10th International Spring Seminar on Nuclear Physics: New Quests in Nuclear Structure

Journal of Physics: Conference Series Volume 267

Vietri sul Mare, Italy
21 - 25 May 2010

ISBN: 978-1-61782-369-5
ISSN: 1742-6588

Printed from e-media with permission by:

Curran Associates, Inc.
57 Morehouse Lane
Red Hook, NY 12571

CURRAN ASSOCIATES INC. proceedings.com

Some format issues inherent in the e-media version may also appear in this print version.

Printed by Curran Associates, Inc. (2011)

For permission requests, please contact the Institute of Physics at the address below.

Institute of Physics
Dirac House, Temple Back
Bristol BS1 6BE UK

Phone: 44 1 17 929 7481
Fax: 44 1 17 920 0979

techtracking@iop.org

Additional copies of this publication are available from:

Curran Associates, Inc.
57 Morehouse Lane
Red Hook, NY 12571 USA
Phone: 845-758-0400
Fax: 845-758-2634
Email: curran@proceedings.com
Web: www.proceedings.com

TABLE OF CONTENTS

NUCLEAR STRUCTURE FAR FROM STABILITY: NEW ADVANCES AND PERSPECTIVES

FROM NUCLEAR FORCES TO NUCLEAR STRUCTURE

EXPLORING NUCLEAR SHELL STRUCTURE: EXPERIMENT AND THEORY

NEW ASPECTS OF COLLECTIVE NUCLEAR MOTION

SPECIAL TOPICS

Foreword

LIST OF PARTICIPANTS

V. Abrosimov	Institute for Nuclear Research 47 Prospect Nauki, 03028 Kiev, UKRAINE abrosim@kinr.kiev.ua
Y. Alhassid	Center for Theoretical Physics, Sloane Physics Laboratory Yale University, New Haven, Connecticut 06520, USA yoram.alhassid@yale.edu
G. de Angelis	INFN Laboratori Nazionali di Legnaro Via Romea 4, 35020 Legnaro (Padova), ITALY deangelis@lnl.infn.it
N. Auerbach	School of Physics and Astronomy Tel Aviv University, Tel Aviv, 69978, ISRAEL auerbach@post.tau.ac.il
T. Aumann	GSI Darmstadt Planckstrasse 1, 64291 Darmstadt, GERMANY t.aumann@gsi.de
D. Balabanski	INRNE, Bulgarian Academy of Sciences Tsarigradsko Schosse Blvd. 72, Sofia 1784 , BULGARIA dimiter.balabanski@unicam.it
B. Balantekin	University of Wisconsin Department of Physics, 1150 University Avenue, Madison, WI 53706, USA baha@physics.wisc.edu
G. Ball	TRIUMF 4004 Wesbrook Mall, Vancouver B.C., V6T 2A3, CANADA ball@triumf.ca
B. Barrett	University of Arizona Dept of Physics, PO Box 210081, Tucson, Arizona, 85721-0081, USA bbarrett@physics.arizona.edu
N. Benczer-Koller	Department of Physics and Astronomy, Rutgers University New Brunswick, NJ 08903, USA nkoller@physics.rutgers.edu
D. Bianco	Dipartimento di Scienze Fisiche, Università di Napoli Federico II Complesso Universitario di Monte S.Angelo, Via Cintia, 80126 Napoli, ITALY bianco@na.infn.it
P. G. Bizzeti	Dipartimento di Fisica Università di Firenze and INFN, Via G. Sansone 1, 50019 Sesto Fiorentino, ITALY bizzeti@fi.infn.it
J. de Boer	University of München Sektion Physik, Moehlstrasse 12A, Müunchen, Bavaria, D-81675, GERMANY deboerv@aol.com
M. Böyükata	Physiscs Department Faculty of Science and Arts, Kirikkale, TURKEY mboyukata@yahoo.com
B. A. Brown	NSCL Michigan State University, East Lansing, MI 48824-1321, USA brown@nscl.msu.edu
F. Catara	Dipartimento di Fisica ed Astronomia Università di Catania and INFN, Via S. Sofia 64, Catania, 95100, ITALY catara@ct.infn.it

J. Cederkall	Physics Department Lund University, Box 117, 221 00, Lund, SWEDEN oakim.cederkall@cern.ch
J. Clark	Argonne National Laboratory 9700 S. Cass Avenue, Argonne, IL 60439, USA jclark@phy.anl.gov
G. Co'	Dipartimento di Fisica Università del Salento, 73100 Lecce, ITALY gpco@le.infn.it
L. Coraggio	Istituto Nazionale di Fisica Nucleare, Sezione di Napoli Complesso Universitario di Monte S. Angelo, Via Cintia, 80126 Napoli, ITALY coraggio@na.infn.it
A. Covello	Dipartimento di Scienze Fisiche, Università di Napoli Federico II Complesso Universitario di Monte S. Angelo, Via Cintia, 80126 Napoli, ITALY covello@na.infn.it
N. D. Dang	RI-Beam Factory Project Office, RIKEN 2-1 Hirosawa, Wako City 351-0198 Saitama, JAPAN dang@rikaxp.riken.go.jp
C. H. Dasso	University of Seville Facultad de Fsica, Av. Reina Mercedes, E-41012 Sevilla, SPAIN dasso@us.es
V. De Donno	Dipartimento di Fisica Università del Salento, 73100 Lecce, ITALY viviana.de.donno@le.infn.it
A. Dellafiore	Istituto Nazionale di Fisica Nucleare, Sezione di Firenze Via G.Sansone, 50019 Sesto Fiorentino, ITALY della@fi.infn.it
A. Di Nitto	Dipartimento di Scienze Fisiche, Università di Napoli Federico II Complesso Universitario di Monte S. Angelo, Via Cintia, 80126 Napoli, ITALY dinitto@na.infn.it
J. Dudek	University of Strasbourg and the IPHC Strasbourg, 23 rue du Loess, F-67037 Strasbourg Cedex2, FRANCE Jerzy.Dudek@ires.in2p3.fr
A. Faessler	University of Tübingen Inst. Theor. Physik, Auf der Morgenstelle 14, D-72076 Tübingen, GERMANY amand.faessler@uni-tuebingen.de
L. S. Ferreira	CFIF and Departamento de Física, Instituto Superior Técnico Av. Rovisco Pais, 1049-001, Lisboa, PORTUGAL flidia@ist.utl.pt
B. Fornal	Niewodniczanski Institute of Nuclear Physics Radzikowskiego 152, 31-342 Kraków, POLAND bogdan.fornal@ifj.edu.pl
M. Gai	University of Connecticut 1084 Shennecossett Rd., Groton, CT 06340, USA moshe.gai@yale.edu
S. Gales	GANIL Bd Henri Becquerel, BP55027, F-14076 Caen Cedex, FRANCE gales@ganil.fr
D. Gambacurta	Dipartimento di Fisica ed Astronomia Università di Catania and INFN, Via S. Sofia, 64, 95100 Catania, ITALY catara@ct.infn.it

A. Gargano — Istituto Nazionale di Fisica Nucleare, Sezione di Napoli
Complesso Universitario di Monte S. Angelo, Via Cintia, 80126 Napoli, ITALY
gargano@na.infn.it

J. Ginocchio — Los Alamos National Laboratory, MS B283
Los Alamos, NM 87545, USA
gino@lanl.gov

J. M. Gómez — Departamento de Fisica Atómica, Molecular y Nuclear
Facultad de Ciencias Fisicas, Universidad Complutense, 28040 Madrid, SPAIN
gomezk@nuc1.fis.ucm.es

P. M. Gore — 629 Country Club Ln, Nashville, TN 37205, USA
p.m.gore.phys@earthlink.net

M. Grasso — IPN-Orsay
Groupe de Physique Théorique, Institut de Physique Nucléaire,
15 rue G. CLEMENCEAU, 91406 Orsay Cedex, FRANCE
grasso@ipno.in2p3.fr

P. Guazzoni — Dipartimento di Fisica
Università di Milano, Via Celoria 16, 20133 Milano, ITALY
paolo.guazzoni@mi.infn.it

G. Hagen — ORNL
Physics Division, Bldg. 6025, M.S. 6373, Oak Ridge, TN 37831-6373, USA
hageng@ornl.gov

I. Hamamoto — Matematisk Fysik
LTH, University of Lund, P.O.Box 118, Lund S-22100, SWEDEN
ikuko@matfys.lth.se

M. Hass — Department of Particle Physics and Astrophysics, The Weizmann Institute of Science
Rehovot 76100, ISRAEL
michael.hass @weizmann.ac.il

A. Heusler — Max-Planck-Institut für Kernphysik
Gustav-Kirchhoff-Strasse 7/1, 69120 Heidelberg, GERMANY
A.Heusler@mpi-hd.mpg.de

J. Holt — ORNL
Physics Division, Bldg. 6025, M.S. 6373, Oak Ridge, TN 37831-6373, USA
jholt31@utk.edu

N. Itaco — Dipartimento di Scienze Fisiche, Università di Napoli Federico II
Complesso Universitario di Monte S. Angelo, Via Cintia, 80126 Napoli, ITALY
itaco@na.infn.it

E. F. Jones — 629 Country Club Ln, Nashville, TN 37205, USA
e.f.jones.phys@earthlink.net

P. Kienle — Technische Universität München
Albrechtstrasse 51, D-80636, München, GERMANY
Paul.Kienle@ph.tum.de

F. Knapp — Istituto Nazionale di Fisica Nucleare , Sezione di Napoli
Complesso Universitario di Monte S. Angelo, Via Cintia, 80126 Napoli, ITALY
knapp@ipnp.troja.mff.cuni.cz

T. T. S. Kuo — Physics Department, State University of New York at Stony Brook
Stony Brook, NY 11794-3800, USA
Thomas.Kuo@stonybrook.edu

J. Kvasil — Institute of Particle and Nuclear Physics
Charles University, V Holesovickach 2, CZ-18000 Prague, CZECH REPUBLIC
kvasil@ipnp.troja.mff.cuni.cz

S. Lenzi — Dipartimento di Fisica, Università di Padova

Via Marzolo 8, 35131 Padova, ITALY
lenzi@pd.infn.it

A. Leviatan
The Hebrew University
Racah Institute of Physics, The Hebrew University, Jerusalem, 91904, ISRAEL
ami@phys.huji.ac.il

N. Lo Iudice
Dipartimento di Scienze Fisiche, Università di Napoli Federico II
Complesso Universitario di Monte S. Angelo, Via Cintia, 80126 Napoli, ITALY
loiudice@na.infn.it

S. Lunardi
Dipartimento di Fisica, Università di Padova
Via Marzolo 8, 35131 Padova, ITALY
lunardi@pd.infn.it

R. Machleidt
University of Idaho
Department of Physics, University of Idaho, Moscow, ID 83843, USA
machleid@uidaho.edu

L. Moretto
Lawrence Berkeley National Laboratory
BLDG88, Berkeley, CA 94720, USA
lgmoretto@lbl.gov

R. Moro
Dipartimento di Scienze Fisiche, Università di Napoli Federico II
Complesso Universitario di Monte S.Angelo, Via Cintia, 80126 Napoli, ITALY
moro@na.infn.it

D. Morrissey
Michigan State University - NSCL, W208 Cyclotron Lab
East Lansing, MI 48824, USA
morrissey@nscl.msu.edu

K. Nomura
Department of Physics, University of Tokyo
7-3-1 Hongo, Bunkyo-ku, Tokyo 113-0033, JAPAN
nomura@nt.phys.s.u-tokyo.ac.jp

F. Nowacki
IPHC
23 rue du Loess, BP 28, 67037, Strasbourg Cedex 2, FRANCE
frederic.nowacki@ires.in2p3.fr

R. Okamoto
Kyushu Institute of Technology
Tobtaa-ku, Kitakyushu, 804-8550, JAPAN
okamoto@mns.kyutech.ac.jp

M. G. Pellegriti
LNS-INFN and Università di Catania
Via S. Sofia, 64, 95125 Catania, ITALY
pellegriti@ct.infn.it

A. Poves
Universidad Autónoma de Madrid, Departamento de Fisica Teórica,
Avenida Tomas y Valiente 7, 28049, Madrid, SPAIN
alfredo.poves@uam.es

A. V. Ramayya
Department of Physics and Astronomy, Vanderbilt University
VU Station B 351807, Nashville, TN 37235, USA
a.v.ramayya@vanderbilt.edu

D. Rifuggiato
INFN Laboratori Nazionali del Sud
Via S. Sofia, 62, 95123 Catania, ITALY
rifuggiato@lns.infn.it

P. Ring
Physik-Department der Technischen Universität München
James Franck Str. 1, D-85748 Garching, GERMANY
ring@ph.tum.de

D. J. Rowe
University of Toronto
Dept. of Physics, Univ. of Toronto, Toronto, ON M5S 1A7, CANADA
rowe@physics.utoronto.ca

G. Royer
Laboratoire Subatech

Subatech-EMN, 4 rue Alfred Kastler, La Chantrerie, 44 Nantes, FRANCE
royer@subatech.in2p3.fr

F. Sammaruca
University of Idaho
Department of Physics, University of Idaho, Moscow, ID 83843, USA
fsammarr@uidaho.edu

N. Sandulescu
Institute of Physics and Nuclear Engineering
P.O.Box MG-6, RO-76900, Bucharest-Magurele, ROMANIA
sandulescu@theory.nipne.ro

S. Sarkar
Bengal Engineering and Science University
Department of Physics, Shibpur, Howrah- 711103, INDIA
ss@physics.becs.ac.in

V. Scuderi
INFN Laboratori Nazionali del Sud
Via S. Sofia, 62, 95123 Catania, ITALY
scuderi@lns.infn.it

K. Sieja
IPHC
23 rue du Loess, BP 28, 67037, Strasbourg Cedex 2, FRANCE
kamila.sieja@ires.in2p3.fr

G. Simpson
LPSC, Université de Joseph Fourier 1, CNRS/IN2P3
Institut National Polytechnique de Grenoble, F-38026 Grenoble Cedex, FRANCE
simpson@lpsc.in2p3.fr

A. M. Sona Bizzeti
Dipartimento di Fisica
Università di Firenze and INFN, Via G. Sansone 1, 50019 Sesto Fiorentino, ITALY
Annamaria.Sona@fi.infn.it

O. Sorlin
GANIL
15 bD Henri Becquerel, BP 55027, F14076 Caen Cedex 5, FRANCE
sorlin@ganil.fr

Ch. Stoyanov
Institute for Nuclear Research and Nuclear Energy
72 Bvd. Tzarigradsko Chaussee, Sofia 1784, BULGARIA
stoyanov@inrne.bas.bg

T. Suda
Research Center for Electron Photon Science, Tohoku Univ.
1-2-1 Mikamine, 982-0826 Sendai, JAPAN
suda@lns.tohoku.ac.jp

B. Szpak
The Niewodniczanski Institute of Nuclear Physics, Polish Academy of Sciences
ul. Radzikowskiego 152, Kraków, PL31-342, POLAND
Bartlomiej.Szpak@ifj.edu.pl

I. Talmi
Department of Physics, The Weizmann Institute of Science
Rehovot 76100, ISRAEL
igal.talmi@weizmann.ac.il

D. Tonev
Institute for Nuclear Physics and Nuclear Energy, Bulgarian Academy of Sciences
GB-1784, Sofia, BULGARIA
mitko@lnl.infn.it

N. Tsunoda
University of Tokyo, Department of Physics
7-3-1 Hongo, Bunkyo-ku, Tokyo 113-0033, JAPAN
tsunoda@nt.phys.s.u-tokyo.ac.jp

J. Uusitalo
Department of Physics
Survontie 9, 40500, Jyväskylä, FINLAND
juha.uusitalo@phys.jyu.fi

J. J. Valiente Dobón
INFN Laboratori Nazionali di Legnaro
Via Romea 4, 35020 Legnaro (Padova), ITALY
Javier.Valiente@lnl.infn.it

P. Van Duppen
Instituut voor Kern- en Stralingsfysica

K.U.Leuven, Celestijnenlaan 200 D, B-3001 Leuven, BELGIUM
piet.vanduppen@fys.kuleuven.be

N. Van Giai — Institut de Physique Nucléaire
Université Paris-Sud, F-91406 Orsay Cedex, FRANCE
nguyen@ipno.in2p3.fr

P. Veselý — Department of Physics, P.O. Box 35 (YFL)
University of Jyväskylä, V Holesovickach 2, 180 00 Prague 8, CZECH REPUBLIC
vesely@ipnp.troja.mff.cuni.cz

A. Vitturi — Dipartimento di Fisica, Università di Padova
Via Marzolo 8, 35131 Padova, ITALY
vitturi@pd.infn.it

A. Voinov — Flerov Laboratory of Nuclear Reactions
Joint Institute for Nuclear Research
Joliot-Curie st. 6, 141980 Dubna, Moscow Region, RUSSIA
voinov@sungns.jinr.ru

V. Voronov — Bogoliubov Laboratory of Theoretical Physics
Joint Institute for Nuclear Research
141980 Dubna, Moscow Region, RUSSIA
voronov@thsun1.jinr.ru

W. B. Walters — University of Maryland
Department of Chemistry and Biochemistry, University of Maryland,
College Park, MD 20742, USA
wwalters@umd.edu

L. Zetta — Dipartimento di Fisica
Università di Milano, Via Celoria 16, 20133 Milano, ITALY
luisa.zetta@mi.infn.it

Status of the FRIB project with a new fragment separator

2011 J. Phys.: Conf. Ser. 267 012001

(http://iopscience.iop.org/1742-6596/267/1/012001)

Status of the FRIB project with a new fragment separator

D.J. Morrissey

National Superconducting Cyclotron Laboratory, and Department of Chemistry, Michigan State University, East Lansing, Michigan, USA 48824

E-mail: morrissey@nscl.msu.edu

Abstract. The surprising variation of the cross sections for the production of the most neutron-rich nuclei by the fragmentation of ^{48}Ca and ^{76}Ge beams at Michigan State University (MSU) continues to show the importance of determining the properties of exotic nuclei. Future detailed studies of such exotic nuclei are being assured through the establishment of the Facility for Rare Isotope Beams (FRIB), a cutting-edge research facility to advance understanding of rare nuclear isotopes and the evolution of the cosmos, by the US Department of Energy at MSU. The new facility is expected to take something less than a decade to design and build and will provide a broad range of research opportunities for an international community. The facility is based on a superconducting-RF driver linear accelerator that will provide a maximum beam power of 400 kW for all beams ranging from uranium accelerated to 200 MeV/u, to lighter ions with increasing energy, down to protons at 600 MeV/nucleon. Rare isotope beam production at a beam power of 400 kW has never been done and presents a number of large challenges. The heart of the FRIB facility will be a next-generation three-stage projectile fragment separator specifically designed to handle the very intense primary and secondary beams.

1. Introduction

One of the important challenges facing nuclear physics is to push the study of neutron-rich isotopes to higher atomic number. The benchmark in this exploration is to find the maximum number of neutrons that can be bound for each atomic number. Often the delineation of the heavy limit of stability, called the neutron drip-line, itself has yielded surprises. Some years ago the heaviest bound oxygen isotope was shown to be ^{24}O while many theories predicted that the much heavier and putatively doubly-magic $^{28}O_{20}$ should be stable. Recently we found that heavy isotopes of aluminum are likely more bound than predicted [1]. Continuing this work, we report here the next step in this challenging exploration in the region about calcium.

The neutron drip line is only confirmed by experiment at present up to $Z = 8$ at ($^{24}O_{16}$) through years of work at various projectile fragmentation facilities. The neutron drip line has been found to rapidly shift to higher neutron numbers at $Z = 9$, i.e., $^{31}F_{22}$ has been observed several times, see e.g. [1]. The nuclide $^{30}F_{21}$ has been shown not to exist and $^{32}F_{23}$ is thought to be unbound based on systematics while the particle stability of $^{33}F_{24}$ remains an open question. The shift in stability of especially neutron-rich isotopes is predicted to continue at higher masses [2, 3, 4] and makes the search for the neutron drip line in this region even more challenging. The fragmentation of $^{48}Ca_{28}$ projectiles has produced a number of heavier nuclei in this region including $^{40}Mg_{28}$ and $^{42}Al_{29}$, but no clear limit has been established yet. [5] On the other hand,

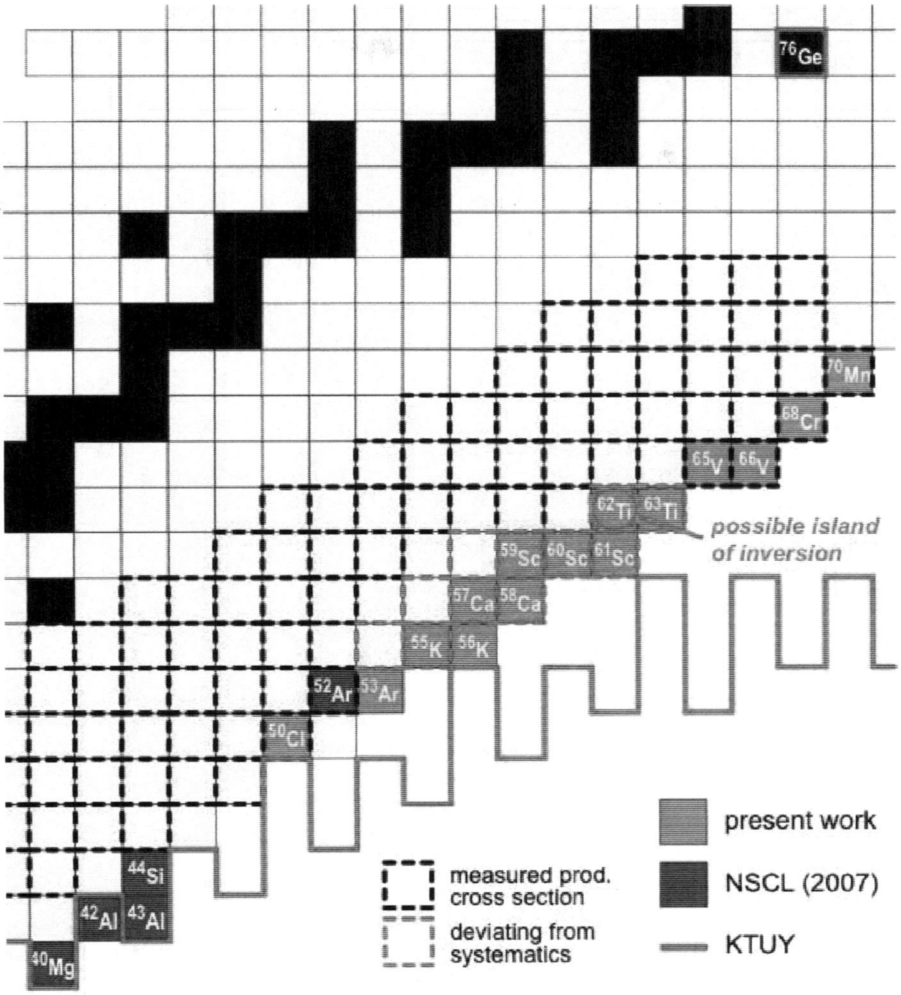

Figure 1. The region of the chart of nuclides under investigation. The solid line is the limit of bound nuclei from the KTUY mass model [4]. Nuclei in the green squares were recently discovered [5, 1, 6], those in red squares are new in this work, and some of the cross sections for those in dashed boxes are shown in Fig. 2. The center of the new proposed island of inversion at ^{62}Ti [8] is highlighted.

all nuclei up to $Z = 12$ with $A = 3Z + 3$ have been shown to be unbound. In the short term, the fragmentation of heavier stable beams such as ^{76}Ge in which ^{52}Ar was observed [6] is necessary in order to go beyond the previous work.

In the longer term, the production of the most neutron-rich nuclei will be an important component of the broad scientific programs at the next generation facilities such as **FAIR** in Germany, **RIBF** in Japan, and the recently announced Facility for Rare-Ion Beams, **FRIB**, in the United States. Michigan State University was recently selected by the US Department of Energy to be the site of FRIB. The proposed facility will provide primary beams at power levels up to 400 kW, compared to approximately 4 kW at the present National Superconducting Cyclotron Laboratory (NSCL) facility. In addition, all stable beams will be available at these

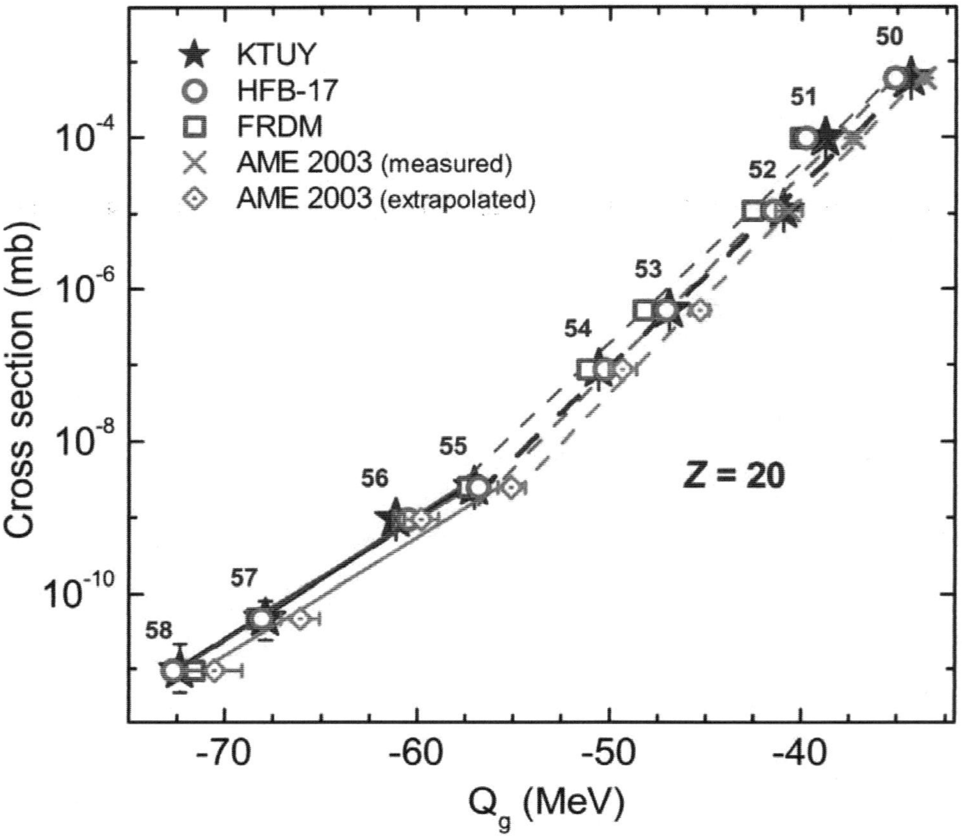

Figure 2. The cross sections for the calcium isotopes are shown as a function of Q_g using a variety of mass models, see text. Points are labeled by mass number.

high power levels and a new higher acceptance separator will provide significantly higher secondary beam rates. An overview of some of these features is given below.

2. Production of Neutron-rich Nuclei

In the most recent work at the NSCL a primary beam of ^{76}Ge was fragmented and a search for new neutron-rich isotopes above ^{40}Mg was carried out using the tandem fragment separator technique [1]. A 132 MeV/u ^{76}Ge primary beam from the NSCL was used to irradiate a series of ^{9}Be targets and finally a thick tungsten target located at the target position of the A1900 fragment separator [7]. The average beam intensity for the measurements of the most exotic fragments was 32 pnA. The A1900 fragment separator was combined with the S800 analysis beam line to form a two-stage separator system as described in Ref. [1]. Such a two-stage separator provides a high degree of rejection of unwanted reaction products and allows the identification of each fragment of interest. During the search for the most exotic fragments, a Kapton wedge (20.2 mg/cm^2) was used at the center of the A1900 to reject less exotic fragments at the A1900 focal plane by an 8 mm aperture. The transmitted fragments were identified by event-by-event momentum analysis and particle identification. The momentum acceptance of the A1900 was set to $\Delta p/p = \pm 0.05\%$, $\pm 0.5\%$, $\pm 1\%$ and $\pm 2.5\%$ as the production rate of the increasingly exotic nuclei decreased, always with an angular acceptance of 8.2 msr. The simultaneous measurement of multiple ΔE signals, the magnetic rigidity, the total energy, and the TOF's of each particle provided an unambiguous identification of the atomic number, charge state and mass number of each isotope.

Tarasov et al. [5] have shown that the cross sections for projectile fragments in this region have an exponential dependance on Q_g (the difference in mass-excess of the beam particle and the observed fragment that is independent of the target in contrast to the older Q_{gg} four-body analysis applied to low energy reactions) and deviations from the predicted yield may be used to identify anomalies in the mass surface such as the new island of inversion near ^{62}Ti predicted by Brown [8] similar to the original island of inversion near ^{31}Na.

Figure 2 shows that the Q_g function using masses from a wide range of mass models and extrapolations, including Ref. [4] and Ref. [9], give similar results since the differences among mass models are small compared to the variation of Q_g. The figure shows that the logarithm of the cross sections for each isotopic chain falling on an approximately straight line except for the heaviest isotopes. On the other hand, most of isotopic chains with $15 \leq Z \leq 24$ can be fit with a single exponential slope of 1/1.8 MeV [10, 11]. The heaviest members of the isotopic chains with Z=19, 20, 21 and 22 break away from the uniform fit and the heaviest four or five isotopes have a shallower slope or enhanced cross sections. Recall that the masses of these most neutron-rich nuclei are not known but rather represent extrapolations. The observed increase in cross section might indicate that the excited precursors that give rise to the observed nuclei are more bound (i.e., less negative Q_g) than current mass models predict. One reason for a stronger binding can be deformation. In a shell-model framework, the wave functions of the ground and low-lying excited states of nuclei in the new island of inversion around ^{62}Ti would be dominated by neutron particle-hole intruder excitations across the $N = 40$ sub-shell gap, leading to deformation and shape coexistence.

3. FRIB: the Facility for Rare-Ion Beams

The US Department of Energy announced that Michigan State University has been selected to design and establish the Facility for Rare Isotope Beams (FRIB), a cutting-edge research facility to advance understanding of rare nuclear isotopes and the evolution of the cosmos. The new facility is expected to take about a decade to design and build, and to cost an estimated $550 million and will provide a broad range of research opportunities for an international community. The facility is based on a superconducting-RF driver linear accelerator that will provide a maximum beam power of 400 kW for all beams ranging from uranium accelerated to 200 MeV/u, to lighter ions with increasing energy, down to protons at 600 MeV/nucleon. In addition, space will be provided in the linac tunnel and shielding in the production area to allow a future upgrade of the driver linac energy to 400 MeV/u for uranium and 1000 MeV for protons without significant interruption of the future science program. Space will also be provided in the production target area for other potential upgrades such as ISOL or even a second fragment separator. A schematic layout of the proposed FRIB facility and its connection to the existing NSCL facility is shown in Fig. 3. This facility layout has several important advantages, for example it allows the continued operation of the Coupled Cyclotron Facility (CCF) using fast, stopped and reaccelerated beams, it allows a relatively rapid connection of the new accelerator complex to the existing facility, and reuse of the experimental equipment.

Rare isotope beam production at a beam power of 400 kW has never been done and presents a number of large challenges. The production targets must be able to sustain very high power densities with the consequent thermal stress and radiation damage. Isotope production requirements include a maximum target thickness of ≈50 mm with a 1-mm diameter beam spot, the ability to absorb (and dissipate) a continuous power of up to 200 kW, and a lifetime of more than one week (anticipated duration of typical FRIB experiment). Research and Development work to establish a target design that has these features is under way; a rotating multi-disk

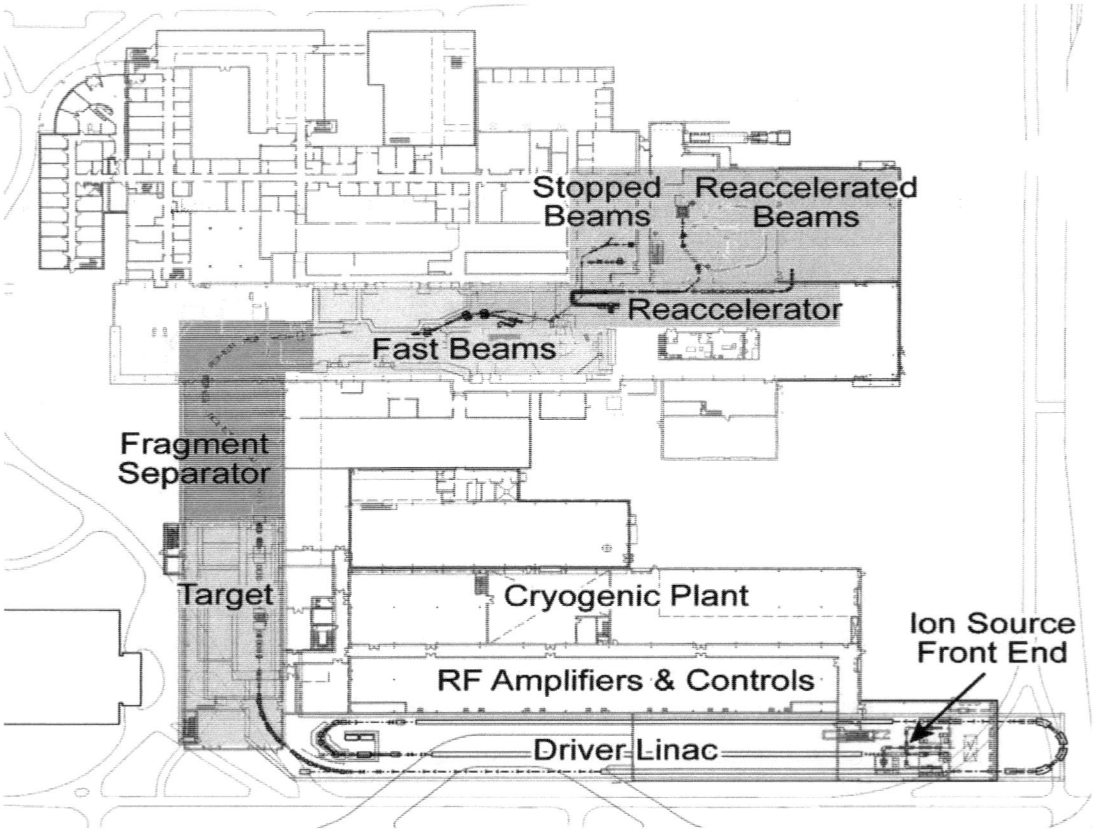

Figure 3. A schematic view of the major components of the FRIB facility in relation to the existing NSCL facility.

graphite target with radiative cooling is under active study. In addition, a windowless liquid lithium target is being considered for use with the heaviest beams. Note that the beam dump for collecting the primary beam after the first separation stage has to be able to absorb up to the full 400 kW beam power although the beam spot size will be larger than that at the target. Research and Development on different beam dump concepts including a water-filled rotating cylinder is also under way. Nuclear interactions in the target and the dump will create very high radiation fields that will reduce the lifetime of the magnets that make up the fragment separator so that radiation resistant superconducting magnets will be needed.

The heart of the FRIB facility will be a next-generation three-stage projectile fragment separator specifically designed to handle the very intense primary and secondary beams. The conceptual design of the FRIB projectile fragment separator is shown in Fig. 4. The beam purity from the new separator will be at least two orders of magnitude higher than that possible from even the most modern one-stage separators. The first stage of the fragment separator, called the pre-separator, provides the initial rough-cut on the broad distribution of fragments, hosts a well-defined primary beam dump, and delivers the secondary fragments to the object point of the main separator without significant aberrations. The projectile fragments pass from the pre-separator into two subsequent high-resolution stages, based on a reconfiguration of the existing A1900 fragment separator used in the work described above.

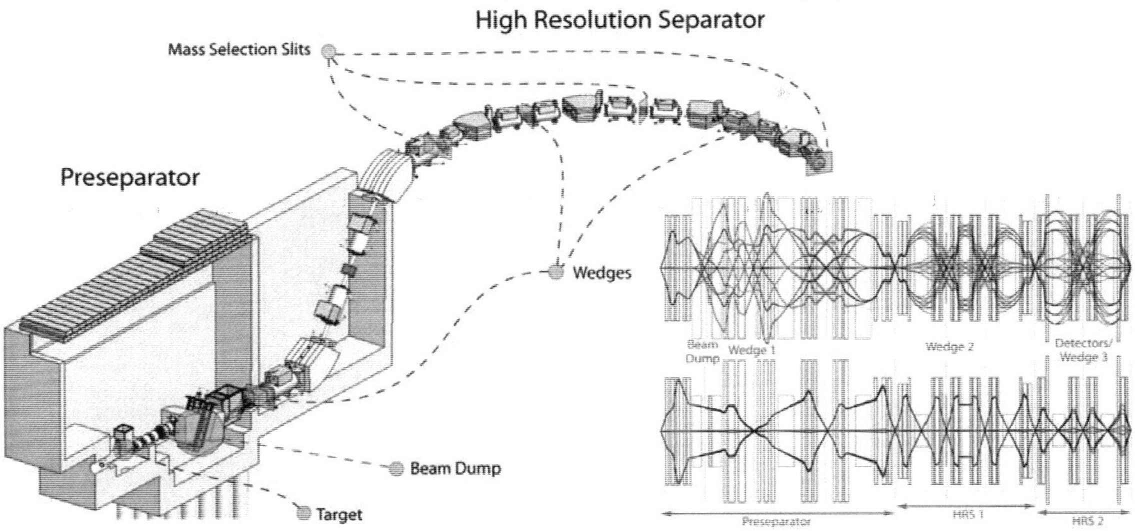

Figure 4. The present conceptual design of the fourth generation fragment separator for the FRIB facility. The ion-optical layout is shown in the inset.

There is the opportunity for a pre-FRIB science program using the existing in-flight separated beams from the existing CCF and the new ReA3 reaccelerator presently under construction. Users will be able to mount and test equipment and techniques and push nuclear science forward with a variety of thermalized and reaccelerated fragmentation beams so that the science program will seamlessly continue on when FRIB is complete. This transition will allow for a continually evolving science program during the time FRIB is under construction that will form the basis of the research program at FRIB.

Acknowledgments
It is important to note that the work reported here was carried out by a large group of people at the National Superconducting Cyclotron Laboratory. The reader is referred to the author list in Ref.[10] for the names of the participants in the projectile fragmentation study. The FRIB project is made up from an even larger group with a leadership team headed by C.K. Gelbke (director) and T. Glasmacher (project manager), with G. Bollen, B.M. Sherrill, and R.C. York. The design and extensive calculations for the performance of the next-generation fragment separator described above come from M. Hausmann, B.M. Sherrill, and M. Portillo.

References

[1] Baumann, T, *et al.*, 2007 *Nature* **449**, 1022.
[2] Heenan, P-H, 2007 *Nature* **449**, 992.
[3] Samyn, M, Goriely, S, Bender, M, and Pearson, J M, 2004 *Phys. Rev.*C **70**, 044309.
[4] Koura, H, *et al.*, 2005 *Prog. Theo. Phys.* **113**, 305.
[5] Tarasov, O B, *et al.*, 2007 *Phys. Rev.* C **75**, 064613.
[6] Mantica, P F, *et al.*, 2008 *Bull. Am. Phys. Soc.* **53**, 64.
[7] Morrissey, D J, *et al.*, 2003 *Nucl. Instrum. Meth. Phys. Res.* B **204**, 90.
[8] Brown, B A, 2001 *Prog. Part. Nucl. Phys.* **47**, 517.
[9] Goerily, S, Chamel, N, and Pearson, J M , 2009 *Phys. Rev. Lett.* **102**, 152503.
[10] Tarasov, O B, *et al.*, 2009 *Phys. Rev. Lett.* **102**, 142501.
[11] Tarasov, O B, *et al.*, 2009 *Phys. Rev.* C **80**, 034609.

Expansion of the radioactive ion beam program at Argonne

2011 J. Phys.: Conf. Ser. 267 012002

(http://iopscience.iop.org/1742-6596/267/1/012002)

Expansion of the radioactive ion beam program at Argonne

J A Clark

Physics Division, Argonne National Laboratory, Argonne, IL 60439, USA

E-mail: `jclark@phy.anl.gov`

Abstract. The Argonne Tandem Linear Accelerator System (ATLAS) at Argonne National Laboratory (ANL) provides a wide range of stable ion beams and radioactive beams which have contributed to our understanding of nuclear structure and reactions. Until now, most radioactive ion beams at ATLAS were produced in flight using light-ion reactions such as (p, n), (d, n), (d, p), $(d,^{3}\mathrm{He})$, and $(^{3}\mathrm{He},n)$. Within the next few months, the radioactive ion beam program at ATLAS will acquire much extended, new capabilities with the commissioning of a new facility: the CAlifornium Rare Isotope Breeder Upgrade (CARIBU). CARIBU will supply ion beams of ^{252}Cf fission fragments, which are thermalized in a gas catcher. The singly- and doubly-charged ions extracted from the gas catcher will be mass-separated and either delivered to a low-energy experimental area, or charge bred with a modified ECR source and subsequently reaccelerated by the ATLAS facility. Properties of hundreds of these neutron-rich nuclides will be investigated using ion traps, decay stations, the newly commissioned HELical Orbit Spectrometer (HELIOS), and other available experimental equipment such as Gammasphere and the FMA. HELIOS was constructed to take advantage of rare ion beams, such as those provided by CARIBU, through light-ion transfer reactions in inverse kinematics, and represents a new approach to the study of direct reactions in inverse kinematics which avoids kinematic broadening. Experiments are currently being conducted with HELIOS, and first results with the $d(^{28}\mathrm{Si},p)$ and $d(^{12}\mathrm{B},p)$ reactions have shown excellent energy resolution.

1. Introduction

Radioactive ion beams have played an important role for studies of nuclear structure and astrophysics. We have been surprised many times when our models of nuclei that are based on the properties of stable nuclei do not accurately characterize the properties of nuclei far from stability. Indeed, studies of the single-particle strength near closed shells, pair correlations, and collective aspects of nuclear structure have provided valuable insight, and much has been learned about explosive stellar phenomena and the origin of the elements from experiments using radioactive ion beams.

Many more radioactive ion beams are available today than in the past. The earliest radioactive ion beams accelerated by the ATLAS facility at ANL used a two-accelerator method in which one accelerator produced the radioactive isotope, such as ^{18}F produced at the University of Wisconsin, which was then transported to Argonne and used as an ion source for the ATLAS accelerator [1, 2, 3]. This method provided radioactive ion beams for only a few long-lived isotopes; whereas, the majority of radioactive ion beams available at Argonne today are produced via an in-flight method and can have lifetimes on the order of milliseconds. This latter method uses a primary beam from the ATLAS accelerator which hits a cryogenically cooled gas target of either protons, deuterons, or ^{3}He, and the reaction products are subsequently focussed, rebunched, and transported in flight to the experimental areas. [4]

The number of radioactive ion beams will be increased significantly in the very near future with the addition of CARIBU. This new ion source will provide neutron-rich isotopes from a 1-Ci ^{252}Cf source which can accelerated by the current ATLAS facility. ^{252}Cf has a 3% fission branch, a 2.6 year half-life, and the nuclides available from ^{252}Cf are quite complementary to existing facilities, including those which use particle-induced fission of uranium as a source of radioactive ions.

2. CARIBU

CARIBU, shown in Fig. 1, is a new source of ions that will make hundreds of neutron-rich nuclei available for acceleration by the ATLAS facility [5]. A key component of CARIBU is the gas catcher, which is composed of three parts: the ion source, the stopping volume, and the extraction cone. As part of the ion source, 1 Ci of ^{252}Cf was electrodeposited onto a flat 0.1"-thick Ta backing covering an area of 5 cm in diameter. This ^{252}Cf source sits inside the gas catcher and is covered by a gold foil for protection and confinement. Fission fragments from the ^{252}Cf source pass through the gold foil and an additional aluminum degrader such that the residual energy of the fragments is lost to collisions with approximately 150 Torr of purified helium gas inside the 50-cm diameter stopping volume of the gas catcher. Although the flow of helium gas through the extraction cone of the gas catcher will carry fission fragment ions, this process is slow and inefficient for the extraction of short-lived isotopes. Therefore, the stopping volume and extraction cone consist of parallel plate electrodes which permit the application of electric fields: a DC gradient along the length of the gas catcher reduces the time the ions spend inside, and a $\sim$ 2-MHz RF field applied such that adjacent plates with opposite phases form an RF *carpet* which helps to recenter the ions, thereby increasing the efficiency of the device. In the end, about 45% of the fission fragment ions are extracted within 5-20 ms, and since the fission fragment ions are thermalized in helium gas, most of the neutron-rich nuclides are extracted from the gas catcher in a singly- or doubly-charged state due to the high ionization potential of helium.

Once the ions are extracted from the gas catcher, they pass through a radio-frequency quadrupole (RFQ) structure that forms a differentially pumped system. Here, the ions are guided with a DC gradient along the length of the RFQ and confined with RF fields applied in a similar fashion as those on the gas catcher. Since the mass of helium is significantly different from those of the fission fragments, the inherent mass selection of the RFQ allows the fission fragment ions to be guided along the length of the RFQ while the helium gas is pumped away. Due to the cooling of the ions through gas collisions in the RFQ, the transverse emittance of the ions is 3 π mm mrad, and the ions have an energy spread of only 1 eV. After two sections of RFQ structures, the residual gas pressure is sufficiently low for electrostatic acceleration of the ions by 50 kV for injection into the beam transport section.

Following the RFQ structures, a particular ^{252}Cf fission fragment is selected with a compact isobar separator consisting of two 60-degree magnets with a 50-cm bending radius which provides a mass resolving power of 20,000. Since the fission fragments are far from stability, the masses of nearby isobars for a particular fission fragment are sufficiently different that a mass resolving power of 20,000 is quite adequate in most cases.

Once a particular fission fragment has been selected with the isobar separator, the fragment can then be directed to one of two beamlines. The fragment is either delivered to a modified ECR source, where the ions are charge bred to a higher charge state and subsequently reaccelerated through the ATLAS facility, or the fragment is sent through a low-energy beamline to a 'stopped' beam experimental area where mass measurements, lifetime measurements, laser spectroscopy, and decay studies can be performed. The low-energy beamline itself consists of a gas-filled

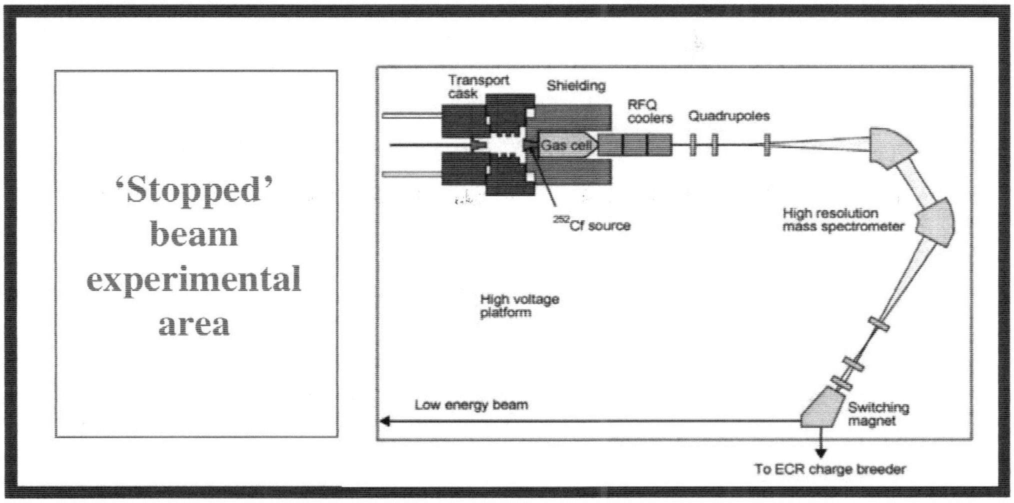

Figure 1. Layout of the CARIBU facility at Argonne. Fission fragments from a 1-Ci ^{252}Cf source are thermalized in a gas catcher, extracted, and guided through RFQ coolers to an isobar separator where a particular nuclide is selected. This nuclide can then be delivered to either the low-energy beamline towards a 'stopped' beam experimental area, or a modified ECR source for charge breeding.

RFQ structure at 50 kV which serves three purposes: to stop and cool the ions from the isobar separator, to accumulate the ions, and to bunch the continuous stream of ions injected into the RFQ. Before the ion bunches are sent to the stopped beam experimental area, they pass through a pulsed tube to convert the 50 kV beam to a lower potential as required for the particular experimental apparatus.

The first of three ^{252}Cf fission sources has been installed in the CARIBU gas catcher. Once sufficient testing has been completed with the 2.5-mCi source, this source will be replaced with the 100-mCi, and eventually the 1-Ci, source. Initial experiments in the stopped beam experimental area are expected to begin by the end of the summer 2010 with the 100-mCi source, and the full experimental program with the 1-Ci source is anticipated by the end of the year. Further information about CARIBU, including expected yields, can be found online at: `http://www.phy.anl.gov/atlas/facility/caribu_beams.html` .

3. HELIOS

Ion traps, decay stations, the newly commissioned HELIOS, and other available experimental equipment such as Gammasphere and the Fragment Mass Analyzer (FMA) will be able to take advantage of the ion beams provided by CARIBU. For example, one of these devices, the Canadian Penning Trap (CPT) mass spectrometer, makes precise mass measurements of ions using a Penning trap located in the bore of a 5.9 T superconducting magnet. The results from previous measurements made by the CPT with a 100-μCi source show a two-neutron separation energy which decreases with neutron excess more quickly than predicted by the 2003 atomic mass evaluation. If this trend were to continue with increasing neutron number, the location of the neutron dripline and the path of the astrophysical r process would be closer to stability than previously thought. With CARIBU, more than 100 neutron-rich nuclides whose masses have never been measured will be accessible with the CPT.

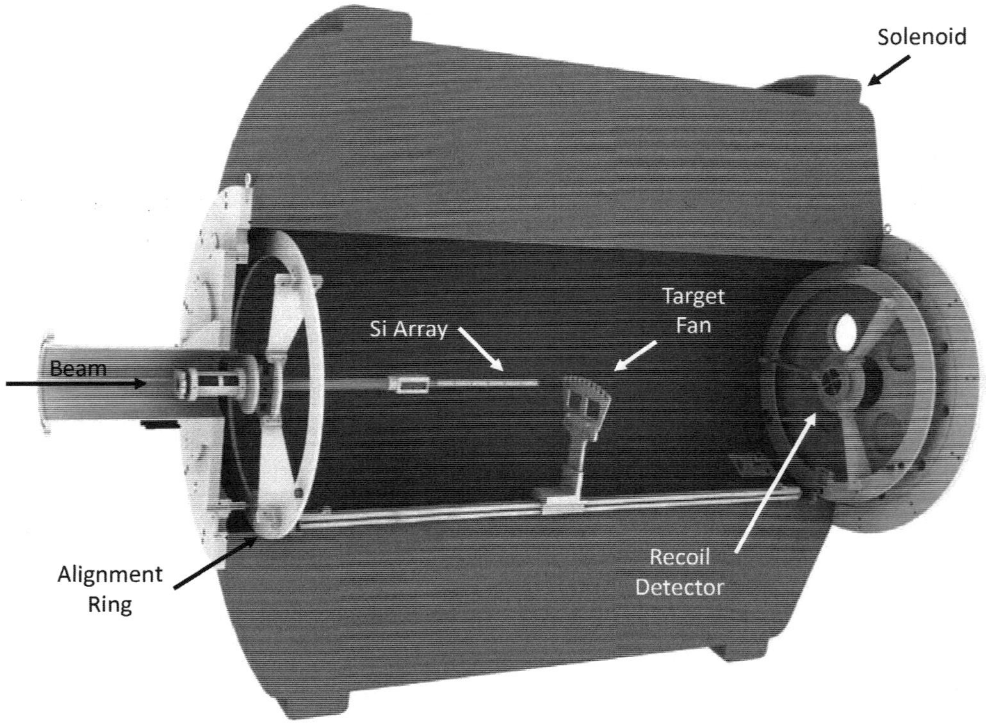

Figure 2. Schematic of the HELIOS concept. From the left, ion beams travel along the magnetic field axis through a hollow Si detector towards a target. Charged particles emitted from the target undergo helical trajectories and reach the detector array in a time equal to one cyclotron period. If necessary, coincidences with recoils detected in a forward detector may be used.

Many properties of nuclei close to stability have been studied through transfer reactions such as (d,p), (^{3}He,d), and (α,t) in 'normal' kinematics where light ion beams strike heavy, stable targets; however, using similar reactions for the study of the nuclei further from stability requires radioactive targets, which are difficult to produce. Instead, targets of light ions are used together with heavy radioactive ion beams to study the properties of nuclei through reactions in 'inverse kinematics'. Traditional spectrometers, which measure the ejected particles as a function of angle, encounter difficulties in separating adjacent nuclear states in the inverse kinematics regime; much of the information contained in the forward angles in the center-of-mass system is compressed into a small energy interval at backward angles in the lab frame. This kinematical compression affects the energy resolution that can be obtained and makes particle identification difficult.

HELIOS is a new concept which avoids the kinematic compression from reactions in inverse kinematics. As shown in Fig. 2, the concept involves a beam of ions which enters the spectrometer along the magnetic field axis of a large-bore, homogeneous, superconducting magnet and strikes a target. Particles emitted from the target undergo helical trajectories and return to the magnetic field axis after one cyclotron period. With a hollow silicon detector array positioned along the axis of the magnetic field, the position along the field axis from the target, the flight time with respect to the pulsed beam, and the energy are measured. Since the cyclotron period is unique for every mass-to-charge ratio, m/q, a measurement of the flight time

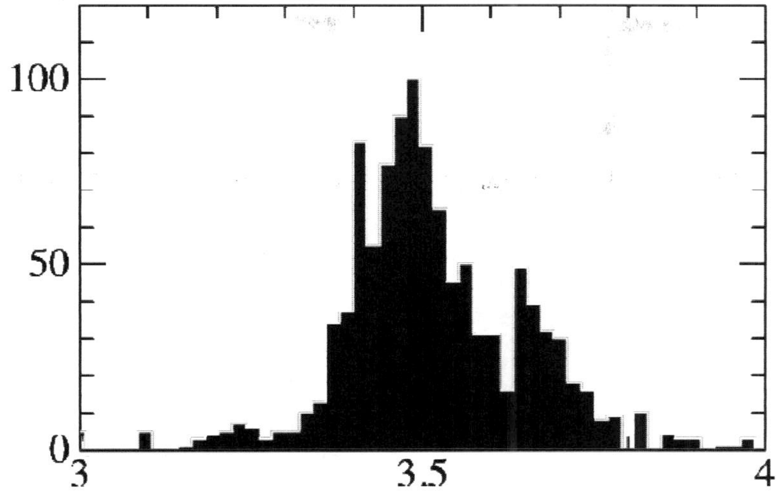

Figure 3. Spectrum showing the resolution of the 3.48/3.68-MeV doublet in ^{13}B.

of the particles serves as a means of particle identification, while measurements of the energy and position on the array determine the center-of-mass energy and scattering angle. [6]

The HELIOS superconducting magnet has a maximum field strength of 2.86 Tesla, and the bore of the magnet is 92.5 cm in diameter and 235 cm long. Housed within the magnet is the Si detector array which consists of 24 position-sensitive detectors where the position-sensitive axis is along the axis of the magnetic field. The array itself forms an open square, with six 700-μm thick, 12mm $\times$ 56 mm silicon wafers along each of the four sides. Tests of the array have shown the intrinsic energy resolution to be < 50 keV (FWHM) for protons.

A study of the excited states in ^{29}Si via the $d(^{28}\text{Si},p)^{29}$Si reaction served as a commissioning experiment for HELIOS. [7]. Targets of deuterated polyethylene were bombarded by a ^{28}Si beam accelerated to 6 MeV/u by the ATLAS facility with an average beam current of 25 ppA. With the detector array fixed in position throughout the experiment, the range in center-of-mass angles was achieved by placing the target at three different positions. For the ground-state transition in $^{28}\text{Si}(d,p)$, this angular range corresponded to 51°. Particles emitted from the target were easily identified by their flight time relative to the 82-ns period of the ATLAS beam RF structure.

The first radioactive beam experiment with HELIOS studied the properties of low-lying states in ^{13}B populated via the $d(^{12}\text{B},p)^{13}$B reaction. Using the existing in-flight capability, a radioactive 75-MeV ^{12}B secondary beam of up to 10^5 ions/s was produced by bombarding a cryogenic deuterium gas cell [4] with a primary ^{11}B beam accelerated to 81 MeV with the ATLAS facility. The ^{12}B beam then passed through the Si detector array inside HELIOS before striking a 73-μg/cm^2 CD_2 target. The outgoing protons were detected in the Si array, and backgrounds were reduced by requiring a coincidence between these protons and recoiling ^{13}B ions detected at forward angles in a Si telescope. A resolution of $\sim$ 100 keV (FWHM) in the excitation energies of ^{13}B was observed which was sufficient to resolve the 3.48- and 3.68-MeV positive-parity doublet in ^{13}B as shown in Fig. 3. Angular distributions for these two states were also obtained, and combined with relative spectroscopic factors, the spin and configuration of these states were constrained. Further details are provided in Ref. [8].

4. Summary

Radioactive ion beams have provided valuable insight into nuclear structure and astrophysics. The first such beams were available at the ATLAS facility in the mid 1990s with long-lived isotopes produced via the two-accelerator method. Since then, many more radioactive ion beams have become possible with in-flight techniques. The latest technological advance in the production of radioactive ion beams is the gas catcher, a device which is a key component of the new CARIBU facility that will provide neutron-rich isotopes from a 1-Ci ^{252}Cf source. Experimental equipment will be able to take advantage of the ion beams provided by CARIBU, including HELIOS, a powerful new tool designed to study transfer reactions in inverse kinematics while avoiding the effects of kinematical compression that plague conventional spectrometers. As more radioactive ion beams become available, the properties of many more nuclei will be accessible to experimental studies.

Acknowledgments

This work was supported by the U.S. Department of Energy, Office of Nuclear Physics, under Contracts No. DE-AC02-06CH11357 (Argonne National Laboratory) and No. DE-FG02-04ER41320 (Western Michigan University).

References

[1] Rehm K E *et al.* 1995 *Phys. Rev. C* **52** R460
[2] Rehm K E *et al.* 1996 *Phys. Rev. C* **53** 1950
[3] Rehm K E *et al.* 1997 *Phys. Rev. C* **55** R566
[4] Harss B *et al.* 2000 *Rev. Sci. Instrum.* **71** 380
[5] Savard G *et al.* 2008 *Nucl. Instrum. Methods Phys. Res., Sect. B* **266** 4086
[6] Wuosmaa A H *et al.* 2007 *Nucl. Instrum. Methods Phys. Res., Sect. A* **580** 1290
[7] Lighthall J C *et al.* 2010 *Nucl. Instrum. Methods Phys. Res., Sect. A* **622** 97
[8] Back B B *et al.* 2010 *Phys. Rev. Lett.* **104** 132501

Future Perspectives of the Legnaro National Laboratories: The SPES project

2011 J. Phys.: Conf. Ser. 267 012003

(http://iopscience.iop.org/1742-6596/267/1/012003)

Future Perspectives of the Legnaro National Laboratories: The SPES project

G. de Angelis, A.Andrighetto, G.Bassato, L.Biasetto,L.Calabretta*,M.Comunian, A.Galata', M.Giacchini, F.Gramegna, A.Lombardi, M.Manzolaro, G.Prete, L.Sarchiapone, D.Scarpa, D.Zafiropoulos and the SPES collaboration

INFN, Laboratori Nazionali di Legnaro,
Viale dell'Universita' 2, I-35020 Legnaro (Pd), Italy
* INFN, Laboratori Nazionali del Sud, Via S. Sofia 2, Catania, Italy

E-mail: giacomo.deangelis@lnl.infn.it

Abstract. The aim of the SPES (Selective Production of Exotic Species) project is to provide high intensity and high-quality beams of neutron-rich nuclei to perform forefront research in nuclear structure, reaction dynamics and interdisciplinary fields like medical, biological and material sciences. SPES is a second generation ISOL radioactive ion beam facility, part of the INFN Road Map for the Nuclear Physics and supported by the italian national laboratories LNL (Legnaro) and LNS (Catania). It represents an intermediate step toward the future generation European ISOL facility EURISOL. It is based on the ISOL method with an UCx Direct Target able to sustain a power of 10 kW. The primary proton beam will be delivered by a Cyclotron accelerator with an energy of more then 40 MeV and a beam current of 200 μA. Neutron-rich radioactive ions will be produced by Uranium fission at an expected fission rate in the target of the order of 10^{13} fissions per second. The exotic isotopes will be re-accelerated by the ALPI superconducting LINAC up to energies of 10-13 AMeV, for masses in the region of A=130 amu, with an expected rate on the secondary target of 10^8 pps.

1. Introduction

Presently our knowledge about the structure of nuclei is mostly limited to nuclei close to the valley of stability or nuclei with a deficiency of neutrons. Only recently the availability of beams of unstable ions has given access to unexplored regions of the nuclear chart, especially on the neutron rich side. Starting from a nucleus on the stability line and adding successively neutrons, one observes that the binding energy of the last neutron decreases steadily until it vanishes decaying by neutron emission. The position in the nuclear chart where this happens defines the neutron drip line. It lies much farther away from the valley of stability than the corresponding drip line associated with protons, owing to the absence of electrical repulsion between neutrons. The location of the neutron drip line is largely unknown as experimental data are available only for nuclei with mass up to around 30. The interest in the study of nuclei with large neutron excess is not only focused on the location of the drip line but also on the investigation of the density dependence of the effective interaction between the nucleons for exotic N/Z ratios. In fact, changes of the nuclear density and size in nuclei with increasing N/Z ratios are expected to lead to different nuclear symmetries and new excitation modes. While in the case of some very light nuclei a halo structure has been identified, for heavier nuclei the formation of a neutron skin has been predicted. The nuclear properties towards the neutron drip line depend on how the shell structure changes as a function of neutron excess. These changes have consequences on the ground state properties of the nuclei and on the single-particle and collective excitations. In particular, studies of neutron-rich nuclei beyond the doubly magic ^{132}Sn are of key importance to investigate the single-particle structure above the N=82 shell closure and find out how the effective interaction between valence nucleons behaves far from stability. New modes of collective motion are also expected in connection with the formation of a neutron skin, namely oscillation of the skin against the core, similar to the soft dipole mode already identified in the case of very light halo nuclei. Presently, neither the thickness nor the detailed properties of the neutron skin of exotic nuclei are known. This information is needed to enable a quantitative description of compact systems like neutron stars, where exotic nuclei forming a Coulomb lattice are immersed in a sea of free neutrons, a system which is expected to display the properties of both finite and infinite (nuclear matter) objects. At the beam energies of SPES (Selective Production of Exotic Species), the radioactive beam project of the Legnaro National Laboratories, it will be possible to address questions related to the properties of neutron rich matter from the perspective of nuclear forces, level density, viscosity, barrier, neutron pairing and collective modes.

Figure 1: The Laboratori Nazionali di Legnaro.

The LNL accelerator facilities

In order to better underline the framework in which the SPES project is going to be developed, a short description of the Legnaro National Laboratory (LNL - figure 1) will be briefly outlined in the following. The LNL heavy-ion accelerator complex is based on a 16 MV Tandem XTU, a Superconductive LINAC (ALPI) and a superconductive radiofrequency quadrupole (RFQ) heavy ion injector (PIAVE). The Tandem XTU is operating stand alone or as an injector to ALPI. The super-conducting RFQ injector PIAVE is based on an ECR Ion Source (placed on a 350 kV platform) and on a super-conducting RFQ able to accelerate ions with $A/q < 8.5$ up to an energy/nucleon of 1.2 MeV/A. The ALPI accelerator is a superconducting heavy ion LINAC, composed of the RFQ section and three quarter wave resonator (QWR) sections for a total of 80 cavities installed. It operates routinely at an equivalent voltage of 50 MV. The LINAC is constructed in a bended configuration: it is composed by two branches connected by an achromatic and isochronous U-bend. It uses three different kinds of cavities: Low Beta, Medium Beta and High Beta cavities, according to the different velocity along the acceleration path. In the last years the cavities of the medium energy QWR section were upgraded using a new Nb sputtered coating in substitution to the original Pb sputtered layer. An upgrade program is on the way, to improve the accelerating fields of the present QWRs and adding more cavities in the Low Beta section. The final equivalent voltage is expected to exceed 70 MV in optimized conditions (all the resonators operating at the designed voltage, normalized transit time factor and synchronous phase taken into account). A further energy improvement has been tested recently installing a stripper station before the U-bend. In this condition the energy/nucleon increased by 20% (from 6.8 to 8.1 MeV/A for ^{136}Xe) with the drawback of a reduced transmission (30%).

2. The SPES Project

SPES [1] is designed to provide neutron-rich radioactive nuclear beams (RIB) of final energies in the order of 10 - 13 MeV/A for nuclei in the A= 80-130 mass regions. The radioactive ions will be produced with the ISOL technique using the proton induced fission on a Direct Target of UCx [2,3] and subsequently reaccelerated using the PIAVE ALPI accelerator complex. An Uranium fission rate of 10^{13} fission/s is foreseen. As proton driver accelerator will be used a variable energy (30-70 MeV) Cyclotron with a maximum current of 0.750 mA rowing two exit ports. Only a maximum beam current of 200 μA will be used for RIB production. The proton beams delivered by the Cyclotron will also be used to produce neutrons for a Boron Neutron Capture Therapy (BNCT) installation, to perform researches dedicated to cancer therapy, and for an irradiation-facility (LENOS) designed for material research studies and cross section measurements. The expected thermal (after moderation) neutrons fluency is estimated as 10^9 n cm^{-2} s^{-1} while the rate of fast neutrons is estimated to be 10^{14} n s^{-1}. The SPES lay-out is shown in figures 2 and 3.

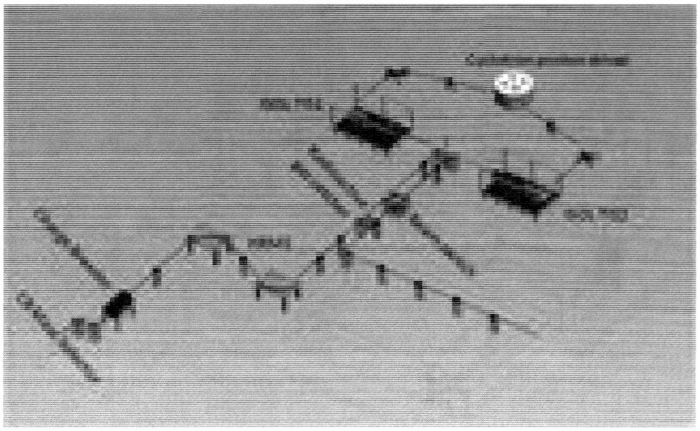

Figure 2: SPES beam transfer layout.

Figure 2 shows schematically the transfer line for the exotic beam. The general configuration of the SPES layout follows the one of the EXCYT facility, the ISOL facility for proton-rich nuclei in operation at LNS (Catania, Italy). The production target and the first mass selection element will be housed in a high radiation bunker and mounted on a high voltage platform. Before the High Resolution Mass Spectrometer a cryopanel will be installed to prevent the beam line to be contaminated by radioactive gasses. After passing through the High Resolution Mass Spectrometer (HRMS), the selected isotopes will be stopped inside the Charge Breeder and extracted with increased charge. Before reaching the PIAVE-ALPI accelerator, a final mass selector (CB_MassSelector) is designed to clean the beam from the contaminations introduced by the Charge Breeder itself. In figure 3 the ISOL facility is located in the white area, housing the cyclotron proton driver, the two RIB targets, the High Resolution Mass Spectrometer (HRMS) and the transfer lines. For safety reasons the

ISOL facility is designed to be constructed 5 meter below ground level [4]. The target development laboratory (not shown in figure 3) will be constructed at ground level above the ISOL facility. Two laboratories for applied physics and other applications are planned: one at the same level of the ISOL facility, which makes use of the Cyclotron proton beam, and another at ground level.

The most critical element of the SPES ISOL facility is the Direct Target. The proposed target represents an innovation in term of capability of sustaining the primary beam power. The design is carefully oriented to optimize the irradiative cooling, taking advantage of the target system high operating temperature, which is in the order of 2000°C.

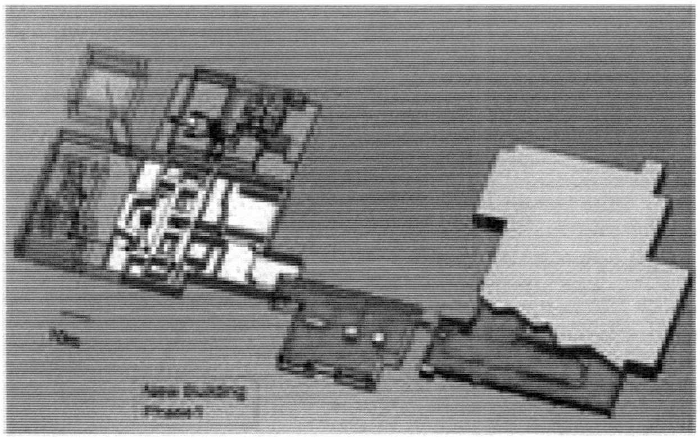

Figure 3: SPES building layout and connection to PIAVE-ALPI laboratory.

An extensive simulation of the target behaviour for thermal and release properties is at the bases of the target-ion-source design. Experimental work to bench mark the simulations was carried out in collaboration with HRIBF, the Oak Ridge National Laboratory ISOL facility (USA). The production target is designed following the ISOLDE (CERN) and EXCYT (LNS, Catania) projects, devoting special care to the system for safety and radiation protection. According to the estimated level of activation in the production target area (10^{13} Bq) special infrastructures needs to be designed. The use of up-to-date techniques of nuclear engineering will result in a high security level of the installation: the control system will integrate radiation management and survey as well as facility operation and safety infrastructures. Redundancies and fault tolerant PLC will be adopted in the low level layer of the control system while EPICS and LabView will be used in the general architecture and user front-end. The interaction of the proton beam with the UCx target will produce fission fragments of neutron-rich isotopes that will be extracted by thermal motion and ionized at 1^+ charge state by a source directly connected with the production target. Several kinds of sources are foreseen according to the beam of interest. A laser ion source for producing a beam as pure as possible will be implemented in collaboration with INFN-Pavia. The selection and the transport of the low intensity exotic beam at low energy is a challenging task. Techniques already applied to the EXCYT beam are of reference for SPES; they include the High Resolution Mass Spectrometer, the online identification station and several systems for low current beam diagnostics. Before the injection in the PIAVE-ALPI Linac, the Charge Breeder is an essential element for an effective reacceleration as it increases the charge state from 1+ to n+. The SPES Charge Breeder is based on ECR method and aims to produce ions with A/q less than 6 for A~130. An experimental program is under way to characterize the SPES target-ion source system. The initially expected beam-on-target intensities are of the order of 10^8 pps for ^{132}Sn and ^{90}Kr, and of about 10^6-10^5 pps for ^{134}Sn and ^{95}Kr, considering a total efficiency of 2% for the transmission from the 1^+ source to the experimental target. Figure 4 shows the beam on target intensities expected for the final stage of the project. A crucial task for the experimental use of radioactive beams is not only the beam intensity but also the beam quality. Special efforts have been dedicated to design a mass spectrometer with an effective mass resolution of at least 1/20000. Such design takes advantage of the 260 keV beam energy obtained with the HV platforms. Such high selectivity results in an advantage also for the safety issue, reducing the problems of contaminations along the beam transport areas and in the target location. The in-target beam intensities at SPES have been estimated from fission fragments production yields calculated with the MCNPX [5] Monte Carlo code in which the target geometry is included. The diffusion and effusion of the exotic species inside the target was evaluated with both GEANT4 [6] and RIBO [7] Monte Carlo codes. The calculations have been tuned using the available experimental data from ISOLDE, ORNL and PNPI taking into account the complete target geometry. Finally, source ionization and extraction, charge breeding, beam transport and re-acceleration efficiencies have been considered. The final estimated beam intensities are shown in figure 4 for some selected species.

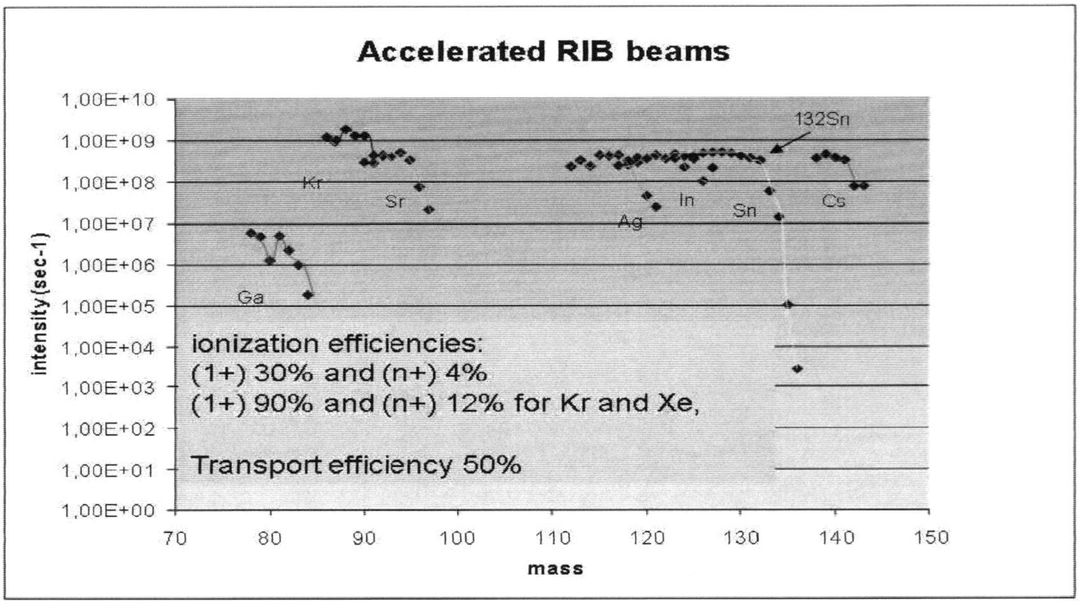

Figure 4: Expected on-target intensities calculated considering emission, ionization and acceleration efficiencies for different isotopes.

[1] SPES Technical Design Report 2008 ed. G. Prete and A. Covello *INFN-LNL 223* http://www.lnl.infn.it/spes/TDR2008/tech_design08_index.htm.

[2] Andrighetto A *et al.* 2010 *Nucl. Phys.* A **834** 754c.

[3] Prete G *et al.* 2009 *J. Phys. Conf. Ser.* **168** 012022.

[4] Gramegna F. *et al.* 2007 VII Latin American Symposium on nuclear physics and applications, *AIP Conf. Proc.* 947, 422.

[5] LA-CP-05-0369 2005 MCNPXTM User Manual, Version 2.5.0 ed. Denise B. Pelowitz.

[6] GEANT4: CERN Program Library W5013 Geant detector description and simulation tool.

[7] Litner S., 2005, *MUCP-ETSEIB / CERN,* Ph.D. thesis http://ribo.web.cern.ch/rib.

Probing nuclei at the limits of nuclear landscape

J Uusitalo[1] for the Gamma, IGISOL, and RITU collaborations
University of Jyväskylä, Department of Physics
P. O. Box 35, FI-40014, Jyväskylä, Finland

E-mail: juha.uusitalo@phys.jyu.fi

Abstract. Light- and heavy ion beams from the K=130 cyclotron are used to produce nuclei far from stability at the Accelerator Laboratory of the Department of Physics of the University of Jyväskylä(JYFL), Finland. Novel instruments at the target and focal-plane areas of a gas-filled recoil separator for in-beam tagging measurements are available for probing structures of very neutron deficient and very heavy nuclei. Bunched and cooled radioactive beams from the IGISOL system are available for studies of nuclear ground-state properties by employing Penning traps and laser systems. Recent results from measurements at these two facilities are presented. Future plans will also be introduced.

1. Introduction

The Accelerator Laboratory at JYFL has been nominated to be the Finnish Centre of Excellence in Nuclear and Accelerator Based Physics by the Academy of Finland for the period of 2006-2011. The status covers research and development work as well as ion beam applications at the Accelerator Laboratory of JYFL and theoretical nuclear physics activities carried out at JYFL. The K130 Cyclotron accelerates light and heavy ion beams, provided by two ECR sources and one multi-cusp source for heavy and light ions, respectively. The total beam time on target has exceeded 6000 hours a year for many years. In this report some recent highlights related to experimental nuclear structure studies performed by the combined Gamma and RITU groups (Nuclear Spectroscopy Group) and by the IGISOL group will be presented.

2. Some highlights from the Nuclear Spectroscopy Group

Figure 1. JUROGAMII+RITU+GREAT facility

[1] To whom any correspondence should be addressed.

The main activity of the Nuclear Spectroscopy Group has been on making use of the powerful JUROGAMII+RITU+GREAT facility (see figure 1). JUROGAMII is a γ-ray array placed at the target position of RITU. RITU is an in-flight gas-filled recoil separator which separates the fusion products from other recoils. At the focal plane a GREAT spectrometer measures recoils, decay products (α's, β's and protons), γ-rays and electrons. Spectroscopic studies have successfully been performed covering a broad area of the nuclear landscape. The nuclei at the N=Z line below A=100, the neutron-deficient rare earth nuclei (A = 117-160), the proton drip-line nuclei in the vicinity of Z = 82 and the heaviest elements (nobelium) have been examined. In these studies the Recoil Gating (RG), Recoil Decay Tagging (RDT) and Recoil Isomer Tagging (RIT) methods have been utilized.

The program to explore nuclear structure at the proton drip line in the lead region has been extensive. One intriguing question is the survival of shape coexistence when approaching the proton drip line. Following the successful development of Zr beams by the ECR group, the JUROGAM II array at RITU was employed in an RDT experiment to observe prompt γ rays from ^{180}Pb. Excited states were populated via the cold ^{92}Mo(^{90}Zr,2n)^{180}Pb reaction. The measured

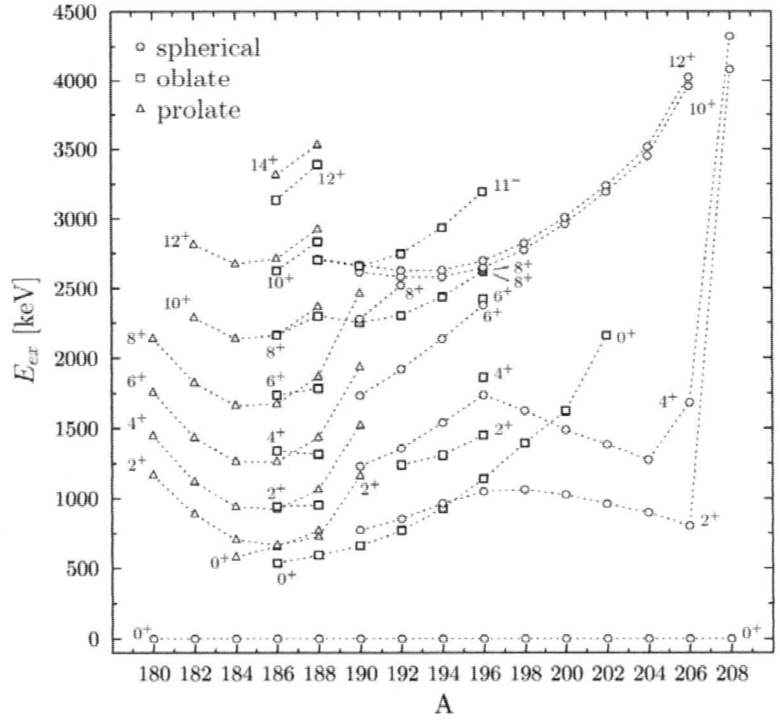

Figure 2. Level systematic of lead nuclei with A < 208. The data are taken from the works [1-4]

production cross-section was only 10 nb. All the excited states of ^{180}Pb are proton unbound. In figure 2 is shown level systematic of the neutron deficient lead isotopes. From the systematics it can be seen that the intruding prolate structures still persist in the very neutron deficient ^{180}Pb.

At JYFL persistent work has been performed in the Nobelium region [5-10]. In in-beam γ-ray spectroscopy studies one of the limiting factors has been the permitted total counting rates at the target position. The combination of digital electronics with JUROGAMII enables low cross-section in beam experiments at high total γ-ray counting rates for heavy nuclei. A successful in-beam spectroscopy run was carried out, for the first time, for ^{246}Fm produced in the ^{208}Pb(^{40}Ar,2n)^{246}Fm reaction with a cross-section of approximately 10 nb, which is the lowest cross-section so far to be studied in-beam in the heavy-element region (figure 3) [11].

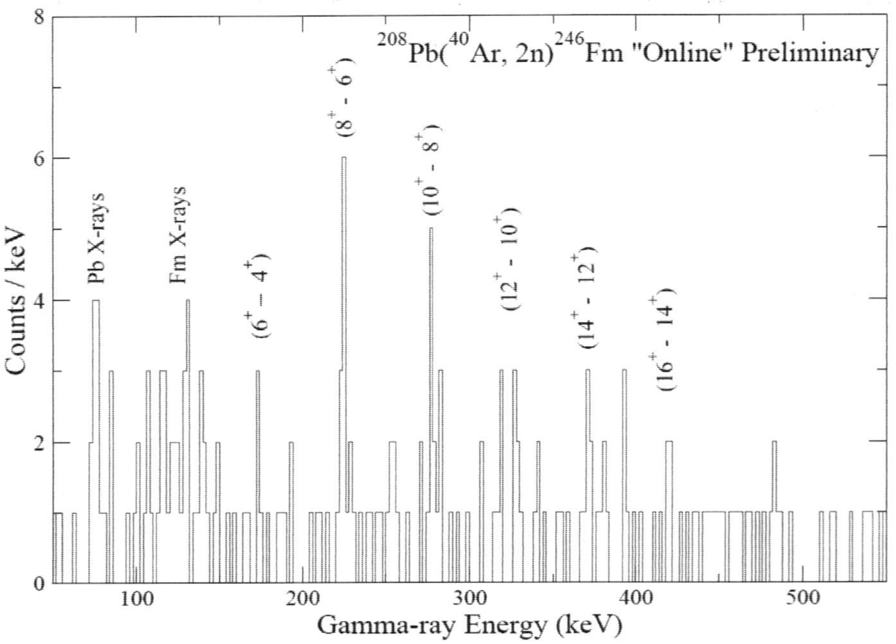

Figure 3. Gamma-ray spectrum obtained for ^{246}Fm

2. Some highlights from the IGISOL group

One of the main research tools at JYFL is the IGISOL isotope separator facility (figure 4). Neutron-rich nuclei are produced using light ion induced fission reactions whereas the proton-drip line nuclei are probed employing the heavy ion induced fusion evaporation reactions. After ion guide the products are mass separated and transported to the radiofrequency quadrupole ion cooler and buncher which is followed by the JYFLTRAP tandem Penning trap. The ion cooler is used to improve the ion beam quality and bunch the beam for laser spectroscopy and the Penning trap. The Penning trap can be used for isobaric purification of the ion beams for nuclear spectroscopy experiments. It can also be employed for very precise (down to 1 keV/c²) mass measurement of the ions. These mass measurements have recently provided many data points for the CVC hypothesis and CKM unitarity tests [12, 13]. A new development is resonant laser ionization of atoms in the helium gas jet leaving the ion guide.

The collinear laser spectroscopy set-up, owned and operated by the Universities of Birmingham and Manchester, is connected to the IGISOL and can be considered as a part of the facility. In the experiments, the isotope shifts, nuclear electric quadrupole and magnetic dipole moments for short-lived isotopes of refractory elements such as hafnium, niobium, molybdenum and zirconium have been determined, an achievement that would have not been possible anywhere else in the world.

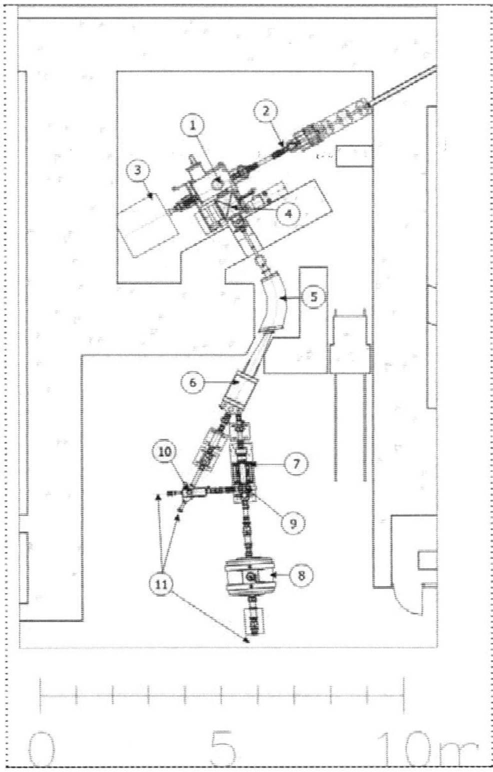

Figure 4. The IGISOL facility floor level layout after the front end upgrade [14]. (1) Target chamber (2) K-130 cyclotron beam line (3) light ion beam dump (4) extraction chamber (5) dipole magnet (6) switchyard (7) radiofrequency cooler (8) JYFLTRAP (9) mini-quadrupole beam deflector (10) electrostatic deflector and beam line upstairs to collinear laser experiments and (11) beam lines for experimental setups.

IGISOL group has studied extensively the evolution of shell gaps in different parts of chart of nuclei [15-18]. As an example in figure 5 two-neutron separation energies around neutron number 50 and covering the elements from cobalt up to zirconium are shown. At proton number Z = 40 the shell gap N = 50 is pronounced. When approaching the Z = 28 and ^{78}Ni it can be noticed that N = 50 shell gap is again pronounced. Although more data points are needed this could be a sign of that the double magic structure persists in very neutron rich ^{78}Ni.

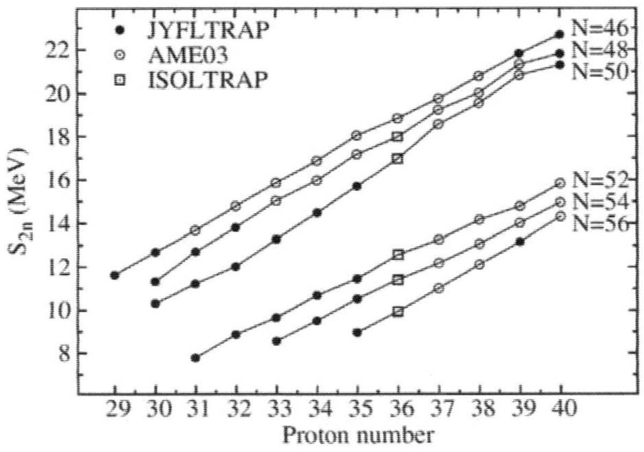

Figure 5. Two-neutron separation energies as a function of proton number [15].

In figure 6 is shown results from the collinear laser spectroscopy studies and from the mass measurements for the neutron rich elements of $36 \leq Z \leq 42$ [15-22]. At N = 60 a sudden increase of the obtained charge radii for higher neutron numbers for elements around zirconium can be seen. This has been interpreted as sign of rapid change from β soft, weekly oblate to static strongly prolate ground state structures. This shape change as a function of neutron number is much smoother for Mo isotopes. The effect caused by this shape change can be seen as anomaly behavior in the two-neutron separation energies around neutron number 60.

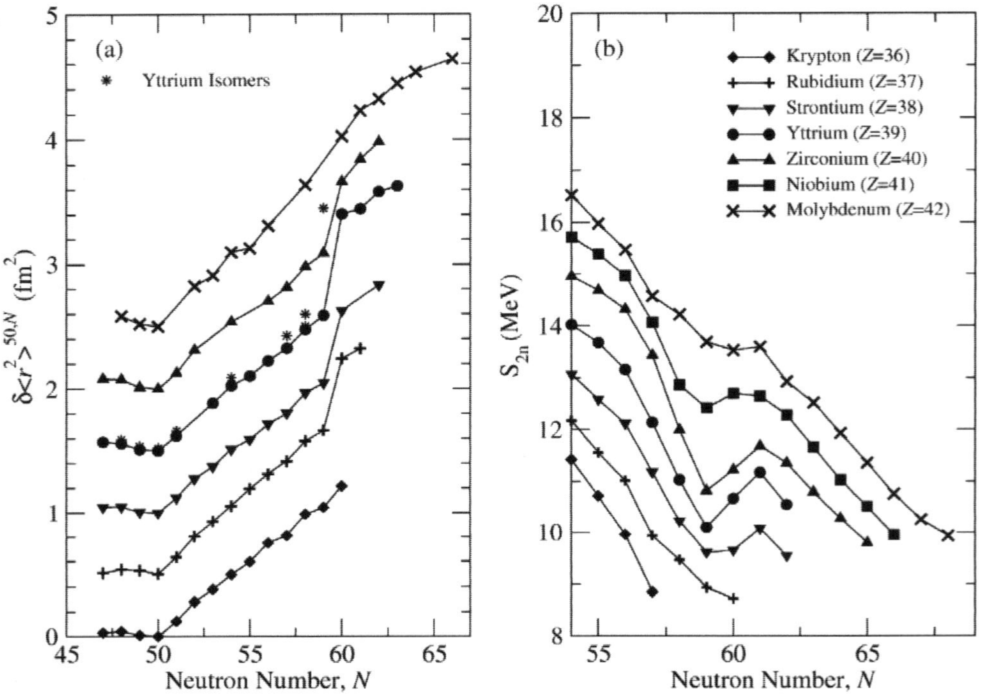

Figure 6. Charge radii and mass measurements in the neutron-rich A ~ 100 region: Figure taken from [19].

3. Future plans

The highlight of the year 2009 was the advent of the new light-ion MCC30/15 cyclotron from NIIEFA, St. Petersburg and the commissioning of the new extension of the Accelerator Laboratory constructed in 2008-2009 (figure 7). First beams from the cyclotron were delivered on 11th November 2009. In the future the use of the new light-ion MCC30/15 cyclotron will release beam time for basic research, tests and service at the K130 cyclotron. The IGISOL facility is foreseen to move in to the new site. The IGISOL group is likely to be the main user of the new accelerator; however the IGISOL facility will also have access to the heavy ion beams provided by the K130 cyclotron through a connection from the old beam line to the new one. The move is expected to take place on the second half of 2010. This move will allow updating and improving the performance of the IGISOL facility.

The SAGE (combined electron and γ spectrometer) and LISA (charge particle spectrometer) spectrometers were recently commissioned at JYFL. The SAGE spectrometer will allow to look for E0 transitions and to perform more comprehensive in-beam studies in the heaviest element region where the low energy transitions are highly converted. The LISA spectrometer will be used to search for fast charge particle decays. The planned vacuum-mode mass separator MARA will be used for spectroscopic studies in the A ~ 80 region. The magnetic components for the MARA separator have been purchased but the most expensive part, the electric dipole, is waiting for funding.

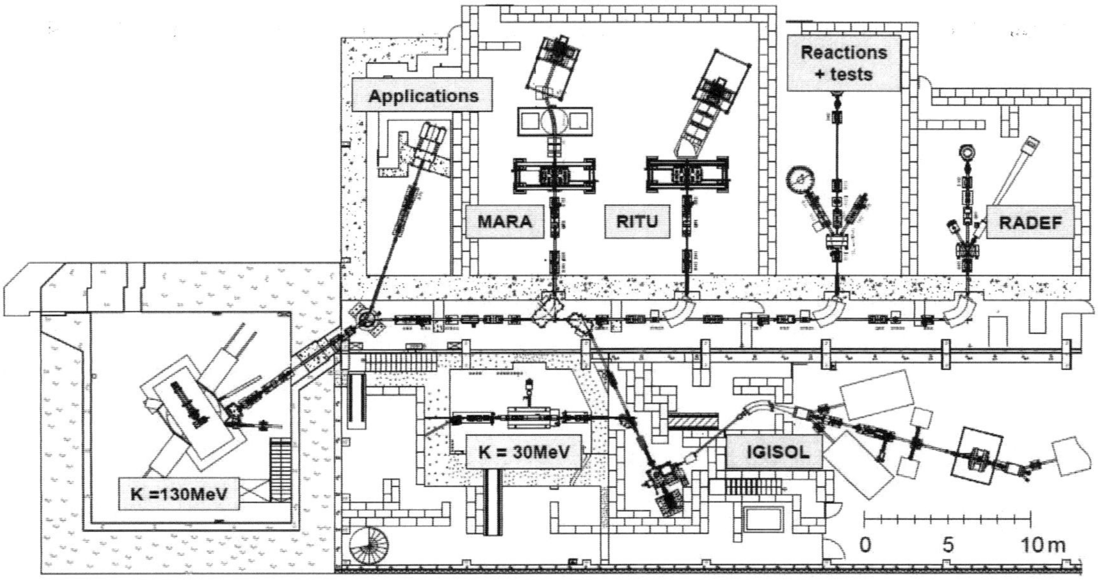

Figure 7. The future layout of the JYFL accelerator laboratory. The new extension part is shown in the right lower corner. The IGISOL facility will move to this new extension. The new MARA separator will be placed alongside the existing RITU separator.

[1] Rahkila P et al., *Phys. Rev.* C **82** (2010) 011303
[2] Jenkins D G et al., *Phys .Rev.* C **62** (2000) 021302
[3] Julin R *J. Phys.* G **27** (2001) R109
[4] http://www.nndc.bnl.gov/ensdf/, database version of October 27, 2009.
[5] Greenlees P T et al., *Phys. Rev.* C **78** (2008) 021303
[6] Chatillon A et al., *Phys. Rev. Lett.* **98** (2007) 132503
[7] Chatillon A et al., *Eur. Phys. J.* A **30** (2006) 397
[8] Bastin J E et al., *Phys .Rev.* C **73** (2006) 024308
[9] Ketelhut S et al., *Phys. Rev. Lett.* **102** (2009) 212501
[10] Herzberg R D et al., *Nature(London)* **442**, (2006) 896
[11] Greenlees et al., to be submitted
[12] Eronen T et al., *Phys. Rev. Lett.* **97** (2006) 232501
[13] Eronen T et al., *Phys. Rev. Lett.* **100** (2008) 132502
[14] Karvonen P et al., *Nucl. Instrum. Methods Phys. Res., Sect.* B **266** (2008) 4454
[15] Hakala J et al., *Phys. Rev. Lett.* **101** (2008) 052502
[16] Hager U et al. *Phys. Rev. Lett.* **96** (2006) 042504
[17] Hager U et al., *Nucl. Phys.* A **793** (2007) 20
[18] Rahaman S et al., *Eur. Phys. J.* A **32** (2007) 87
[19] Charlwood F C et al., *Phys. Lett.* B **674** (2009) 23
[20] Cheal B et al., *Phys. Rev. Lett.* **102** (2009) 222501
[21] Cheal B et al., *Phys. Lett.* B **645** (2007) 133
[22] Campbell P et al., *Phys. Rev. Lett.* **89** (2002) 082501

Shell structure of weakly-bound and resonant neutrons

2011 J. Phys.: Conf. Ser. 267 012005

(http://iopscience.iop.org/1742-6596/267/1/012005)

Shell structure of weakly-bound and resonant neutrons

Ikuko Hamamoto
Division of Mathematical Physics, LTH, University of Lund, Sweden
E-mail: ikuko@MATFYS.LTH.SE

Abstract. The systematic change of shell structure in both weakly bound and resonant neutron one-particle levels in nuclei towards the neutron drip line is exhibited, solving the coupled equations derived from the Schrödinger equation in coordinate space with the correct asymptotic behaviour of wave functions for $r \to \infty$. The change comes from the behaviour unique in the one-particle motion with low orbital angular momenta compared with that with high orbital angular momenta. The observed deformation of very neutron-rich nuclei with $N \gtrsim 20$ in the island of inversion is a natural result of this changed shell structure, while a possible deformation of neutron-drip-line nuclei with $N \approx 51$, which are not yet observed, is suggested.

1. Introduction

The study of one-particle motion in deformed potentials is the basis for understanding the structure of deformed drip-line nuclei. Since the Fermi level of drip line nuclei lies close to the continuum, both weakly-bound and positive-energy one-particle levels play a crucial role in the many-body correlation of those nuclei. Among an infinite number of one-particle levels at a given positive-energy, only some selected levels related to one-particle resonant levels will be important to bound states of drip-line nuclei. The motion of weakly-bound or low-energy resonant protons is not so different from that of well-bound protons except for very light nuclei, due to the presence of the Coulomb barrier. Therefore, the present talk is devoted to the change of shell structure of one-neutron levels.

The change of the shell structure in weakly-bound and resonant one-neutron levels is exhibited, which comes from the behavior unique in the one-particle motion with low orbital-angular-momenta (ℓ). For spherical shape the centrifugal barrier is absent for ℓ=0 orbits while the barrier is low for ℓ=1. For deformed shape a given one-particle wave-function contains components with various ℓ values, however, the minimum value of possible ℓ for the wave-function components plays a crucial role in the present problem. For example, in the case of axially-symmetric quadrupole deformation ℓ_{min}= 0 for $\Omega^\pi = 1/2^+$ and ℓ_{min}=1 for both $\Omega^\pi = 3/2^-$ and $1/2^-$, where Ω expresses the component of the particle angular-momentum along the symmetry axis. For bound one-particle levels the $\ell = 0$ component becomes always overwhelming in one-particle wave-functions with $\Omega^\pi = 1/2^+$, as the binding energy approaches zero. Namely, in the limit of small binding energy the halo phenomena which are created exclusively by the $\ell = 0$ neutron wave-function appear for both spherical and deformed shapes. On the other hand, the decay of $\Omega^\pi = 1/2^+$ one-particle resonances goes most easily through the $s_{1/2}$ channel because of the absence of the centrifugal barrier. Therefore, it is in general difficult for an Ω^π = $1/2^+$ Nilsson level to survive as a resonant level in the positive-energy region. Nevertheless, we note that a given one-particle resonance with $\Omega^\pi = 1/2^+$ may survive even at several MeV excitation-energy, if the higher-ℓ components are predominant in the wave function inside the potential. In such cases the existence of the resonance is made possible by the presence of the higher-ℓ components while the width of the resonance is determined by the small $\ell = 0$ component.

One-particle resonant levels for spherical shape are defined in terms of phase shift as described in standard textbooks. On the other hand, the definition of one-particle resonance for deformed potentials is obtained using the eigenphase formalism [1, 2, 3]. The definition is a natural extension of the definition of one-particle resonance for spherical potentials in terms of phase shift.

Taking the three bound one-particle levels for neutrons in the *sd* shell, in [4] it is shown how the order of the three levels varies as a function of binding energies: from the bottom to the top in energy, (i) the relative level order is $1d_{5/2}$, $1d_{3/2}$ and $2s_{1/2}$ for strongly bound cases,; (ii) $1d_{5/2}$, $2s_{1/2}$ and $1d_{3/2}$ for about 10 MeV binding and (iii) $2s_{1/2}$, $1d_{5/2}$ and $1d_{3/2}$ for very weak bindings. Namely, the $\ell = 0$ level shifts downwards relative to the $\ell = 2$ levels, as the potential strength becomes weaker or the binding energy becomes smaller. The level order in (ii) is known as the usual energy splitting in stable *sd*-shell nuclei, which is used in the shell model. In the present talk we concentrate on the shell structure of bound and resonant neutron levels in the *pf*-shell as well as in the $50 < N \leq 82$ major shell, taking examples of a stable *sd*-shell nucleus (figure 1), a known deformed neutron-drip-line nucleus (figure 2), and a not yet observed neutron-drip-line nucleus in the region of medium mass (figure 3).

2. Model and numerical examples

In the description of one-particle resonances the formalism using complex energy is sometimes popular in the literature. However, it is equally possible to formulate the resonances using real energy. I prefer to choosing the latter, since the treatment of certain observable quantities such as matrix-elements or wave-function components is much simpler with real variables.

Solving the coupled equations derived from the Schrödinger equation in coordinate space with the correct asymptotic behaviour of wave functions for $r \to \infty$, the shell structure in axially symmetric quadrupole-deformed Woods-Saxon potentials is studied. Whereas one-particle bound levels in the deformed potential are obtained by solving the well-known eigenvalue problem, for positive-energy one-particle levels all eigenphases for a given Ω^π are calculated, and the possible one-particle resonant level for a given energy ε_Ω and a given Ω^π value may be identified, which is associated with a particular eigenphase among all possible eigenphases. One-particle resonance is not obtained if none of calculated eigenphases do not increase through $\pi/2$ as the one-particle energy increases at the given energy. The relevant formulation can be found in [2, 3].

The parameters of the Woods-Saxon potential are chosen as follows: the diffuseness a= 0.67 fm and the radius $R = r_0 A^{1/3}$ with r_0=1.27 fm, while the depth which depends on the neutron excess in respective nuclei is taken approximately from the expression (2-182) in [5]. The neutron potential for nuclei with a neutron excess is shallower than that for $N = Z$ nuclei.

2.1. Nilsson diagram for a stable nucleus $^{25}_{12}Mg_{13}$

The analysis of observed spectroscopic properties of low-lying states in light deformed mirror nuclei, $^{25}_{12}\mathrm{Mg}_{13}$ and $^{25}_{13}\mathrm{Al}_{12}$, in terms of one-particle motion in a deformed potential is very successful [6] using the deformation parameter $\delta_{osc} \approx 0.4$. In [6] the Nilsson diagram based on a modified oscillator potential is used. Since the neutron separation energy S_n of $^{25}\mathrm{Mg}$ is 7.3 MeV, the properties of low-lying levels could well be described using modified oscillator potential

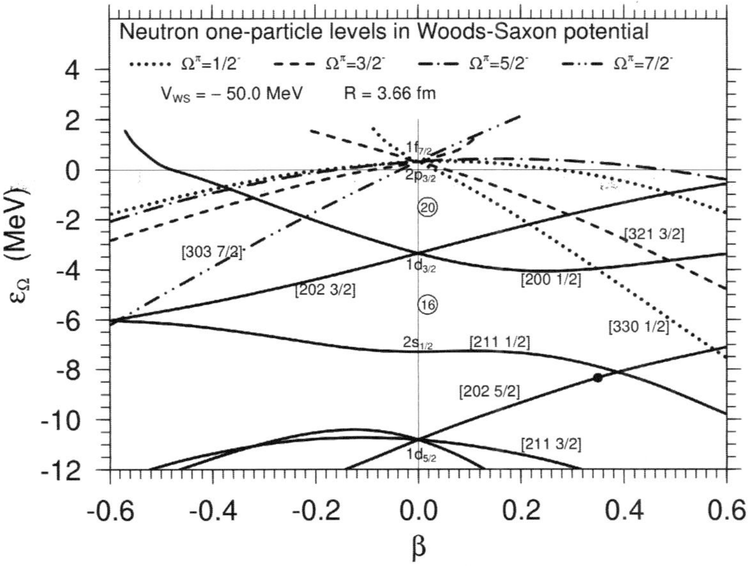

Figure 1. Neutron one-particle levels as a function of axially-symmetric quadrupole deformation. The depth and radius of the Woods-Saxon potential are designed approximately for ${}^{25}_{12}\mathrm{Mg}_{13}$. The possible occupancy of the 13th neutron in the ground state of ^{25}Mg is indicated by the filled circle. Some one-particle levels are denoted by the asymptotic quantum numbers, $[Nn_z\Lambda\Omega]$. One-particle levels plotted for $\varepsilon_\Omega > 0$ are resonant levels obtained by using the eigenphase formalism, of which the widths are not plotted for simplicity. Note that one-particle resonant $p_{3/2}$ and $f_{7/2}$ levels, of which the resonant energies are about 0.3 MeV, are almost degenerate.

when the related parameters were properly adjusted. In figure 1 the Nilsson diagram based on a deformed Woods-Saxon potential is shown. Observed spectroscopic properties of low-lying states are equally well understood also using the Nilsson diagram in figure 1.

At β=0 of figure 1 the $2s_{1/2}$ level appears around the middle of the $1d_{5/2}$ and $1d_{3/2}$ level, as usually assumed. A striking difference of figure 1 from the Nilsson diagram based on the modified oscillator potential (or from one-particle energies used in the conventional shell model) is seen in the energies of the $2p_{3/2}$ and $1f_{7/2}$ levels. Calculated one-particle resonant energies are $\varepsilon_{res}(p_{3/2}) = +0.31$ MeV and $\varepsilon_{res}(f_{7/2}) = +0.32$ MeV. Namely, the $p_{3/2}$ level lies even lower than the $f_{7/2}$ level. In other words, the N=28 energy gap totally disappears in the low-energy resonance spectra.

2.2. Nilsson diagram for a deformed neutron-drip-line nucleus ${}^{31}_{10}Ne_{21}$

In figure 2 the Nilsson diagram appropriate for neutrons in ${}^{31}_{10}\mathrm{Ne}_{21}$, of which the measured neutron separation energy S_n is 0.29±1.64 MeV, is shown. At β=0 the $f_{7/2}$ one-neutron resonance is found at 2.04 MeV, while neither $p_{3/2}$ nor $p_{1/2}$ resonance defined by the eigenphase formalism is obtained. However, the $p_{3/2}$ resonance lying clearly lower than the $f_{7/2}$ resonance is found if we use a slightly more attractive Woods-Saxon potential. The next low-lying one-particle resonant level for β=0 is the $f_{5/2}$ level at 9.50 MeV.

The measured very low excitation-energies of the first 2^+ state of both ^{30}Ne and ^{32}Ne indicate that the Ne isotope with the neutron numbers is deformed. Furthermore, the large Coulomb breakup cross section reported in [7] clearly suggests the halo nature of the ground state of ^{31}Ne.

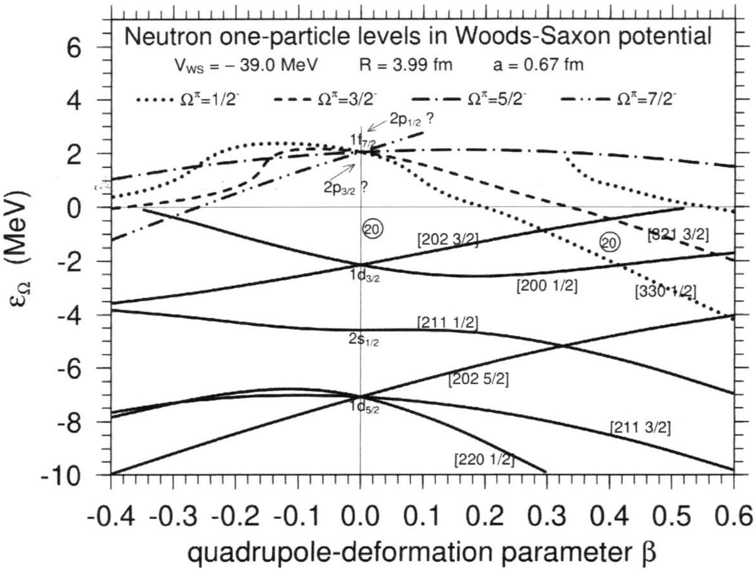

Figure 2. Neutron one-particle levels as a function of axially-symmetric quadrupole deformation. The depth and radius of the Woods-Saxon potential are designed approximately for $^{31}_{10}\mathrm{Ne}_{21}$. The neutron number 20, which is obtained by filling in all lower-lying levels, is indicated with a circle. The approximate positions of the $2p_{3/2}$ and $2p_{1/2}$ levels for $\beta = 0$ are indicated with "?", which are extrapolated from the resonant energies obtained by using a slightly more attractive Woods-Saxon potential. For the potential with the present strength the $p_{3/2}$ and $p_{1/2}$ resonant levels are not obtained for $\varepsilon_\Omega \gtrsim 1.3$ MeV, due to the low centrifugal barrier for ℓ=1 orbits. The calculated widths of one-particle resonant levels are not plotted, for simplicity.

In the conventional shell model the 21st neutron for a spherical shape occupies the $1f_{7/2}$ shell and, thus, the related halo phenomena are impossible to obtain. If ^{31}Ne is prolately deformed, from figure 2 we obtain the bound Nilsson level which is to be occupied by the 21st neutron having a halo structure: [330 1/2] for $0.20 \lesssim \beta \lesssim 0.30$ and [321 3/2]for $0.40 \lesssim \beta \lesssim 0.58$. In both cases the ground state should have $3/2^-$ and the p-wave neutron halo together with the deformed core ^{30}Ne. The measurement of S_n with an accuracy which is an order of magnitude better than the presently available one will clarify the intrinsic configuration of the ground state of ^{31}Ne [8].

Examining figure 2 we may also note that the presence of ^{31}Ne inside the neutron drip line is possibly realized by the deformation, which is created by the Jahn-Teller effect due to the near degeneracy of $1f_{7/2}$, $2p_{3/2}$ and $2p_{1/2}$ shells in the continuum for the spherical shape. Furthermore, it is remarked that the measured magnetic moment of the isotone nucleus $^{33}_{12}\mathrm{Mg}_{21}$ [9] is in good agreement with the estimated value for the prolately deformed shape [10].

2.3. Nilsson diagram for neutron-drip-line nuclei with $A \approx 75$ and $N \approx 50$

In figure 3 the Nilsson diagram appropriate for neutron-drip-line nuclei with $N \approx 51$ (for example, $^{75}_{24}\mathrm{Cr}_{51}$) is shown. Since in this example the $1f_{7/2}$, $2p_{3/2}$, $1f_{5/2}$ and $2p_{1/2}$ levels are well bound, the presence of the magic number $N = 28$ which is established for stable nuclei is clearly seen. Moreover, the level order of four levels in the $1f - 2p$ major shell is exactly what is used in traditional shell-model calculations. In contrast, we note the peculiar order of weakly-bound

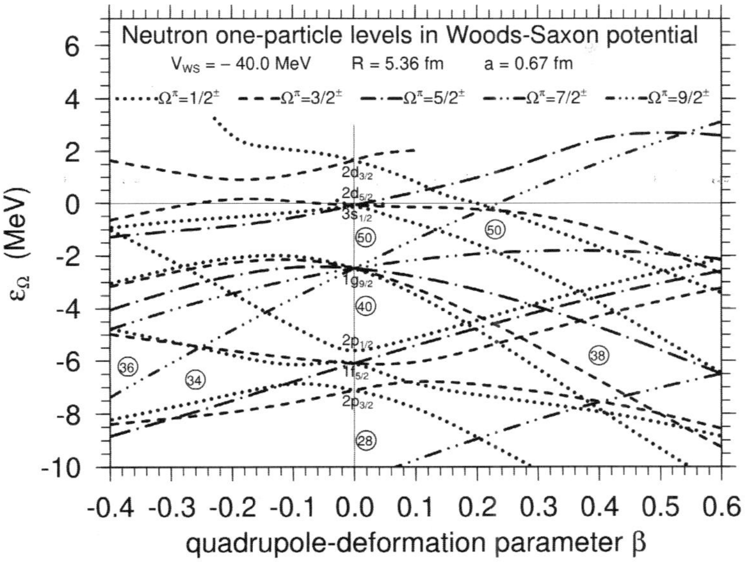

Figure 3. Neutron one-particle levels as a function of axially-symmetric quadrupole deformation. The depth and radius of the Woods-Saxon potential are designed approximately for $^{75}_{24}Cr_{51}$.

and resonant one-particle levels which belong to the $50 < N \leq 82$ major shell. At $\beta = 0$ the very weakly bound $3s_{1/2}$ level is almost degenerate with (but slightly lower than) the $2d_{5/2}$ level, and the one-particle resonant levels in the $50 < N \leq 82$ major shell which are not plotted in the figure are: $\varepsilon_{res}(1g_{7/2}) = +3.44MeV$ and $\varepsilon_{res}(1h_{11/2}) = +5.48MeV$. It is noted that the level order in the major shell consisting of the $3s_{1/2}$, $2d_{3/2}$, $2d_{5/2}$ $1g_{7/2}$, and $1h_{11/2}$ levels is totally different from what is known from stable nuclei.

A similarity of the level bunching around N=21 in figure 2 to that around N=51 in figure 3 as a function of deformation is seen especially for $\beta \geq 0$. Considering the fact that the N=21 neutron-drip-line nuclei, ^{31}Ne and ^{33}Mg, are deformed, the nucleus such as $^{75}_{24}Cr_{51}$ may be also deformed, if it indeed lies inside the neutron drip line. In other words, there might be $N = 51$ nuclei such that they may remain inside the neutron drip line due to the possible deformation. Furthermore, instead of having $I^{\pi} = 7/2^{+}$ or $5/2^{+}$ coming from the occupancy of $1g_{7/2}$ or $2d_{5/2}$ neutron orbits, respectively, by the 51st neutron the neutron-drip-line nuclei with $N = 51$ have a good chance to have the ground or very low-lying $I^{\pi} = 1/2^{+}$ state, irrespective of spherical or deformed shape.

3. Summary and discussions

It is known that the presence of the surface in realistic potentials of stable nuclei pushes down one-particle levels with higher ℓ relative to lower ℓ levels among the levels belonging to a given major shell of the harmonic-oscillator potential. On the other hand, weakly-bound neutrons with small ℓ have an appreciable probability of being outside the potential compared with those with larger ℓ and thus are insensitive to the strength of the potential. Consequently, those small-ℓ levels are pushed down relative to high-ℓ levels, as the binding energy becomes small. The resulting shell structure, which is quite different from what we are accustomed to in stable nuclei, would appear in weakly-bound and resonant neutron one-particle levels and affect the nuclear many-body problem in neutron drip line nuclei. In the present talk I have chosen to demonstrate the shell structure of one-particle levels in the $1f$-$2p$ shell as well as in the $50 < N \leq 82$ major shell, when the levels become weakly-bound or resonant. Examining the obtained shell structure one may understand the possible origin of the deformation of nuclei in the island of inversion ($N \gtrsim 20$) as well as a possible deformation of neutron drip line nuclei with $N \approx 51$.

The systematic change of shell structure described in the present article is strictly related to the characteristic behavior of the one-particle motion with small orbital angular-momenta. The change cannot be estimated by using harmonic-oscillator potential (or wave function). In the present meeting the change of shell structure of neutrons (protons) as the proton (neutron) number varies is discussed, considering the tensor force between protons and neutrons using harmonic-oscillator wave functions. Though some of the shell-structure change looks numerically similar to that shown in the present work, the relevant physics is essentially different.

References

[1] For example, see; Newton R G 1966 *Scattering Theory of Waves and Particles* (New York: McGraw-Hill)
[2] Hamamoto I 2005 *Phys. Rev.* C **72** 024301
[3] Hamamoto I 2006 *Phys. Rev.* C **73** 064308
[4] Hamamoto I 2010 *J. Phys. G: Nucl. Part. Phys.* **37** 055102
[5] Bohr A and Mottelson B R 1969 *Nuclear Structure* vol 1 (New York: Benjamin)
[6] Bohr A and Mottelson B R 1975 *Nuclear Structure* vol 2 (New York: Benjamin)
[7] Nakamura T et al 2009 *Phys. Rev. Lett.* **103** 262501
[8] Hamamoto I 2010 *Phys. Rev.* C **81** 021304(R)
[9] Yordanov D T et al 2007 *Phys. Rev. Lett.* **99** 212501
[10] Yordanov D T, Blaum K, De Rydt M, Kowalska M, Neugart R, Neyens G and Hamamoto I 2010 *Phys. Rev. Lett.* **104** 129201

Giant and Pygmy Dipole Resonances in neutron-rich nuclei: their excitation via Coulomb and nuclear fields

2011 J. Phys.: Conf. Ser. 267 012006

(http://iopscience.iop.org/1742-6596/267/1/012006)

Giant and Pygmy Dipole Resonances in neutron-rich nuclei: their excitation via Coulomb and nuclear fields

A Vitturi (1,2), **EG Lanza** (3), **MV Andrés** (4), **F Catara** (3,5) **and D Gambacurta** (3,5)

(1) Dipartimento di Fisica, Padova, Italy
(2) INFN, Sezione di Padova, Italy
(3) INFN, Sezione di Catania, Italy
(4) Departamento de FAMN, Facultad de Física, Sevilla, Spain
(5) Dipartimento di Fisica, Catania, Italy

E-mail: `vitturi@pd.infn.it`

Abstract. We study the excitation of low-lying dipole states in very-neutron rich nuclei (so called Pygmy Dipole Resonances). The states are described within the Hartree-Fock plus RPA formalism, using different parametrizations of the Skyrme interaction. We show how the combined information from reaction processes involving the Coulomb and different mixtures of isoscalar and isovector nuclear interactions can provide a clue to reveal the characteristic features of these states.

1. Introduction

A great deal of interest has been raised in the last few years by the presence of dipole strength at low excitation energy in neutron-rich nuclei. This strength (typically of the order of few per cent of the corresponding Energy Weighted Sum Rule) has been often associated to the possible existence of a collective mode of new nature, corresponding to the oscillation of the valence neutron skin against the proton plus neutron core (Pygmy Dipole Resonance, PDR).

The presence of this low-lying strength is predicted by almost all microscopic models, ranging from Hartree-Fock plus RPA with Skyrme interactions to relativistic Hartree-Bogoliubov plus relativistic quasiparticle RPA. All these approaches [1, 2, 3, 4] predict similar amounts of strength, but often disagree on the collective (or not) nature of these states, on their fragmentation and on their isoscalar/isovector contents. Information on the precise nature of these states must come from experimental data [5]. Heavy-ion high-energy Coulomb excitation processes provide values of the multipole B(Eλ) transition rates, being instead unable to provide further information on wave functions and transition densities. We suggest in this talk that a part of this information can be obtained by resorting to reactions where the nuclear part of the interaction is involved. By tuning masses, charges, bombarding energies and scattering angles one can alter the relative role of the nuclear and Coulomb components, as well as of the isoscalar and isovector contributions.

Along this line, we present the predictions for the excitation of the low-lying (PDR) and high-lying (GDR) dipole states in the neutron-rich ^{132}Sn by different projectiles (α, ^{40}Ca, ^{48}Ca) at different bombarding energies. The dipole states, their wave functions and the corresponding transition densities have been obtained within the Hartree-Fock plus discrete RPA with Skyrme interactions. Inelastic formfactors have then been obtained by double-folding the M3Y nucleon-nucleon interaction with the microscopic densities and transition densities and then used to describe the quantal evolution of the system along classical ion-ion trajectories. The excitation probabilities turn out to be sensitive to the details of the transition densities (and not simply to the B(E1) values) and therefore these can be probed by comparing different processes.

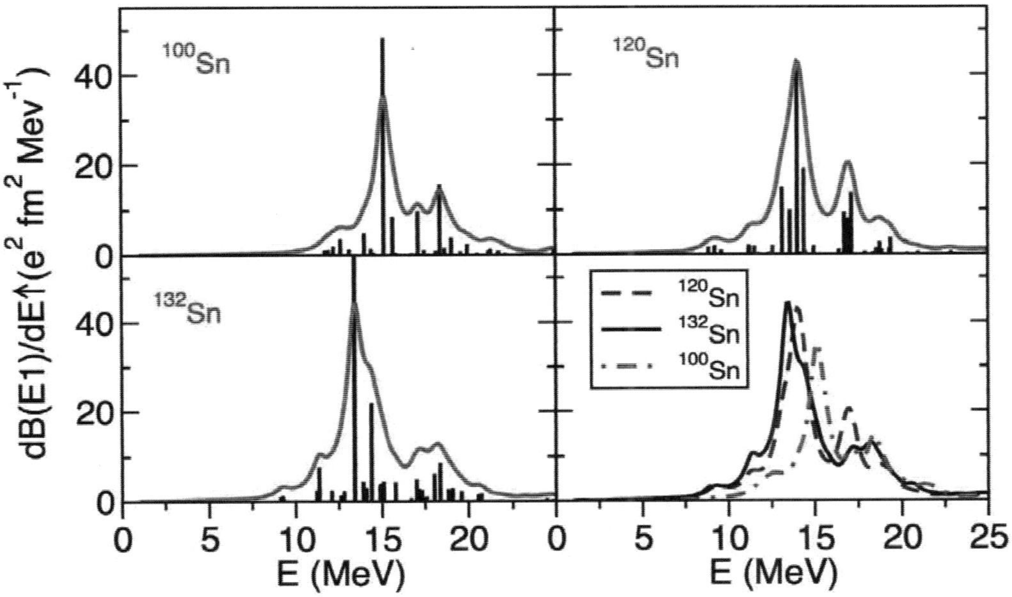

Figure 1. Isovector strength distributions for dipole states for tin isotopes calculated with the SGII interaction. The bars are the RPA B(E1) values. The solid curves represent $dB(E1)/dE$ as obtained by adopting a smoothing procedure described in the text. In the lower right figure we report the same continuous lines shown in the other frames.

2. RPA dipole strength, transition densities and formfactors

Dipole states and corresponding dipole response have been calculated using the simple discrete non-relativistic RPA approach with Skyrme interactions. An example of the evolution of the dipole response with the neutron number is shown in Fig. 1, where the three isotopes ^{100,120,132}Sn are separately considered (from Ref.[4]). Note that the bars are the discrete response of the RPA calculation while the continuous lines are generated by a smoothing procedure using a Lorentzian with a 1 MeV width. For a better comparison the three distributions of dipole strength are shown together in the lower right frame. Together with the usual lowering of the energies of the dominant Giant Dipole Resonance with increasing mass number, we can notice the appearance of some low-lying strength (carrying a fraction of the EWSR of the order of few per cent) below 10 MeV. These are precisely the states that are candidates to be interpreted as Pygmy Dipole Resonances, associated with the occurrence of a neutron skin in the nuclear densities and its oscillation with respect to the proton+neutron core.

More precise information on the specific nature of the states is embedded in the corresponding transition densities. As an example we show in Fig. 2 the RPA transition densities associated with the GDR (right frame) and with a state in the "PDR" region (left frame) in ^{132}Sn. Neutron and proton components of the transition densities are separately shown, together with their isoscalar and isovector combinations. The two cases clearly display very different behaviours. The one associated with the GDR shows the usual opposite-phase behaviour of the proton and neutron components, leading to the usual predominant isovector character. The situation is rather different in the case of the other state at lower energy. Here neutron and proton components seem to oscillate in phase in the interior region, while having a pure neutron oscillation (with opposite phase) in the surface region. This may favor the interpretation of this state as a Pygmy Dipole Resonance, macroscopically described as the oscillation of the neutron skin with respect to the proton+neutron core, although a full interpretation in terms of the PDR is not obvious (cf. e.g. ref.[4]).

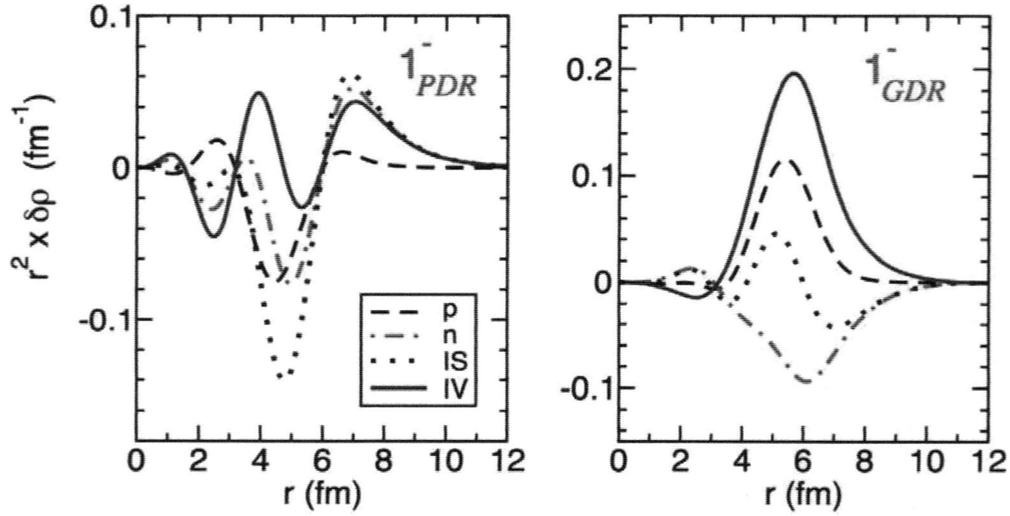

Figure 2. Transition densities for the low-lying dipole state (PDR) (left) and for the GDR (right) for the ^{132}Sn isotope calculated with the SLY4 interaction. We show the proton, neutron, isoscalar and isovector components.

The transition densities are the basic ingredients to construct the nuclear formfactors describing nuclear excitation processes. These formfactors can be obtained by doublefolding the transition densities with the density of the reaction partner and the nucleon-nucleon interaction (M3Y interaction, in our case), including both isoscalar and isovector terms. The relative role of isoscalar and isovector formfactors depends on the corresponding mixture in the transition density of the specific state but also on the nature of the specific reaction partner (being purely isoscalar in the case of α-particle or other $N = Z$ nuclei, and both isoscalar and isovector for $N \neq Z$ nuclei). The change of reaction (and of the bombarding energy), with the consequent change of the relative role of nuclear and Coulomb components as well as of the isoscalar/isovector nuclear components) will have therefore the consequence of altering the relative population of the different states. We will consider here the excitation of the Pygmy and Giant dipole states in ^{132}Sn by different partners: α, ^{40}Ca and ^{48}Ca. The square of the corresponding formfactors (nuclear and Coulomb) at the surface for the excitation of different

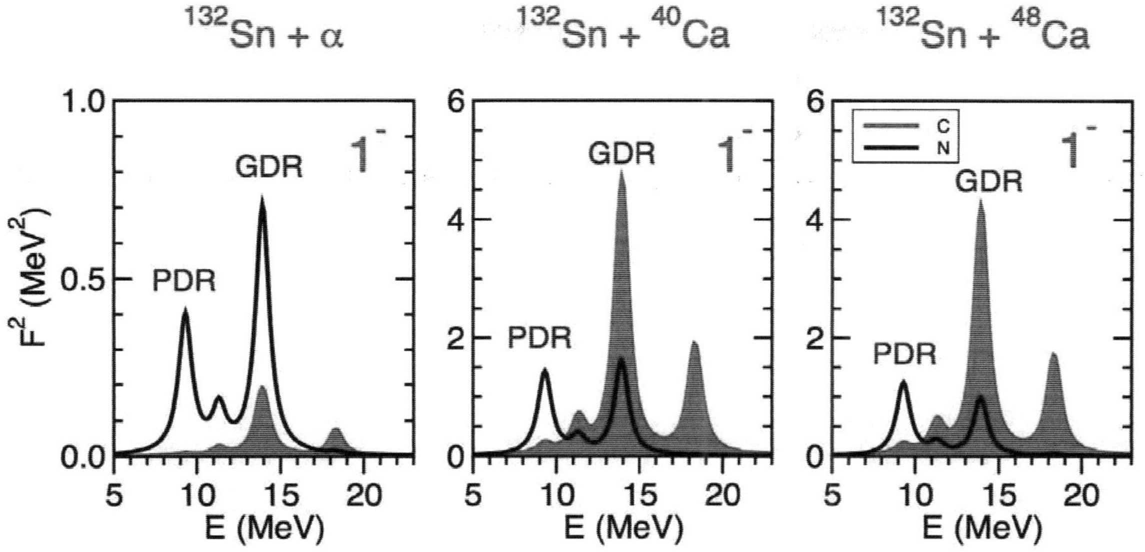

Figure 3. Square of the formfactor around the surface for different systems ^{132}Sn + α, ^{40}Ca, ^{48}Ca (the surface has been assumed to be at 7.7 fm, 11.0 fm and 11.3 fm for the three cases). The continuous lines are obtained with the same smoothing procedure used previously. Coulomb and nuclear contributions to the total formfactors are separately shown.

states and shown in Fig. 3. One can see how the different reactions and the different interplay of isoscalar and isovector contributions alter the relative "intensities" of PDR and GDR states.

3. Cross sections

With ion-ion potential and formfactors we can now proceed to the next step and calculate the inelastic cross sections. This is done within a semiclassical model where the relative motion is treated classically. These assumptions are usually well justified for grazing collisions. The Hamiltonian describing the colliding nucleus can be separated into two parts: the internal Hamiltonian describing the structure of the nucleus and an external field which is responsible of the excitation of one of the partner of the collision through the mean field of the other one. The time dependent Schrödinger equation is cast into a set of coupled differential equations for the different excitation amplitudes which are integrated along the classical relative motion trajectories. For each state considered the associated cross section is obtained by integrating the corresponding excitation probability over the whole impact parameters range modulated by the transmission coefficient [6]. The resulting energy differential total cross sections for ^{132}Sn on different targets (α, ^{40}Ca and ^{48}Ca) at E=10 MeV/u are shown in Fig. 4. The different contributions from Coulomb and nuclear formfactors are separately show. It is clear that the balance between PDR and GDR changes in the different reactions. The ratios can be further modified by selecting different scattering angles, that are associated to different ranges of impact parameters. For example nuclear contributions are known to be enhanced at grazing angles, corresponding to "surface" impact parameters. An alternative possibility is to consider the total cross sections but varying the bombarding energy, so altering the relative role of nuclear and Coulomb contributions. This is better shown in Fig. 5 where we show the cross sections for the case ^{132}Sn + ^{40}Ca at different bombarding energies.

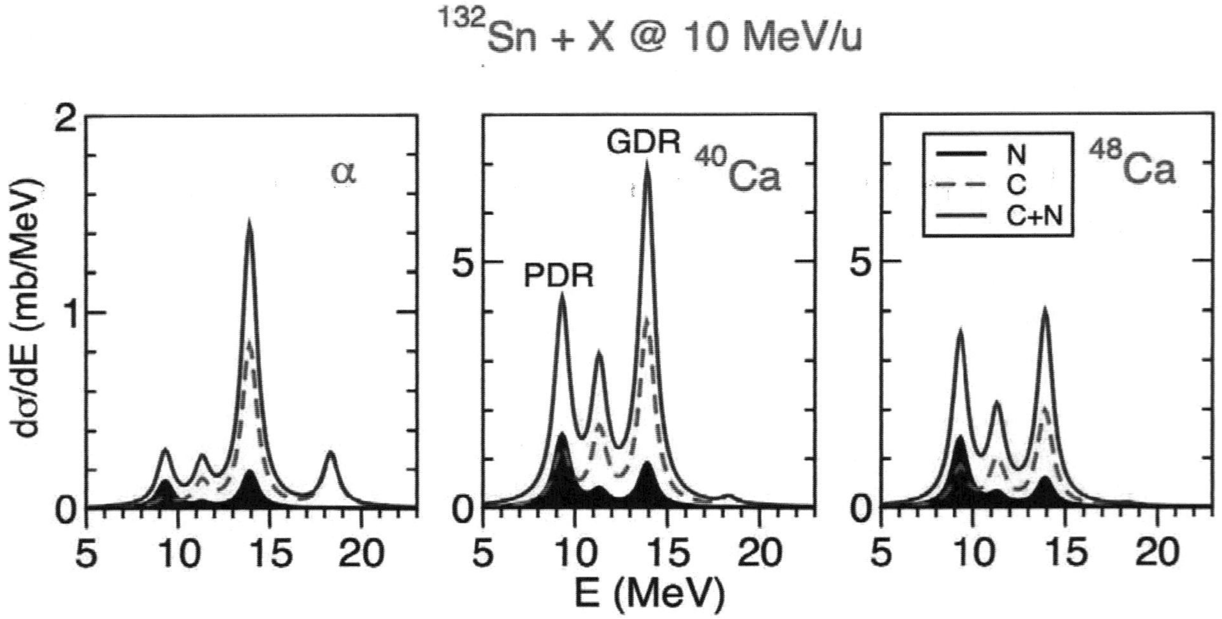

Figure 4. Differential cross sections as function of the excitation energy for the systems ^{132}Sn + α, ^{40}Ca, ^{48}Ca at 10 MeV per nucleon. Coulomb, nuclear and total contributions are separately shown.

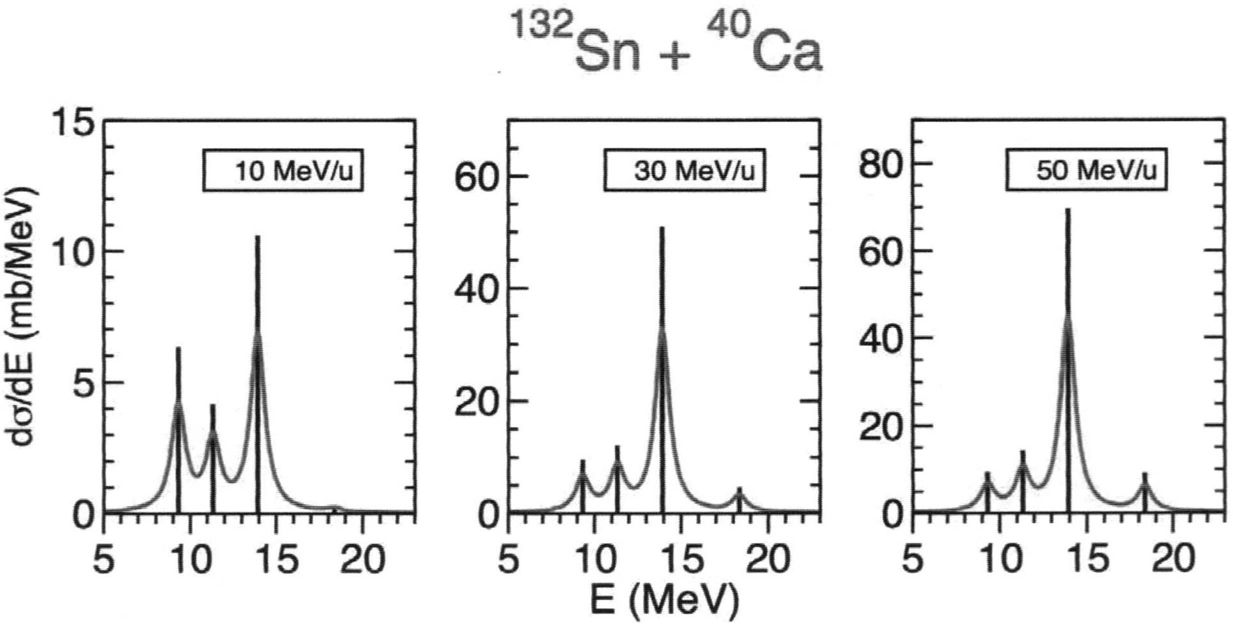

Figure 5. Differential cross sections as function of the excitation energy for the systems ^{132}Sn + ^{40}Ca at different bombarding energies.

4. Conclusions

The interpretation of the low-lying dipole strength in very-neutron rich nuclei as a pygmy dipole state of collective nature needs to be carefully checked. Valuable information on the nature of these states can be obtained by excitation processes involving the nuclear part of the interaction, which are sensitive to the shape of the transition densities. The use of different bombarding energies, of different combinations of colliding nuclei involving different mixture of isoscalar/isovector components, together with the mandatory use of microscopically constructed formfactors, can provide the clue towards the solution of the problem.

[1] Catara F, Lanza E G, Nagarajan M A and Vitturi A *Nucl. Phys. A* **614** 86 (1997); *Nucl. Phys. A* **624** 449 (1997).

[2] Paar N, Vretenar D, Khan E and Colò G, *Rep. Prog. Phys.* 70, 691 (2007) and referencs therein.

[3] Tsoneva N and Lenske H, *Phys. Rev. C* **77** 024321 (2008).

[4] Lanza E G, Catara F, Gambacurta D, Andres M V and Chomaz Ph, *Phys. Rev. C* **79** 054615 (2009).

[5] Adrich P, et al. (LAND-FRS Collaboration), *Phys. Rev. Lett.* **95** 132501 (2005); Klimkiewicz A, et al. (LAND-FRS Collaboration), *Nucl. Phys. A* **788** 145 (2007); Klimkiewicz A, et al. (LAND Collaboration) *Phys. Rev. C* **76** 051603(R) (2007).

[6] Andrés M V, Catara F, Lanza E G, Chomaz Ph, Fallot M and Scarpaci J A, *Phys. Rev. C* **65** 014608 (2001).

Radioactive ion beam facilities at INFN LNS

2011 J. Phys.: Conf. Ser. 267 012007

(http://iopscience.iop.org/1742-6596/267/1/012007)

Radioactive ion beam facilities at INFN LNS

D Rifuggiato*, L Calabretta, L Celona, F Chines, L Cosentino, G Cuttone, P Finocchiaro, A Pappalardo, M Re and A Rovelli

Istituto Nazionale di Fisica Nucleare, Laboratori Nazionali del Sud, Catania, Italy

*E-mail: Rifuggiato@lns.infn.it

Abstract. Radioactive ion beams are produced at INFN- Laboratori Nazionali del Sud (LNS) by means of the two operating accelerators, the Tandem and the Superconducting Cyclotron (CS), originally designed to accelerate stable beams. Both the ISOL (Isotope Separation On Line) and the IFF (In-Flight Fragmentation) methods are exploited to produce RIBs in two different ways at different energies: in the first case, the Cyclotron is the primary accelerator and the Tandem accelerates the secondary beams, while in the second case radioactive fragments are produced by the Cyclotron beam in a thin target with energies comparable to the primary beam energy. The ISOL facility is named EXCYT (Exotics at the Cyclotron and Tandem) and was commissioned in 2006, when the first radioactive beam (^{8}Li) has been produced. The IFF installation is named FRIBs (in Flight Radioactive Ion Beams), and it has started to produce radioactive beams in 2001, placing a thin target in the extraction beam line of the Cyclotron. The development of both facilities to produce and accelerate radioactive ion beams at LNS, is briefly described, with some details on the future prospects that are presently under consideration or realization.

1. Introduction

Low and intermediate energy radioactive beams far from the valley of beta stability, offer new opportunities in nuclear physics and nuclear astrophysics research, allowing a deeper exploration of nuclear structure and reaction mechanisms. Furthermore, using radioactive ion beams in nuclear reactions, it is possible to obtain compound nuclei with N/Z values very interesting for studying the behaviour of super heavy elements (at beam energies near the coulomb barrier) or rather the influence of the isospin degree of freedom in the dynamics of nuclear collision (at intermediate energies) [1,2]. At LNS – INFN Catania we have developed two complementary facilities for RIBs production, exploiting the techniques known as ISOL and In-Flight Fragmentation . The former, called EXCYT, allows to produce radioactive beams with energy up to 7 AMeV; the latter, called FRIBs, produces beams with energy up to 60 AMeV.

Within EXCYT a primary ion beam is accelerated by a Superconducting Cyclotron, (CS, K=800, Energy range 8 - 80 AMeV, P_{max}=500W) and it is stopped inside a thick carbon target, that is the core of the target-ion-source (TIS) complex. The radioactive isotopes, produced as fragments of nuclear reactions, are extracted by means of a high voltage (some tens of kV) and then selected in mass by means of a two-stage high resolving mass separator [3]. The beam, whose energy along such a separator can range between 10 keV and 300 keV, is finally transported to the injection stage of a Tandem Van De Graaf (V_{max} = 15 MV), figure 1, for the post-acceleration.

The radioactive beams produced by means of the FRIBs facility exploit the primary beam accelerated by the CS, that reacts within a thin target placed in the extraction beam line, before a beam line composed by a couple of magnetic dipoles. The nuclear fragments with a kinetic energy comparable to the primary beam are selected in magnetic rigidity (p/q) by setting the suitable magnetic field in the dipoles, as given by the LISE code, in order to select the nuclei of interest within an acceptance in momentum of $\pm$ 1%, [4].

In comparison to EXCYT, the FRIBs beams are not composed by just one nuclear species: in fact they consist of a nuclear admixture that depends on the setting and on the acceptance of the magnetic dipoles. Such a beam is then transported to the experimental apparatus where the particles interact with the final target, thus producing reactions with several projectiles at the same beam time session. Such a technique requires therefore an efficient system of tagging detectors, in order to identify event by event the projectile producing the nuclear reaction, with very high rates (at least 10^5 Hz) without changing the performances of the system.

2. The EXCYT facility.

The production yield of the radioactive nuclear beam (RNB) depends on many factors, such as the primary beam parameters (element, energy and intensity), the target ion-source properties (nuclear cross-sections, diffusion and effusion mechanism, etc), the charge exchange efficiency, the transport efficiency and the acceleration efficiency. The beam we have produced and post-accelerated, as approved by the LNS Scientific Committee, is ^{8}Li. Taking into account the CS operational diagram, the production of the radioactive ions has been performed by injecting a $^{13}C^{4+}$ primary beam of 45 AMeV on a graphite target, with a beam power up to 150 W (~ 1 eμA). The selection of the target material has been done following the criteria of high porosity, small grain size, high thermal conductivity, high chemical purity, high melting point and low vapour pressure. The chosen UTR146 graphite has been selected, since it is the best target material for such a purpose [5]. The recoils, produced and diffused on the target surface, effuse through the transfer tube to the Tungsten positive surface ionizer and then extracted by an acceleration voltage up to 50 kV. The charge of the exiting ions is +1 and it must be converted to -1 by means of the charge exchange cell (CEC), in order to be accelerated by the Tandem. The beam transport has been followed by means of several diagnostic devices, densely distributed along the whole beam pipe, that allow to visualize the beam profile, to measure the RIB intensity and to identify the radionuclide by means of beta and gamma spectroscopy, figure 2 [6]. The measured transport efficiency through the two stages of the isobaric mass separation is close to 100%, whereas the acceleration transmission at the Tandem is of the order of 50%, a value that is lower than for Li stable beams produced with their current source.

The maximum intensity of ^{8}Li that up to now has been reached on the final target of the experiment, is $5 \cdot 10^4$ pps. With the same positive ion source, also other elements can be produced, in particular ^{9}Li and ^{21}Na. Different activities are under way in order to achieve a higher intensity of ^{8}Li , reasonably corresponding to $5 \cdot 10^5$·pps. In order to obtain such a result, it is necessary to increase the primary beam intensity (possibly by a factor three) from the Superconducting Cyclotron, to gain efficiency by the CEC (up to now the main bottleneck) from the measured value of 2.8% to the expected one of 4%, and to increase the transmission through the Tandem. In our current planning a lot of efforts are focused on the optimisation of the TIS assembly and to improve the beam injection into the Tandem, by installing a new quadruplet. The TIS system is presently affected by a number of technological problems related to the current flow of the heater, which encounters many discontinuities due to bad electrical contacts. A new design of the system is planned to be accomplished in the next future.

Not less important than the upgrading of the facility, many efforts have recently been put on safety and reliability issues. In particular, the front-end of the source has been completely renovated, and as a consequence the robot is now able to perform all the remote handling manipulations for which it was realized.

Finally, in order to start developing a new radioactive beam, the installation of a different source type, able to ionise a big number of ion species, is planned in the near future: at the moment a possible choice might be ^{15}O.

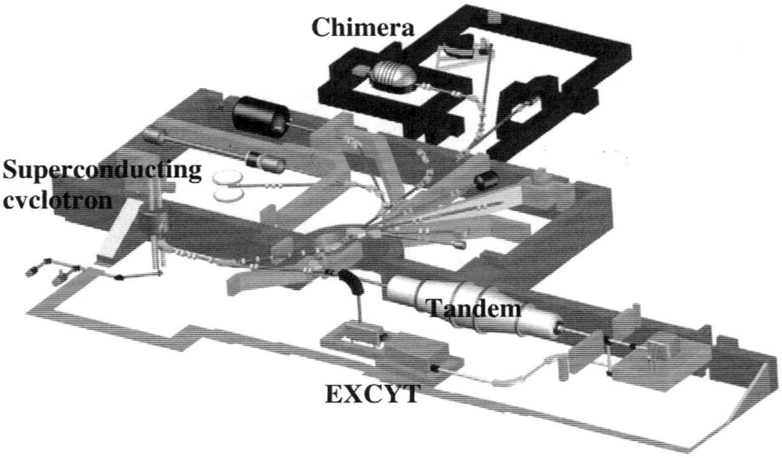

Figure 1. The layout of INFN-LNS with the Excyt facility for the production of radioactive beams with the ISOL technique (left). The new target geometry adopted recently (right).

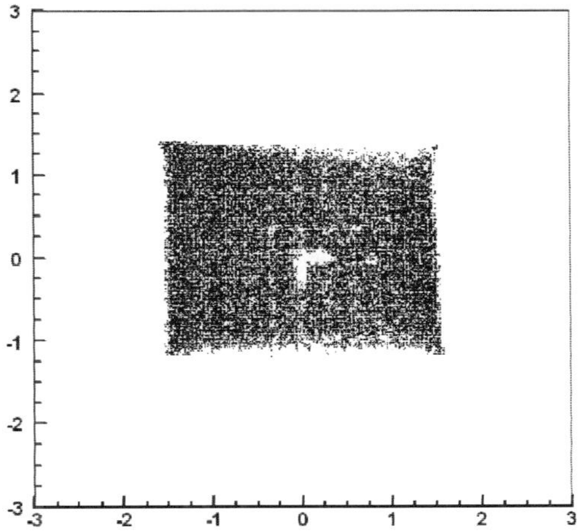

Figure 2. A diagnostic device based on a PSSD (Position Sensitive Silicon Detector), used for very low intensity beam imaging. The picture on the right corresponds to a transversal profile of a radioactive beam.

3. FRIBs.

Fragmentation beams have been produced at LNS for many years. Several production experiments have been performed, in order to produce proton and neutron rich light nuclei, whose identification has been performed by using tagging detectors, suitably developed for the experiments that have exploited such beams, such as the HODO and the CHIMERA detectors. The first experiment performed with a ^{18}Ne beam by the Prof. Raciti team, has allowed to highlight the two proton decay [7].

By exploiting the LISE code, in which the beam line properties have been included, the magnetic separator is set in order to select the fragments of interest produced by the interaction of the primary beam with the production target. The particles composing the beam can be tagged one by one, in order to identify them before they react with the final target [8]. Therefore in the data acquired during the experiments, the events related to a particular class of projectiles can be thus selected, allowing to study different reactions in the same beam time. To this aim it is necessary to have a good isotopic resolution for the selected fragments by means of 2D plots ΔE/Tof, choosing suitable detectors for such a purpose, that are able to work at very high rates, at least 10^5 Hz, for all the time of the measurements, without any deterioration of their performances. Besides, since the beam obtained by fragmentation is poorly focused, its trajectory should be reconstructed, by using position sensitive detectors. The beam tagging devices should therefore identify the fragments, measure their energy and their trajectory (if possible), and also they should perturb as less as possible the secondary beams. Besides, the reactions that occur in the tagging detector should not reach the detectors or should be simply discriminated by the reactions on target.

Taking for instance the CHIMERA tagging system, it consists of a large area microchannel plate (MCP) used as start detector for time of flight (TOF) measurements and a double side silicon strip detector (DSSSD), consisting of 16*16 strips providing the stop signal for TOF measurement, energy loss (ΔE) and position of the impinging beam. For the test of beam production, several primary beams have been used. The last beams are ^{13}C, ^{18}O and ^{16}O, with an energy of 55 AMeV. The first two beams were used to produce neutron rich nuclei, while the ^{16}O was used for the production of nuclei on the proton rich side. One of the secondary beam used in last experiment with CHIMERA, consisted to set the acceptance window of the separator in the region of ^{11}Be, with the help of LISE code that has established the most suited primary beam to be ^{13}C accelerated at 55AMeV, interacting on a 1.5mm ^{9}Be target. With such a combination of primary beam and target, a quite intense ^{11}Be has been produced (of the order of 10^4 pps with a 100 W of $^{13}C^{5+}$ beam), with a high purity (around 50% of the produced beam was ^{11}Be). The measured identification scatter plot (ΔE-TOF) is reported in figure 3 (left). The same isotopic region can be produced by using 55 AMeV $^{18}O^{7+}$, also if with lower intensity (10^3 pps with 100 W) and with much more contaminants. However most of the contaminants produced with ^{18}O are quite useful neutron rich beams. Therefore also this choice is good for some experiments. In figure 3 (right) the ΔE-TOF identification plots obtained with ^{18}O as primary beam is shown.

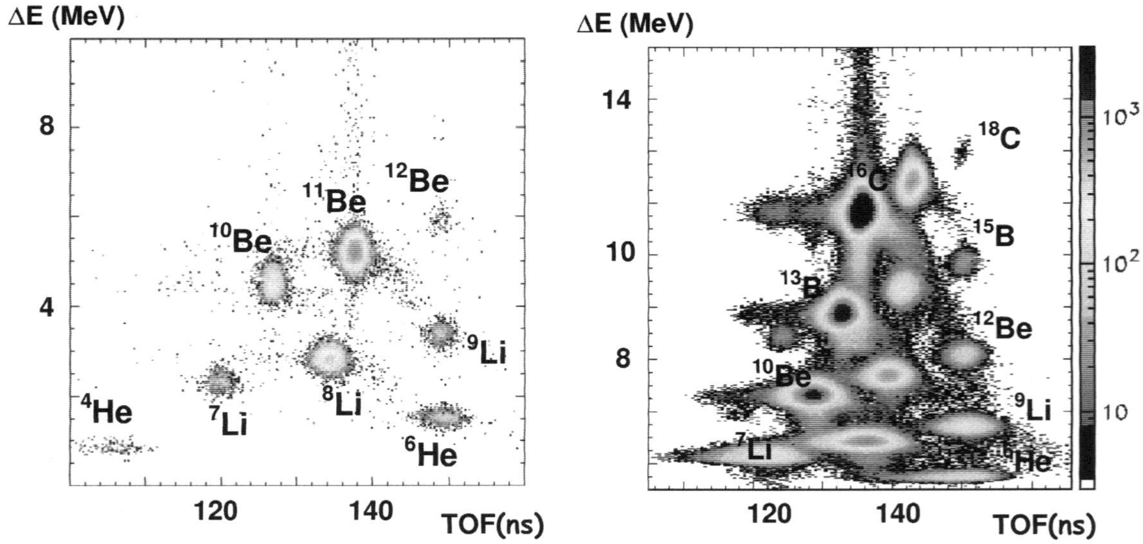

Figure 3. Identification scatter plot for the fragmentation beams obtained using a primary beam of ^{13}C 55 MeV/A on a 1.5mm ^{9}Be target (left). Identification scatter plot for the fragmentation beams obtained using a primary beam of ^{18}O 55 MeV/A on a 1.5mm ^{9}Be target (right).

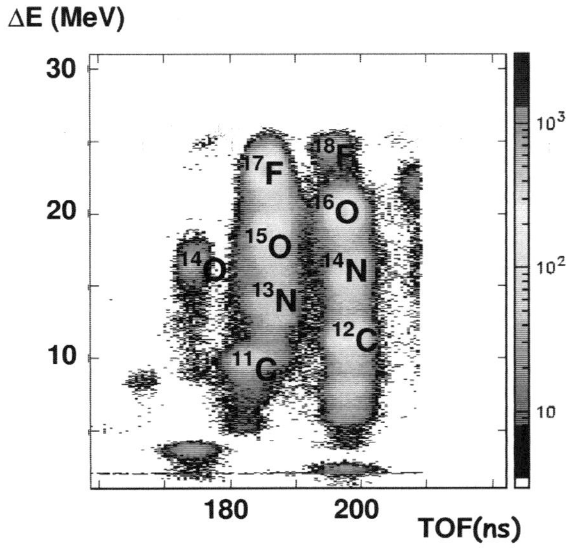

Figure 4. Identification scatter plot for the fragmentation beams obtained using a primary beam of ^{16}O 55 MeV/A on a 2.5mm ^{9}Be target. A 70 μm thick DSSSD detector was used to measure the beam energy loss.

The 55 AMeV ^{16}O beam was used as a primary beam in order to investigate the production of proton rich nuclei. The main aim was to observe the production yield for nuclei in the region of ^{13}O. In order to prepare a pilot beam for the beam line setting, a 2.1 mm Al degrader has been used. $^{15,14}O$ beams are produced with reasonable rate, whereas a large yield has been measured for ^{17}F, which confirms the presence of proton pickup reactions from the ^{9}Be target.

In figure 4 one of the identification scatter plot obtained with ^{16}O beam on a 2.5 mm thick ^{9}Be target is shown.

In this case a DSSSD detector 70μm thick has been used for the ΔE measurement, whose performances are worse than with the 140μm DSSSD detector used for the ^{13}C and ^{18}O primary beams. Such a trouble is due to a non-uniformity of the 70 μm thick detector. A correction that takes it into account is necessary, in order to overcome this problem.

An increase of the beam transmission is expected to come from the new beam monitoring system developed for the transport along the CHIMERA and MAGNEX beam line, that has recently been installed and will be used for the next beam run [9].

Moreover, a significant upgrade of the FRIBs line from the Cyclotron exit to the beam distribution point will be realized at the end of 2010. A new magnetic configuration has been designed that will imcease the transverse and momentum acceptance of the existing line after the production target.

As a consequence of this upgrading, an enhacement factor of 10-20 is expected in the experimental halls making use of FRIBs beams, i.e. the CHIMERA and the 20° beam lines. Radioactive beams far from the stability valley, already produced in the present configuration with a poor intensity, like ^{14}Be, ^{8}B, ^{8}He, will benefit from the above described upgrading and it will be possible to use them for experiments.

Table 1. Characteristics of the produced fragmentation beams. Intensity is normalized to a primary beam of 100 Watt.

Primary beam (55 MeV/A)	^{9}Be production target thickness (mm)	Secondary beam and energy	Intensity (kHz)	Abundance
^{13}C	1.5	^{11}Be - 48 MeV/A	10	50%
^{13}C	1.5	^{12}B - 47 MeV/A	160	80%
^{16}O	1.5	^{15}O - 31 MeV/A	5	10%
^{16}O	1.5	^{14}O - 36 MeV/A	0.7	1.5%
^{16}O	2.5	^{17}F - 24 MeV/A	10	20%
^{18}O	1.5	^{16}C - 49.9 MeV/A	9	30%
^{18}O	1.5	^{13}B - 52 MeV/A	4.5	15%
^{18}O	1.5	^{11}Be - 48 MeV/A	3	10%
^{18}O	1.5	^{10}Be - 56 MeV/A	4.5	15%
^{18}O	1.5	^{8}Li - 51 MeV/A	3	10%

References

[1] Li B, Chen L, Ming C and Ko C 2008 *Physics Reports* **464** 113

[2] Baran V, Colonna M, Greco V and Di Toro M 2005 *Physics Reports* **410** 335

[3] Cuttone G, Calabretta L, Celona L, Chines F, Cosentino L, Finocchiaro P, Pappalardo A, Re M, Rifuggiato D and Rovelli A 2008 *Nucl. Instr. and Meth.* B **266** 4108

[4] Raciti G, Rapisarda E, De Napoli M, Amorini F, Calabretta L, Cardella G, Cosentino G, Sfienti C and Shchepunov V 2008 *Nucl. Instr. and Meth.* B **266** 4632

[5] Re M, Cirrone G, Cosentino L, Cuttone G, Finocchiaro P, Lojacono P, Ma Y, Vynck P, Volckaerts B, Vervaeke M, Van Erps J, Hermanne A and Thienpont H *Proc. PAC* 2005 Knoxville 898

[6] Cappello S, Cosentino L and Finocchiaro P 2002 *Nucl. Instr. and Meth.* A **479** 243

[7] Raciti G, Cardella G, De Napoli M, Rapisarda E, Amorini F and Sfienti C 2008 *Phys. Rev. Lett.* **100** 192503

[8] Lombardo I et al 2009 *LNS - INFN Report*

[9] Cosentino L, Finocchiaro P and Pappalardo A 2009 *LNS - INFN Report*

Present status of the RIKEN Radioactive Ion Beam Factory, RIBF

2011 J. Phys.: Conf. Ser. 267 012008

(http://iopscience.iop.org/1742-6596/267/1/012008)

Present status of the RIKEN Radioactive Ion Beam Factory, RIBF

T Suda

Research Centre for Photon Electron Science, Tohoku University

2-1 Mikamine, Sendai, 982-0826, Japan

suda@lns.tohoku.ac.jp

Abstract. The present operation status of the RIKEN Radioactive Ion Beam Factory (RIBF) and recent research highlights are reported. One of the major devices of the RIBF, an electron scattering facility for short-lived nuclei, is described in detail. This facility will open a completely new research field of structure studies of short-lived nuclei.

1. RIKEN Radioactive Ion Beam Factory

The RIKEN RIBF [1] is a next-generation radioactive-beam facility that provides intense exotic beams for various basic sciences. The RIBF consists of an heavy-ion linac injector followed by four cyclotrons to boost the primary beam energy up to 350 AMeV. The intensity goal of primary beams ranging from protons to uranium is 1 pμA (particle micro-Ampere).

BigRIPS [2] is a large acceptance fragment separator consisting of two stages. They are the production stage where the primary beam is converted to radioactive ion (RI) beams by the in-flight fission of a uranium beam as well as projectile fragmentation of other primary beams, and the tagging stage to tag the produced secondary beam on an event-by-event basis for experiments using these RI downstream.

After completion of the facility construction in 2006, continuous efforts have been made for improving the facility performances; primary beam species, beam intensity,and its stability etc... So far, uranium beams with an intensity of 10^9 /sec and the world's most intense ^{48}Ca beam with the intensity of 200 pnA have been provided for nuclear physics experiments.

1. 1. Major experimental devices

There are several large-scale experimental devices to enhance the research capability of this facility.

The major devices in operation today are the ZeroDegree spectrometer [2] and a high-resolution spectrometer SHARAQ [3]. The ZeroDegree is a large-acceptance beam-line spectrometer for particle identification of the reaction products at the secondary target that is bombarded by the tagged secondary beam. SHARAQ, in connection with a long beam transport line for dispersion matching purpose, provides a unique opportunity to carry out high resolution studies using secondary beams; such as missing mass spectroscopy with secondary beams as a "probe".

Devices under construction are SAMURAI (**S**uperconducting **A**nalyzer for **MU**lti particles from **RA**dio **I**sotope beams) [4] and an electron scattering facility for short-lived nuclei based on a novel SCRIT concept.

SAMURAI is a versatile large acceptance spectrometer having a large gap that allows the detection of neutrons emitted forward in coincidence with other charged particles for spectroscopic studies of unbound states excited by nuclear reactions. It will be completed in FY2011. As for the electron scattering facility, it will be described in details in the next section of this report.

Additional unique devices currently under budget requests are SLOWRI [5] and the Rare Isotope ring [6]. SLOWRI aims to provide slow radioactive beams after stopping a fast radioactive beam by a gas catcher. The Rare Isotope ring measures the mass of rare isotopes with a mass precision of 10^{-6}.

1. 2. Recent results from RIBF experiments

As of today, an intense, stable, ^{48}Ca primary beam, ~ 200 pnA, has been regularly provided for several BigRIPS experiments, and a uranium beam of 0.3 pnA has been used for new isotope search measurements. Updated information of the primary beam intensity is always available at the RIBF web site: http://www.nishina.riken.jp/UserGuide/accelerator/tecinfo.html .

Campaign-type experiments using the intense ^{48}Ca beam have been conducted in 2008. Three independent experiments were successfully performed. The key of the success resulted from obtaining secondary beam intensities of very neutron-rich isotopes, which were unexpectedly high compared with those based on our experience with lower energy RIPS experiments. As an example, the ^{31}Ne beam intensity was 400 times more intense, a result which is attributed to the larger production cross section at higher beam energy in addition to a thicker production target and a larger acceptance of the separator.

The experiments performed were the measurements of the reaction cross-section [7], including Coulomb breakup [8] and inelastic scattering by detecting de-excitation gamma rays [9]. These measurements that ^{31}Ne is a halo nucleus, the heaviest halo nucleus discovered so far.

The intensity of U, typically ~0.3 pnA, is still far from the goal intensity of 1 pμA. Nevertheless, many new isotopes were discovered in one-shot experiments thanks to the large acceptance of the separator [10,11]. A higher intensity U beam is expected in the next few years resulting from charge-stripper developments, and medium-heavy exotic nuclei will soon be available as targets for studies of nuclear structure and reactions.

2. Electron scattering facility based on SCRIT scheme at the RIBF

An electron scattering facility dedicated to the structure studies of short-lived nuclei is being under construction at the RI Beam Factory. Although electron scattering is well known to be a very powerful tool for structure study of atomic nuclei, it has never been never used for short-lived nuclei.

2. 1. Electron scattering

Electron scattering provides the most reliable structure information of atomic nuclei by virtue of the fact that the electron is a point particle, and probes a target nucleus through the fairly weak and well-understood electromagnetic interaction. Since this scattering process is perfectly described by QED, one can unambiguously extract structure information of the target nucleus from experimental data.

Thus, as is well known, electron scattering has consistently played a key role in nuclear structure studies. It has been, however, strictly limited to stable nuclei except for a few exceptions, and no experiments for highly unstable (short-lived) nuclei has been ever conducted.

Once electron scattering becomes feasible for short-lived nuclei, one may start with the measurements of elastic scattering, since its cross section is the largest in low momentum transfer region and the charge density distribution, one of the fundamental ground state properties of nuclei, is determined. According to the results of simulations for elastic scattering off ^{132}Sn which is our first target for electron scattering, a luminosity of 10^{27} /cm^2/s is required to determine the size and surface diffuseness with an accuracy better than a few %.

2. 2. A novel technique for electron scattering experiments

We have proposed a novel experimental technique [12] to open up this new research field, namely the structure study of exotic nuclei by electron scattering. It is named as SCRIT (**S**elf **C**onfining **RI** **T**arget), which uses "ion trapping" notoriously known at electron ring facilities. Since ion trapping reduces the performance of electron rings by inducing beam instability and shorter beam lifetime, much effort has been paid so far to remove these effects. Our idea is to grab exotic nuclei of interest on the electron beam using this ion trapping technique.

Ions (exotic nuclei) of interest will be externally injected and trapped by a circulating electron beam. Once they are trapped on the electron beam, they can not escape. Therefore, a rather high collision luminosity is expected with a small number of trapped ions. Since the ions are trapped on the electron beam, electron scattering off the target nucleus takes place automatically.

To confine the trapped ions longitudinally also, electrodes are placed in the electron beam direction to form a mirror potential. Controlling the mirror potential, one can inject the target ions from the external ion source, trap them for a certain duration depending on the lifetime of the target nuclei, and extract them from the trap region for the next injection to keep the purity of the target.

2. 3. Feasibility studies

In order to study the achievable luminosity by this novel SCRIT approach, a prototype was constructed and installed at an existing electron ring, KSR, of Kyoto University. Stable Cs ions are used instead of exotic nuclei for this study. Figure 1 shows the SCRIT prototype installed on the 2-m straight line of KSR.

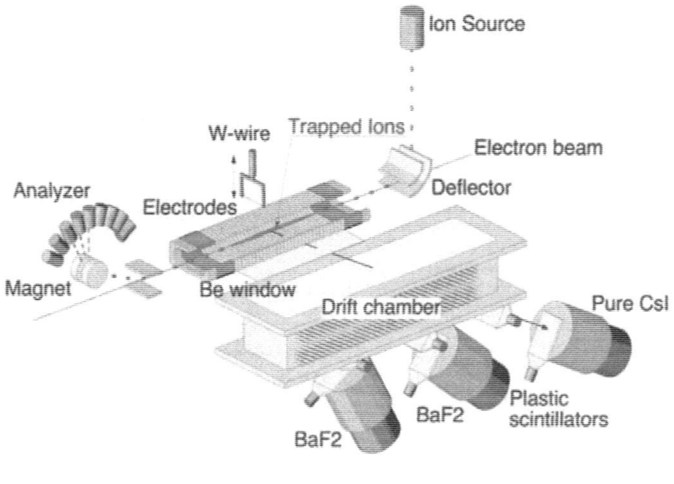

Figure 1. The SCRIT prototype installed at KSR for feasibility studies.

Cs ions extracted from an ion source are transported and merged with the electron beam at the deflector. They are transversely trapped by the electron beam, and are longitudinally confined by a mirror potential created by electrodes. By adjusting the mirror potential shape the ion-trapping length was set to 260 mm. The scattered electrons emerging to the air through a thin Be window are detected by an electron detection system consisting of a drift chamber for scattering angle measurement and calorimeters for energy measurements.

Using 120 MeV electron beams whose stored current is 70 mA with a beam lifetime of 100 s, we measured the scattered electron spectrum with the ion injection-trap-release cycle by controlling the ion source and the mirror potential. In order to mimic short-lived nuclei, the trapping time was set to be 50 msec. Clear signals of elastically scattered electrons from ions trapped in the trapping region are observed, and their angular distribution clearly shows that they are from the trapped Cs ions as shown in figure 2 [13,14]. Detailed analysis shows that the luminosity of $1x10^{26}/cm^2/s$ is achieved with 10^6 trapped ions. Since much improved luminosities are expected with larger stored electron-beam current and improved ion-injection technique, we conclude that a luminosity of $1x10^{26}/cm^2/s$ is achievable, and the SCRIT technique is a way to realize never-yet-performed electron scattering of rarely-produced short-lived nuclei.

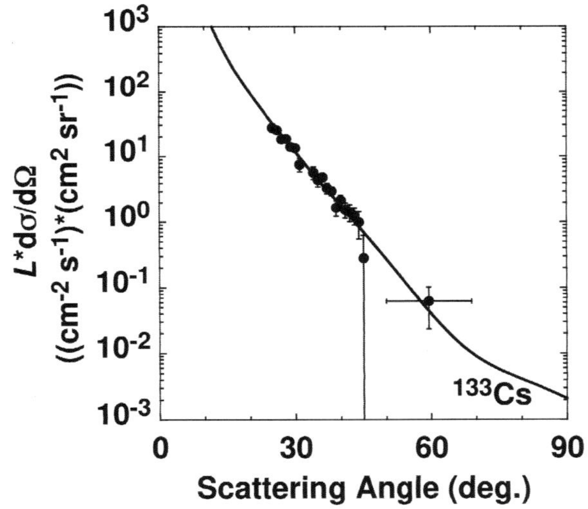

Figure 2. Angular distribution of electron elastic scattering. Solid circles show the elastic events, and the solid line the results of a distorted wave calculation by DREPHA for ^{133}Cs.

2. 4. Electron scattering facility at RIKNE RI Beam Factory

Based on the success of the feasibility studies described above, we started to plan the construction of an electron scattering facility at the RIBF. This has been, fortunately, boosted up by three fortunate recent events:

- Donation of a 700-MeV electron storage ring system from SHI (Sumitomo Heavy Industries), who terminated its use as a light source facility
- FY2008 supplementary budget from the government that enabled us to install the ring at the RIBF
- Grant-in-Aid for Scientific Research (S) has been approved in FY2010 for construction of the electron detection system

Figure 3 shows a layout of the electron scattering facility at the RIBF. The facility consists of an electron accelerator, an ISOL system and an electron detection system. The donated electron ring system has already been installed and is now in operation.

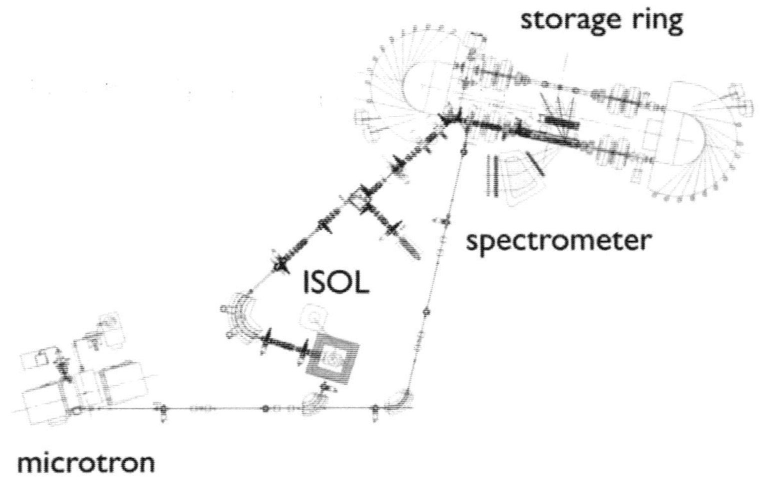

Figure 3. A layout of the electron scattering facility at the partly completed RIBF. The facility consists of an electron accelerator, an ISOL and an electron detection system.

2. 4. 1. Electron storage ring system

The system consists of a 150-MeV injector microtron, the beam transport line and 700-MeV electron storage ring. The basic configuration is the same as that of the SR light source facility of Hiroshima University, HiSOR. Commissioning is underway at the highest electron energy 700 MeV and, as of today, the stored beam current is higher than 400 mA with a beam lifetime of nearly 2 hours.
The much improved electron-beam conditions of the new facility compared with those of KSR, such as stored current, stability and beam lifetime etc..., will result in a much higher luminosity for the same number of trapped ions.

2. 4. 2. ISOL

Production of short-lived nuclei, specifically Sn isotopes including ^{132}Sn as the first targets, will be carried out using 150 MeV electron beam bombarding on an uranium carbide target, $^{238}UC_x$. Photo-fission (and electro-fission) of uranium is known to be a very efficient way to produce neutron rich Sn isotopes, especially ^{132}Sn [15]. Photo-fission of uranium is mainly induced by an excitation of the giant dipole resonance of uranium. Numerical simulations show that the number of photo-fission per unit beam power for the optimized target geometry becomes nearly constant, an order of 10^8 fission/s/watt, when the incident electron energy is larger than 100 MeV.

Since a long beam lifetime (a few hours) of the storage ring will allow us to operate the injector microtron mostly for isotope production, we decided to build an ISOL system using the photo-fission process of uranium. At the electron beam power of 1 kW, the production rate of ^{132}Sn will be 10^9 /s based on the ^{132}Sn yield ratio in photo-fission process being 1 % [15]. Detailed design of the ISOL system including an element-separator, mass separator and ion pulsing for the SCRIT injection after accumulation is underway.

2. 4. 3. Electron spectrometer

In order to identify elastic scattering, an electron spectrometer must have an energy resolution at least better than 1 MeV to resolve elastic and inelastic scatterings. The electron beam energy will need to be 300 MeV from considerations of the momentum transfer range required for the measurement of the charge density distribution, and the geometrical limitation of where the spectrometer can be placed. Thus, the momentum resolution must be better than 3×10^{-3}.

Due to the fact that the SCRIT provides a spatially extended target, 30-40 cm, any focusing-type high-resolution magnetic spectrometer can not be employed. In addition, the solid angle of such high-resolution spectrometer, typically a few 10 msr, is obviously insufficient to deal with the small yield from low luminosity experiments.

A non-focusing magnetic spectrometer with tracking detectors as shown in figure 3 will be employed. This spectrometer covers a wide scattering angular range of 30-60 deg., and has a solid angle of the order of 100 msr.

2. 4. 4. Time schedule

In the next three years, the construction for the ISOL system and the electron detection system will be carried out. The first collisions of electrons and the trapped Sn nuclei are planned for 2014.

3. Summary

The RIKEN RIBF has started its full-scale operation with the world's most intense ^{48}Ca beams of 350 AMeV since 2008. In addition to efforts to improve the facility performance including targeting a primary-beam intensity goal of 1 pμA, the physics program using the full potential of major experimental devices will be continuously conducted. An electron scattering facility for short-lived nuclei, which is quite unique, is under construction based on a new experimental technique, named SCRIT. The first collision between electrons and short-lived nuclei will take place in 2014.

References

[1] Yano Y 2007 *Nucl. Instrum. and Methods* B **261** 1009-1013

[2] Kubo T 2003 *Nucl. Instrum. and Methods* B **204** 97

[3] Uesaka T, Shimoura S, Sakai H, Berg G.P.A, Nakanishi K, Sasamoto Y, Saito A, Michimasa S, Kawabata T and Kubo T 2008 *Nucl. Instrum. and Methods* B **266** 4218-4222

[4] Yoneda K, Contribution to the 8th International Conference on Radioactive Nuclear Beams, Grand Rapids, USA, May, 2009.

[5] Wada M *et al.* 2004 *Nucl. Instrum. Methods* A **532** 40

[6] Yamaguchi Y *et al.* 2008 *Nucl. Instrum. Methods* B**266** 4575

[7] Takechi M *et al.* 2010 *Nucl. Phys.* A **834** 388c-393c

[8] Nakamura T *et al.* 2009 *Phys. Rev. Lett.* **103** 262501

[9] Doornenbal P *et al.* 2009 *Phys. Rev. Lett.* **103** 032501

[10] Ohnishi T *et al.* 2008 J. Phys. Soc. Japan 77 083201

[11] Ohnishi T *et al.* in press

[12] Wakasugi M, Suda T and Y Yano 2004 *Nucl. Instrum. Method* A **278** 216

[13] Wakasugi M *et al.* 2008 *Phys. Rev. Lett.* **100** 164801

[14] Suda T *et al.* 2009 *Phys. Rev. Lett.* **102** 102501

[15] De Frenne D *et al. Phys. Rev.* C **29** 1908

GANIL-SPIRAL2: a new era

2011 J. Phys.: Conf. Ser. 267 012009

(http://iopscience.iop.org/1742-6596/267/1/012009)

GANIL-SPIRAL2: a new era

S Gales

GANIL (DSM-CEA/IN2P3-CNRS)

Blvd. Henri Becquerel, F-14076 Caen cedex, France

E-mail:**gales@ganil.fr**

Abstract. GANIL presently offers unique opportunities in nuclear physics and in many other fields that arise from not only the provision of low-energy stable beams, fragmentation beams and re-accelerated radioactive species, but also from the availability of a wide range of state-of-the-art spectrometers and instrumentation. A few examples of recent highlights are presented. With the construction of SPIRAL2 over the next few years, GANIL is in a good position to retain its world-leading capability. As selected by the ESFRI committee, the next generation of ISOL facility in Europe is represented by the SPIRAL2 project to be built at GANIL (Caen, France). SPIRAL 2 is based on a high power, CW, superconducting LINAC, delivering 5 mA of deuteron beams at 40MeV (200KW) directed on a C converter+ Uranium target and producing therefore more 10^{13} fissions/s. The expected radioactive beams intensities in the mass range from A=60 to A=140, will surpass by two order of magnitude any existing facilities in the world. These unstable atoms will be available at energies between few KeV/n to 15 MeV/n. The same driver will accelerate high intensity (10μA to 1 mA), heavier ions (Ar up to Xe) at maximum energy of 14 MeV/n. Under the 7FP program of European Union called *Preparatory phase*, the SPIRAL2 project has been granted a budget of about 4M€ to build up an international consortium around this new venture. The scientific pillars of the future facility, the status of the construction of SPIRAL2 accelerator and associated physics instruments in collaboration with EU and International partners will be presented.

1. Introduction

Since its beginning in 1983, GANIL [1] has been involved in active research in the field of exotic nuclei. With the LISE spectrometer and its successor LISE3 some 50 new nuclei have been observed. In addition, in 1994, the physicists developed SISSI, a dual superconducting solenoid system for producing high-energy beams of exotic nuclei. At GANIL the available stable ion beams range from ^{12,13}C to Kr isotopes with intensities up to 2pμA for ^{13}C at 95MeV/n and 0.1pμA for Kr at 50 MeV/n (beam power between 3 to 6 KW on target). For heavier ions (Lead and Uranium) an energy range between 5-25 MeV/n with 10^{10-11}pps can be achieved. Light and medium mass ion beams raised to high energy are fragmented on a thin target, giving rise through the "fragmentation process" to light exotic nuclei. The successful program of the physics of "exotic " nuclei produced by the so-called "in-flight " method was extended recently towards new possibilities offered by high- quality, low-energy Radioactive Ions Beams (RIB) at the SPIRAL facility [1]using the "ISOL" production method. This SPIRAL facility, which became operational in September 2001, can be summarized as follows: the high energy, high power extracted beam from the three GANIL accelerators C01 (2)-CSS1-CSS2 is sent to a carbon target. The incident nuclei are fragmented by nuclear reactions in this target, generating a population of radioactive nuclei which diffuses out of this hot target (2000°C). After ionization in an ion source, the particles are injected into the CIME cyclotron to be accelerated to energies from 1.7 to 25 MeV per nucleon. Since its start-up, various ions have been produced and accelerated in the SPIRAL facility, ranging from exotic nuclei as light as ^{8}He to heavier nuclei like ^{76}Kr. About 200 shifts per year are devoted to experiments with SPIRAL beams. At GANIL, the rather unique capabilities, using both "In flight" and "ISOL" techniques, to produce a large variety of light and medium "exotic" species have boosted the scientific program.

2. Exotic nuclei at GANIL: recent highlights

As intensities of secondary radioactive beams are usually four or five orders of magnitude lower than typical intensities of stable beams, physics with secondary RIBs requires very specific detection devices. Reactions and Structure studies of "exotic" nuclei are carried out using intensively the high energies of the fragmented "in flight" RIB from GANIL, the post accelerated ISOL- SPIRAL1 RIB and low energy (5-10 MeV/n) intense Uranium beams. The beams are coupled to spectrometers like LISE, VAMOS and/or 4Π γ-array called EXOGAM, charged particle detectors like MUST and TIARA. Those combinations are ideal tools for the studies of nuclear structure and reactions. In the following we will present a few recent highlights of the physics results achieved in the last few years.

2.1Reaction studies with neutron rich dilute matter are also of strong interest.

A modern variation of the Rutherford experiment to probe the tunneling of exotic nuclear matter from the measurement of the residues formed in the bombardment of ^{197}Au by extremely neutron-rich ^{8}He nuclei was carried out. Using a novel off-beam technique the most precise and accurate measurements of fusion and neutron transfer involving reaccelerated unstable beams are reported. The results are displayed in Fig.1 and reveal an unusual behavior of the tunneling of ^{8}He compared to that for lighter helium isotopes, highlighting the role of the intrinsic structure of composite many-body quantum systems and pairing correlations [4].

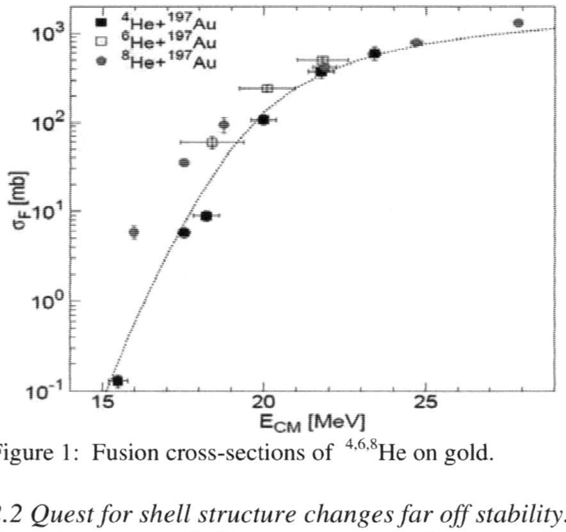

Figure 1: Fusion cross-sections of ^{4,6,8}He on gold.

2.2 Quest for shell structure changes far off stability.

Experiments on shell structure evolution around magic number 8,20 and 28 have been recently investigated using both "in flight" and ISOL RIB via knock out or transfer reactions associated to gamma spectroscopy.

For the N=28 neutron shell ,the energies of the excited states in very neutron-rich ^{42}Si and $^{41-43}$P have been measured using in-beam γ–ray spectroscopy from the fragmentation of secondary beams of $^{42-44}$S at 39 MeV/n. The low 2^{+} energy of ^{42}Si, 770(19) keV, together with the level schemes of $^{41-43}$P provide evidence for the disappearance of the Z=14 and N=28 spherical shell closures, which is ascribed mainly to the action of proton-neutron tensor forces [5]. Recent investigation, of the neighboring nuclei N=29, ^{43}S have led to the discovery of a low lying 7/2- isomeric state, very weakly connected to the deformed 3/2- ground state demonstrating a shape coexistence at low excitation energy in ^{43}S, [6] in line with the previous conclusion, namely a shell erosion of the N=28 gap in that region of mass.

2.3 Prompt gamma spectroscopy of fully identified fission fragments.

Last but not least, thanks to the high intensity and low energy ^{238}U beams coupled to the VAMOS spectrometer and the gamma array EXOGAM, it has been possible for the first time to carry out the study of fission fragments production and spectroscopy using the fusion-fission reactions of the type ^{12}C (^{238}U, ^{134}Xe)Ru [7]. The measurements with high precision of the magnetic rigidity, the energy loss, the total energy and time of flight have allowed unique determination of the atomic charge Z and the mass A of each fission fragment detected in the focal plane of VAMOS spectrometer. Resulting two dimensional plots for ΔE, E and Mass identification are shown in Figure 2.

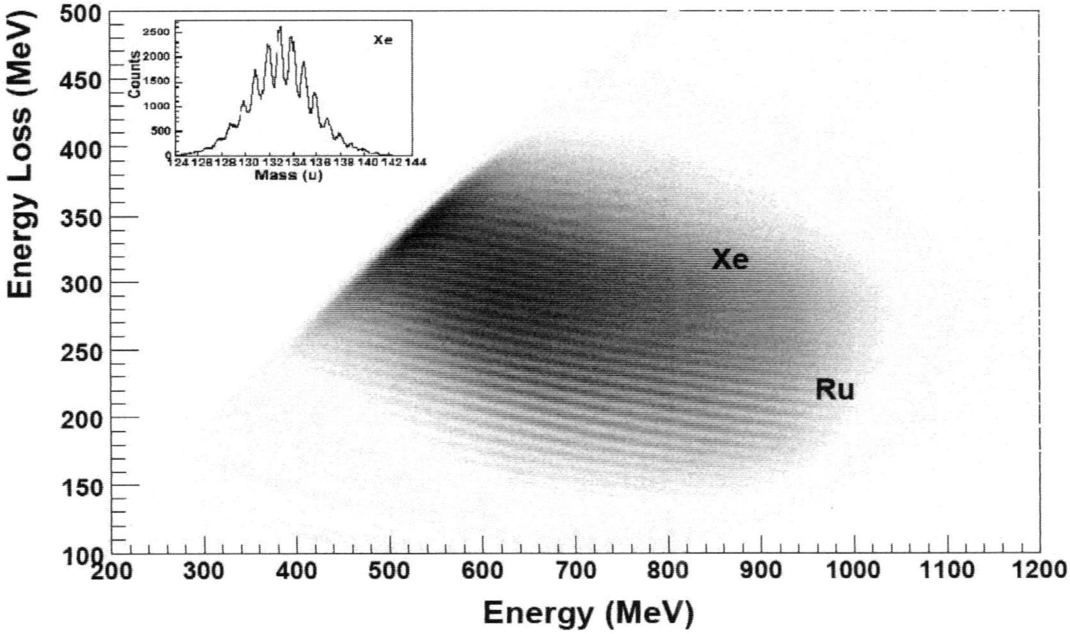

Figure 2 : Two dimensional spectrum of ΔE vs. E_{total} in the ^{238}U+^{12}C system at beam energy of 1.45 GeV. The inset shows the associated mass spectrum for the Xe isotopes.

Measurements of prompt Doppler-corrected de-excitation γ-rays from uniquely identified fragments formed in the above mentioned reaction was also achieved through gamma-fission fragment coincidences. These very promising results open a new window in the spectroscopy of very neutron rich fission fragments [7, 8].

3. SPIRAL2: Status and Perspectives

Since the beginning of the SPIRAL project it was proposed to enlarge the range of RIB produced through the fission of uranium target and the delivery of high intensity neutron –rich fission fragments. This idea is now being implemented in the SPIRAL2 project [9].

The importance of the RIB's has been underlined by NuPECC [10] (Nuclear Physics European Collaboration Committee - an expert committee of the European Science Foundation) which recommend the design and construction of the ultimate-generation RIB ISOL facility EURISOL [11].

Because of the time-line for EURISOL (2020), NuPECC recommends the construction of intermediate-generation facilities that will benefit the EURISOL project in terms of R&D and that will give the community opportunities to perform research and applications with RIBs. Among the proposed intermediate facilities SPIRAL2 at GANIL (Caen, France), meets in terms of physics potential, site and size of the investments, the criteria of European dimension as it was recognized recently in the ESFRI (European Strategy Forum on Research Infrastructures) roadmap [12]. The construction of SPIRAL 2 started in the middle of 2005 and is supported by the EU FP7 through the Preparatory Phase contract since 2008. The first beams are expected in the years 2012-2013. As an added bonus, the facility will also produce over 10^{15} neutrons/s, making it the world's most powerful source of fast neutrons for several years.

3.1 The Science Drivers of SPIRAL2

A rather complete presentation of the scientific case of SPIRAL2 facility can be found in the White Book of SPIRAL2, Letters of Intent and technical proposals for the facility (see Ref [9]). In the following only few examples of a rich and exciting scientific program will be shortly illustrated.

3.1.1 N=Z Proton Drip Line and Super-Heavy Elements

Very intense stable heavy ions beams will be available at the exit of the LINAG accelerator. These heavy ions (from He to Ca), will reach intensities from 1pμA up to 1pmA using new PHOENIX ion sources in combination with A/Q=1/3 RFQ injector and later A/Q=1/6 for heavier ions (Kr to U).
Reaching such beams intensities will open a wide range of physics experiments not accessible to day. Super Heavy Elements synthesis, spectroscopy and chemistry, spectroscopy of N=Z nuclei or beyond the proton drip line, multi-nucleon transfer and deep inelastic scattering, production and study of isomers , access to many short-lived isotopes or isotopes of refractory elements which are difficult to produce with ISOL technique. For example, a production of up to $8x10^4$ atoms of ^{80}Zr per second using a 200 μA $^{24}Mg^{8+}$ beam on a ^{58}Ni target should be possible.
A new device, S3 (for Super Separator Spectrometer, see Ref [13] has been proposed for experiments with the very high intensity stable beams of SPIRAL2 coming from the LINAG accelerator. Most of the physics cases listed above can be seen as either the study of rare processes or as the study of a secondary reaction with exotic ions produced in a first step reaction.

3.1.2 Physics of "Very Neutron Rich Isotopes" and Nuclear Astrophysics

The SPIRAL2 facility at GANIL will produce radioactive isotopes ranging from the lightest to very heavy elements beyond uranium. Different production mechanisms will be utilized: fission of ^{238}U to produce medium-mass, neutron-rich isotopes, fusion evaporation for medium-mass, transfer and deep-inelastic reactions for light to heavy nuclei closer to the line of stability. Thanks to very high intensities, SPIRAL2 will give access to a whole range of experiments which were not feasible with the first generation of RIB facilities. New exotic shapes and excitation modes, halo-like and molecular structures, single particle properties and new magic numbers, new modes of nuclear decay have been recently observed, while studies of fundamental symmetries, tests of the Standard Model and of fundamental interactions in very exotic nuclei have all a huge potential of discovery.
A fundamental challenge in Nuclear Astrophysics is our lack of understanding of the formation of trans-ferric elements. The site for the production of many of the elements heavier than iron, including gold, platinum and uranium is still unknown. Nuclear physics is needed to provide the data required to understand the underlying reaction processes. The future SPIRAL2 facility will contribute prominently to several areas of active research in nuclear astrophysics, such as explosive hydrogen burning, s-process and r-process nuclear synthesis, which are linked to astrophysical observations (novae, X-ray bursts, type2 supernovae, etc...). The (d,p) reaction can be used to simulate the (n,γ) capture on medium-mass nuclei at SPIRAL2 near the r-process possible paths. The neutron rich isotopes of Sn and Cd nuclei are excellent test cases.

3.1.3 Neutrons for Science at SPIRAL2

In addition, let us stress that, thanks to the high energy and high intensity neutron flux available at SPIRAL2, the facility will offer a unique opportunity for material irradiations and cross-section measurements both for fission (notably accelerator driven systems (ADS) and Gen-IV fast reactors) and fusion related research, tests of various detection systems and of resistance of electronics components to irradiations, etc. The scientific opportunities discussed above are just a few examples among many exciting avenues which can be investigated with SPIRAL2.

4. Layout, performances and construction of the SPIRAL2 facility

The SPIRAL 2 facility (Figure 3.) is based on a high power, superconducting driver LINAC, which will deliver a high intensity, 40 MeV deuteron beam as well as a variety of heavy-ion beams with mass over charge ratio equals to 3 and energy up to 14.5 MeV/n. Using a carbon converter, the 5 mA deuteron beam and a uranium carbide target, fast-neutron induced fission is expected to reach a rate of up to 10^{14} fissions/s. The RIB intensities in the mass range from A=60 to A=140 will surpass by one or two order of magnitude any existing facilities in the world. A direct irradiation of the UC_x target with beams of deuterons, $^{3,4}He$, $^{6,7}Li$, or ^{12}C can be used if higher excitation energies were to lead to higher production rate for a nucleus of interest.

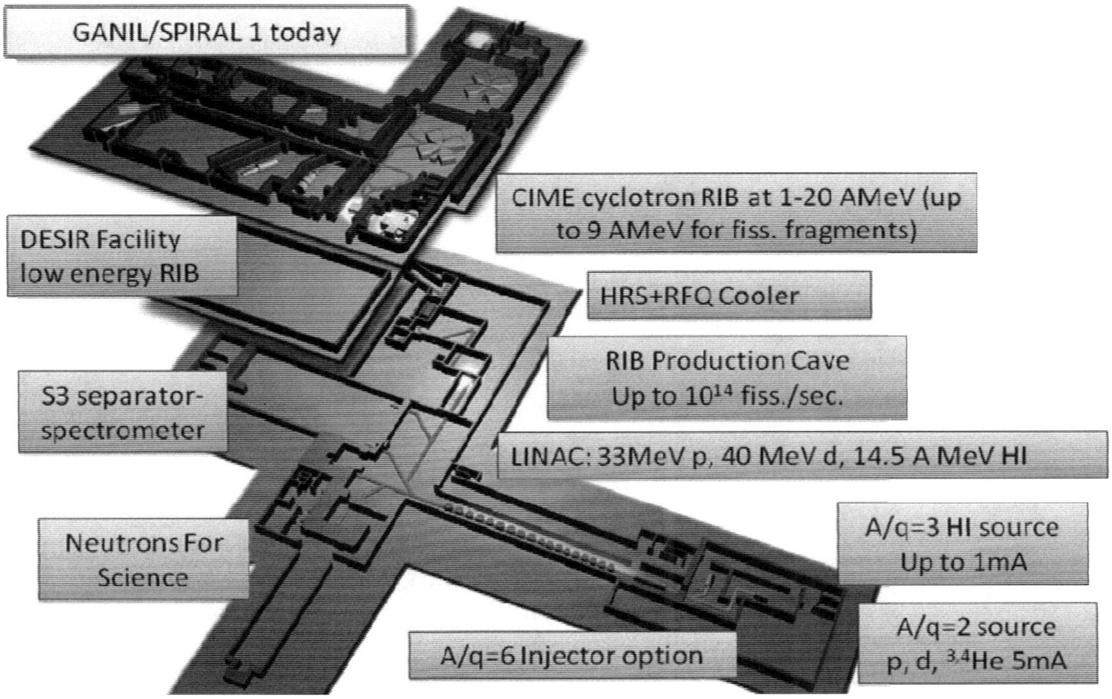

Figure 3: Schematic layout of the SPIRAL 2 facility at GANIL

The stable heavy-ions available at the exit of the linac accelerator will reach intensities from 1pμA to 1pmA.These heavy ions beams will be used to induce fusion and deep inelastic processes leading to the production of neutron and proton rich isotopes. Similarly, the heavy- and light-ion beams from the linac driver on different production targets can be used to produce high-intensity light RIB with ISOL technique. The extracted RIBs will be subsequently accelerated to energies of up to 20 MeV/n, (typically 6-7 MeV/n for fission fragments), by the existing CIME cyclotron. Thus using different

production mechanisms and techniques SPIRAL 2 would allow performing experiments in a wide range of neutron- and proton-rich nuclei far from the line of stability (see Figure 1).
One of the important features of the future GANIL/SPIRAL/SPIRAL 2 facility will be a possibility to deliver up to five stable or radioactive beams simultaneously in the energy range from KeV to several tens of MeV/n.
SPIRAL2 has also a remarkable potential for neutron-based research both for fundamental physics and various applications.
The civil construction of SPIRAL2 has started. It is divided into two phases. Phase 1 includes the LINAC building and associated experimental areas (AEL) and is planned to be completed by 2012 with the commissioning of first beams from LINAC. The second phase (RIB production building and dedicated area low energy ISOL facility, DESIR) will start in 2012 aiming to be ready by 2014. Moreover all the components of the LINAC are under construction including serial production of super-conducting RF cavities, ion sources, injection and other beam-line components.

5. New experimental areas and detectors for SPIRAL2

A process leading to the definitions of the new experimental areas, detectors and associated collaborations was initiated in March 2006 by the Scientific Advisory Committee (SAC) of SPIRAL2. A call for Letters of Intents was launched and the corresponding evaluation took place at the end of 2006. A first targeted call for full technical proposals related to two news experimental vaults S3 and NFS and using directly intense heavy ions and p,d beams from the linac was launched in 2008.

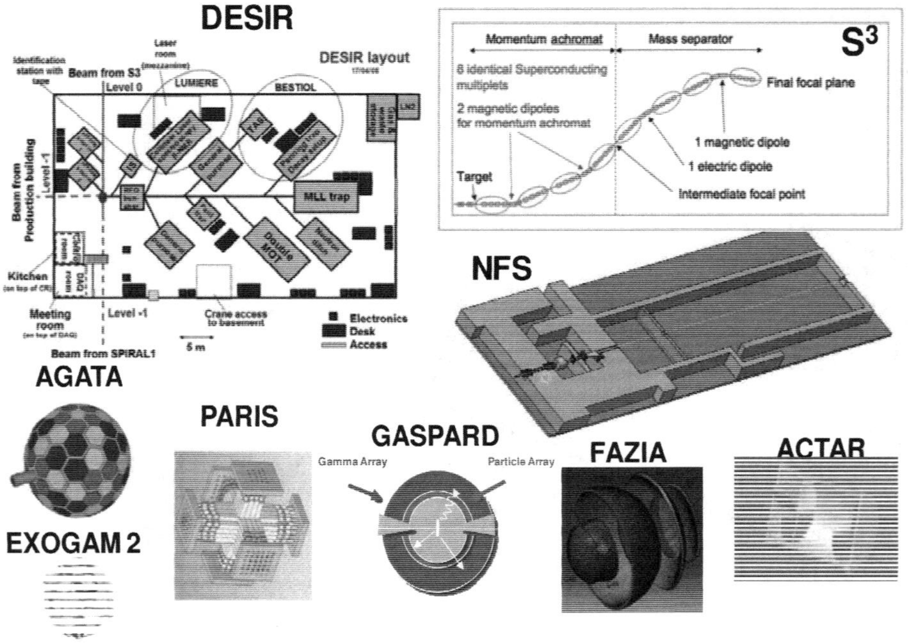

Figure 4: New experimental areas and detectors to be used with GANIL/SPIRAL1/SPIRAL2 facility

Within the SPIRAL2 project these two new experimental areas (Aires Experimentales du LINAC-AEL) will host, respectively, the Neutron for Science facility (Neutron converter and time of flight hall) and the Super Separator Spectrometer (S3) devoted to in flight production of exotic nuclei and to Super Heavy Elements (SHE) using intense heavy ions linac beams (see figs 3 and 4).
Another experimental area devoted to the low energy RIBs has been proposed by the DESIR (Decay, Excitation and Storage of Radioactive Ions) collaboration. At SPIRAL2 and its low-energy beam

facility DESIR, there is an opportunity of building a diversity of apparatus to try to cover the whole nuclear chart.

In addition to these new experimental areas, the Letters of Intent process has induced the formation of new collaborations around detectors the 4Π-γarray EXOGAM2 and AGATA, the active target gas detector ACTAR, the large charge particle arrays FAZIA and GASPARD and the gamma calorimeter PARIS (see Figure 4). These new instruments dedicated to SPIRAL2, for the R&D phase, are supported by the EU FP7 trough the Preparatory Phase contract. Signatures of Memorandums of Understanding (MoU) related to the construction phase are expected by 2010-2011.

6. Conclusions

Nuclear physics has been revolutionized by the recent development of the ability to produce accelerated beams of radioactive nuclei. For the first time it will be possible to study reactions to produce the 6000 to 7000 nuclei we believe exist.

SPIRAL2 is a major expansion of the SPIRAL facility at GANIL, which will help maintain European leadership in ISOL (Isotope Separation On Line) development, aiming at two orders of magnitude increase of the secondary beams available for nuclear physics studies. The technical challenges of the high power accelerator, targets and experimental equipments will provide essential knowledge on the road toward EURISOL, the most advanced nuclear physics research facility presently imaginable and based on the ISOL principle. The scientific program, proposes the investigation of the most challenging nuclear and astrophysics questions aiming at the deeper understanding of the nature of matter.

The construction cost of SPIRAL2 amounts to 200 M€ (including personnel and contingency), and will be shared by French funding agencies CNRS/IN2P3 and CEA/DSM, the Region of Basse Normandie and international partners. The first beams are expected to be delivered by SPIRAL2 in 2012. The full GANIL/SPIRAL1/SPIRAL2 facility will be operational by the end of 2013 and will serve a community of about 800 users.

Acknowledgments

I would like to thank my colleagues from the physics group at GANIL, N.Alahri, M. Jacquemet, SPIRAL2 project leader and M. Lewitowicz Scientific Coordinator of SPIRAL2, together with all the members of SPIRAL2 project for allowing me to use some of the results of their work in this paper.

References

[1] *www.**ganil**-spiral2.eu/*

[2] Rodríguez *et al., Nucl. Inst. and Methods* A **565** (2006) 876-889

[3] Fléchard X. *et al., Phys. Rev. Lett.*, **101**(2008) 212504 *and references herein*

[4] Lemasson A.*et al., Phy. Rev. Lett.* **103** 232701 (2009)

[5] Bastin B.*et al., Phys.Rev.Lett* **101**(2008) 092501

[6] Gaudefroy L.*et al .,Phys. Rev. Lett* **102**, 092501 (2009)

[7] Shrivastava A. *et al., Phys. Rev. C* **80,** 051305 (2009)

[8] Caamaño M. *et al., Proceeding of the 4th International Workshop on Fission And Fission Products Spectroscopy* (Cadarache) May 2009

[9] *www.ganil.fr/research/developments/spiral2/*

[10] *www.nupecc.org*

[11] *www.eurisol.org*

[12] *Cordis.europa.eu esfri/publications.html*

[13] Savajols H. *et al. S3, SPIRAL2 Technical Proposal* *www.ganil.fr/research/developments/spiral2/collaborations.html*

Mass predictions of exotic nuclei within a macroscopic-microscopic model

2011 J. Phys.: Conf. Ser. 267 012010

(http://iopscience.iop.org/1742-6596/267/1/012010)

Mass predictions of exotic nuclei within a macroscopic-microscopic model

G. Royer, A. Onillon, M. Guilbaud and A. Auzizeau

Subatech, Université-IN2P3/CNRS-Ecole des Mines, 44307 Nantes, France

E-mail: royer@subatech.in2p3.fr

Abstract. Different Liquid Drop Model mass formulae have been studied. They include a Coulomb diffuseness correction Z^2/A term and pairing and shell energies of the Thomas-Fermi model. The influence of the selected charge radius, the curvature energy and different forms of the Wigner term has been investigated. Their coefficients have been determined by a least square fitting procedure to 2027 experimental atomic masses. The different fits lead to a surface energy coefficient of 17-18 MeV. A large equivalent rms radius ($r_0 = 1.22 - 1.24$ fm) or a shorter central radius may be used. A rms deviation of 0.54 MeV can be reached between the experimental and theoretical masses. The remaining differences come from the determination of the shell and pairing energies. Mass predictions are given for exotic nuclei.

1. Introduction

Predictions of masses of exotic nuclei close to the proton and neutron drip lines and in the superheavy element region must still be pursued and improved. Beyond the Bethe-Weizsäcker formula [1, 2] and beside the statistical Thomas-Fermi model [3] and the microscopic Hartree-Fock self-consistent mean field approaches [4], the ability and accuracy of different versions of the macro-microscopic Liquid Drop Model mass formula and nuclear radii have been studied and compared [5].

2. Different possible mass formulae

Different subsets of the following expansion of the nuclear binding energy in powers of $A^{-1/3}$ and the relative neutron excess $I = (N - Z)/A$ have been studied :

$$B = a_v \left(1 - k_v I^2\right) A - a_s \left(1 - k_s I^2\right) A^{\frac{2}{3}} - a_k \left(1 - k_k I^2\right) A^{\frac{1}{3}} - \frac{3}{5}\frac{e^2 Z^2}{R_0} + f_p \frac{Z^2}{A} - E_{pair} - E_{shell} - E_{Wigner}. \quad (1)$$

The first term is the volume energy. $I^2 A$ is the asymmetry energy of the Bethe-Weizsäcker mass formula. The second term is the surface energy. It takes into account the deficit of binding energy of the nucleons at the nuclear surface. The third term gives the curvature energy which is a correction to the surface energy resulting from the mean local curvature (see Ref. [5]). This term is considered in the TF model [3] but not in the Finite Range LDM [6]. The fourth term is the usual Coulomb energy. Different formulae may be assumed for the charge radius. The Z^2/A term is the diffuseness correction to the basic sharp radius Coulomb energy term (called also the proton form-factor correction). The shell and pairing energies of the recent Thomas-Fermi model [3, 5] have been chosen and four versions of the Wigner term have been introduced: $W_1 = |I|$, $W_2 = |N-Z| \times e^{-(A/50)^2}$, $W_3 = |N-Z| \times e^{-A/35}$ and $W_4 = e^{-80I^2}$. To obtain the coefficients of the selected expansions by a least square fitting procedure, the masses of the 2027 nuclei verifying the two conditions : N and Z higher than 7 and the one standard deviation uncertainty on the mass lower than 150 keV have been used [7].

As examples, three possible accurate possible formulae are given here.

$$B = 15.4133\left(1-1.7962I^2\right)A - 17.3079\left(1-1.7858I^2\right)A^{\frac{2}{3}} - 0.6\frac{e^2Z^2}{1.2318A^{\frac{1}{3}}} + 0.8956\frac{Z^2}{A}$$
$$-0.4838|N-Z| \times e^{-(A/50)^2} + 2.2 \times e^{-80I^2} - E_{pair} - E_{shell}. \quad (2)$$

$$B = 15.6096\left(1-1.8543I^2\right)A - 18.1132\left(1-2.0021I^2\right)A^{\frac{2}{3}} - 0.6\frac{e^2Z^2}{1.28A^{\frac{1}{3}} - 0.76 + 0.8A^{-\frac{1}{3}}}$$
$$+1.8086\frac{Z^2}{A} - 0.47|N-Z| \times e^{-(A/50)^2} + 2.4954 \times e^{-80I^2} - E_{pair} - E_{shell}. \quad (3)$$

$$B = 15.3848\left(1-1.7837I^2\right)A - 17.1947\left(1-1.8204I^2\right)A^{\frac{2}{3}} - 0.6\frac{e^2Z^2}{1.2257A^{\frac{1}{3}}}$$
$$+1.1035\frac{Z^2}{A} - 16.606|I| - E_{pair} - E_{shell}. \quad (4)$$

The rms deviations between the theoretical and experimental masses are respectively : 0.543, 0.558 and 0.584 MeV. In the formula (2) the reduced radius r_0 is determined by the adjustment to the experimental masses. The combination of two Wigner terms allows to reach a very good accuracy. In the formula (3) the assumed radius is the radius proposed in Ref. [8]. It corresponds to a central or equivalent sharp radius. In the last formula (4) the radius $R_0 = 1.2257\ A^{1/3}$ fm has been obtained previously by an adjustment on 782 ground state charge radii [9]. So it it possible to obtain accurate mass formulae with a large constant reduced radius r_0 or with a more sophisticated central radius corresponding to a smaller value of r_0 increasing with the mass. On the other hand a constant value $r_0 = 1.16$ fm does not allow to obtain a rms deviation better than 0.72 MeV. The formula (2) is more precise for the light nuclei while the formula (4) is the most appropriate for the heaviest elements.

The difference between the theoretical masses obtained with the formula (4) and the experimental masses of the 2027 nuclei used for the adjustment of the coefficients is indicated in Figure 1 (formulas (2) and (3) lead about to the same figures). The more the colour is dark the more the accuracy is high. The errors are slightly larger for the light nuclei. The same behaviour is encountered by all the mass models. Nevertheless the error is very rarely higher than 2 MeV.

3. Predictability of the formulae

Since the last mass evaluation [7] other masses have been newly or more precisely obtained. The predictions given by the formula (4) (not readjusted) for 161 new masses are compared with the experimental data in Figure 2. The accuracy is correct in the whole mass range showing the predictability of such formulae.

Table 1. Theoretical mass excess (in MeV) predicted with the formula (3) and 2003 AME values for heavy exotic nuclei.

Nucleus	E_{th}	E_{AME}	Nucleus	E_{th}	E_{AME}	Nucleus	E_{th}	E_{AME}
^{152}La	-49.34	-50.07	^{152}Ce	-59.26	-59.11	^{152}Lu	-33.18	-33.42
^{153}Ba	-36.48	-37.62	^{153}La	-46.02	-46.93	^{153}Ce	-54.97	-55.35
^{153}Yb	-47.09	-47.06	^{153}Hf	-26.82	-27.3	^{154}La	-41.3	-42.38
^{154}Ce	-52.25	-52.7	^{154}Lu	-39.88	-39.57	^{154}Hf	-32.87	-32.73
^{155}La	-37.71	-38.8	^{155}Ce	-47.65	-48.4	^{155}Pr	-55.37	-55.78
^{155}Nd	-62.38	-62.47	^{155}Hf	-34.18	-34.1	^{155}Ta	-24.38	-23.67
^{156}Ce	-44.6	-45.4	^{156}Pr	-51.28	-51.91	^{156}Ta	-26.2	-25.8
^{157}Ce	-39.66	-40.67	^{157}Pr	-48.29	-48.97	^{157}Nd	-56.36	-56.79
^{157}Hf	-38.54	-38.75	^{158}Pr	-43.86	-44.73	^{158}Nd	-53.94	-54.4
^{158}Ta	-31.1	-31.02	^{158} W	-23.81	-23.7	^{159}Pr	-40.68	-41.45
^{159}Nd	-49.6	-50.22	^{159}Pm	-56.5	-56.85	^{159} W	-24.99	-25.23
^{160}Nd	-46.91	-47.42	^{160}Pm	-52.73	-53.1	^{160}Sm	-60.5	-60.42
^{160}Eu	-63.14	-63.37	^{160}Re	-16.95	-16.66	^{161}Nd	-42.36	-42.96
^{161}Pm	-50.12	-50.43	^{161}Sm	-56.88	-56.98	^{161}Eu	-61.57	-61.78
^{161}Ta	-38.7	-38.73	^{161} W	-30.08	-30.41	^{162}Pm	-46.07	-46.31
^{162}Sm	-54.8	-54.75	^{162}Eu	-58.43	-58.65	^{162}Re	-22.54	-22.35
^{162}Os	-14.13	-14.5	^{163}Pm	-43.07	-43.15	^{163}sm	-50.83	-50.9
^{163}Eu	-56.47	-56.63	^{163}Gd	-61.17	-61.49	^{163}Os	-15.9	-16.12
^{164}Sm	-48.36	-48.18	^{164}Eu	-53.02	-53.1	^{164}Gd	-59.76	-59.75
^{164}Re	-27.54	-27.64	^{164}Ir	-7.55	-7.27	^{165}Eu	-50.66	-50.56
^{165}Gd	-56.41	-56.47	^{165}Tb	-60.27	-60.66	^{165}Os	-21.43	-21.65
^{165}Ir	-11.32	-11.63	^{166}Eu	-46.68	-46.6	^{166}Gd	-54.52	-54.4
^{166}Re	-31.99	-31.85	^{166}Ir	-13.32	-13.21	^{166}Pt	-4.35	-4.79
^{167}Eu	-43.65	-43.59	^{167}Gd	-50.67	-50.7	^{167}Tb	-55.69	-55.84
^{167}Re	-34.75	-34.84	^{167}Pt	-6.17	-6.54	^{168}Gd	-48.17	-48.1
^{168}Tb	-52.35	-52.5	^{168}Ir	-18.73	-18.74	^{169}Gd	-43.71	-43.9
^{169}Tb	-49.99	-50.1	^{169}Pt	-12.03	-12.38	^{169}Au	-1.71	-1.79
^{170}Tb	-46.13	-46.34	^{170}Dy	-53.49	-53.66	^{170}Ir	-23.4	-23.32
^{170}Au	-3.7	-3.61	^{171}Tb	-43.48	-43.5	^{171}Dy	-49.76	-50.11
^{171}Hg	3.57	3.5	^{172}Dy	-47.65	-47.73	^{172}Ho	-51.09	-51.4
^{172}Ir	-27.48	-27.52	^{172}Au	-9.06	-9.28	^{173}Dy	-43.62	-43.78
^{173}Ho	-49.11	-49.1	^{173}Er	-53.54	-53.65	^{173}Hg	-2.31	-2.57
^{174}Ho	-45.65	-45.5	^{174}Er	-52.09	-51.95	^{174}Au	-13.87	-14.2
^{175}Ho	-43.25	-42.8	^{175}Er	-48.73	-48.65	^{176}Er	-46.89	-46.5
^{176}Au	-18.14	-18.54	^{176}Tl	0.87	0.55	^{177}Er	-43.06	-42.8
^{177}Tm	-47.76	-47.47	^{178}Tm	-44.44	-44.12	^{178}Tl	-4.38	-4.75
^{179}Tm	-42.19	-41.6	^{179}Yb	-46.88	-46.42	^{179}Pb	2.34	2
^{180}Yb	-45.15	-44.4	^{180}Tl	-9.07	-9.4	^{181}Yb	-41.58	-40.85
^{181}Lu	-45.29	-44.74	^{182}Lu	-42.24	-41.88	^{183}Lu	-40.21	-39.52
^{184}Lu	-37.01	-36.41	^{184}Bi	1.88	1.05	^{185}Hf	-39.06	-38.36
^{185}Bi	-1.47	-2.21	^{186}Hf	-37.44	-36.43	^{187}Hf	-33.9	-32.98
^{187}Ta	-37.73	-36.77	^{188}Hf	-31.73	-30.88	^{188}Ta	-34.59	-33.81
^{189}Ta	-32.55	-31.83	^{190}Ta	-29	-28.66	^{191} W	-31.72	-31.11
^{192} W	-29.98	-29.65	^{192}Re	-32.1	-31.71	^{193}Re	-30.48	-30.3
^{194}Re	-27.15	-27.55	^{198}Ir	-25.39	-25.82	^{202}Pt	-22.66	-22.6

Nucleus	E_{th}	E_{AME}	Nucleus	E_{th}	E_{AME}	Nucleus	E_{th}	E_{AME}
^{204}Au	-20.5	-20.75	^{205}Au	-18.8	-18.75	^{208}Hg	-13.62	-13.1
^{209}Hg	-8.54	-8.35	^{210}Hg	-5.14	-5.11	^{211}Tl	-5.61	-6.08
^{212}tl	-0.84	-1.65	^{215}Pb	5.67	4.48	^{217}Bi	9.78	8.82
^{218}Bi	14.28	13.34	^{219}Po	13.72	12.8	^{220}Po	16.51	15.47
^{220} U	22.6	23.03	^{221}At	17.4	16.81	^{221} U	24.05	24.59
^{222}At	21.58	20.8	^{222}Pa	21.27	22.12	^{222} U	24.17	24.3
^{223}At	24.39	23.46	^{226}Np	32.44	32.74	^{228}Fr	34.05	33.28
^{228}Np	33.5	33.7	^{231}Fr	43.04	42.33	^{231}Am	42.09	42.44
^{232}Fr	46.99	46.36	^{232}Np	37.42	37.36	^{232}Am	43.16	43.4
^{233}Ra	45.18	44.77	^{233}Ac	41.91	41.5	^{233}Am	43.11	43.17
^{234}Ra	47.81	47.23	^{234}Ac	45.49	45.1	^{234}Am	44.43	44.53
^{235}Ac	48.09	47.72	^{235}Am	44.62	44.66	^{235}Cm	47.98	47.91
^{235}Bk	52.44	52.7	^{236}Ac	51.75	51.51	^{236}Th	46.8	46.45
^{236}Am	46.25	46.18	^{236}Cm	47.75	47.89	^{236}Bk	53.32	53.4
^{237}Th	50.42	50.2	^{237}Am	46.79	46.57	^{237}Cm	49.36	49.28
^{237}Bk	53.04	53.1	^{237}cf	57.73	57.82	^{238}Th	52.88	52.63
^{238}Bk	54.27	54.29	^{238}Cf	57.05	57.2	^{239}Pa	53.64	53.34
^{239}Cm	51.4	51.19	^{239}Bk	54.28	54.29	^{239}Cf	58.24	58.15
^{240}Pa	57.27	56.8	^{240}Bk	55.66	55.67	^{240}Cf	57.85	58.03
^{240}Es	63.9	64.2	^{241} U	56.49	56.2	^{241}Bk	56.12	56.1
^{241}Cf	59.16	59.36	^{241}Es	63.39	63.84	^{242} U	58.8	58.62
^{242}Bk	57.91	57.74	^{242}Es	64.44	64.97	^{242}Fm	67.88	68.4
^{243}Np	59.98	59.88	^{243}Cf	60.72	60.95	^{243}Es	64.44	64.78
^{243}Fm	68.88	69.26	^{244}Np	63.5	63.2	^{244}Es	65.53	66.03
^{244}Fm	68.45	69.01	^{245}Es	65.82	66.44	^{245}Fm	69.49	70.22
^{245}Md	74.78	75.29	^{246}Es	67.39	67.9	^{246}Md	75.71	76.28
^{247}Pu	69.11	69	^{247}Am	66.85	67.15	^{247}Es	67.96	68.61
^{247}Fm	70.89	71.58	^{247}Md	75.27	76.04	^{248}Am	70.43	70.56
^{248}Bk	67.76	68.08	^{248}Es	69.79	70.3	^{248}Md	76.37	77.15
^{248}No	79.81	80.66	^{249}Am	73.26	73.1	^{249}Es	70.62	71.18
^{249}Fm	72.82	73.62	^{249}Md	76.46	77.33	^{249}No	80.9	81.82
^{250}Es	72.57	73.23	^{250}Md	77.84	78.64	^{250}No	80.58	81.52
^{251}Md	78.15	79.03	^{251}No	81.91	82.91	^{251}Lr	86.76	87.9
^{252}Cm	79.3	79.06	^{252}Bk	78.59	78.53	^{252}Md	79.67	80.63
^{252}Lr	87.72	88.84	^{253}Bk	80.96	80.93	^{253}Md	80.42	81.3
^{253}No	83.38	84.47	^{253}Lr	87.48	88.69	^{253}Rf	92.61	93.79
^{254}Bk	84.74	84.39	^{254}Md	82.89	83.51	^{254}Lr	88.69	89.85
^{254}Rf	92	93.32	^{255}Cf	84.81	84.81	^{255}Lr	88.81	90.06
^{255}Rf	93.03	94.4	^{255}Db	98.47	100.04	^{256}Cf	87.11	87.04
^{256}Es	87.01	87.19	^{256}Lr	90.89	91.87	^{256}Db	99.14	100.72
^{257}Es	89.14	89.4	^{257}Lr	91.64	92.74	^{257}Rf	94.86	95.93
^{257}Db	98.94	100.34	^{258}Es	92.54	92.7	^{258}Fm	89.92	90.43
^{258}No	90.83	91.48	^{258}Lr	93.83	94.84			

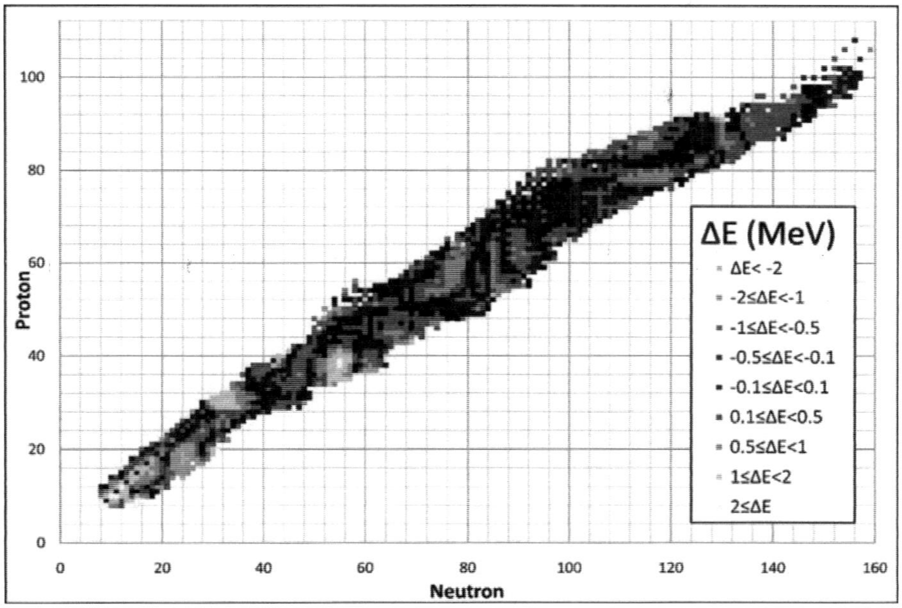

Figure 1. Difference between the theoretical masses obtained with the formula (4) and the experimental masses of the 2027 selected nuclei.

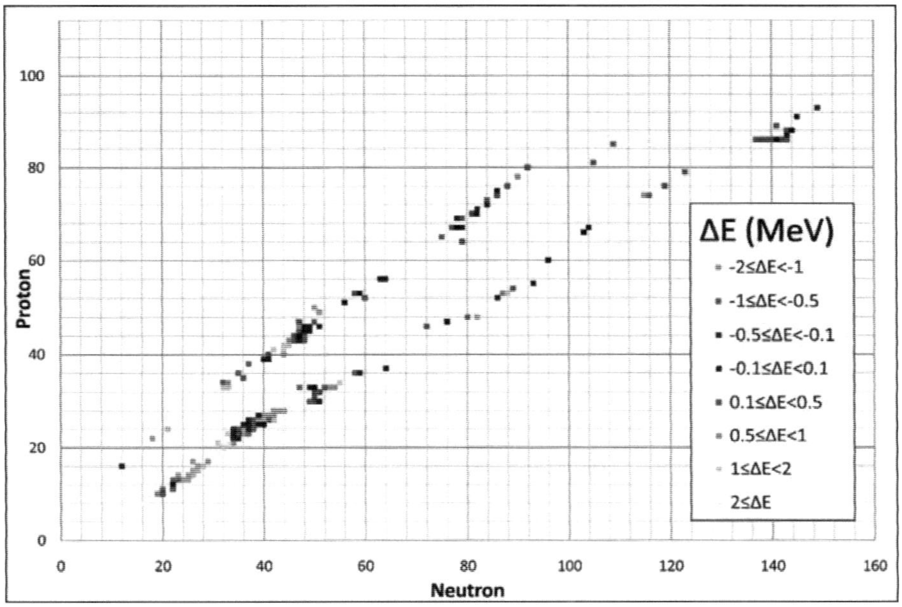

Figure 2. Difference between the theoretical masses obtained with the formula (4) and 161 new experimental masses.

4. Mass of exotic nuclei

Finally, the predictions given by the formula (4) for 656 other nuclei for which the mass is still unknown are compared in Figure 3 to the extrapolations given in Ref. [7] with an

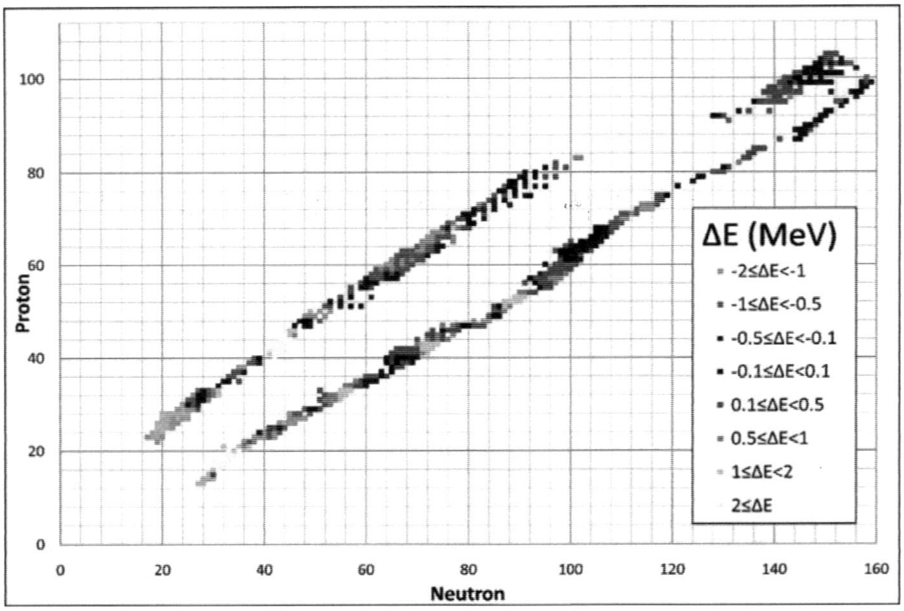

Figure 3. Difference between the theoretical masses obtained with the formula (4) and 656 extrapolated masses.

assumed uncertainty often higher than 500 keV. Without readjustment the formula (4) leads to $\sigma = 0.748$ MeV for the 2844 nuclei.

The theoretical mass excesses predicted with the formula (3) (which uses the radius adopted in our generalized liquid drop model) and 2003 AME values are given and compared in the table for heavy elements.

5. Conclusion

The coefficients of different macro-microscopic Liquid Drop Model mass formulae have been determined by an adjustment to 2027 experimental atomic masses. A rms deviation of 0.54 MeV can be reached. The remaining small differences come probably mainly from the determination of the shell and pairing energies (Strutinsky procedure and Thomas-Fermi model [3]). A large constant coefficient $r_0 = 1.22 - 1.23$ fm or a small value increasing with the mass can be used. Extrapolations are compared to 161 new experimental masses and to 656 mass evaluations of exotic nuclei. The different fits lead always to a surface energy coefficient of around 17-18 MeV.

References

[1] von Weizsäcker C F 1935 *Z. Physik* **96** 431
[2] Bethe H A and Bacher R F 1936 *Rev. Mod. Phys.* **8** 82
[3] Myers W D and Swiatecki W J 1996 *Nucl. Phys. A* **601** 141
[4] Goriely S, Hilaire S, Girod M and Pru S 2009 *Phys. Rev. C* **109** 242501
[5] Royer G 2008 *Nucl. Phys. A* **807** 105, 2010 *Nucl. Phys. A* in press
[6] Möller P, Nix J R, Myers W D and Swiatecki W J 1995 *At. Data Nucl. Data Tables* **59** 185
[7] Audi G, Wapstra A H and Thibault C 2003 *Nucl. Phys. A* **729** 337
[8] Blocki J, Randrup J, Swiatecki W J and Tsang C F 1977 *Ann. of. Phys.* **105** 427
[9] Angeli I 2004 *At. Data Nucl. Data Tables* **87** 185

Alpha structure of ^{12}B studied by elastic scattering of ^{8}Li EXCYT beam on ^{4}He thick target

2011 J. Phys.: Conf. Ser. 267 012011

(http://iopscience.iop.org/1742-6596/267/1/012011)

Alpha structure of ^{12}B studied by elastic scattering of ^{8}Li EXCYT beam on ^{4}He thick target

M G Pellegriti[1,2], D Torresi[1,2], L Cosentino[1], A Di Pietro[1], C Ducoin[3], M Lattuada[1,2], T Lonnroth[4], P Figuera[1], M Fisichella[1,5], C Maiolino[1], A Musumarra[1,6], M Papa[3], M Rovituso[2], V Scuderi[1,7], G Scalia[a,b], D Santonocito[1], M Zadro[8]

1) INFN Laboratori Nazionali del Sud, Catania, Italy
2) Dipartimento di Fisica ed Astronomia, Università di Catania, Catania, Italy
3) INFN, Sezione di Catania, Italy
4) Abo Academy, Turku, Finland
5) Dipartimento di Fisica, Università di Messina
6) Dipartimento di Metodologie Fisiche e Chimiche per l'Ingegneria, Università di Catania, Catania, Italy
7) CSFNSM, Catania, Italy
8) Ruđer Boskovic Institute, Zagreb, Croatia

E-mail: pellegriti@lns.infn.it

Abstract. The ^{8}Li elastic scattering on ^{4}He in inverse kinematics has been studied in order to search for ^{8}Li-α cluster configurations in excited states of ^{12}B. The experimental technique, the inverse kinematic thick target method (TTIK), involved a ^{8}Li beam provided by the EXCYT facility at an initial energy of E_b=30.6 MeV which is continuously degraded down to zero in the passage through a helium extended thick target. After the interaction, the recoil alpha particles were detected by using four ΔE-E silicon detector telescopes. The explored ^{8}Li-α centre-of-mass energy ranges from E_{CM}=2.7 MeV up to E_{CM}=9.6 MeV. A micro-channel plate, placed before the gas chamber, is used to measure the time interval between the entrance of the ^{8}Li into the chamber and the arrival of the recoil alpha particle on the detectors. This "time interval" measurement improves the TTIK technique allowing the discrimination of elastic from inelastic events. In this paper, the experimental technique will be described and the preliminary results for the elastic cross section will be shown.

1. Introduction

The availability of radioactive ion beams is allowing for the study of nuclei in the region far from the stability valley. In particular, the search for cluster structure in neutron rich nuclei is now possible. Kanada En'yo et al. [1] report a study on the spatial density distribution for Boron isotopes in their ground states. A microscopic calculation for the ground states of Boron isotopes, based on Antisymmetrized Molecular Dynamics (AMD) model, shows the change in nuclear shape and properties when the neutron number increases. In particular ^{11}B ground state is described by a

deformed cluster state α-^{7}Li whilst ^{13}B (N=8) has a more spherical shape. When N increases in ^{15}B, ^{17}B and ^{19}B a cluster structure with a prolate shape is again present. A specific theoretical study of ^{12}B and its excited states has been performed by Descouvemont [2]. This study, based on the Generator Coordinate Method (GCM), predicts the presence of cluster states above the alpha decay threshold in ^{12}B (E_x~ 10 MeV).
Experimental signatures of ^{8}Li-α cluster configuration in the excited states of ^{12}B can be the large reduced width for alpha decay (see ref. [2]).

2. Resonant elastic scattering in inverse kinematics

The resonant elastic scattering method [4-5] has been used in order to investigate the α-^{8}Li structure of ^{12}B from E_x~12.7 MeV up to E_x~19.6 MeV.
The resonant elastic scattering method in inverse kinematics (study of the reaction *a(A,a)A*) makes use of the sensitivity of the detected recoil yield to the presence of a resonant state in the intermediate system (*a+A*). By using the beam energy loss in the target it is possible to induce the elastic scattering in a given continuous range of E_{CM} energies (see Fig. 1). The inverse kinematics (*A* (projectile) heavier than *a* (target)), gives forward focused recoiled particles (α particles in the present case). In this context a crucial role is played by the smaller dE/dx of the recoils with respect of that of the beam particles.

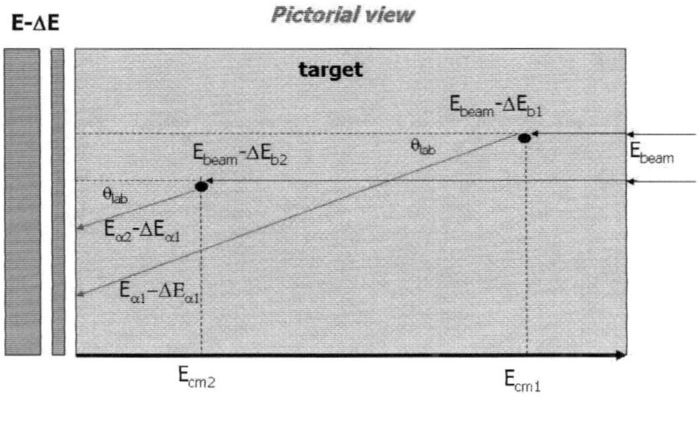

Figure 1. Pictorial view of elastic scattering induced at E_{beam} and at a degraded beam energy E_{beam}-ΔE_{b1}. Different reaction points inside the gas target will correspond to different E_{cm} energies.

In particular the beam particles are stopped inside the target while the recoils can reach the detectors placed around θ_{lab}=0° which correspond at backward angles in centre-of-mass system. At these angles the elastic cross section is more sensitive to the interference effects due to the presence of both nuclear and Coulomb contribution.

An R-matrix analysis [4] of the elastic cross section can provide precise information on the energy position of the levels and their reduced particle widths.
An experimental limitation of the thick target method is represented by the contribution of other possible reaction mechanisms to the final elastic scattering spectrum. For example, if the centre-of-mass energy is high enough to open the inelastic cannel (see Fig. 2, right side), we can have the superposition of elastic events generated at a given beam energy (E_{beam}-ΔE_2) onto inelastic events generated at higher beam energy (E_{beam}-ΔE_1, where $\Delta E_2 > \Delta E_1$).

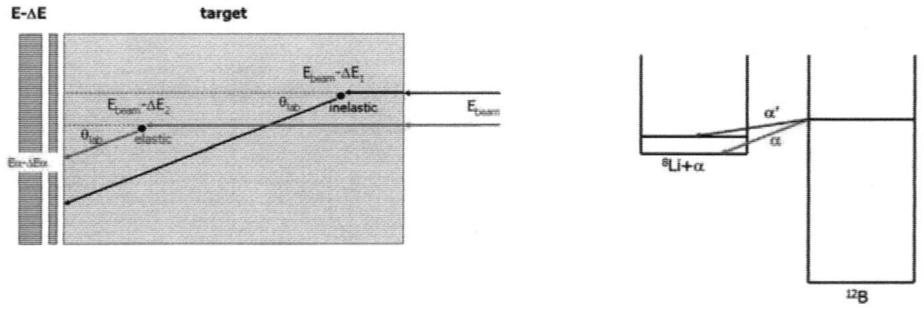

Figure 2. Pictorial view of elastic scattering (Q=0) induced at E_{beam}-ΔE_1 and inelastic scattering (Q≠0) induced at E_{beam}-ΔE_2. Due to a compensation of the energy loss of beam particles and recoils, they give the same alpha final energy $E=E_{\alpha}-\Delta E_{\alpha}=E_{\alpha'}-\Delta E_{\alpha'}$.

This means that recoil particles reaching the detectors at the same energy ($E=E_{\alpha}-\Delta E_{\alpha}=E_{\alpha'}-\Delta E_{\alpha'}$) can be generated by different events (see Fig. 2, left side). In order to bypass this problem it would be necessary to disentangle the different events by knowing the position inside the target where these events have taken place. A way to perform this analysis will be described in the following paragraph.

3. Experimental set-up

The ^{8}Li beam was produced by using the EXCYT facility at INFN-LNS, Catania (see [5,6]). The EXCYT production-acceleration system uses the ISOL technique. It is based on a K-800 superconducting cyclotron, as a driver for stable heavy ion beams (up to 80 MeV/amu, 1 pμA), and on a 15 MV tandem for post-accelerating the radioactive ion beams produced in a target-ion source (TIS) assembly. The production of the radioactive ions was performed by injecting a $^{13}C^{4+}$ primary beam of 45 MeV/amu on a graphite target up to a beam power of 150 W. For ^{8}Li, the higher extraction efficiency from the TIS is obtained by positive ionisation achieved by a tungsten positive surface ioniser. After passing through a charge exchange cell (CEC) to obtain negative ions, the beam is injected into the tandem. The tandem accelerates the $^{8}Li^{3+}$ beam at a final energy of 30.6 MeV.
The ^{8}Li beam impinges on the aluminized Mylar foil (0.8 μm thick) of a Micro Channel Plate (MCP) detector. The foil is placed at 45° with respect to the beam direction (see Fig. 3, upper part). The MCP signal provides a reference for time of flight measurements and at the same time allows counting of the beam intensity, (~$5\cdot 10^{4}$ pps). The ^{8}Li then enters the CT2000 scattering chamber which is filled with ^{4}He enriched gas. The gas chamber is isolated from the beam line (which is at a pressure of 10^{-6} mbar) by using a kapton foil (13μm thick). Pressure (~700mbar) and temperature (~300K°) inside the chamber are continuously monitored. A system of four Silicon ΔE-E telescope detectors is placed at forward angles and at distance of 80 cm with respect to the centre of the chamber. The ΔE detectors

are four quadrant Silicon Pad detectors (50-60 μm thick) whilst the residual energy is measured by Double Sided Silicon Strip detectors, 16x16 strips, (500-1000 μm thick) (see Fig. 3).
The detectors energy calibration was performed by using the known alpha energy peaks from a ^{241}Am-^{244}Cu-^{239}Pu source and the elastic diffusion of a ^{7}Li beam on a gold target.
Stopping power measurements of ^{7,8}Li in ^{4}He have also been preliminarily performed. As it has been underlined by Zadro et al. [7], this is essential to correctly recover the beam energy which induces a given scattering event.
During the measurement, the micro-channel plate, placed before the gas chamber, is used to measure the "time interval", Δt, between the passage of the ^{8}Li through the foil of the micro-channel and the arrival of the recoil alpha particles in the detectors. This "time interval" measurement improves the TTIK technique for infinite target allowing the discrimination of elastic from inelastic events.
During the measurement, the acquisition trigger is generated by events from ΔE detectors. Δt is measured by using the signal produced by the ΔE's as a start. The stop is given by a delayed signal coming from the MCP (inverse logic). The time resolution is ~ 1 ns.

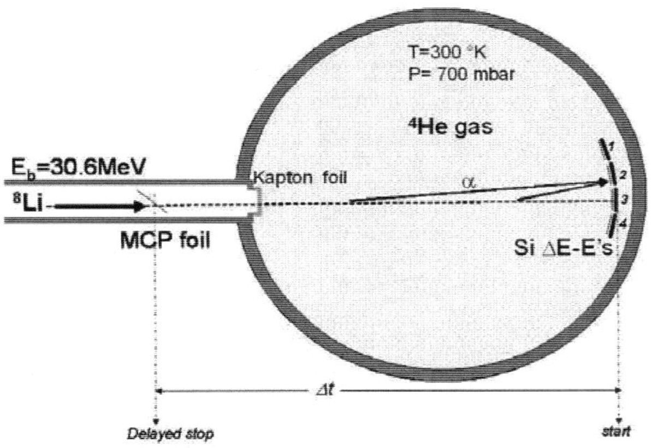

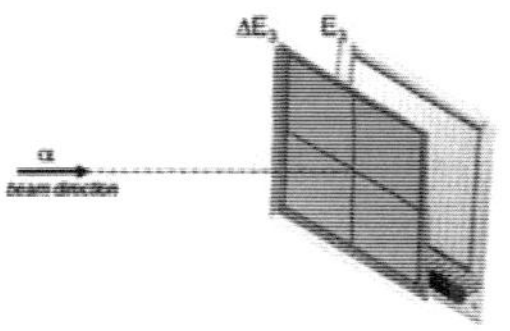

Figure 3. Experimental set-up. In the bottom the ΔE_3-E_3, designed from Micron Semiconductors, used at 0°.

4. Preliminary Results

In the following we will focus the attention on the preliminary results obtained from the most forward detector ΔE-E, placed at $\theta_{lab}=0°$ with respect to the beam direction (see Fig.3, bottom part).
For this telescope, the ΔE nominal thickness is 47 μm. In Fig. 4, the ΔE-E plot is shown. We can distinguish alpha particles from hydrogen isotopes. We can notice the presence of a large yield of tritons, coming most probably from ^{8}Li break-up.

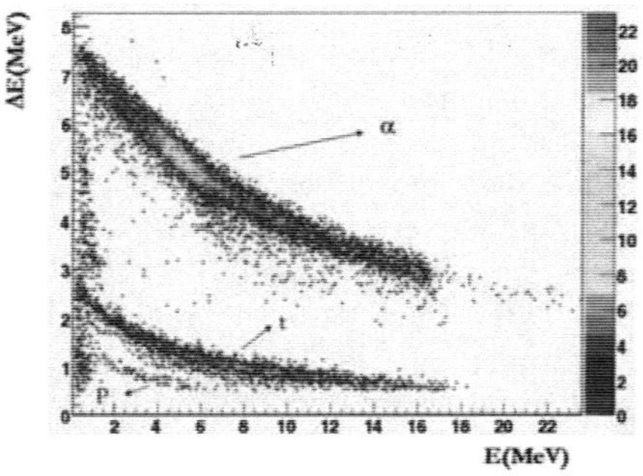

Figure 4. ΔE-E spectrum.

In Fig. 5, the time interval Δt is shown as a function of the energy loss in the ΔE detector. Since inverse time logic has been used, large time intervals (~220ns) correspond to events which occur at the beginning of the chamber, while smaller intervals (~175ns) corresponds to events which take place closer to the ΔE detectors. The α-particle punch-through energy is around 8MeV.

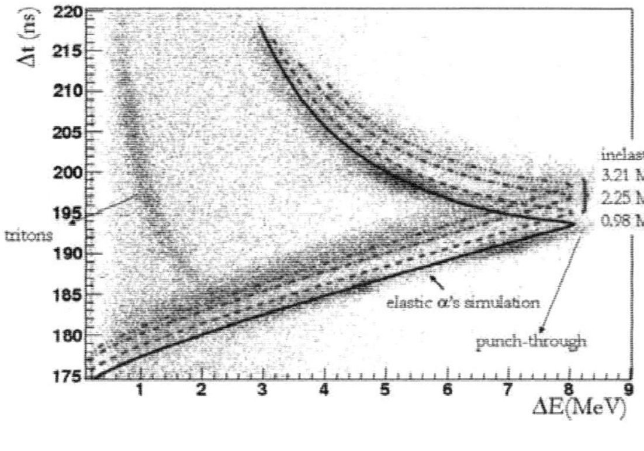

Figure 5. Δt v.s. ΔE spectrum. In Δt axis, the "zero" has been arbitrarily chosen.

As one can observe, the elastic events are clearly distinguished from the inelastic events. The time interval Δt has been calculated as a function of the detected energy both for the elastic and inelastic events (^{8}Li first three excited levels). The calculation includes the kinematics and energy loss of ^{8}Li and α in ^{4}He gas. It can be seen that the calculated locus of the elastic events is in agreement with the data. As regards the inelastic events, we can notice that the 2^+ level is scarcely populated.
In the plot one can also observe a large quantity of tritons as well as alpha background events (for ΔE<5MeV). The first are due, as previously mentioned, to ^{8}Li break-up, while the last, which are uncorrelated in time, are due to the ^{8}Li decay (^{8}Li(β^-) ^{8}Be* 2α). The elastic events which stop in the ΔE detector, have been selected by a graphical contour has been done on the ΔE-E plot. For these events the centre-of-mass energy has been reconstructed and the preliminary cross section has been plotted in Fig.6.

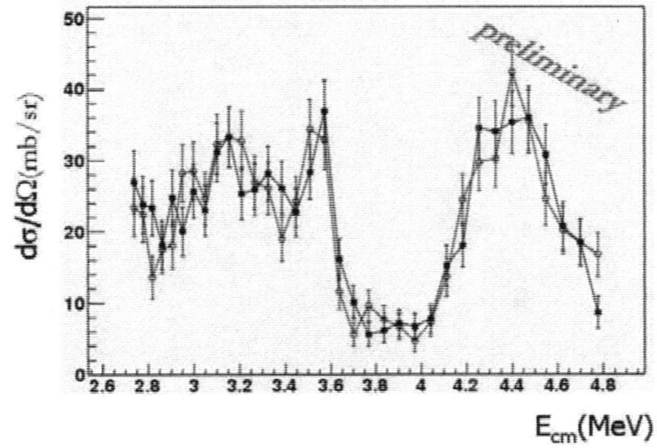

Figure 6. Elastic differential cross section for alpha particles stopped in the ΔE_3 detector. As example, the curves corresponding to two of the four different ΔE_3 quadrants are plotted.

In Fig. 6, the ^{8}Li+^{4}He elastic scattering cross section obtained for E_{cm}=2.7-4.8MeV (events stopped in ΔE detector) is plotted, as example, just for two quadrants of the ΔE_3-E_3 telescope. The two data sets are in agreement within the experimental error bars. The cross-section shows evidences of large peaks.

5. Conclusions

The ^{8}Li-α elastic scattering cross section has been studied by using the TTIK method. The improvement of the technique, concerning the tagging of the events by using time interval measurements, leads to the discrimination of elastic scattering from other reaction mechanisms. The preliminary results of the elastic cross section for E_{CM}=2.7-4.8 MeV have been shown. A more precise reconstruction of E_{CM}, including the beam size, angular and energy straggling for beam particles and alpha recoils, is in progress. Moreover, before extracting information from the data, the background subtraction for ΔE<5 MeV is also needed.

6. References

[1] Kanada En'yo Y et al. 1995 *Prog. Theor. Phys.* **93** 115

[2] Descouvemont P 1996 *Nucl. Phys.* A **596** 285

[3] Goldberg V Z et al. 2004 *Phys. Rev.* C **69** 024602 and references therein

[4] Angulo C et al. 2003 *Phys. Rev.* C **67** 014308

[5] Ciavola G et al. 2002 *Nucl. Phys.* A **701** 54c

[6] Rifuggiato D *this volume*

[7] Zadro M et al. 2007 *Nucl. Instr. Meth.* B **259** 836

Acknowledgments

The authors wish to acknowledge the EXCYT technical staff for the smooth operation of the facility.

Structure effects in the reactions $^{9,10,11}Be+^{64}Zn$ at the Coulomb barrier

2011 J. Phys.: Conf. Ser. 267 012012

(http://iopscience.iop.org/1742-6596/267/1/012012)

Structure effects in the reactions $^{9,10,11}Be+^{64}Zn$ at the Coulomb barrier

V Scuderi[1,2,3], A Di Pietro[2], L Acosta[4], F Amorini[1,2], M J G.Borge[5], P Figuera[2], M Fisichella[1,2], L M Fraile[6†], J Gomez-Camacho[7], H Jeppesen[6§], M Lattuada[1,2], I Martel[4], M Milin[8], A Musumarra[2,9], M Papa[2], M G Pellegriti[1,2], F Perez-Bernal[4], R Raabe[10],G Randisi[1,2‡], F Rizzo[1,2], D Santonocito[2], G Scalia[1,2], O Tengblad[5], D Torresi[1,2], A M Vidal[5], M Zadro[11]

[1]Dipartimento di Fisica ed Astronomia Università di Catania, Catania, Italy
[2]INFN-Laboratori Nazionali del Sud and Sezione di Catania, Italy
[3]Centro Siciliano di Fisica Nucleare e Struttura della Materia, Catania, Italy
[4]Departamento de Fisica Aplicada Universidad de Huelva, Huelva, Spain
[5]Instituto de Estructura de la Materia CSIC, Madrid, Spain
[6]CERN, Geneva, Switzerland
[7]Departamento de Fisica Atomica Molecular Nuclear Universidad de Sevilla, Sevilla, Spain
[8]Department of Physics Faculty of Science University of Zagreb, Zagreb, Croatia
[9]Dipartimento di Metodologie Fisiche e Chimiche per l'Ingegneria Università di Catania,Catania, Italy
[110]Instituut voor Kern-en Stralingsfysica University of Leuven, Leuven, Belgium
[11]Division of Experimental Physics Ruder Bo·skovic Institute, Zagreb, Croatia

E-mail: scuderiv@lns.infn.it

Abstract. Elastic scattering and direct reactions have been studied for the collisions induced by the three Beryllium isotopes $^{9,10,11}Be$, on a medium mass ^{64}Zn target at energies around the Coulomb barrier. The elastic scattering angular distributions, measured for the three systems at the same center of mass energy, were analyzed within the Optical Model and reaction cross-sections were deduced from optical model calculations. For the ^{11}Be induced reaction the transfer/break-up angular distribution was also extracted.

1. Introduction

In the last years scattering of halo nuclei has been studied to some extent in the attempt to investigate both the influence of the projectile halo structure and low binding energy and the effect of the break-up coupling on the reaction dynamics. Since elastic-scattering is a peripheral process it probes the tail of the wave function and hence one can learn about surface properties, such as size of nuclei and surface diffuseness. Therefore, elastic-scattering is an ideal tool to study peculiar nuclear structures as for example the nuclear halo. As it is well known, the nuclear halo structure originates when very

---

†present address: Departamento de Fysica Atomica, Molecular y Nuclear, Universidad Complutense, Madrid, Spain
§present address: Nuclear Science Division, Lawrence Berkeley National Laboratory, Berkeley, USA
‡present address: LPC-ENSICAEN, IN2P3-CNRS and Universitè de Caen, France

weakly bound valence nucleon(s) can tunnel into the classically forbidden region, giving rise to a diffuse tail surrounding a well bound core. The weakly bound halo structure may affect also the reaction mechanisms in collisions induced by such nuclei especially at energies around the Coulomb barrier where couplings between the entrance channel and the continuum [1-5], as well as to the various reaction channels [6-9] are expected to be very important. Direct reactions, such as break-up or transfer, may be favored owing to the low binding energy, the extended tail of halo nuclei and the large Q-values for selected transfer channels.

From an experimental point of view, high quality data are necessary; however, due to the low intensities of the radioactive beams, these are not always available. Since ^{6}He beams are available in different facilities with intensities up to 10^{7} pps, in a wide range of energies, several experiments have been performed so far using mainly this two neutron halo nucleus on different targets. A common feature, observed in the ^{6}He induced elastic scattering, is a reduction of the cross-section in most of the measured angular range [10,11]. As a consequence of the elastic cross-section suppression, the total reaction (TR) cross-section is found to be much larger, a factor of 2 in some cases, than the one observed in reactions induced by non-halo nuclei [10,12].

Optical Model (OM) calculations [11-13] have shown that the optical potential of halo and weakly bound nuclei has a long range part due to the coupling to non-elastic channels and that a large diffuseness parameter of the imaginary potential is needed in order to take into account the diffuse structure of the ^{6}He halo and to describe the data. Continuum Discretized Coupled Channel calculations (CDCC) of the ^{6}He elastic scattering on different target masses (e.g. [2-4]) have been recently published. According to these calculations, the suppression of the elastic cross-section at forward angles is mostly originated by the coupling to states into the continuum. Besides coupling to bound and unbound states, also coupling to one and two neutron stripping can be of importance in describing the low-energy scattering of a halo nucleus. This has been investigated in (e.g. [6-7]) and according to the authors it has a significant effect on the elastic scattering of ^{6}He and should not be neglected. Moreover, coupling effects of break-up or transfer depend upon nuclear structure properties of projectile and/or target as well as break-up thresholds or beam energies. Therefore drawing general conclusions about channel coupling effects on a limited sets of reactions induced by ^{6}He could be misleading. Moreover, due to the complexity of modelling reactions with the 2n-halo ^{6}He it would be of great help to have detailed elastic scattering data of 1n-halo systems, such as ^{11}Be.

To our knowledge, only two attempts to measure elastic-scattering and reaction mechanisms around the barrier with n-halo nuclei different than ^{6}He are reported in literature. ^{11}Be+^{209}Bi was measured by Mazzocco *et al.* [15-16], using a fragmentation beam degraded in energy. The quasi elastic cross-section, which included inelastic excitations up to 2.6 MeV, was measured. The extracted TR cross-section was found to be similar to the one of ^{9}Be+^{209}Bi, measured by the same group [17]. The authors concluded that the breakup process has comparable strengths in both ^{9,11}Be nuclei and that the differences in binding energies and radii of the two Be isotopes do not play an important role. This result is not in agreement with what observed by different authors in ^{6}He induced collisions. In [18] ^{11}Be+^{120}Sn quasi-elastic-scattering angular distribution was measured, the ^{11}Be(1/2-) inelastic excitation was included, but in a very limited angular range.

Recently, we have measured the elastic scattering angular distributions for the collisions induced by the three different Be isotopes, ^{9,10,11}Be, on a medium mass ^{64}Zn target in a wide angular range and with small angular step at $E_{c.m.}$= 24.5 MeV, close to the coulomb barrier [19]. By comparing the elastic-scattering angular distributions for these three systems, the separate effect of the weak binding and halo structure can be investigated. Moreover, in the case of the halo ^{11}Be nucleus the transfer/break-up cross-section has been measured, as well.

2. Experimental Set-up

The experiment with the stable ^{9}Be beam was performed at Laboratori Nazionali del Sud (LNS) in Catania. The ^{9}Be beam was delivered by the SMP 14MV Tandem of LNS and was impinging on a 550 μm/cm^{2} ^{64}Zn. The light charged particles, emitted in the reaction, have been detected using five ΔE (~

10 μm thick) - E (~ 150 μm thick) Si detector telescopes each consisting of a surface barrier detector. The five telescopes were placed on a rotating arm allowing the measurement of the elastic scattering angular distribution up to 110°.

The data with the radioactive beams were obtained, in the same experiment, using the new post accelerated $^{10,11}Be$ beams of REX-ISOLDE at CERN. The detection system, shown in fig. 1, consisted of 6 ΔE (~ 50 μm thick) - E (~ 1500 μm thick) Si detector telescopes, each one with a surface of 50x50mm^2, surrounding a thin ^{64}Zn target (550 μg/cm^2 and 1000 μg/cm^2 for the $^{10,11}Be$ beams, respectively). The target was tilted at 45° to allow the measurement in the angular region around 90°. The ΔE stage consists of a Double sided Silicon Strip Detector (DsSSD) segmented in 16 both on the front side and on the rear side, allowing a 256 pixels for each detector. The E stage consists of a Silicon Single Pad detector. The detectors were placed very close to the target in order to have a large polar angle ($10°<\theta<150°$) and solidangle coverage. The average beam intensity was 10^6 pps and 10^4 pps for ^{10}Be and ^{11}Be, respectively.

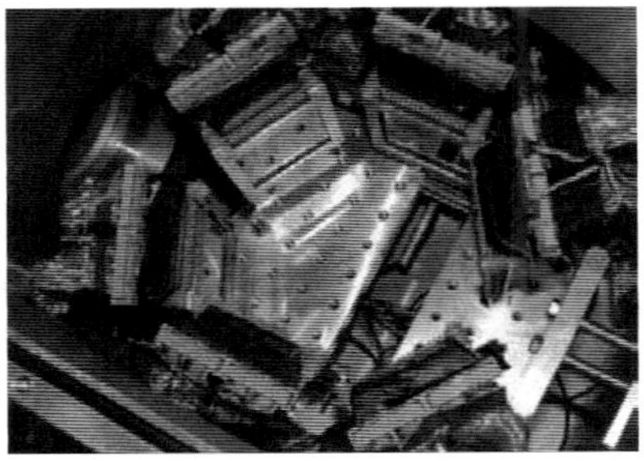

Figure 1. Picture of the detectors set-up inside the scattering chamber

The effective solid angles covered by each detector pixel have been evaluated by using a Monte Carlo simulation code. However, due to the very compact geometry of the detection system, small variations of the beam position onto the target resulted in a large variation of the detector angles. Therefore, polar angle and solid angle determination has been verified by measuring the Rutherford scattering for the $^{12}C+^{197}Au$ reaction at E_{lab} = 28 MeV as well as the elastic scattering angular distribution for the collision $^{10}Be+^{197}Au$ at E_{lab} = 29.4 MeV. With the adopted analysis procedure, the expected Rutherford behaviour of the cross-sections for the two reactions were obtained, indicating that the geometry was properly determined.

3. Results and Discussion

In Fig. 2 the angular distributions for the scattering of $^{9,10,11}Be+^{64}Zn$ are shown in linear scale. We must say that, as in [15], also in the present case the $^{11}Be+^{64}Zn$ scattering cross-section is the quasi elastic cross-section, since the inelastic contribution of the *1*st excited state of ^{11}Be at E_x=320 keV is included. However, the possible contribution to the scattering cross-section of the inelastic channel has been evaluated by performing OM calculations using as dipole strength $B(E1)=0.115e^2fm^2$ [20] and the potential parameters extracted from the fit of the scattering data. According to the these calculations (see ref. [19]), the inelastic contribution to the scattering cross-section is quite small at all measured angles and therefore negligible. For this reason we refer to the angular distribution for $^{11}Be+^{64}Zn$ as elastic scattering angular distribution.

As one can see from fig. 2, in spite of the very different binding energies of ^9Be and ^{10}Be, their elastic-scattering angular distributions are similar. On the contrary, as one can see in fig. 2, ^{11}Be scattering shows a very different pattern and the main feature that one can observe is a strong reduction of the elastic cross-section at forward angles. In [1] the suppression of the scattering cross-section at small angles, in collision induced by weakly bound nuclei, is connected to the coupling with a large Coulomb dipole due to the presence of the low-lying electric dipole strength. The authors studied also the dependence of the suppression of the cross-section from the target charge and beam energy concluding that this effect, close to the Coulomb barrier, should be evident only in scattering with targets having high charge (Z_T ~80) and that measurements with lighter targets (Z_T ~28) are not sensitive to this coupling. This would explain the similar angular distribution for ^{9,10}Be scattering. Indeed, in the ^{6}He induced collisions, a clear reduction of elastic cross-section at small angles has been observed although not as large as in the present case and only for heavy targets [11]. The elastic scattering angular distribution of ^{6}He on a medium mass target ^{64}Zn does not show such feature [10]. Thus, the observation of a strong reduction of elastic cross-section at small angles in the scattering of ^{11}Be which has a strong low-lying continuum dipole strength, as ^{6}He and ^{9}Be, with a light charge target must be due to other mechanisms besides coupling to Coulomb break-up, that could be associated with the halo structure.

The observed difference in ^{6}He and ^{11}Be scattering could be due to the more extended halo distribution in ^{11}Be compared to ^{6}He (e.g. [21]).

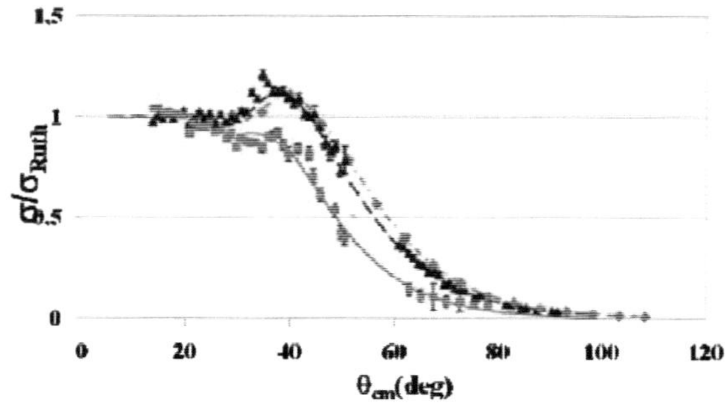

Figure 2. Elastic-scattering angular distributions on ^{64}Zn: ^{9}Be(diamonds), ^{10}Be (triangles) and ^{11}Be (squares). The lines represent the OM calculations for ^{9}Be(dot dashed), ^{10}Be (dashed) and ^{11}Be (full line).

The ^{9,10,11}Be+^{64}Zn elastic-scattering data were analyzed within the OM framework using the code PTOLEMY [22]. For ^{9,10}Be a Woods-Saxon (W-S) form of the real and imaginary potential was used. In the case of ^{11}Be, in order to take into account the diffuse halo structure and the effect of the break-up channel on elastic scattering, a surface term was added to the imaginary volume potential.

Couplings can be represented in the OM by the addition of a complex dynamic polarization potential (DPP) to the OP (e.g. [23]). In [24] the DPP was a surface potential calculated analytically and applied to the ^{11}Be+^{12}C data at 49.3 AMeV. The real part of the DPP was not considered in the calculations since it produced a negligible effect on the elastic scattering fit. The volume potential, responsible for the inelastic core-target interaction, was obtained from the fit of the scattering of the ^{10}Be core on the same target. The calculated surface potential was responsible for peripheral reactions like transfer or break-up. This term has an exponential tail, and a diffuseness parameter of 3.2 fm. A similar result was also obtained in [21]. Then, we used a procedure similar to [24] to deduce the optical potential in the case of ^{11}Be+ ^{64}Zn.

In the present case, all terms of the OP were extracted phenomenologically from a fit of the elastic scattering angular distributions. The fit of ^{11}Be+^{64}Zn was performed using as volume part of the real and imaginary W-S potentials the ones obtained from the fit of the scattering data of the core ^{10}Be on ^{64}Zn and in addition we considered a surface imaginary term having the shape of a W-S derivative. The effect of such term, as one can see from fig. 2, is to reduce the scattering cross-section at all the measured angles and in particular in order to reproduce the behaviour of the cross-section at forward angles we have found that a very large surface diffuseness parameter is needed, a_{si}= 3.5 fm. The surface diffuseness and also the surface potential depth, extracted from the fit, are consistent with [24].

The large diffuseness confirms the presence of long range absorption mechanisms. The suppression of the elastic cross-section at forward angles is then originated by absorption occurring at large impact parameters due to the diffuse halo structure of the ^{11}Be. More details on the OM analysis and the potential parameters are reported in [19].

From the OM analysis we have also extracted TR cross-section for all the reaction studied. The deduced TR cross-sections are $\sigma_{TR} \sim$ 1 b for ^{9}Be, $\sigma_{TR} \sim$ 1.2 b for ^{10}Be and $\sigma_{TR} \sim$ 2.7 b for ^{11}Be. Consequently to the reduction of the elastic cross-section, we have observed a much larger TR cross-section in the halo ^{11}Be nucleus case with respect the other two Be isotopes.

Since the TR cross-section for the ^{11}Be case is double than the one for the ^{9,10}Be cases, it is important to understand if such enhancement could be due to transfer and/or break-up processes as observed in the ^{6}He case [10,25,7]. Indeed, when comparing the ΔEvsE spectra for the reactions induced by ^{10}Be and ^{11}Be, shown in fig. 3, events close to the elastic and consistent with the detection of ^{10}Be coming from transfer and/or break-up processes are present in the ^{11}Be beam case. On the contrary, similar events are not observed in the ^{10}Be case. In fig. 4 the angular distribution for such events is shown. The corresponding integrated cross-section is σ = 1100±150 mb. This cross-section corresponds to about 40% of the TR cross-section measured for ^{11}Be. This result confirms what was found in the ^{6}He experiments (e.g. [10, 12]) but disagrees with the previously published data of ^{11}Be [15,16].

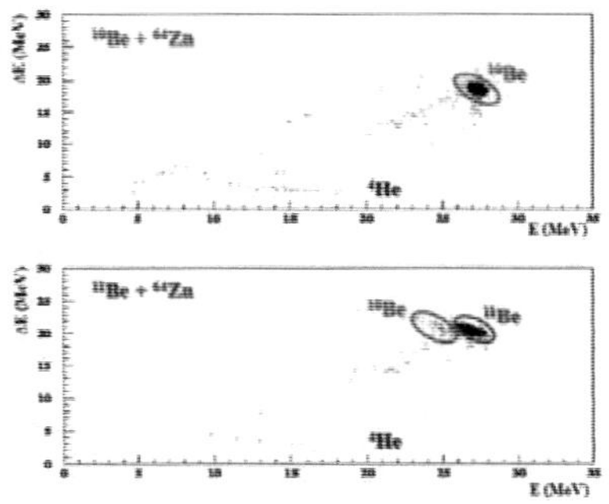

Figure 3. ΔEvs$E$ scatter plots for the reactions ^{10}Be+^{64}Zn (top) and ^{11}Be+^{64}Zn (bottom), at θ=35°

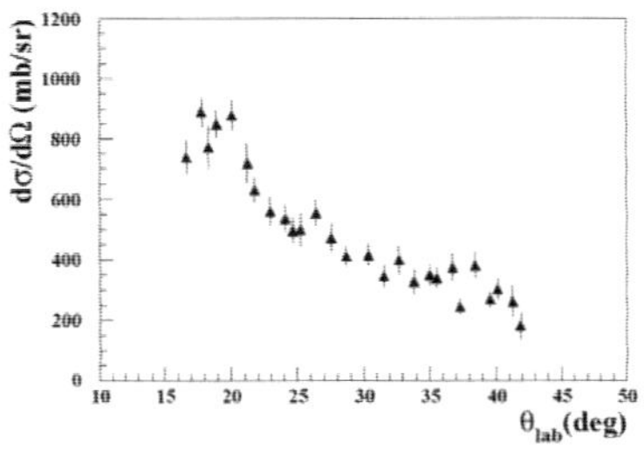

Figure 4. Angular distribution of transfer/break-up events in ^{11}Be+^{64}Zn obtained by selecting ^{10}Be events in the ΔEvs$E$ spectrum.

4. Couclusion

In summary, we have studied the collisions ^{9,10,11}Be+^{64}Zn at energy close to the Coulomb barrier using high quality post-accelerated beams. The scattering of the ^{11}Be halo nucleus shows extremely different features with respect the ^{9}Be and ^{10}Be. The ^{11}Be elastic cross-section is strongly suppressed and absorption occurs at much smaller scattering angles than for the other two Be isotopes.

All the scattering data were interpreted within the OM. In order to take into account the ^{11}Be diffuse halo structure and to reproduce the scattering data a surface imaginary potential with a very large diffuseness parameter was needed. Along with the OM parameters the TR cross-section were also deduced and we have found that in the ^{11}Be halo nucleus case the TR cross-section is much larger than the ones for ^{9,10}Be induced reaction.

Moreover, from the analysis of ^{10}Be events, present in the collision induced by the ^{11}Be halo nucleus, it was found that about 40% of the TR cross-section can be attributed to transfer and/or break-up processes.

These results show a strong effect on nuclear reaction mechanisms around the barrier due to the ^{11}Be halo structure and to its weak binding and represents an independent confirmation that similar effects observed for the ^{6}He are also due to its halo nature.

References

[1] Kucuk Y et al., 2009 *Phys. Rev.* C 79, 067601.
[2] Moro A et al., 2007 *Phys. Rev.* C 75, 064607.
[3] Matsumoto T et al., 2006 *Phys. Rev.* C 73, 051602R.
[4] Rodriguez-Gallardo M et al., 2008 *Phys. Rev.* C 77, 064609.
[5] Rusek K et al., 2005 *Phys. Rev.* C 72, 037603.
[6] Keeley N and Alamanos N, 2008 *Phys. Rev.* C 77, 054602.
[7] Chatterjee A et al., 2008 *Phys. Rev. Lett.* 101, 032701.
[8] Keeley N et al., 2009 *Prog. Part. Nuc. Phys.* 63, 396.
[9] Keeley N et al., 2007 *Prog. Part. Nuc. Phys.* 59, 579.
[10] Di Pietro A et al., 2004 *Phys. Rev.* C 69, 044613.
[11] Sanchez-Benitez A M et al., 2008 *Nucl. Phys.* A 803, 30.
[12] Aguilera E et al., 2001 *Phys. Rev.* C 63, 061603(R).
[13] Kakuee O et al., 2003 *Nucl. Phys.* A 728, 339.
[14] Moro A et al., 2007 *Phys. Rev.* C 75, 064607.
[15] Mazzocco M et al., 2006 *Eur. Phys. J.* A 28, 295.
[16] Mazzocco M et al., 2007 *Eur. Phys. J.* S. T. 150, 37.
[17] Signorini C et al., 2002 *Nucl. Phys.* A 701, 23c.
[18] Acosta L et al., 2009 *Eur. Phys. J* A 42, 461.
[19] Di Pietro A et al., 2010 *Phys. Rev. Lett.* 105, 022701.
[21] Hassan M et al., 2009 *Phys. Rev.* C 79, 064608.
[22] Rhoades-Brown M et al., 1980 *Phys. Rev.* C 21, 2417.
[23] Satchler G, 1991 *Phys. Rep.* 199, 147.
[24] Bonaccorso A and Carstoiu F, 2002 *Nucl. Phys. A* 706, 322.
[25] Young P D et al., 2005 *Phys. Rev.* C 71, 051601R.

A novel method for fundamental interaction studies with an electrostatic ion beam trap

2011 J. Phys.: Conf. Ser. 267 012013

(http://iopscience.iop.org/1742-6596/267/1/012013)

A novel method for fundamental interaction studies with an electrostatic ion beam trap

M Hass[1], S Vaintraub[1,2], O Aviv[1], K Blaum[3], O Heber[1], I Mardor[2], M Rappaport[1], A Wolf[3] and D Zajfman[1]

[1] Weizmann Institute of Science, Rehovot 76100, Israel
[2] Soreq NRC, Yavne 81800, Israel
[3] Max Planck Institute for Nuclear Physics, Saupfercheckweg 1, 69117, Heidelberg, Germany

Abstract. We present the first steps towards the realization of a novel experimental scheme to measure β−ν correlations in the β decay of ^{6}He, for the purpose of Fundamental Interaction studies. Our method is based on a novel use of an ion trapping device, the Electrostatic Ion Beam Trap (EIBT) coupled to a d+t neutron generator. The EIBT which has not been previously considered for Fundamental Interaction studies, exhibits potentially very significant advantages over other state of the art experimental schemes aimed at precision measurements of the β−ν angular correlation coefficient.

1. Introduction

Trapped radioactive atoms present exciting opportunities for the study of fundamental interactions and symmetries. For example, detecting beta decay in a trap can probe the minute experimental signal that originates from possible tensor or scalar terms in the weak interaction. Such scalar or tensor terms affect, e.g., the angular correlation between a neutrino and an electron in the beta-decay process, thus probing new physics of "beyond-the-standard-model" nature [1].

We propose an innovative experimental scheme for beta-decay studies using a practically "table-top" experiment, featuring an Electrostatic Ion Beam Trap (EIBT) [2]. Such a trap has not been previously considered for Fundamental Interaction studies and exhibits potentially very significant advantages over other experimental schemes (e.g. Magneto Optical Traps []). The long (~seconds) storage time of the ion beam is perfectly suitable for beta decay studies for relatively short lived species. Other advantages include high injection efficiency of the radionuclide under study, an extended field-free region, ion-beam kinematics for better efficiency, ease-of-operation and the potential for a much larger solid angle for the electron and recoiling atom counters.

In the following we briefly present the theory of beta decay formalism, with special emphasis on the case of radioactive ^{6}He (half life 807 ms). The subsequent sections include a brief method description and preliminary simulation results. We conclude with a brief summary.

2. Theory

The β-decay transition rate W (inverse lifetime) in case of non-oriented nucleus is given by [3]

$$dW \propto \xi\left(1+a\frac{\vec{p}_e\cdot\vec{p}_\nu}{E_eE_\nu}\right)=\xi\left(1+a\frac{p_e}{E_e}\cdot\cos\theta_{e\nu}\right)$$

a is the beta decay coefficient (more coefficients also exist in case of polarized nuclei [3]). In our recent study we concentrate on the beta-neutrino correlation coefficient - a, which can be measured completely unbiased in case of ^{6}He beta decay, as will be shown below. The following analysis can be very similarly attuned to others beta decay coefficients and radio-nuclei.

The ^{6}He nuclide has a half-life of 807 ms and is decaying into ^{6}Li, electron and electron anti-neutrino

$$^6He \rightarrow {}^6Li + e^- + \bar{\upsilon}_e$$

The most general way to write the beta decay Hamiltonian [3] is:

$$H_\beta{}^{^6He} = \sum_{i=S,P,V,A,T} \overline{(^6Li)}\hat{O}_i(^6He)[\bar{e}\hat{O}_i(C_i + C'_i\gamma_5)\upsilon_e] + h.c.$$

where S,P,V,A and T stands for Scalar, Pseudo-scalar, Vector, Axial-vector and Tensor respectively. The $\beta-\nu$ correlation coefficient can be expressed by means of the Gamow-Teller (GT) and Fermi (F) matrix elements together with eight coupling constants C, C' [3]:

$$a\xi = |F|^2\left[|C_V|^2 - |C_S|^2 + |C'_V|^2 - |C'_S|^2\right] + \frac{|GT|^2}{3}\left[|C_T|^2 - |C_A|^2 + |C'_T|^2 - |C'_A|^2\right]$$

and

$$\xi = |F|^2\left[|C_V|^2 + |C_S|^2 + |C'_V|^2 + |C'_S|^2\right] + |GT|^2\left[|C_T|^2 + |C_A|^2 + |C'_T|^2 + |C'_A|^2\right]$$

The ^{6}He beta decay is a $J^\pi=0^+$ to $J^\pi=1^+$ pure Gamow-Teller decay. In that case the beta-neutrino correlation coefficient can be expressed only by Axial and Tensor coupling constants in the Hamiltonian, since the Fermi matrix element is zero. In the Standard Model (SM) only the V-A interactions can occur, therefore only the axial coupling constants remain [3]:

$$|C_T|^2 = |C'_T|^2 = 0 \quad |C_A|^2 = |C'_A|^2 = 1$$

Consequently the transition rate for pure Gamow-Teller decays in the SM frame has the following simple form (with a=-1/3),

$$dW_{SM}(\theta_{e\nu}) \propto \left(1 - \frac{1}{3}\frac{p_e}{E_e}\cos\theta_{e\nu}\right)$$

where the squared GT matrix element can be brought out of brackets, hence its absolute value is not important. On the contrary, in a general case, beta decay may be neither a pure Gamow-Teller nor a pure Fermi decay. As a result any beta decay coefficient in the transition rate equation would be matrix elements depended, thus nuclear structure theory depended.

3. Method

The main purpose of the experiment is a **high-precision measurement** of the beta decay coefficients (a). In the following we present a novel approach – that of using an Electrostatic Ion Beam Trap (EIBT). Following the pioneering work at the Weizmann Institute of Science [2], the EIBT is being used worldwide (Germany, France, India, Japan and USA) in atomic and molecular physics experiments and even a cryogenic EIBT operated at 2K has been recently commissioned [4]. However, injecting and trapping radioactive nuclei for Fundamental Interaction studies in an EIBT device is an entirely novel direction that has never been tried before. This direction offers significant improvements in detection efficiency (paramount consideration when working with radioactive beams of limited intensity) and ease of-operation.

The EIBT was developed at WI in 1996 [2] for storing ions at typical energy of few keV. The principle of operation of an EIBT [5] is based on the analogy with the classical optical resonator: it is possible to store a beam of photons between two spherical mirrors separated by distance *L* if the focal length *f* fulfils the stability condition, namely: $L/4 < f < \infty$ and the potential on the mirrors is larger than the ion beam kinetic energy. A scheme of the EIBT setup is shown in figure 1. The basic EIBT consist of 2 sets of eight electrodes both acting as an electrostatic mirror and as a lens by producing a retarding field which reflects the beam along its path and focuses it on the lateral direction. The typical distance between the electrostatic mirrors is 500 mm.

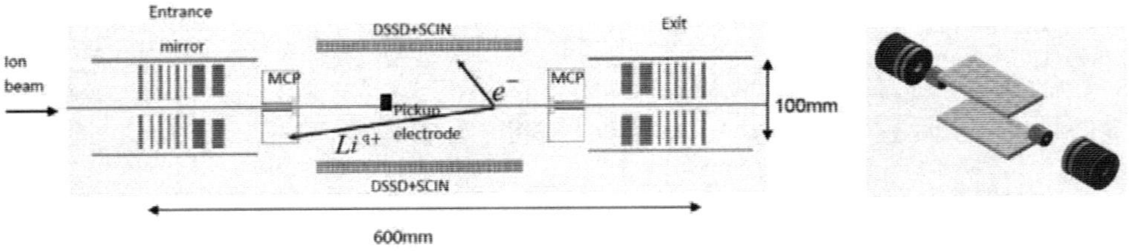

Figure 1. A schematic view of the EIBT for β-decay studies. The radioactive ion, like ^6He, moves with E_k~10 keV between the reflecting electrodes. The β electrons are detected in position sensitive counters while the recoiling ions, due to kinematical focusing, are detected with very high efficiency in either one (determined by the instantaneous direction) of the annular MCP counters.

Ions with kinetic energy of ~10 keV are injected through the grounded entrance mirror, once the ions fill the trap; potentials on the entrance mirror are quickly raised (< 10 ns) so that the ions bounce back and forth between the mirrors. For the $^6He^+$, the revolution time is ~4 microseconds. Shortly after injection the trapped beam is bunched using RF fields. A pickup electrode, located in the trap center is used to continuously monitor the bunch location (in the new set up two pickup electrodes will be located as a part of the ground electrode in the mirrors). Therefore, the position of the decay is known. The products of the decay (recoil Li nuclei and electron) are detected by a set of dedicated detectors. Two annular position sensitive Multi-Channel-Plate (MCP) detectors are situated along the main trap axis with the purpose of counting the recoil Li. These MCPs have central holes which allow the stable trajectories of the stored ions (such technology was demonstrated before [6]). The electrons are counted by two detectors located above and below the optical axis. The detectors are composed of plastic scintillators coupled to double sided silicon strip detectors (DSSD), thereby obtaining both energy and position of the beta particle.

4. Preliminary simulations results

Simulations are crucial to demonstrate the concept validity and, later on in the project itself, to optimize both the bunch dynamics and the detection geometry. The construction of the dedicated set-up of an EBIT for beta-decay studies should naturally follow such simulations and take advantage of the gained insight. This process is currently ongoing and we present below first and preliminary results. The observable parameters (recoil time of flight, position of the ion bunch, position of particles on detectors, electron energy), allow the full reconstruction of the momentum of the undetected neutrino particle in the beta decay. Thus the angular correlation between electron and the neutrino upon decay can be studied.

The accuracy of the neutrino momentum is mainly limited by the certainty of the position of the decay, which can occur anywhere inside the ion bunch. While only the location of the bunch is measured. Previous works on the subject show that under special trapping conditions ("self bunching") [7], the ion bunch size can be reduced to few cm. We plan to investigate this issue further in the hope of achieving bunch size of typically 1 cm. Moreover, a study of the influence of the bunch size (e.g. the uncertainty of the decay within the bunch) on the sensitivity of our measurement is now underway and preliminary results are shown in figure 2 for different bunch sizes. As we can see small bunch sizes (~<2 cm) are almost identical to a point-like beta decay occurrences. Further simulations, aimed at a better understanding of the detailed characteristics of the particles in the bunch, as far as the correct labeling of the corresponding energy and momentum (and hence the ensuing $\beta-\nu$ angle) are presently being carried out.

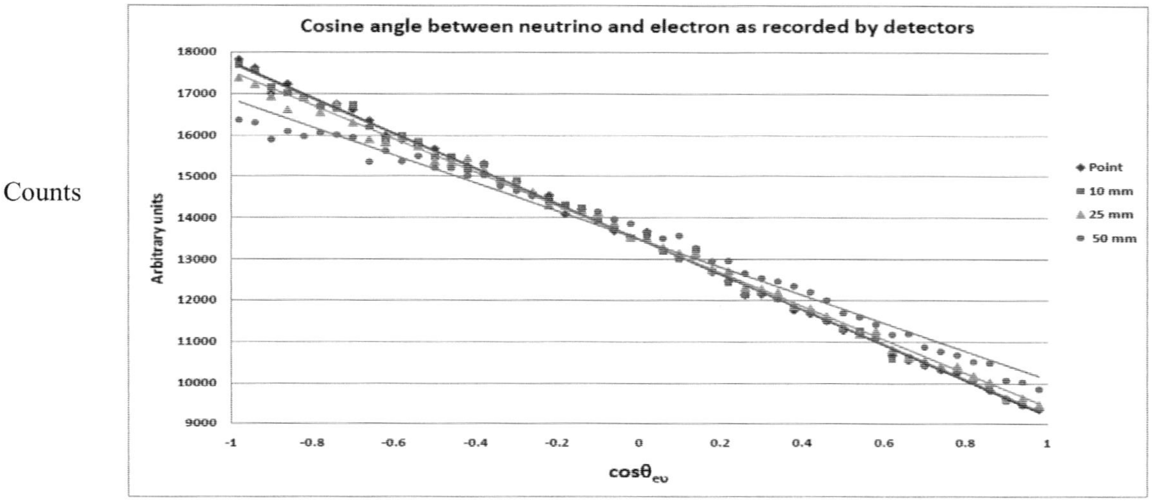

Figure 2. GEANT4 [8] simulation of the beta-neutrino correlation with different bunch sizes.

As a consequence of these simulations, we are presently embarking on a R&D effort to investigate the efficient bunching of the ion beam as an input to the EIBT. For that purpose, we are using ^{4}He stable ions from an rf source (shown in figure 3).

Figure 3. A picture of the initial experimental setup, with the rf source, beam-lines and analyzing magnet. The d+t neutron generator will be placed next to the ion source

5. Summary and future plans

This research direction is also closely related in turn to the intensive R&D efforts that are currently under way in our laboratory to facilitate the production and extraction of such radioactive nuclei (^{6}He and ^{8}Li, for example) in record yields, using neutron-induced reactions [9]. We are planning to use an intense d+t commercial neutron generator that provides ~14 MeV neutrons at a fluxes reaching 10^9 - 10^{10} n/s. The purchase of such a generator has recently been approved by the WI. At a later stage, we envisage taking advantage of uniquely high yields of radioactive nuclei that will become available at the new and modern SARAF accelerator at Soreq.

It is also imperative to use an appropriate buncher-cooler device, such as an RFQ buncher-cooler as a pre-injection device into the EBIT. Simulations regarding design option to this effect are presently being examined in detail. Another subject under initial considerations is the electron and ion detection system, consisting of a positive sensitive Micro Channel Plate (MCP) device and for the recoiling ^{6}Li ions and a set of silicon strip detectors for the electrons' position and energy determination.

Though the research program is only at the beginning, it seems robust and promising in the light of the recent calculations and our growing understanding of the system.

References

[1] Fléchard X 2008 Paul trapping of radioactive $^{6}He^{+}$ ions and direct observation of their β decay *Phys. Rev. Lett.* **101**, 212504; and references therein.

[2] Dahan M 1998 A new type of electrostatic ion trap for storage of fast ion beams *Rev. Sci. Ins.* **69**, 76

[3] Jackson J D, Treiman S B and Wyld H W 1957 Possible Tests of Time Reversal Invariance in Beta Decay *Phys. Rev.* **106** 517; 1957 Coulomb corrections in allowed beta transitions *Nucl. Phys.* **4**, 206

[4] Suzuki T and Yamauchi Y 2006 Electrostatic linear ion trap for low-energy ion beam storage *Nucl. Inst. Meth. A* **562** 53; Lange M 2010 *Rev. Scient. Instrum.* **81**, 055105

[5] Zajfman D 1997 Electrostatic bottle for long-time storage of fast ion beams *Phys. Rev. A* **55** R1577

[6] Aviv O 2008 A bent electrostatic ion beam trap for simultaneous measurements of fragmentation and ionization of cluster ions *Rev. Sci. Ins.* **79** 083110

[7] Pedersen H B 2001 Ion Motion Synchronization in an Ion-Trap Resonator *Phys. Rev. Lett.* **87** 055001

[8] *www.geant4.web.cern.ch/geant4*

[9] Hass M 2008 Light radio-isotopes for nuclear astrophysics and neutrino physics *J. Phys. G* **35** 014042

Nuclear forces from chiral EFT: The unfinished business

2011 J. Phys.: Conf. Ser. 267 012014

(http://iopscience.iop.org/1742-6596/267/1/012014)

Nuclear forces from chiral EFT: The unfinished business

R. Machleidt

Department of Physics, University of Idaho, Moscow, Idaho 83843, USA

E-mail: machleid@uidaho.edu

Abstract. In recent years, there has been substantial progress in the derivation of nuclear forces from chiral effective field theory (EFT). Accurate two-nucleon forces have been constructed at next-to-next-to-next-to-leading order (N^3LO) and applied (in some cases together with three-nucleon forces at NNLO) to nuclear few and many-body systems—with a fair deal of success. This may suggest that the 80-year old nuclear force problem has finally been cracked. Not so! Some pretty basic issues are still unresolved, like, the proper renormalization of the two-nucleon potential and subleading many-body forces. In this contribution, I will focus on the latter issue.

1. Introduction

The fundamental goal of nuclear structure physics is to understand the properties of atomic nuclei and their reactions in terms of the basic forces between the constituents. During the past half century, a large variety of phenomenological forces has been developed and applied in microscopic nuclear structure calculations with some success. But in the long run phenomenology is not good enough and, ultimately, we need nuclear interactions that are based upon proper theory. Since the nuclear force is a manifestation of strong interactions, any serious derivation has to start from quantum chromodynamics (QCD). However, the well-known problem with QCD is that it is non-perturbative in the low-energy regime characteristic for nuclear physics. For many years this fact was perceived as the great obstacle for a derivation of nuclear forces from QCD—impossible to overcome except by lattice QCD. The effective field theory (EFT) concept has shown the way out of this dilemma. One has to realize that the scenario of low-energy QCD is characterized by pions and nucleons interacting via a force governed by spontaneously broken approximate chiral symmetry. This chiral EFT allows for a systematic low-momentum expansion known as chiral perturbation theory (ChPT) [1]. Contributions are analyzed in terms of powers of small external momenta over the large scale: $(Q/\Lambda_\chi)^\nu$, where Q is generic for an external momentum (nucleon three-momentum or pion four-momentum) or pion mass and $\Lambda_\chi \approx 1$ GeV is the chiral symmetry breaking scale ('hard scale'). The early applications of ChPT focused on systems like $\pi\pi$ [2] and πN [3], where the Goldstone-boson character of the pion guarantees that the expansion converges.

The past 15 years have also seen great progress in applying ChPT to nuclear forces [4, 5, 6, 7, 8, 9, 10, 11, 12, 13, 14, 15, 16, 17, 18]. As a result, nucleon-nucleon (NN) potentials of high precision have been constructed, which are based upon ChPT carried to next-to-next-to-next-to-leading order (N^3LO) [15, 17, 18], and applied in nuclear structure calculations with

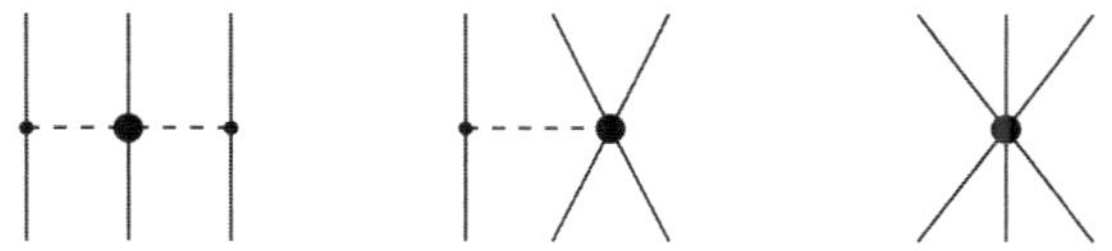

Figure 1. The three-nucleon force at NNLO. From left to right: 2PE, 1PE, and contact diagrams. Solid lines represent nucleons and dashed lines pions. Small dots denote vertices with $\Delta_i = 0$ and large solid dots are $\Delta_i = 1$.

great success [19, 20, 21, 22, 23].

However, in spite of this progress, we are not done. Due to the complexity of the nuclear force issue, there are still many subtle and not so subtle open problems, like

- the proper renormalization of chiral nuclear potentials and
- subleading chiral few-nucleon forces.

This contribution is devoted to the latter issue. The renormalization problem is discussed in Refs. [24, 25].

2. Few-nucleon forces and what is missing

Nuclear three-body forces in ChPT were initially discussed by Weinberg [5]. The 3NF at NNLO, was derived by van Kolck [7] and applied, for the first time, in nucleon-deuteron scattering by Epelbaum *et al* [26]. The leading 4NF (at N^{3}LO) was recently constructed by Epelbaum [27] and found to contribute in the order of 0.1 MeV to the ^{4}He binding energy (total ^{4}He binding energy: 28.3 MeV) in a preliminary calculation [28], confirming the traditional assumption that 4NF are essentially negligible. **Therefore, the focus is on 3NF.**

The power of a 3NF diagram is given by

$$\nu = 2 + 2L + \sum_i \Delta_i \,, \tag{1}$$

where L denotes the number of loops in the diagram and

$$\Delta_i = d_i + \frac{n_i}{2} - 2 \tag{2}$$

with d_i the number if derivatives or pion-mass insertions and n_i the number of nucleon fields (nucleon legs) involved in vertex i. The sum in (1) runs over all vertices contained in the diagram under consideration. We will use (1) to analyze 3NF contributions order by order. The lowest possible power is obviously $\nu = 2$ (NLO), which is obtained for no loops ($L = 0$) and only leading vertices ($\sum_i \Delta_i = 0$). This 3NF happens to vanish [5, 29, 30]. The first non-vanishing 3NF occurs at NNLO.

2.1. The 3NF at NNLO

The power $\nu = 3$ (NNLO) is obtained when there are no loops ($L = 0$) and $\sum_i \Delta_i = 1$, i.e., $\Delta_i = 1$ for one vertex while $\Delta_i = 0$ for all other vertices. There are three topologies which fulfill this condition, known as the two-pion exchange (2PE), 1PE, and contact graphs (figure 1).

The 3NF at NNLO has been derived (without the $1/M_N$ corrections) [7, 26] and applied in calculations of few-nucleon reactions [26, 31, 32, 33], structure of light- and medium-mass nuclei [21, 34, 35], and nuclear and neutron matter [23, 36] with a fair deal of success. However, the famous 'A_y puzzle' of nucleon-deuteron scattering is not solved [26, 31], and there exists

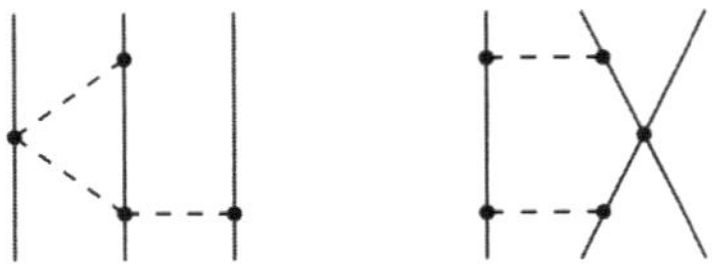

Figure 2. The 3NF at N^3LO: Two examples of one-loop graphs. Notation as in figure 1.

a similar problem with the analyzing power in p-^{3}He scattering [37, 38, 39]. Furthermore, the spectra of light nuclei leave room for improvement [21].

We note that there are further 3NF contributions at NNLO, namely, the $1/M_N$ corrections of the NLO 3NF diagrams. Part of these corrections have been calculated by Coon and Friar in 1986 [30]. These contributions are believed to be very small.

In summary, because of various unresolved problems in microscopic nuclear structure, the 3NF beyond NNLO is very much in need. In fact, it is no exaggeration to state that the 3NF at sub-leading orders is presently one of the most important outstanding issues in the chiral EFT approach to nuclear forces.

2.2. The 3NF at N^3LO

According to (1), the value $\nu = 4$, which corresponds to N^3LO, is obtained for the following classes of diagrams.

3NF loop diagrams at N^3LO. For this group of graphs, we have $L = 1$ and, therefore, all Δ_i have to be zero to ensure $\nu = 4$. Thus, these one-loop 3NF diagrams can include only leading order vertices, the parameters of which are fixed from πN and NN analysis. We show two samples of this very large class of diagrams in figure 2. One sub-group of these diagrams ("2π exchange graphs") has been calculated by Ishikawa and Robilotta [40], and two other topologies (2π-1π and ring diagrams) have been evaluated by the Bonn-Jülich group [41]. The remaining topologies, which involve a leading order four-nucleon contact term (e.g., second diagram of figure 2), are under construction by the Bonn-Jülich group. The N^3LO 2π-exchange 3NF has been applied in the calculation of nucleon-deuteron observables in [40] producing very small effects.

The smallness of the 2π loop 3NF at N^3LO is not unexpected. It is consistent with experience with corresponding 2NF diagrams: the NLO 2PE contribution to the NN potential, which involves one loop and only leading vertices, is also relatively small.

By the same token, one may expect that also all the other N^3LO 3NF loop topologies will produce only small effects.

3NF tree diagrams at N^3LO. The order $\nu = 4$ is also obtained for the combination $L = 0$ (no loops) and $\sum_i \Delta_i = 2$. Thus, either two vertices have to carry $\Delta_i = 1$ or one vertex has to be of the $\Delta_i = 2$ kind, while all other vertices are $\Delta_i = 0$. This is achieved if in the NNLO 3NF graphs of figure 1 the power of one vertex is raised by one. The latter happens if a relativistic $1/M_N$ correction is applied. A closer inspection reveals that all $1/M_N$ corrections of the NNLO 3NF vanish and the first non-vanishing corrections are proportional to $1/M_N^2$ and appear at N^4LO. However, there are non-vanishing $1/M_N^2$ corrections of the NLO 3NF and there are so-called drift corrections [42] which contribute at N^3LO (some drift corrections are claimed to contribute even at NLO [42]). We do not expect these contributions to be sizable. Moreover, there are contributions from the $\Delta_i = 2$ Lagrangian [43] proportional to the low-energy constants d_i. As

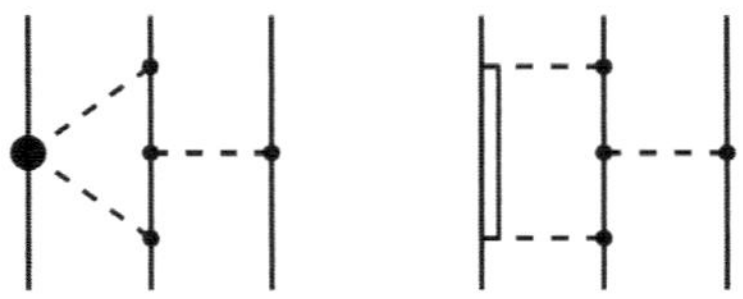

a) *b)*

Figure 3. (a) One-loop 3NF at N^4LO of the Δ-less theory. (b) Corresponding diagram of the Δ-full theory which contributes at N^3LO. Double lines represent Δ isobars; other notation as in figure 1.

it turns out, these terms have at least one time-derivative, which causes them to be Q/M_N suppressed and demoted to N^4LO.

Thus, besides some minor $1/M_N^2$ corrections, there are no tree contributions to the 3NF at N^3LO.

Summarizing the entire N^3LO 3NF contribution: For the reasons discussed, we anticipate that this 3NF is weak and will not solve any of the outstanding problems. In view of this expectation, we have to look for more sizable 3NF contributions elsewhere.

2.3. The 3NF at N^4LO of the Δ-less theory

The obvious step to be taken is to proceed to the next order, N^4LO or $\nu = 5$, of the Δ-less theory which is the one we have silently assumed so far. (The Δ-full theory will be introduced and discussed below.) Some of the tree diagrams that appear at this order were mentioned already: the $1/M_N^2$ corrections of the NNLO 3NF and the trees with one d_i vertex which are $1/M_N$ suppressed. Because of the suppression factors, we do not expect sizable effects from these graphs. Moreover, there are also tree diagrams with one vertex from the $\Delta_i = 3$ πN Lagrangian [44, 45] proportional to the LECs e_i. Because of the high dimension of these vertices and assuming reasonable convergence, we do not anticipate much from these trees either.

However, we believe that the loop contributions that occur at this order are truly important. They are obtained by replacing in the N^3LO loops (figure 2) one vertex by a $\Delta_i = 1$ vertex [with LEC c_i]. We show one symbolic example of this large group of diagrams in figure 3(a). This 3NF is presumably large and, thus, what we are looking for.

The reasons, why these graphs are large, can be argued as follows. Corresponding 2NF diagrams are the three-pion exchange (3PE) contributions to the NN interaction. In analogy to Figs. 2 and 3(a), there are 3PE 2NF diagrams with only leading vertices and the ones with one (sub-leading) c_i vertex (and the rest leading). These diagrams have been evaluated by Kaiser in Refs. [46] and [47], respectively. Kaiser finds that the 3PE contributions with one sub-leading vertex are about an order magnitude larger then the leading ones.

2.4. N^3LO 3NF contributions in the Δ-full theory

The above considerations indicate that the Δ-less theory exhibits, in some cases, a bad convergence pattern. The reason for the unnaturally strong subleading contributions are the large values of the $\Delta_i = 1$ LECs, c_i. The large values can be explained in terms of resonance saturation [48]. The $\Delta(1232)$-resonance contributes considerably to c_3 and c_4. The explicit inclusion of the Δ takes strength out of these LECs and moves this strength to a lower order, thus improving the convergence [49, 6, 10, 50, 51]. Figure 3 illustrates this fact for the 3NF under consideration: the diagram of the Δ-less theory shown in (a) is (largely) equivalent to diagram (b) which includes one Δ excitation. Note, however, that diagram (a) is N^4LO, while diagram (b) is N^3LO. Moreover, there are further N^3LO one-loop diagrams with two and three Δ excitations, which correspond to diagrams of order N^5LO and N^6LO, respectively, in the Δ-less theory.

This consideration clearly shows that the inclusion of Δ degrees of freedom in chiral EFT makes the calculation of sizable higher-order 3NF contributions much more efficient.

2.5. Summarizing the open 3NF business

To make a complicated story short, this is the bottom line concerning 3NF:

- The chiral 3NF at NNLO is insufficient. Additional *sizable* 3NF contributions are needed.
- The chiral 3NF at N^3LO (in the Δ-less theory) most likely does *not* produce sizable contributions.
- Sizable contributions are expected from one-loop 3NF diagrams at N^4LO of the Δ-less or N^3LO of the Δ-full theory (figure 3). *These 3NF contributions may turn out to be the missing pieces in the 3NF puzzle and have the potential to solve the outstanding problems in microscopic nuclear structure.*[1]

3. Conclusions and Outlook

The past 15 years have seen great progress in our understanding of nuclear forces in terms of low-energy QCD. Key to this development was the realization that low-energy QCD is equivalent to an effective field theory which allows for a perturbative expansion that has become know as chiral perturbation theory. In this framework, two- and many-body forces emerge on an equal footing and the empirical fact that nuclear many-body forces are substantially weaker than the two-nucleon force is explained automatically.

In spite of the great progress and success of the past 15 years, there are still some unresolved issues that will need our attention in the near future. For this contribution, we picked the issue of the few-nucleon forces beyond NNLO ("sub-leading few-nucleon forces") which are needed to hopefully resolve some important outstanding nuclear structure problems. We believe that we identified correctly where these forces will emerge within the systematic scheme of ChPT.

If the open issues discussed in this paper will be resolved within the next few years, then, after 80 years of desperate struggle, we may finally claim that the nuclear force problem is essentially solved. The greatest beneficiary of such progress will be the field of *ab initio* nuclear structure physics.

Acknowledgment

This work was supported in part by the US Department of Energy under Grant No. DE-FG02-03ER41270.

References

[1] Weinberg S 1979 *Physica* **96A** 327
[2] Gasser J and Leutwyler H 1984 *Ann. Phys. (N.Y.)* **158** 142
[3] Gasser J Sainio M E and Švarc A 1988 *Nucl. Phys.* **B307** 779
[4] Weinberg S 1990 *Phys. Lett.* B **251** 288
Weinberg S 1991 *Nucl. Phys.* **B363** 3
[5] Weinberg S 1992 *Phys. Lett.* B **295** 114
[6] Ordóñez C *et al* 1994 *Phys. Rev. Lett.* **72** 1982
Ordóñez C *et al* 1996 *Phys. Rev.* C **53** 2086

[1] Note that the Illinois 3NF [52] includes two one-loop diagrams with one and two $\Delta(1232)$-isobars. The deeper reason for this may be in arguments we are presenting.

[7] van Kolck U 1994 *Phys. Rev.* C **49** 2932
[8] van Kolck U 1999 *Prog. Part. Nucl. Phys.* **43** 337
[9] Kaiser N, Brockmann R and Weise W 1997 *Nucl. Phys.* **A625** 758
[10] Kaiser N, Gerstendörfer S and Weise W 1998 *Nucl. Phys.* **A637** 395
[11] Epelbaum E *et al* 1998 *Nucl. Phys.* **A637** 107
Epelbaum E *et al* 2000 *Nucl. Phys.* **A671** 295
[12] Bedaque P F and van Kolck U 2002 *Ann. Rev. Nucl. Part. Sci.* **52** 339
[13] Entem D R and Machleidt R 2002 *Phys. Lett.* B **524** 93
[14] Entem D R and Machleidt R 2002 *Phys. Rev.* C **66** 014002
[15] Entem D R and Machleidt R 2003 *Phys. Rev.* C **68** 041001
[16] Machleidt R and Entem D R 2005 *J. Phys. G: Nucl. Part. Phys.* **31** S1235
[17] Epelbaum E, Glöckle W and Meißner U-G 2005 *Nucl. Phys.* **A747** 362
[18] Machleidt R 2007 Nuclear forces from chiral effective field theory *Preprint* arXiv:0704.0807 [nucl-th]
[19] Coraggio L *et al* 2002 *Phys. Rev.* C **66** 021303
Coraggio L *et al* 2005 *Phys. Rev.* C **71** 014307
Coraggio L *et al* 2010 *Preprint* arXiv:1005.2896 [nucl-th]
[20] Fujii S *et al* 2004 *Phys. Rev.* C **69** 034328
Fujii S *et al* 2009 *Phys. Rev. Lett.* **103** 182501
[21] Navratil P *et al* 2007 *Phys. Rev. Lett.* **99** 042501
[22] Hagen G *et al* 2008 *Phys. Rev. Lett.* **101** 092502
Hagen G *et al* 2009 *Phys. Rev.* C **80** 021306
Hagen G *et al* 2010 *Preprint* arXiv:1005:2627 [nucl-th]
[23] Bogner S K *et al* 2005 *Nucl. Phys.* **A763** 59
Bogner S K *et al* 2009 Nuclear matter from chiral low-momentum interactions *Preprint* arXiv:0903.3366 [nucl-th]
[24] Entem D R *et al* 2008 *Phys. Rev.* C **77** 044006
[25] Machleidt R *et al* 2010 *Phys. Rev.* C **81** 024001
[26] Epelbaum E *et al* 2002 *Phys. Rev.* C **66** 064001
[27] Epelbaum E 2006 *Phys. Lett.* B **639** 456
Epelbaum E 2007 *Eur. Phys. J.* **A34** 197
[28] Rozpedzik D *et al* 2006 *Acta Phys. Polon.* **B37** 2889 *Preprint* arXiv:nucl-th/0606017
[29] Yang S N and Glöckle W 1986 *Phys. Rev.* C **33** 1774
[30] Coon S A and Friar J L 1986 *Phys. Rev.* C **34** 1060
[31] Kalantar-Nayestanaki N *et al* 2007 The three-nucleon system as a laboratory for nuclear physics: the need for 3N forces *Preprint* arXiv:nucl-th/0703089, and references therein.
[32] Marcucci L C, Kievsky A, Girlanda L, Rosati S and Viviani M 2009 *Phys. Rev.* C **80** 034003
[33] Kievsky A, Viviani M, Girlanda L and Marcucci L C 2010 *Phys. Rev.* C **81** 044003
[34] Hagen G *et al* 2007 *Phys. Rev.* C **76** 034302
[35] Otsuka T *et al* 2009 Three-body forces and the limit of oxygen isotopes *Preprint* arXiv:0908.2607
[36] Hebeler K and Schwenk A 2009 Chiral three-nucleon forces and neutron matter *Preprint* arXiv:0911.0483 [nucl-th]
[37] Fisher B M *et al* 2006 *Phys. Rev.* C **74** 034001
[38] Deltuva A and Fonseca A C 2007 *Phys. Rev. Lett.* **98** 162502
[39] Viviani M *et al* 2010 *Preprint* arXiv:1004.1306 [nucl-th]
[40] Ishikawa S and Robilotta M R 2007 *Phys. Rev.* C **76** 014006
[41] Bernard V, Epelbaum E, Krebs H, and Meißner U-G 2008 *Phys. Rev.* C **77** 064004
[42] Robilotta M R 2006 *Phys. Rev.* C **74** 044002
[43] Fettes N, Meißner U-G and Steininger S 1998 *Nucl. Phys.* **A640** 199
[44] Fettes N *et al* 2000 *Ann. Phys. (N.Y.)* **283** 273, Erratum 2001 *Ann. Phys. (N.Y.)* **288** 249
[45] Fettes N and Meißner U-G 2000 *Nucl. Phys.* **A676** 311
[46] Kaiser N 2000 *Phys. Rev.* C **61** 014003
Kaiser N 2000 *Phys. Rev.* C **62** 024001
[47] Kaiser N 2001 *Phys. Rev.* C **63** 044010
[48] Bernard V, Kaiser N, and Meißner U-G 1997 *Nucl. Phys.* **A615** 483
[49] Ordóñez C and van Kolck U 1992 *Phys. Lett.* B **291** 459
[50] Krebs H, Epelbaum E and Meißner U-G 2007 *Eur. Phys. J.* **A32** 127
[51] Krebs H, Epelbaum E and Meißner U-G 2008 *Nucl. Phys.* **A806** 65
[52] Pieper S C *et al* 2001 *Phys. Rev.* C **64** 014001

Effects of Brown-Rho scalings in nuclear matter, neutron stars and finite nuclei

T T S Kuo and Huan Dong

Department of Physics and Astronomy,
State University of New York at Stony Brook, Stony Brook, NY 11794-3800, USA

E-mail: thomas.kuo@stonybrook.edu

Abstract. We have carried out calculations for nuclear matter, neutron stars and finite nuclei using NN potentials with and without the medium-dependent modifications based on the Brown-Rho (BR) scalings. Using the V_{low-k} low-momentum interactions derived from such potentials, the equations of state (EOS) for symmetric and asymmetric nuclear matter, for densities up to $\sim 5\rho_0$, are calculated using a RPA method where the particle-particle hole-hole ring diagrams are summed to all orders. The medium effects from both a linear BR scaling (BR_1) and a non-linear one (BR_2) are considered, and they both are essential for our EOSs to reproduce the nuclear matter saturation properties. For densities ρ below ρ_0, results from BR_1 and BR_2 are close to each other. For higher densities, the EOS given by BR_2 is more desirable and is well reproduced by that given by the interaction (V_{low-k}+TBF) where V_{low-k} is the unscaled low-momentum interaction and TBF is an empirical Skyrme three-body force. The moment of inertia of neutron stars is ~ 60 and $\sim 25 M_\odot \mathrm{km}^2$ respectively with and without the inclusion of the above BR_2 medium effects. Effects from the BR scaling are important for the long half-life, ~ 5000yrs, of the $^{14}\mathrm{C} -^{14}\mathrm{N}$ β-decay.

1. Introduction

In microscopic nuclear many-body problems, nuclear properties are commonly described by the the many-body Schroedinger equation $H\Psi = E\Psi$ where $H = T + V_{NN}$; T being the kinetic energy operator and V_{NN} a realistic nucleon-nucleon (NN) potential such as the CDBonn [1], Nijmegen [2], Argonne V18 [3] and BonnA [4] potentials. As is well known, this approach has had both many successes and a number of shortcomings. For example, in NCSM (no core shell model) calculations [5] the use of V_{NN} alone is not capable in providing adequate binding energies for light nuclei, and this shortcoming can be remedied by the inclusion of the three-body force V_{3N}. Another well-known difficulty concerns the empirical saturation properties of symmetric nuclear matter: practically all nuclear matter calculations using the two-body free-space V_{NN} alone can not simultaneously reproduce the empirical nuclear matter binding energy $E_0/A \simeq -16$MeV and saturation density $n_0 \simeq 0.16\mathrm{fm}^{-3}$.[6]

The above results indicate that the free-space two-body interaction V_{NN} alone is inadequate for describing nuclear properties. To satisfactorily describe nuclear properties, certain modifications concerning the NN interaction may be necessary. Generally speaking, there are two approaches for such modification. As mentioned earlier, one approach is the inclusion of many-body forces (3-body, 4-body,..), in addition to the 2-body interaction V_{NN}. The Brown-Rho (BR) scaling [7, 8, 9] leads to another approach. According to this scaling, the masses of mesons in nuclear medium are suppressed (dropped) compared to those in free space. And this suppression is particularly important for the vector mesons, for which there is now evidence from both theory [8, 10, 11] and experiment [12, 13] that the masses may decrease by approximately 10 − 15% at normal nuclear matter density and zero temperature. For densities below that of normal nuclear matter, a linear approximation for the in-medium mass decrease has been suggested [8], namely

$$\frac{m_V^*}{m_V} = 1 - C\frac{n}{n_0}, \tag{1}$$

where m_V^* is the vector meson mass in-medium, n is the local nuclear matter density and n_0 the nuclear matter saturation density. C is a constant of value $\sim 0.10 - 0.15$. The above linear scaling will be referred to as the BR_1 scaling. Current NN potentials are mostly based on meson exchanges. The above in-medium modifications of meson masses can significantly alter such NN interactions, making them dependent on the density n of the nuclear medium where the interacting pair of nucleons reside.

We have carried out several preliminary studies about the effects of the BR scaling on nuclear matter [14, 15], neutron stars [15] and the $^{14}C - ^{14}N$ β-decay [16]. In the following we would like to report and discuss some of our results.

2. Nuclear matter equation of state with BR scalings

As mentioned earlier, it has been difficult for nuclear matter calculations using realistic NN potentials V_{NN} to reproduce simultaneously the ground state energy per particle E_0/A and saturation density n_0 of nuclear matter [6]. Would nuclear matter calculations using NN potentials with BR scalings give better results in this respect? To answer this question, we have performed nuclear matter calculations using the BonnA potential [4] with and without the BR scaling. This potential is a one-boson exchange potential and it is relatively convenient to scale its meson parameters. In a previous work [14] we have carried nuclear matter calculations using the Nijmegen one-boson potential with and without BR scalings.

Our nuclear matter calculations are based on a low-momentum ring-diagram framework [14, 15]. We first obtain the low-momentum NN potential V_{low-k} by integrating out the $k > \Lambda$ momentum components of V_{NN}, Λ being a decimation scale. This "integrating-out" procedure is carried out by way of a T-matrix equivalence approach whose details have been described in [17, 18, 19, 20]. For nuclear matter calculations we need to employ a cut-off scale of $\Lambda \sim 3$ fm^{-1} [14, 15].

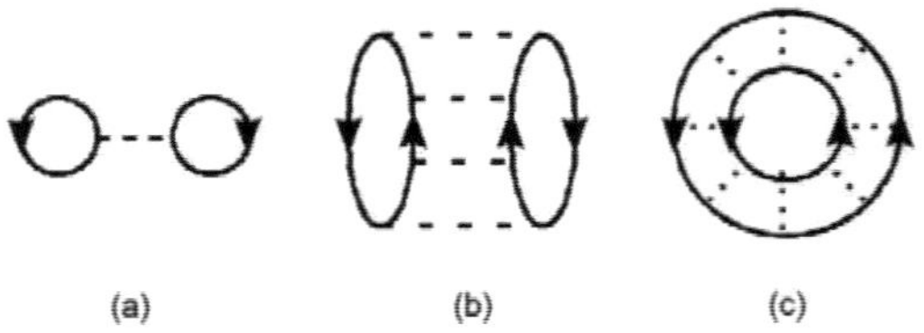

Figure 1. Diagrams included in the all-order *pphh* ring-diagram summation for the EOS of nuclear matter. Each dashed line represents a V_{low-k} vertex.

We employ a ring-diagram method to calculate the nuclear matter equation of state (EOS). In this method we include the all-order sum of the *pphh* ring diagrams as illustrated in Fig. 1, where (a), (b) and (c) are respectively 1st-, 4th- and 8th-order such diagrams. Then the ground-state energy energy shift ΔE_0, defined as $(E_0 - E_0^{free})$ where E_0 is the true ground state energy and E_0^{free} that for the non-interacting system, is given by

$$\Delta E_0(n,\alpha) = \int_0^1 d\lambda \sum_m \sum_{ijkl<\Lambda} Y_m(ij,\lambda) Y_m^*(kl,\lambda) \langle ij|V_{low-k}|kl\rangle. \tag{2}$$

The transition amplitudes Y are obtained by solving a RPA equation [14, 15]. The above EOS includes the effect from the particle-hole ground-state correlations generated by the all-order sum of *pphh* ring diagrams. Such long-range correlations are not included in BHF and DBHF calculations which include only diagrams first-order in the G-matrix.

Our results of ring-diagram calculations for symmetric nuclear matter using the BonnA potentials (without BR scaling) are presented in Fig. 2, labelled as 'V_{low-k} alone'. As seen, the energy per particle at saturation E_0/A and the saturation density n_0 are both rather far from the corresponding empirical values of -16 MeV and 0.16fm^{-3} ($k_F \simeq 1.34$fm^{-1}).

Can nuclear matter calculations using BR-scaled NN potentials give improved nuclear matter saturation properties? To explore this possibility, we have carried out nuclear matter calculations using BR-scaled NN potentials. We have chosen to employ the BonnA potential with its vector mesons ρ and ω scaled according to the linear BR scaling (BR_1) of Eq.(1). We also scaled its σ meson which is an effective meson representing multi-meson exchanges. The scaling parameters C_ρ, C_ω and C_σ are adjusted, to see if satisfactory saturation properties may be obtained.

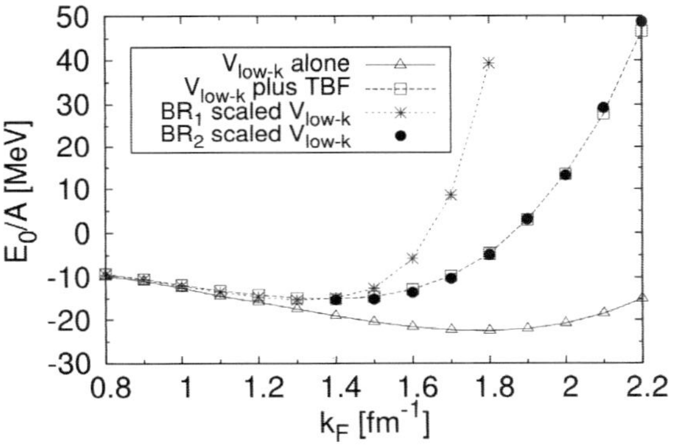

Figure 2. Ring-diagram EOSs for symmetric nuclear matter given by V_{low-k} alone, V_{low-k} with linear (BR_1) and nonlinear (BR_2) scalings, and V_{low-k} plus the three-body force (TBF) V_{3b} of Eq.(8). Λ=3.5fm^{-1} is used for all cases. See text for other explanations.

In Fig.2, our results of ring-diagram calculations for symmetric nuclear matter with the linear BR_1 scaling of Eq.(1) are also presented (labelled 'BR_1 scaled-V_{low-k}'). In this calculation, the mesons (ρ, ω, σ) of the BonnA potential are slightly scaled using Eq.(1) with $C_\rho = 0.113$, $C_\omega = 0.128$ and $C_\sigma = 0.102$. The cutoffs for these mesons are also scaled according to Eq.(1) with the same C parameters. We applied a larger momentum cutoff $\Lambda = 3.5$fm^{-1} than the usual 2.1 fm^{-1} used in finite nuclei. As explained in [14], momenta $k \sim 2-3$fm^{-1} are necessary for symmetric nuclear matter saturation, so we have applied $\Lambda = 3-4$fm^{-1} and found that the results are nearly the same [15]. In order to calculate the higher density region in neutron star and retain the similarity of V_{low-k}s from different NN potentials, we have chosen $\Lambda = 3.5$fm^{-1}. With these values the calculated E_0/A and saturation Fermi momentum k_F^0 are in satisfactory agreement with the empirical values, as shown in the figure.

A main effect of the above scaling is making the effective NN interaction density dependent, it becoming more repulsive at higher densities and thus providing a mechanism for nuclear matter saturation. So there arises an important question: Should the effective NN interaction be generally density dependent? In fact density dependent effective interactions have been successfully employed in nuclear theory since long time ago. A well-known example is the Skyrme effective interaction [21]. This interaction has a three-body force (TBF)

$$V_{3b}(i,j,k) = t_3\delta(\vec{r}_i - \vec{r}_j)\delta(\vec{r}_j - \vec{r}_k) \tag{3}$$

which is equivalent to a density-dependent two-body interaction usually written as

$$V_n(1,2) = \frac{1}{6}(1 + x_3 P_\sigma)t_3\delta(\vec{r}_1 - \vec{r}_2)n(\vec{r}_{av}), \tag{4}$$

where P_σ is the spin-exchange operator and $\vec{r}_{av} = \frac{1}{2}(\vec{r}_1+\vec{r}_2)$. t_3 and x_3 are parameters determined by fitting certain experimental data. For nuclear matter, n is uniform and is readily determined (e.g. $n = \frac{2k_F^3}{3\pi^2}$ for symmetric nuclear matter). In the present work, we have used $x_3 = 0$.

We have calculated the EOS for symmetric nuclear matter using the interaction (V_{low-k}+V_{3b}) where V_{low-k} is without BR scaling and V_{3b} is calculated using Eq.(4). The resulting EOS reproduces well the empirical saturation properties when t_3=2000 MeV is used; in fact this EOS , labelled as 'V_{low-k} *plus TBF*' in Fig. 2, is in very good agreement with the BR_1 EOS for densities $<\sim n_0$. But for larger densities, they deviate from each other largely. The EOS given by the BR_1 scaled V_{low-k} seems to be too stiff.

Let us come back to the BR_1 scaling of Eq.(1). This is a linear scaling and is clearly suitable only for low densities. As far as we know, how to extend this scaling to high densities is largely unknown. We feel that an EOS which is softer than the above BR_1 one may be desirable. To explore this possibility we have adopted the following non-linear BR_2 scaling

$$\frac{m^*}{m} = 1 - C(\frac{n}{n_0})^\alpha, \tag{5}$$

with α being a parameter. Using the same C_ρ, C_ω and C_σ parameters as used in BR_1, we have found that the EOS given by the above BR_2 scaling with α= 0.3 is considerably softer. (Recall that BR_1 has α=1.) Our results are also presented in Fig. 2, labelled as 'BR_2 scaled V_{low-k}'. As shown, this BR_2 EOS is practically identical to the 'V_{low-k} plus TBF' one, up to densities of about $5n_0$. These EOSs are both in good agreement with the constraints on nuclear EOS derived by Danielowicz et al. [22] from heavy ion collisions. It is of interest that the density-dependent effects provided by the above non-linear BR_2 scaling are quite similar to those from a Skyrme-type three-body force.

3. Brown-Rho scalings and neutron star properties

As we mentioned earlier, how to extend the BR scaling to high densities is by and large still an open question. As discussed extensively in the literature [23, 24], the properties of neutron stars can be calculated from the nuclear matter EOS using the Tollman-Oppenheimer-Volkov equations. In so doing, however, we need to know the EOS at densities as high as $\sim 8n_0$. The EOSs there are hardly well known, and neutron star properties may provide a useful test for them. We have adopted an empirical approach in extending the BR scaling to high densities. Recall that for $n <\sim 5n_0$ the EOS given by the non-linear BR_2 of Eq.(5) is nearly identical to that given by the interaction 'V_{low-k} (unscaled) plus TBF'. For higher densities, we feel that the density dependence provided by the TBF force may be a reasonable choice. Thus we have used the above combined interaction to calculate the EOS for neutron stars, covering both low and high density regimes. As shown in Fig. 3 the inclusion of 'TBF' is both important and desirable. Without 'TBF', the calculated maximum mass and its radius are both too small comparing with

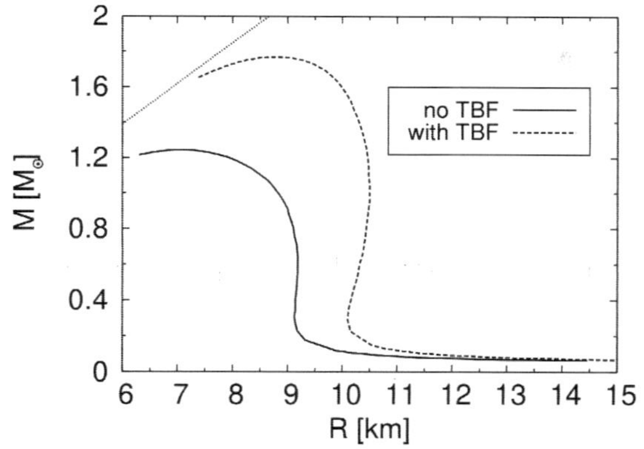

Figure 3. Mass-radius trajectories of pure neutron stars from ring-diagram EOSs given by the CD-Bonn V_{low-k} interaction with and without the three-body force (TBF) V_{3b}. Only stars to the right of maximum mass are stable against gravitational collapse. Causality limit is indicated by the straight line in the upper left corner.

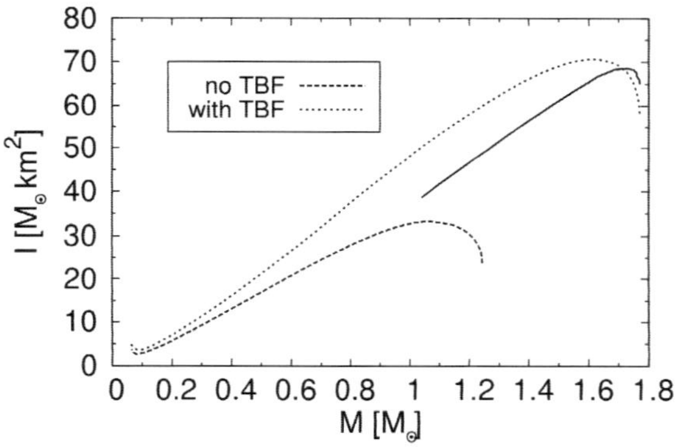

Figure 4. Pure neutron stars' moments of inertia calculated from CDBonn potential with and without three-body force (TBF). Results from the empirical formula (6) for $M \geq 1.0 M_\odot$ with TBF are denoted by the solid line.

the empirical values of $\sim 1.8 M_\odot$ and $\sim$ 9km, respectively. With 'TBF', our results are quite close to the empirical values.

There has been considerable progress in the measurement of moment of inertia of neutron stars [25]. For $M \geq 1.0 M_\odot$, Lattimer and Schutz [26] have determined an empirical formula relating the moment of inertia I of neutron stars to their mass M and radius R, namely

$$I \approx (0.237 \pm 0.008) M R^2 (1 + 4.2 \frac{M}{M_\odot} \frac{\text{km}}{R} + 90 (\frac{M}{M_\odot} \frac{\text{km}}{R})^4). \qquad (6)$$

As displayed in Fig. 4, our 'with TBF' results for $(M,\ R,\ I)$ are in remarkably good agreement with this empirical relation, especially near the maximum neutron star mass. There the inclusion of 'TBF' enhances the moment of inertia by more than a factor of 2.

4. Summary and outlooks

We have seen that the BR scaling has had important and desirable effects on nuclear matter and neutron stars. Although how to scale the in-medium mesons is still not precisely known (e.g. we still do not know how to precisely determine the scaling parameters Cs, especially at high densities), we feel that the physics of BR scalings is both interesting and in the right direction. That is, the NN interaction is most likely medium-dependent or density dependent. In nuclear many-body calculations, the medium corrections to the NN intertactions, which are determined for two nucleons in free space, should be taken into account. Because of 'time and space' limitations, we have not been able to report and discuss the effects of BR scalings in finite nuclei. In a recent work [16], the BR scaling was found to have crucial effects on the $^{14}C - ^{14}N$ β-decay; the BR scaling weakens the tensor force (the contributions to tensor force from π and ρ are of opposite signs) and thus increases the life time of this decay from few hours to few thousand years. Further studies of the effects of BR scaling in finite nuclei are needed and should be of interest. There should be certain connections between BR scalings and three-nucleon forces. In two recent publications [27, 28], density-dependent corrections to the in-medium NN interactions from the leading-order chiral three-nucleon force have been derived. It should be of interest to compare such density-dependent NN interactions with those obtained from the BR scaling.

We sincerely thank R. Machleidt and Jeremy Holt for many helpful discussions and their long time collaboration with us. We are very grateful to Gerry Brown for suggesting the Brown-Rho scaling topic to us and numerous discussions with him. TTSK also want to thank the organizers, particularly Aldo Covello, for their unsurpassed hospitality. This work is supported in part by the U.S. Department of Energy under Grant No. DF-FG02-88ER40388.

[1] Machleidt R, *Phys. Rev.* C **63** (2001) 024001.
[2] Stoks V G J, Klomp R A M, Terheggen C P F and de Swart J J, *Phys. Rev.* C **49** (1994) 2950.
[3] Wiringa R B, Stoks V G Jand Schiavilla R, *Phys. Rev.* C **51**, 38 (1995).
[4] Machleidt R, *Adv. Nucl. Phys.* **19** (1989) 189-376.
[5] Barrett B R, report in this Proceedings.
[6] Li Bao-An, Chen Lie-Wen, Ko Che Ming, *Phys. Rep.* **464**, 113 (2008).
[7] Brown G E, Rho M, *Phys. Rev. Lett.* **66** (1991) 2720.
[8] Hatsuda T and Lee S H, *Phys. Rev.* C **46** (1992) R34.
[9] Brown G E, Rho M, *Phys. Rep.* **396** (2004) 1.
[10] Harada M and Yamawaki K, *Phys. Rept.* **381** (2003) 1.
[11] Klingl F, Kaiser N, and Weise W, *Nucl. Phys.* A **624** (1997) 527.
[12] Trnka D *et al.*, *Phys. Rev. Lett.* **94** (2005) 192303.
[13] Naruki M *et al. Phys. Rev. Lett.* **96** (2006) 092301.
[14] Siu L W, Holt J W, Kuo T T S and Brown G E, *Phys. Rev.* C **79**, 0540004(2009)
[15] Dong H, Kuo T T S and Machleidt R, *Phys. Rev.* C **80**, 065803(2009)
[16] Holt J W, Brown G E, Kuo T T S, Holt J D and Machleidt R, *Phys. Rev. Lett.* **100**, 062501(2008).
[17] Bogner S K, Kuo T T S and Coraggio L, *Nucl. Phys.* A **684**, (2001) 432.
[18] Bogner S K, Kuo T T S, Coraggio L, Covello A and Itaco N, *Phys. Rev.* C **65** (2002) 051301R.
[19] Bogner S K, Kuo T T S and Schwenk A, *Phys. Rep.* **386** (2003) 1.
[20] Holt J D, Kuo T T S and Brown G E, *Phys. Rev.* C **69** (2004) 034329.
[21] Ring P and Schuck P, *The Nuclear Many-Body Problem* (Springer-Verlag, New York, 1980).
[22] Danielewicz P, Lacey R, and Lynch W A, *Science* **298**, 1592(2002).
[23] Lattimer J M and Prakash M, *Phys. Rep.* **442**, 109(2007).
[24] Sedrakian A, *Prog. Part. Nucl. Phys.* **58**, 168(2007).
[25] Bejger M, Haensel P, *Astrophys. J.* **405**, 747 (2003).
[26] Lattimer J M and Schutz B F, *Astrophys. J.* **629**, 979 (2005).
[27] Holt J W, Kaiser N and Weise W, *Phys. Rev.* C **79**, 054331 (2009).
[28] Holt J W, Kaiser N and Weise W, *Phys. Rev.* C **81**, 0240021 (2010).

Ab initio shell model with a core

Ab initio shell model with a core

Bruce Barrett[1], **Alexander Lisetskiy**[1,5], **Michael Kruse**[1], **Petr Navratil**[2], **Ionel Stetcu**[3] **and James Vary**[4]

[1] Department of Physics, PO Box 210081, University of Arizona, Tucson, AZ 85721, USA
[2] Lawrence Livermore National Laboratory, PO Box 808, Livermore, CA 94551, USA
[3] Department of Physics, University of Washington, Seattle, WA 98195, USA
[4] Department of Physics and Astronomy, A515 Zaffarano Hall, Iowa State University, Ames, IA 50011, USA

E-mail: bbarrett@physics.arizona.edu

Abstract. The No Core Shell Model (NCSM) has been successful in describing the properties of light nuclei, $A \leq 16$, starting from the fundamental interactions among the A nucleons, but it is currently difficult to extend the NCSM to heavier nuclei, because of the extremely large model spaces involved in the calculations. We present a new procedure, based on performing two unitary transformations on the Hamiltonian, so as to obtain the core, one-body, two-body and perhaps three-body input necessary for performing Standard Shell Model (SSM) calculations for nuclei. Such SSM calculations can be easily performed for *sd*- and *pf*-shell nuclei. We demonstrate the accuracy of our approach by applying to it nuclei within the $0p$-shell.

1. Introduction

The No Core Shell Model (NCSM) has had considerable success in describing the binding energies, exitation spectra and other physical properties of light nuclei, $A \leq 16$, *e.g.*, [1, 2, 3]. One of the principle reasons for these successes is that one can exactly calculate the effective interactions and operators to be used in a given model space, using a *unitary transformation* approach [4, 5]. The unitary matrix employed in this transformation is composed of the eigensolutions for the interacting system of a few nucleons n, where n is usually only two or three. This is referred to as a *cluster approximation* but is perhaps more clearly defined as an n-body correlation, given by the eigensolutions for the fully correlated system of n nucleons. This unitary transformation is used to truncate the full space solution to a smaller model space. The n-body matrix elements obtained from this truncation are used as input to the many-body Hamiltonian for all A nucleons, which is then diagonalized in the model space. As the size of the model space increases, the *cluster* Hamiltonian approaches the full-space Hamiltonian, yielding the converged, (*i.e.,* exact) solution. This procedure works well for lighter nuclei, $A \leq 16$, whose many-body Hamiltonians can be diagonalized in sufficiently large model spaces, so as to obtain converged results. For heavier mass nuclei, existing computer technology does not allow calculations in the very large model spaces that they require in order to yield converged results. In this contribution, we outline a new approach for obtaining higher-order correlations, *i.e., clusters*, in smaller model spaces, so that microscopic calculations based on the NCSM can be extended to heavier nuclei, *e.g.*, nuclei in the *sd*- and *pf*-shells.

[5] Present address: Mintec Inc., 3544 E. Fort Lowell Road, Tucson, AZ 85716 USA

2. The NCSM Formalism and Its Extension

2.1. The NCSM Approach

One can outline the NCSM procedure as follows:

1. Start with a translationally invariant Hamiltonian for the A-nucleon system, which contains a realistic nucleon-nucleon (NN) potential and perhaps a complementary theoretical NNN potential.

2. Add to this the center-of-mass (COM) harmonic-oscillator (HO) Hamiltonian, so as to bind the COM of the A-nucleon system. This defines a basis, *i.e.*, HO, for performing our calculations and also allows for the exact projection of spurious COM components from our final wave functions.

3. Solve this combined Hamiltonian [Steps 1 and 2] in an extremely large space, *e.g.*, 300 or more HO configurations, so as to obtain fully converged solutions for the two-(or three-)nucleon system.

4. Using a unitary transformation constructed from the components of the eigensolutions in step 3, which span a smaller two-(or three-)nucleon model space, obtain the matrix elements of the effective two-(or three-)body Hamiltonian in the smaller model space.

This procedure is similar to the earlier technique of taking a subset d of the eigensolutions obtained in step 3 and explicitly projecting them into a smaller two-(or three-)nucleon model space of dimension d. In that case the *exact* two-(or three-)nucleon effective interaction in the $d \times d$ model space is given by a sum over d of all terms of the form of a projected eigensolution times its corresponding full-space eigenenergy times the corresponding biorthogonal of the projected eigensolution. The truncating of the unitary matrix in step 4 is a straightforward technique for performing this truncation and summation.

5. Use these effective Hamiltonian matrix elements to construct the A-nucleon Hamiltonian matrix, which is then diagonalized in the A-nucleon model space.

6. Repeat steps 4 and 5 as the A-nucleon model-space size is increased, so as to obtain converged results.

2.2. Our Extension of the NCSM Approach

To extend the NCSM approach to heavier nuclei, for which the A-nucleon model spaces become unmanageable with existing computers, a new technique will be necessary. One such procedure [6] is to reintroduce a core into the NCSM approach, by performing a second unitary transformation, so as to include the effects of *all A* nucleons into a single major shell, as proposed by Navrátil *et al.* [7].

This extension, which we call the "*ab initio* Shell Model with a Core," adds additional steps to the ones given above.

7. Use the converged A-nucleon eigensolutions of step 6 to construct a second unitary transformation matrix, so as to project the *converged* A-nucleon model-space results in step 6 into the space, in which all A nucleons occupy the lowest HO configuration allowed by the Pauli Exclusion Principle.

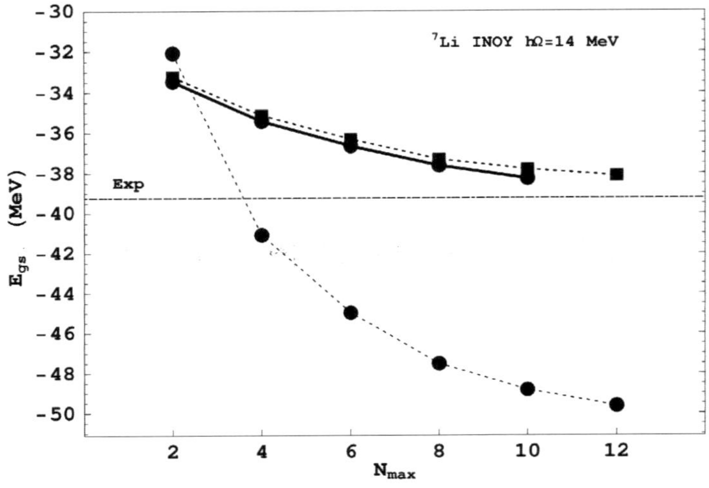

Figure 1. The g.s. energy of ^{7}Li as a function of N_{max}. The dots connected with a solid line are the full NCSM results. The squares connected with a dashed line are the SSM results without the three-body component. The dots connected with a dashed line are SSM results using a constant, bare ^{4}He g.s. energy. See the text for more details.

For example, for ^{6}Li this would be four nucleons in the $0s_{1/2}$-shell and two nucleons in the $0p$-shell. Similar to Standard Shell Model (SSM) terminology, we call this the space of $0\hbar\Omega$ excitation energy, *i.e.,* no energy is available to any of the A nucleons for excitations to any higher HO shells. Thus, in our example of ^{6}Li, the four nucleons in the $0s_{1/2}$-shell are effectively *frozen* into this configuration (like an inert core!) and only the two nucleons in the $0p$-shell can move around in that shell. Consequently, the two nucleons in the $0p$-shell yield a full set of $0p$-shell two-nucleon matrix elements, which *by construction* now contain *all the correlations* of the six-nucleon system. Note that this is equivalent to calculating the six-body cluster! When diagonalized in the $0p$-shell, this six-body-cluster result yields *exactly* the same eigenenergies as the full NCSM calculations for states with parentage predominately within the $0p$-shell.

8. For NCSM solutions involving A nucleons, such that A can be separated into $A-2$ *core* nucleons and two-valence nucleons, *e.g.,* ^{6}Li going to a ^{4}He core and a valence proton and a valence neutron, one can then calculate the core, one-body and two-body parts of the solutions in step 7.

These core, one-body and two-body components are simply the input for a SSM calculation and can then be used to perform SSM calculations for heavier mass nuclei in the same major shell. For example, in the $0p$-shell, the results obtained for the $A = 5, 6$ and 7 nuclear systems in steps 7 and 8 can be used to perform SSM calculations for $A = 8, 9, 10...16$ nuclear systems.

3. Applications to $0p$-shell Nuclei

Figures 1 and 2 compare the results of full NCSM calculations (INOY NN potential [8, 9] as the A-nucleon model-space size, $N_{max}\hbar\Omega$, is varied from $N_{max} = 2$ to 12 with $\hbar\Omega = 14$ MeV) for the ground-state (g.s.) energy and excited-state energies of ^{7}Li, respectively, with the results of SSM calculations, using the core, one-body and two-body components, computed as described above [steps 7 and 8], using the ^{7}Li NCSM two-body matrix elements.

The energy of the ^{4}He core inside the ^{7}Li nucleus is computed to be -63.336 MeV, significantly

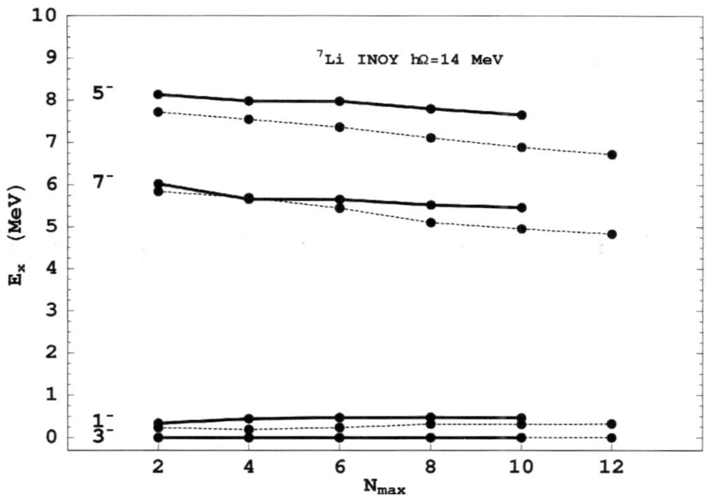

Figure 2. NCSM (solid lines) and SSM without the three-body component (dashed lines) for the excited-state spectrum of ^{7}Li. The states with spin J are marked by 2J.

larger than that for a bare ^{4}He nucleus, being -28.296 MeV. This should not be a surprising result, because the two protons and the two-neutrons of the ^{4}He core are in lower-lying $0s_{1/2}$ levels inside the ^{7}Li mean field (*e.g.*, a phenomenological Woods-Saxon potential), than in the mean field for a bare ^{4}He nucleus.

The single-particle energies are then obtained by calculating the five-body clusters inside the ^{7}Li nucleus, *i.e.*, ^{5}He and ^{5}Li, and subtracting the above ^{4}He core energy. The two-body matrix elements are found by subtracting the previous five-body cluster results from the earlier six-body cluster results, *e.g.*, ^{6}He, ^{6}Li and ^{6}Be. For $A = 7$ it is also possible to obtain the seven-body cluster results, from which effective three-body matrix elements can be obtained.

For the test calculations in the $0p$-shell we have performed such calculations, as described above, for the $A = 6$ and $A = 7$ nuclear systems and determined the core, one-body and two-body (and three-body for $A = 7$) components. For $A = 7$, SSM calculations performed in the $0p$-shell with the core, one-, two- *and* three-body components exactly reproduce (*by construction*) the results of a full NCSM calculation for $A = 7$ for eigenstates predominately in the $0p$-shell.

These components can also be used for performing SSM calculations in the $0p$-shell for nuclear systems with $A > 7$. For example, Fig. 3 shows results for ^{8}He, ^{9}He and ^{10}He calculated with the CD Bonn NN potential [10] for $N_{max} = 6$ and $\hbar\Omega = 20$ MeV. At least for these maximal isospin cases, including the effective three-body contribution improves agreement with the exact NCSM results and appears to leave very limited room for effective four- or higher-body contributions.

The same double unitary transformation approach outlined previously for obtaining the effective components of the shell-model Hamiltonian in a single major shell, *e.g.*, the $0p$-shell, can also be utilized for computing the effective components of any physical operator in the same major shell [11].

It should be noted that the double unitary transformation technique developed here is similar in philosophy to the work of Fujii, *et al.*, [12] using the unitary-model-operator approach. However, in their work the second unitary transformation is to a single-particle basis, while in our case the second unitary transformation is based on a truncation to a *smaller total energy* for the A-nucleon model space.

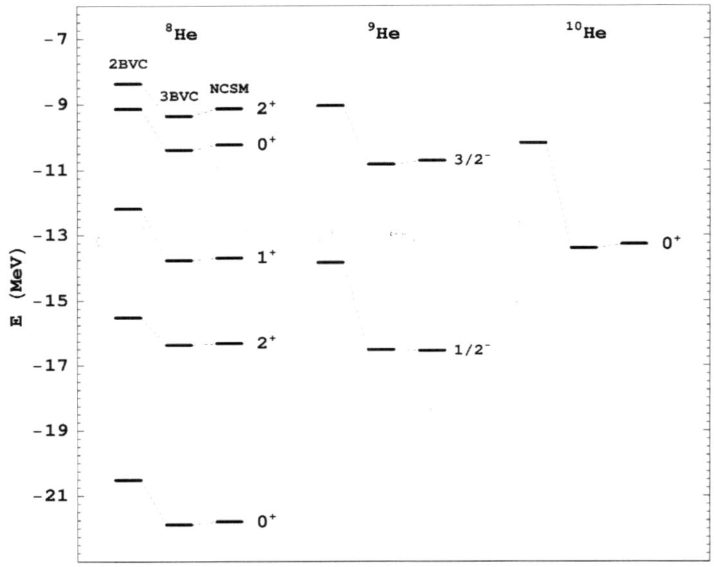

Figure 3. Comparison of spectra for ^{8}He, ^{9}He and ^{10}He from SSM calculations using up to two-body components (2BVC) and up to three-body components (3BVC) with exact NCSM results, all calculated for $N_{max} = 6$ and $\hbar\Omega = 20$ MeV using the CD-Bonn NN potential [10].

4. Conclusions

The purpose of the present investigations was to demonstrate the feasibility of using our double unitary transformation technique for determining the input components for SSM calculations in a single major shell. This was done by performing calculations for nuclei in the $0p$-shell, for which exact NCSM calculations can also be performed, thereby providing a check on our proposed procedure.

The good agreement obtained between our exact NCSM calculations and our SSM calculations encourages us to extend this procedure to $sd-$ and $pf-$shell nuclei. Because the second unitary transformation is always made to the lowest energy model space, *i.e.*, $0\hbar\Omega$, the final results will always look like an inert core plus a few valence nucleons. Work on the extension of this approach to $sd-$ and $pf-$shell nuclei is currently in progress.

Acknowledgments

B.R.B., A.F.L. and M.K.G.K. acknowledge partial support of this work by the NSF under the grants PHY-0555396 and PHY-0854912. P.N. acknowledges partial support from LLNL under Contract DE-AC52-07NA27344, I.S. acknowledges partial support from DE-FC02-07ER41457, and J.P.V. acknowledges partial support from DE-FG02-87ER40371.

References

[1] Navrátil P, Vary J P and Barrett B R 2000 *Phys. Rev. Lett.* **84** 5728
[2] Navrátil P, Vary J P and Barrett B R 2000 *Phys. Rev.* C **62** 054311
[3] Navrátil P, Quaglioni S, Stetcu I and Barrett B R 2009 *J. Phys. G: Nucl. Part. Phys.* **36** No. 8, 083101
[4] Okubo S 1954 *Prog. Theor. Phys.* **12** 603
[5] Suzuki K 1982 *Prog. Theor. Phys.* **68** 246

[6] Lisetskiy A F, Barrett B R, Kruse M K G, Navrátil P, Stetcu I and Vary J P 2008 *Phys. Rev.* C **78** 044302
[7] Navrátil P, Thoresen M and Barrett B R 1997 *Phys. Rev.* C **55** R573
[8] Doleschall P, Borbely I, Papp Z and Plessas W 2003 *Phys. Rev.* C **67** 064005
[9] Doleschall P 2004 *Phys. Rev.* C **69** 054001
[10] Machleidt R, Sammarruca F and Song Y 1996 *Phys. Rev.* C **53** R1483
[11] Lisetskiy A F, Kruse M K G, Barrett B R, Navrátil P, Stetcu I and Vary J P 2009 *Phys. Rev.* C **80** 024315
[12] Fujii S, Okamoto R and Suzuki K 2004 *Phys. Rev.* C **69** 034328, and references therein

A new solution for effective interaction

A new solution for effective interaction

R Okamoto[1], K Suzuki[1], H Kumagai[2], S Fujii[3]

[1]Kyushu Institute of Technology, Kitakyushu, 804-8550 Japan
[2]Fukuoka Institute of Technology, Fukuoka, 813-0000 Japan
[3]Center for Nuclear Study (CNS), University of Tokyo, Wako Campus of RIKEN, Wako 351-0198, Japan

E-mail: okamoto@mns.kyutech.ac.jp

Abstract. A new method is given for the model-space effective interaction. Introducing a new operator in place of the Q-box in the Krenciglowa-Kuo (KK) method, we derive a new equation for the effective interaction. This equation can be viewed as an extension of the KK method. We show that this equation can be solved both in iterative and non-iterative ways. We observe that the iteration procedure brings about fast acceleration of convergence compared to the KK approach. We also find that the non-iterative calculation reproduces successfully any set of the true eigenvalues of the original Hamiltonian. This non-iterative calculation can be made regardless of the magnitudes of the overlaps with the model space and the energy differences between the unperturbed energy and the eigenvalues to be solved.

1. Introduction

In some fields of many-body physics, it is often useful to introduce an effective interaction acting in a truncated model space. Much effort has been made both as regards formal theories and their applications[1].

Among many approaches, we here direct our attention to the Krenciglowa-Kuo (KK)[2] and the Lee-Suzuki (LS)[3,4] methods. These two methods are constructed by the use of the so called Q-box as the building block of formulation. The KK approach has very simple structure and the solution is obtained in an iterative way. If the iteration converges, eigenvalues are given for the states with largest overlaps with the chosen model space. On the other hand, the LS method reproduces eigenvalues for the states which lie closest to the chosen unperturbed energy. The LS method is complicated in structure and higher derivatives of the Q-box with respect to starting energy are necessary if one wants to obtain more accurate solutions.

Both of the two theories yield only certain of the true eigenvalues of the original Hamiltonian. This restriction is not, of course, desirable. In a formal point of view, the Q-box itself contains information regarding all of the true eigenvalues. For a given model space of dimension d, there would be a method of reproducing any d eigenvalues among all the true eigenvalues.

In this report, we want to show that a method exists for reproducing all the true eigenvalues of the original Hamiltonian if the Q-box is given accurately enough.

2. Krenciglowa-Kuo method

We assume that the Hamiltonian H is composed of the unperturbed Hamiltonian H_0 and the perturbation V, i.e., $H = H_0 + V$. The entire Hilbert space is partitioned into the d-dimensional model space (P space) and its complement (Q space) with the projection operators P and Q, respectively. We note some properties, $P + Q = 1, P^2 = P, Q^2 = Q$ and $PQ = QP = 0$. We here assume that H_0 is decoupled between the P and Q spaces as

$$H_0 = PH_0P + QH_0Q. \tag{1}$$

We further assume that the P-space states have a degenerate unperturbed energy E_0, i.e.,

$$PH_0P = E_0P. \tag{2}$$

Thus the P-space eigenvalue equation is written with the P-space effective interaction R and the P-space eigenstate $|\phi_k\rangle$ as

$$(E_0P + R)|\phi_k\rangle = E_k|\phi_k\rangle, \qquad (k = 1, 2, \cdots, d). \tag{3}$$

The effective interaction can be constructed by the use of Q-box which is an operator of energy variable E and is defined as

$$\hat{Q}(E) \equiv PVP + PVQ\frac{1}{E - QHQ}QVP. \tag{4}$$

The Krenciglowa-Kuo (KK) approach leads to a solution for R as

$$\begin{aligned} R_{\mathrm{KK}} &= \sum_{k=1}^{d} \hat{Q}(E_k)|\phi_k\rangle\langle\tilde{\phi}_k| \\ &= \sum_{k=1}^{d} (E_k - E_0)|\phi_k\rangle\langle\tilde{\phi}_k|, \end{aligned} \tag{5}$$

where $\langle\tilde{\phi}_k|$ is the biorthogonal state of the P-space eigenstate $|\phi_k\rangle$ in Eq. (3) defined through $\langle\tilde{\phi}_k|\phi_{k'}\rangle = \delta_{kk'}$. We here note that, as seen in Eq. (5), the determination of the effective interaction R_{KK} is equivalent to solving d eigenvalues $\{E_k\}$. In general, the K-K solution is obtained in an iterative way based on Eqs. (3) and (5). It has been known that if the iteration converges, the KK solution yields d eigenvalues of the states with largest P-space overlaps among the eigenstates of the original Hamiltonian H.

We here want to show that there is another way of obtaining the KK solution. We consider an eigenvalue equation for a given energy E

$$[E_0 + \hat{Q}(E)]|\psi_k\rangle = G_k(E)|\psi_k\rangle, \; (k = 1, 2, \cdots, d), \tag{6}$$

where $\{G_k(E)\}$ and $\{|\psi_k\rangle\}$ are the eigenvalues and the eigenstates, respectively. For a given E, there are d eigenstates because $\hat{Q}(E)$ is a d-dimensional operator. Thus, we have d functions of variable E, which we label in order of energy as $G_1(E) < G_2(E) < \cdots < G_d(E)$.

It may be clear from Eqs. (3) and (6) that the eigenvalues $\{E_k\}$ in Eq. (3) can be given by solving equations

$$E = G_k(E), \quad (k = 1, 2, \cdots, d). \tag{7}$$

We can prove that any solution of Eq. (7) agrees with one of the true eigenvalues of H and the eigenstate $|\psi_k\rangle$ does with $|\phi_k\rangle$ in Eq. (3).

The solutions $\{E_k\}$ to Eq. (7) are obtained as crossing points of two graphs of $y = G_k(E)$ and $y = E$. It should be noted that this procedure of solving eigenvalues $\{E_k\}$ can be made independently of the magnitude of the P-space overlap. Therefore we can, in principle, reproduce all the true eigenvalues of H. However, as we see in Eq. (4) there appear poles in $\hat{Q}(E)$ when E approaches one of the eigenvalues of QHQ. The poles in $\hat{Q}(E)$ induce also poles in $G_k(E)$. Such a situation causes numerical instability in solving Eq. (7) with $\{G_k(E)\}$ around the pole positions.

3. Extension of the Krenciglowa-Kuo method

We define an operator in terms of the Q-box as

$$\hat{Z}(E) \equiv \frac{1}{1-\hat{Q}_1(E)}\left[\hat{Q}(E) - (E-E_0)\cdot\hat{Q}_1(E)\right], \tag{8}$$

where $\hat{Q}_1(E)$ is the energy derivative of the Q-box given as

$$\begin{aligned}\hat{Q}_1(E) &\equiv \frac{d\hat{Q}(E)}{dE} \\ &= -PVQ\frac{1}{(E-QHQ)^2}QVP.\end{aligned} \tag{9}$$

Hereafter we shall refer to $\hat{Z}(E)$ as the Z-box. It should be noted that if $E = E_0$, the Z-box agrees with the first-order recursive solution in the LS method[4].

The Z-box has the following properties:

(i) By using Eqs. (4), (5) and (8), we have

$$\begin{aligned}\sum_{k=1}^{d}\hat{Z}(E_k)|\phi_k\rangle\langle\tilde{\phi}_k| &= \sum_{k=1}^{d}\frac{1}{1-\hat{Q}_1(E_k)}\left[R_{\rm KK} - \hat{Q}_1(E_k)\cdot R_{\rm KK}\right]|\phi_k\rangle\langle\tilde{\phi}_k| \\ &= \sum_{k=1}^{d} R_{\rm KK}|\phi_k\rangle\langle\tilde{\phi}_k| \\ &= R_{\rm KK}.\end{aligned} \tag{10}$$

The above fact means that, replacing $\hat{Q}(E)$ by $\hat{Z}(E)$ in Eq. (5), a new solution for the effective interaction R can be derived as

$$R_{\rm EKK} \equiv \sum_{k=1}^{d}\hat{Z}(E_k)|\phi_k\rangle\langle\tilde{\phi}_k| \tag{11}$$

which we call the extended Krenciglowa-Kuo (EKK) solution.

In a similar way as in Eq. (6) we solve the d-dimensional P-space eigenvalue equation for the Z-box for a given variable E as

$$\left\{E_0 + \hat{Z}(E)\right\}|\chi_k\rangle = F_k(E)|\chi_k\rangle,\ (k=1,2,\cdots,d), \tag{12}$$

where $\{F_k(E)\}$ are the eigenvalues and $\{|\chi_k\rangle\}$ are the eigenstates. We here label $F_k(E)$ in order of energy as $F_1(E) < F_2(E) < \cdots < F_d(E)$. In the same way as in Eq. (7), the true eigenvalues $\{E_k\}$ are given by solving the equations

$$E = F_k(E),\quad (k=1,2,\cdots,d). \tag{13}$$

(ii) We can prove that if E is one of the true eigenvalues $\{E_k\}$ satisfying Eq. (5), the energy derivative of $\hat{Z}(E)$ becomes zero, and we have

$$\frac{dF_k(E)}{dE} = 0. \tag{14}$$

This property of the Z-box plays important roles in two ways:

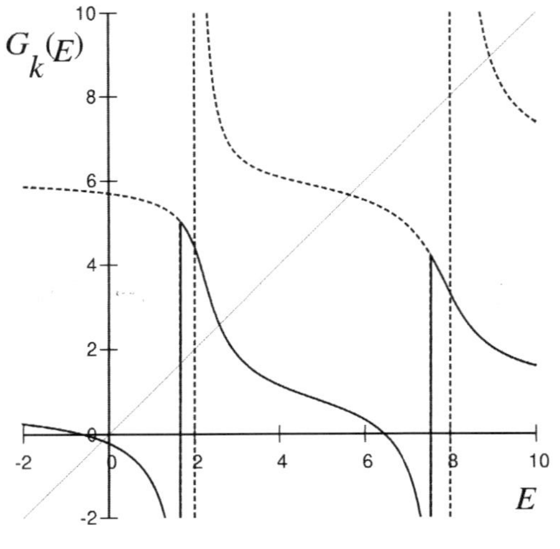

Figure 1. Dependence of $G_k(E)$ on E with $x = 0.2$. The graphs of $y = G_1(E)$ and $y = G_2(E)$ are shown in solid and broken lines, respectively. The direct line denotes the graph of $y = E$.

(a) The two equations (6) and (12) are not identical. The operator $\hat{Z}(E)$ is different from $\hat{Q}(E)$ as a P-space operator. Therefore, Eq. (13) based on the Z-box has possibly excessive solutions, or spurious solutions, different from the true eigenvalues $\{E_k\}$. However, we can remove easily these spurious solutions according to the condition in Eq. (14).

(b) In the same way as in the KK approach we can derive the effective interaction R_{EKK} in an iterative way. The iterative calculation in this approach can be made only by replacing $\hat{Q}(E)$ in the KK method by $\hat{Z}(E)$. It must be pointed out that, due to the property of Eq. (14), the iteration based on $\hat{Z}(E)$ converges quite rapidly as we see in Table 1. The new method can be understood as an application of the Newton-Raphson (NR) method which is used widely for accelerating convergence in solving non-linear equations iteratively.

(iii) In the vicinity of the poles in $\hat{Q}(E)$, $\hat{Q}_1(E)$ has the dominant contribution in $\hat{Z}(E)$, which may be clear from Eqs. (4), (8) and (9). Resultantly $\hat{Z}(E)$ becomes

$$\hat{Z}(E) \approx (E - E_0)P, \tag{15}$$

and thus there appear no poles in $\hat{Z}(E)$. Therefore, the functions $\{F_k(E)\}$ are continuous and differentiable for any E. This fact of the new approach guarantees stability in numerically solving the crossing points of two graphs of $y = F_k(E)$ and $y = E$. We want to emphasize here that the procedure of solving $\{E_k\}$ is regardless of the properties of the eigenstates of the original Hamiltonian H, i.e., the P-space overlap or the position in energy.

4. A schematic model analysis

In order to obtain some assessments of the new approach we consider a model problem for which exact solutions can be derived easily. The model Hamiltonian we adopt here is given with the coupling strength x, the dimension of the entire space $n = 4$, the dimension of the P space $d = 2$ and the unperturbed energy $E_0 = 1$ as

$$PH_0P = \begin{pmatrix} 1 & 0 \\ 0 & 1 \end{pmatrix}, \quad PVP = \begin{pmatrix} 0 & 5x \\ 5x & 25x \end{pmatrix}, \quad PVQ = \begin{pmatrix} -5x & 5x \\ 5x & -8x \end{pmatrix}, \tag{16}$$

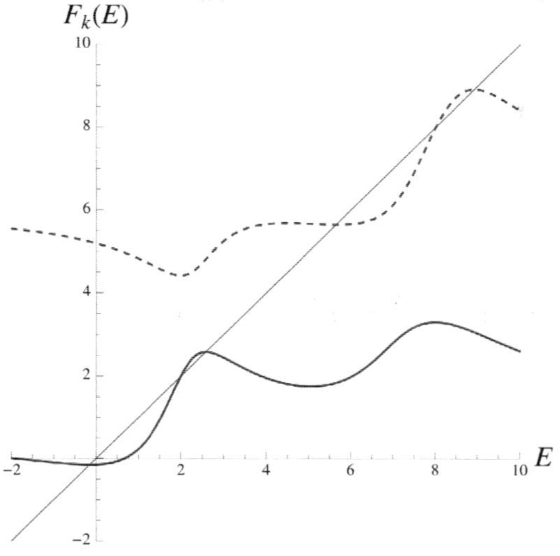

Figure 2. Dependence of $F_k(E)$ on E with $x = 0.2$. The graphs of $y = F_1(E)$ and $y = F_2(E)$ are shown in solid and broken lines, respectively. The direct line denotes the graph of $y = E$.

Table 1. Convergence of the eigenvalue of the lowest-lying state obtained with the KK and EKK methods for the model Hamiltonian with $x = 0.05$. Correct digits with the KK and EKK methods are given in n-th iteration step. The starting energies for (E_1, E_2) are taken to be $(0.0, 0.0)$. The notation c indicates convergence to more than fifteen decimal places. The exact eigenvalue here is 0.8904504858869942.

n	correct digits (KK)	correct digits (EKK)
1	0.9···	0.89···
2	0.89···	0.890450···
3	0.8904···	0.89045048588699···
4	0.890450···	c
5	0.8904504···	c
6	0.89045048···	c
7	0.8904504858···	c

and

$$QVP = \begin{pmatrix} -5x & 5x \\ 5x & -8x \end{pmatrix}, \quad QHQ = \begin{pmatrix} 3-5x & x \\ x & 9-5x \end{pmatrix}. \tag{17}$$

This model Hamiltonian H was introduced many years ago by Hoffmann et al.[5] and the structure of H was investigated in Ref. [6]. We first depict, in Figs. 1 and 2, the dependence of the functions $G_k(E)$ and $F_k(E)$, respectively, on the energy variable E with the strength $x = 0.2$. One may see in Fig. 1 that there are two poles in the graph of $G_k(E)$. On the other hand, poles disappear in the graph of $F_k(E)$ associated with the Z-box. One observes four crossing points between $y = G_k(E)$ and $y = E$ except for pole positions, and six crossing points between $y = F_k(E)$ and $y = E$ for $k = 1, 2$. The spurious solutions in Fig. 2 for $y = F_k(E)$ can be easily removed by the condition that the energy derivative at the crossing points should be zero.

Next we show in Table 1 the results in the iterative method for the KK and EKK solutions. The convergence rate in the EKK method is much higher than that in the KK method. The EKK method reaches the convergence to fourteen decimal places after only three iterations. However, the iterative calculation is not always recommendatory, because convergence conditions are not always clear and we cannot control these conditions. In this respect, it would be recommended to solve directly the crossing points of $y = G_k(E)$ or $y = F_k(E)$ and $y = E$. We could reproduce all the true eigenvalues of H by calculating the crossing points in a non-iterative way, for example, the bisection method.

5. Conclusion

In many cases of calculating the effective interaction, iteration and/or recursion methods, for example, the KK and LS methods, have been employed. However, one has seen that these methods reproduce only certain of the exact solutions. In general, it is quite difficult to control convergence of the iteration and/or the recursion. Moreover, these methods do not work well if a solution to be solved lies in the vicinity of the pole position. Such a case takes place actually in the calculation using the Q-box as in the KK approach.

In the present report we have proposed a new approach by introducing a new operator, the Z-box, and the functions $\{F_k(E)\}$. Characteristics of this approach are as follows: (a) The eigenvalues in the model space can be obtained by calculating the crossing points of two graphs $y = F_k(E)$ and $y = E$. The functions $\{F_k(E)\}$ give us all the information on the eigenvalues of the original Hamiltonian H. The crossing points can be calculated by using a non-iterative method, such as the bisection method. In this case convergence problem does not occur anymore. (b) The solutions can be also calculated by applying an iteration method. In this case, the use of $F_k(E)$ accelerates convergence compared to the usual KK approach. We want to notice that this approach is understood essentially as an application of the Newton-Raphson method for solving a non-linear equation. (c) The function $F_k(E)$ is continuous and differentiable for any of E. Instability or difficulty due to the presence of poles does not occur anymore as long as we use the function $F_k(E)$.

We may conclude that the present formulation on the procedure for obtaining the effective interaction removes some difficulties encountered in the usual approaches, such as the KK and LS methods. This approach makes us free from divergence of iteration or the pole problem, and therefore it may have possibility of applying widely to many-body problems.

6. References

[1] Kuo T T S and Osnes E 1990 *Lecture Notes in Physics*, **364** (Springer-Verlag)
[2] Krenciglowa E M and Kuo T T S 1974 *Nucl. Phys.* A**235** 171
[3] Lee S Y and Suzuki K 1980 *Phys. Lett.* B**91** 173
[4] Suzuki K and Lee S Y 1980 *Prog. Theor. Phys.* **64** 2091
[5] Hoffmann H M, Lee S Y, Richert J and Weidenmueller H A 1973 *Phys. Lett.* B**45** 421
[6] Suzuki K, Okamoto R, Ellis P J and Kuo T T S 1994 *Nucl. Phys.* A**567** 570

From neutron skins to neutron stars with a microscopic equation of state

From neutron skins to neutron stars with a microscopic equation of state

Francesca Sammarruca

Physics Department, University of Idaho, Moscow, Idaho, U.S.A.

E-mail: `fsammarr@uidaho.edu`

Abstract. I will review several issues related to the nuclear matter equation of state, with particular emphasis on isospin-asymmetric matter. I will stress the importance of pursuing a microscopic approach towards a better understanding of dense and exotic matter.

1. Introduction

Isospin-asymmetric nuclear matter (IANM) simulates the interior of a "nucleus" with unequal numbers of protons and neutrons. The equation of state (EoS) of (cold) IANM is then a function of density as well as the relative concentrations of protons and neutrons. The recent and fast-growing interest in IANM stems from its close connection with neutron-rich nuclei, or, more generally, asymmetric nuclei, including the very "exotic" ones known as "halo" nuclei. At this time, the boundaries of the nuclear chart are uncertain, with a few hundreds stable nuclei known to exist and perhaps a few thousands believed to exist. The Facility for Rare Isotope Beams (FRIB) has recently been approved for design and construction at Michigan State University. The facility will deliver intense beams of rare isotopes the study of which can provide crucial information on short-lived elements normally not found on the earth. Thus, this new experimental program will have widespread impact, ranging from the origin of elements to the universe in its evolution.

2. The Microscopic Approach

In this section, we present a discussion of the microscopic (*ab initio*) approach, in general, and the Dirac-Brueckner-Hartree-Fock (DBHF) method, in particular, as pursued by our group in Idaho [1, 2].

By *ab initio* we mean that the starting point is a realistic nucleon-nucleon (NN) interaction which is then applied in the nuclear medium without any additional parameters. Thus the first question to be confronted concerns the choice of the "best" NN interaction. After the development of QCD and the understanding of its symmetries, chiral effective theories [3] were developed as a way to respect the symmetries of QCD while keeping the degrees of freedom (nucleons and pions) typical of low-energy nuclear physics. However, chiral perturbation theory (ChPT) has definite limitations as far as the range of allowed momenta is concerned. For the purpose of applications in dense matter, where higher and higher momenta become involved with increasing Fermi momentum, NN potentials based on ChPT are unsuitable.

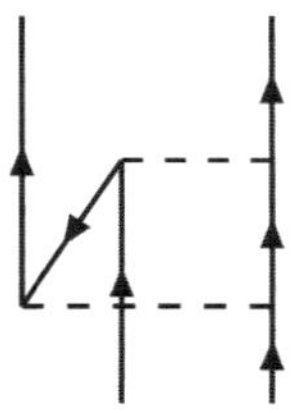

Figure 1. Contribution to the NN interaction from virtual pair excitation. Upward- and downward-pointing arrows represent nucleons and antinucleons, respectively. Dashed lines denote mesons.

Relativistic meson theory is an appropriate framework to deal with the high momenta encountered in dense matter. In particular, the one-boson-exchange (OBE) model has proven very successful in describing NN data in free space and has a good theoretical foundation. Furthermore, we require a potential that uses the pseudovector coupling for the interaction of nucleons with pseudoscalar mesons. With this in mind, as well as the requirement of a good description of NN data, Bonn B [4] is a reasonable choice.

We now move on to our chosen many-body method to investigate symmetric and asymmetric nuclear matter. The main strength of the DBHF approach is in its inherent ability to account for important three-body effects through its density dependence. In Fig. 2 we show a three-body force (TBF) originating from virtual excitation of a nucleon-antinucleon pair, known as "Z-diagram". The main feature of the DBHF method turns out to be closely related to the TBF depicted in Fig. 2, as we will see next. In the DBHF approach, one describes the positive energy solutions of the Dirac equation in the medium as

$$u^*(p,\lambda) = \left(\frac{E_p^* + m^*}{2m^*}\right)^{1/2} \begin{pmatrix} \mathbf{1} \\ \frac{\sigma\cdot\vec{p}}{E_p^*+m^*} \end{pmatrix} \chi_\lambda \,, \tag{1}$$

where the effective mass, m^*, is defined as $m^* = m + U_S$, with U_S an attractive scalar potential. It can be shown that both the description of a single-nucleon via Eq. (1) and the evaluation of the Z-diagram, Fig. 2, generate a repulsive effect on the energy/particle in symmetric nuclear matter which depends on the density approximately as

$$\Delta E \propto \left(\frac{\rho}{\rho_0}\right)^{8/3} \,, \tag{2}$$

and provides the saturating mechanism missing from conventional Brueckner calculations. (Alternatively, TBF are used along with the BHF method in order to achieve a similar result.) Brown showed that the bulk of the desired effect can be obtained as a lowest order (in p^2/m) relativistic correction to the single-particle propagation [6].

Of course, other, more popular, three-body forces (not included in DBHF) need to be addressed as well. Figure 3 shows the TBF that is included in essentially all TBF models, regardless other components; it is the Fujita-Miyazawa TBF [7]. With the addition of contributions from πN S-waves, one ends up with the well-known Tucson-Melbourne TBF [8]. The microscopic TBF of Ref. [9] include contributions from excitations of the Roper resonance (P_{11} isobar) as well.

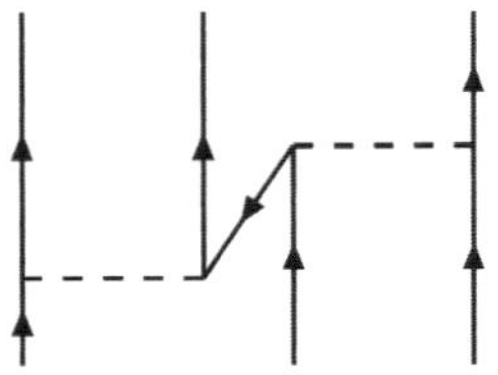

Figure 2. Three-body force due to virtual pair excitation. Conventions as in the previous figure.

Figure 3. Left: three-body force arising from Δ-isobar excitation (thick line). Right: two-meson exchange contribution to the NN interaction involving Δ-isobar excitation.

Now, if diagrams such as the one shown on the left-hand side of Fig. 3 are included, consistency requires that medium modifications at the corresponding two-body level are also included, that is, the diagram on the right-hand side of Fig. 3 should be present and properly medium modified. Large cancellations then take place, a fact that was brought up a long time ago [10]. When the two-body sector is handled via OBE diagrams, the two-pion exchange is effectively incorporated through the σ "meson", which cannot generate the (large) medium effects (dispersion and Pauli blocking on Δ intermediate states) required by the consistency arguments presented above. Thus, caution needs to be exercised when applying TBF in a particular space (of nucleons only, or nucleons and Δ's).

In closing this section, we highlight the most important aspect of the "*ab initio*" approach: namely, the only free parameters of the model (the parameters of the NN potential) are determined by the fit to the free-space NN data and never readjusted in the medium. In other words, the model parameters are tightly constrained and the calculation in the medium is parameter free.

3. The Symmetry Energy and Related Observables

Here we will focus specifically on the symmetry energy and its impact on the structure of neutron-rich nuclei, in particular the neutron skin thickness.

From the experimental side intense effort is presently going on to obtain reliable empirical information for the less known aspects of the EoS. Heavy-ion reactions are a popular way to seek constraints on the symmetry energy, through analyses of observables that are sensitive to the pressure gradient between nuclear and neutron matter. Concerning the lower densities, isospin-sensitive observables can also be identified among the properties of normal nuclei. The neutron skin of neutron-rich nuclei is a powerful isovector observable, being sensitive to the slope of the symmetry energy, which determines to which extent neutrons are "pushed out" to form the skin.

It is our purpose here to systematically examine and discuss the symmetry energy properties in microscopic models and the corresponding neutron skin predictions. We will take the skin of ^{208}Pb as our representative isovector "observable".

Typically, predictions of the symmetry energy at saturation density encountered in the literature are in reasonable agreement with one another, ranging approximately from 26 to 35 MeV. On the other hand, the slope of the symmetry energy is more model dependent and less constrained. This is seen through the *symmetry pressure*, defined as

$$L = 3\rho_0\Big(\frac{\partial E_{sym}(\rho)}{\partial \rho}\Big)_{\rho_0} \approx 3\rho_0\Big(\frac{\partial e_{n.m.}(\rho)}{\partial \rho}\Big)_{\rho_0} . \tag{3}$$

Thus, L is sensitive to the gradient of the energy per particle in neutron matter ($e_{n.m.}$). As to be expected on physical grounds, the neutron skin, $S = \sqrt{< r_n^2 >} - \sqrt{< r_p^2 >}$, is highly sensitive to the same gradient.

Values of L are reported to range from -50 to 100 MeV as seen, for instance, through the numerous parametrizations of Skyrme interactions, all chosen to fit the binding energies and the charge radii of a large number of nuclei. Heavy-ion data impose boundaries for L at 85 ± 25 MeV, with more stringent constraints being presently extracted [12].

Such phenomenological studies are extremely useful, but, ultimately, they must be compared with *ab initio* approaches in order to get true physical insight. The direct connection with the underlying two-body potential will then facilitate the physical understanding, when combined with reliable constraints.

As explained earlier, the DBHF model does not include three-body forces explicitely, but effectively incorporates the class of TBF originating from the presence of nucleons and antinucleons (the "Z-diagrams" in Fig. 2), see previous discussion.

As the other main input of our comparison, we will take the EoS's from the microscopic approach of Ref. [9]. There (and in previous work by the same authors), the Brueckner-Hartree-Fock (BHF) formalism is employed along with microscopic three-body forces. In particular, in Ref. [13] the meson-exchange TBFs are constructed applying the same parameters as used in the corresponding nucleon-nucleon (NN) potentials, which are: Argonne V18 [14] (V18), Bonn B [4] (BOB), Nijmegen 93 [15] (N93). The popular (but phenomenological) Urbana TBF [16] (UIX) is also utilized in Ref. [13]. Convenient parametrizations in terms of simple analytic functions are given in all cases and we will use those to generate the various EoS's. We will refer to this approach, generally, as "BHF + TBF".

In Fig. 4, we display Idaho DBHF predictions for the symmetry energy, solid black curve, along with those from V18, BOB, UIX, and N93. All values of the symmetry energy at the respective saturation densities are between 29 and 34 MeV. A larger spreading is seen in the L parameter, see left panel of Fig. 5, where the values range from about 70 to 106 MeV. The respective neutron skin predictions are shown on the vertical axis.

We notice that all BHF+TBF models predict larger L, and thus larger neutron skins, as compared with DBHF, corresponding to a faster growth of the energy per particle in neutron matter relative to symmetric matter. This can be seen from Fig. 4, especially for the higher densities. The present calculations reveal that there are more subtle, but significant differences at low to medium densities as well.

The next term in the expansion of the symmetry energy is the K_{sym} parameter,

$$K_{sym} = 9\rho_0^2\Big(\frac{\partial^2 E_{sym}(\rho)}{\partial \rho^2}\Big)_{\rho_0} , \tag{4}$$

which is a measure for the curvature of the symmetry energy. The neutron skin *vs.* K_{sym} is shown in the right panel of Fig. 5 for the various models. Although the values of K_{sym} appear

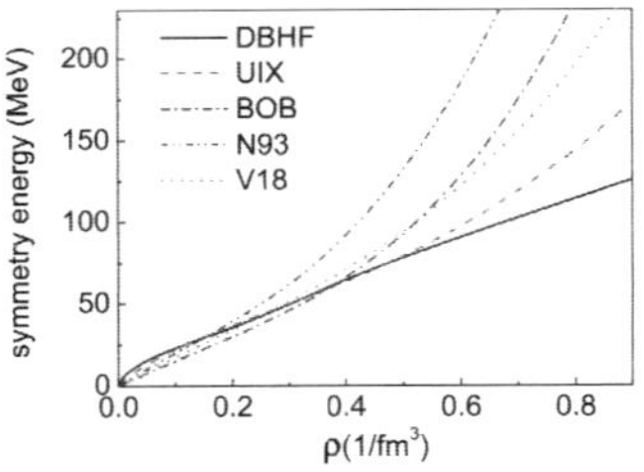

Figure 4. Predictions for the symmetry energy from DBHF and various "BHF + TBF" models considered in the text.

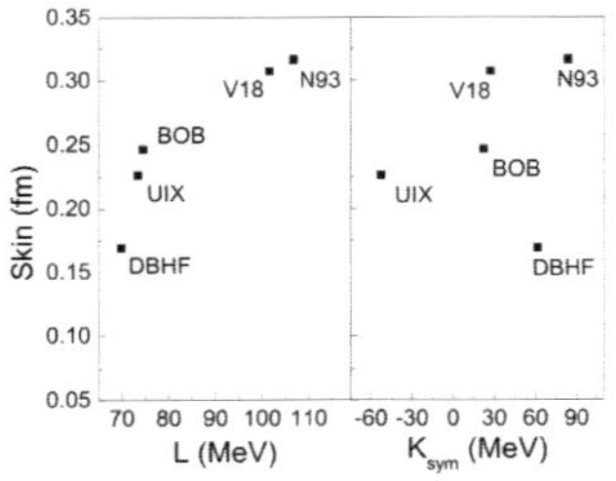

Figure 5. Left panel: neutron skin of ^{208}Pb *vs.* the symmetry pressure for the models considered in the text; Right panel: neutron skin of ^{208}Pb *vs.* the curvature of the symmetry energy, K_{sym}, for the same models.

more spread out, the large negative values obtained with some of the parametrizations of the Skyrme model are not present. Those large negative values (as low as -600 MeV) produced by Skyrme models indicate a strongly downward curvature already at low to medium densities. We also notice from Fig. 5 (right panel) that no clear correlation can be identified between K_{sym} and the neutron skin.

To conclude this section: Although microscopic models do not display as much spread as phenomenological ones, a point that comes out clearly from the present study is that a measurement of the neutron skin of ^{208}Pb with an accuracy of 0.05 fm, which is expected from the electroweak program at the Jefferson Laboratory [17], should definitely be able to discriminate among the EoS's from these microscopic models.

4. Comparing neutron star predictions from microscopic models

As constraints from nuclear physics and/or astrophysics promise to become more stringent, it is important to understand and compare how the nature of the various predictions is related to the features of each model. In microscopic approaches, the tight connection with the underlying forces facilitates physical interpretation of the predictions in terms of the characteristics of the nuclear force and its behavior in the medium. Motivated by these considerations, in this section we calculate the mass-radius relation for static neutron stars, using equations of state based on different microscopic models. These will be the same as those used in the previous section to examine symmetry energy and neutron skin predictions, namely BOB, V18, N93, and UIX, along with Idaho DBHF. There, we compared the predictions of the neutron skin in ^{208}Pb by these models and correlated them with differences in the slope of the symmetry energy. Model differences become larger at high-density and will naturally impact on neutron star predictions.

At subnuclear densities all the EoS considered here are joined with the crustal equations of state from Harrison and Wheeler [18] and Negele and Vautherin [19].

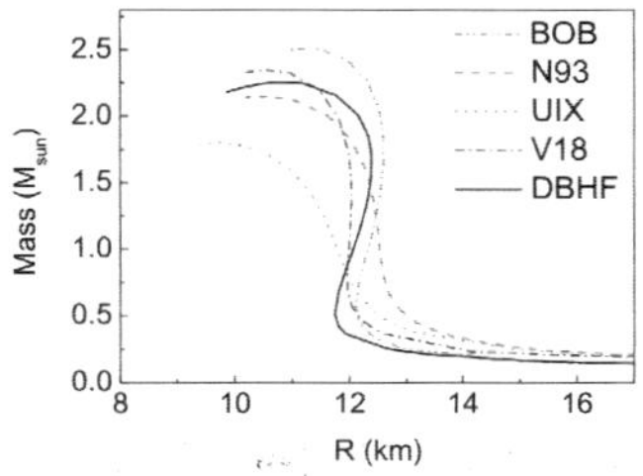

Figure 6. Static neutron star mass-radius relation for the models considered in the text.

In Fig. 6, we show the mass-radius relation for a sequence of static neutron stars as predicted by the various models considered previously.

All models besides DBHF share the same many-body approach (BHF+TBF) but differ in the two-body potential and TBF employed. The resulting differences can be much larger than those originating from the use of different many-body approaches. This can be seen by comparing the DBHF and BOB curves, both employing the Bonn B interaction (although in the latter case the non-reltivistic, r-space version of the potential is adopted). Overall, the maximum masses range from $1.8M_{\odot}$ (UIX) to $2.5M_{\odot}$ (BOB). Radii are less sensitive to the EoS and range between 10 and 12 km for all models under consideration, DBHF or BHF+TBF. Concerning consistency with present constraints, an initial observation of a neutron-star-white dwarf binary system suggested a neutron star mass (PSR J0751+1807) of $2.1\pm0.2M_{\odot}$ [20]. Such observation would imply a considerable constraint on the high-density behavior of the EoS. On the other hand, a dramatically reduced value of $1.3\pm0.2M_{\odot}$ was later reported by Nice, which does not invalidate any of the theoretical models under consideration.

Acknowledgments
Support from the U.S. Department of Energy is gratefully acknowledged.

References

[1] Alonso D and Sammarruca F 2003 *Phys. Rev.* C **67** 054301
[2] Sammarruca F 2010 Int. J. Mod. Phys. **19** 1259.
[3] Weinberg S 1979 *Physica* **96A** 327; 1990 *Phys. Lett.* **B251** 288
[4] Machleidt R 1989 *Adv. Nucl. Phys.* **19** 189
[5] Brown G E 1987 *"Mesons in Nuclei"* vol I ed M Rho and D H Wilkinson (Amsterdam: North-Holland) p 330
[6] Brown G E *et al.* 1987 *Comments Nucl. Part. Phys.* **17** 39
[7] Fujita J and Miyazawa H 1957 *Prog. Theor. Phys.* **17** 360
[8] Ellis R G, Coon S A and McKellar B H J 1985 *Nucl. Phys.* **A438** 631
[9] Li Z H *et al.* 2008 *Phys. Rev.* C **77** 034316
[10] Dickhoff W H, Faessler A and Müther H 1982, *Nucl. Phys.* **A389** 492
[11] Piekarewicz J 2006 *Preprint* nucl-th/0607039
[12] Tsang M B *et al.* 2009 Phys. Rev. Lett. **102** 12701
[13] Li Z H and Schulze H J 2008 *Phys. Rev.* C **78** 028801
[14] Wiringa R B, Stocks V G J and Schiavilla R 1995 *Phys. Rev.* C **51** 38
[15] Stocks V G J, Klomp R A M, Terheggen C P F and de Swart J J 1994 *Phys. Rev.* C **49** 2950
[16] Pieper S C, Pandharipande V R, Wiringa R B and Carlson J 2001 *Phys. Rev.* C **64** 014001
[17] Horowitz C J, Pollock S J, Souder P A and Michaels R 2001 *Phys. Rev.* C **63** 025501
[18] Harrison B K and Wheeler J A 1965 *Gravitation Theory and Gravitational Collapse* ed B K Harrison, K S Thorne, M Wakano and J A Wheeler (Chicago: University of Chicago Press) pp 1-177
[19] Negele J W and Vautherin D 1973 *Nucl. Phys.* **A178** 123
[20] Nice D J *et al.* 2005 *Astrophys. J.* **634** 1242

Shell-model study of exotic Sn isotopes with a realistic effective interaction

Shell-model study of exotic Sn isotopes with a realistic effective interaction

A Covello[1,2], L Coraggio[2], A Gargano[2] and N Itaco[1,2]
[1]Dipartimento di Scienze Fisiche, Università di Napoli Federico II, Complesso Universitario di Monte S. Angelo, I-80126 Napoli, Italy
[2]Istituto Nazionale di Fisica Nucleare, Complesso Universitario di Monte S. Angelo, I-80126 Napoli, Italy

E-mail: covello@na.infn.it

Abstract. We report on a shell-model study of Sn isotopes beyond $N = 82$ employing a realistic effective interaction derived from the CD-Bonn nucleon-nucleon potential renormalized through use of the $V_{\rm low-k}$ approach. At present, the most exotic Sn isotope for which some experimental information exists is ^{134}Sn with an N/Z ratio of 1.68. It is the aim of our study to compare the results of our calculations with the available experimental data and to make predictions for the neighboring heavier isotopes which may be within reach of the next generation of radioactive ion beam facilities. The very good agreement between theory and experiment obtained for ^{134}Sn gives confidence in the predictive power of our realistic shell-model calculations.

1. Introduction

The study of Sn isotopes beyond $N = 82$ is currently the subject of great interest, as it allows to explore for possible changes of the shell-model structure when moving towards the neutron drip line. At present, however, some experimental information is only available for ^{134}Sn. Beyond this nucleus, which has an N/Z ratio of 1.68, lies a neutron-rich *terra incognita* which may become accessible to the next generation of radioactive ion beam facilities. In this perspective, it is challenging to make predictions which may stimulate, and provide guidance to, future experiments. In fact, over the past several years various shell-model studies have been performed [1, 2, 3, 4] predicting spectroscopic properties of hitherto unknown neutron-rich Sn isotopes. In these studies both empirical and realistic effective interactions have been used, the latter being based on the CD-Bonn nucleon-nucleon (NN) potential model [5]. In this connection, it should be mentioned that a main feature of the study of Ref. [2] is that the short-range repulsion of the bare potential is renormalized by use of the $V_{\rm low-k}$ approach [6].

A peculiar property of ^{134}Sn is the position of its first 2^+ state which, lying at 726 keV excitation energy, is the lowest first-excited 2^+ level observed in a semi-magic even-even nucleus over the whole chart of nuclides. This feature as well as the other observed excited states - 4^+, 6^+, and 8^+ - have been successfully reproduced by the realistic calculations of [2]. In that study a very good agreement was also obtained with the experimental $B(E2; 0^+ \to 2_1^+)$ value measured via Coulomb excitation of a radioactive ^{134}Sn beam [7].

Very recently, two sophisticated experimental studies [8, 9] have provided new information of great relevance to the study of Sn isotopes beyond $N = 82$. More precisely, in [8] a high-precision Penning trap mass measurement has revealed a 0.5 MeV discrepancy with respect to previous Q_β measurements. This finding provides clear evidence of the robustness of the $N = 82$ shell closure. In [9], the magic nature of ^{132}Sn has been explored through the study of the single-particle states in ^{133}Sn populated by a (d,p) reaction in inverse kinematics. This study has evidenced the purity of the single-neutron excitations in ^{133}Sn and has identified a strong candidate for the $2p_{1/2}$ single-particle state at 1.363 MeV excitation energy. This is about 300 keV lower than the previous accepted value [10].

Based on the above, we have found it interesting to revisit our previous realistic shell-model calculations for ^{134}Sn and extend them to the heavier isotopes up to $A = 140$. In this paper, we first give a brief description of the theoretical framework in which our calculations are performed and then present and discuss our results which are, for the most part, predictions for possible future experiments. A short summary is given in the last section.

2. Outline of calculations

We have performed shell-model calculations for even tin isotopes beyond the $N = 82$ shell closure, namely, from ^{134}Sn to ^{140}Sn. We have taken doubly magic ^{132}Sn as closed core and let the valence neutrons occupy the six levels $0h_{9/2}$, $1f_{7/2}$, $1f_{5/2}$, $2p_{3/2}$, $2p_{1/2}$, and $0i_{13/2}$ of the $82-126$ shell. Our adopted single-particle (SP) energies relative to the $1f_{7/2}$ level are (in MeV): $\epsilon_{p_{3/2}} = 0.854$, $\epsilon_{p_{1/2}} = 1.363$, $\epsilon_{h_{9/2}} = 1.561$, $\epsilon_{f_{5/2}} = 2.005$, and $\epsilon_{i_{13/2}} = 2.694$. These values are taken from the experimental spectrum of ^{133}Sn [9, 11], with the exception of the $0i_{13/2}$ level which was determined from the position of the 2.434 MeV level in ^{134}Sb assumed to be a 10^+ state of $\pi g_{7/2}\nu i_{13/2}$ nature. As for the $2p_{1/2}$ level, we use the energy reported in the recent experiment of Ref. [9]. It is worth emphasizing that in this experiment the spectroscopic factors of the four levels $1f_{7/2}$, $2p_{3/2}$, $2p_{1/2}$, $1f_{5/2}$ were extracted, which provide a clear confirmation of their SP nature. The absolute energy of the $1f_{7/2}$ level relative to ^{132}Sn was placed at -2.371 MeV, as resulting from the new mass measurement of ^{133}Sn [8].

The shell-model effective interaction has been derived within the framework of perturbation theory [12] starting, as mentioned in the Introduction, from the CD-Bonn NN potential [5] renormalized by way of the so-called $V_{\text{low-k}}$ approach [6]. More precisely, we start by deriving $V_{\text{low-k}}$ with a cutoff momentum of $\Lambda = 2.2$ fm^{-1}. Then, using this potential plus the Coulomb force for protons, we calculate the two-body matrix elements of the effective interaction by means of the $\hat{Q}$-box folded-diagram expansion, with the $\hat{Q}$-box including all diagrams up to second order in the interaction. These diagrams are computed within the harmonic-oscillator basis using intermediate states composed of all possible hole states and particle states restricted to the five proton and neutron shells above the Fermi surface. The oscillator parameter is 7.88 MeV for the $A = 132$ region, as obtained from the expression $\hbar\omega = 45A^{-1/3} - 25A^{-2/3}$. The calculations have been performed by using the OSLO shell-model code [13].

3. Results

We start this section by discussing our results for ^{134}Sn, the only tin isotope beyond $N = 82$ for which some experimental information is available. In figure 1 all the observed levels [11] are reported and compared with the calculated ones up to about 2.5 MeV excitation energy. The experimental spectrum consists of a group of low-lying levels separated by an energy gap of about 1.5 MeV from the highest observed level which has $J^\pi = 8^+$. The former correspond to the four calculated members of the $\nu(f_{7/2})^2$ multiplet while the latter to the maximum-aligned state of the $\nu f_{7/2}h_{9/2}$ configuration and all energies are very well reproduced by the theory. In the energy interval between 1.1 and 2.5 MeV our calculation predicts the existence of other states, namely the members of the $\nu f_{7/2}p_{3/2}$ and $\nu f_{7/2}p_{1/2}$ multiplets, and of the 0^+ state of the $\nu(p_{3/2})^2$ configuration. It is worth noting that in our previous calculation [2] the 3_2^+ and 4_3^+

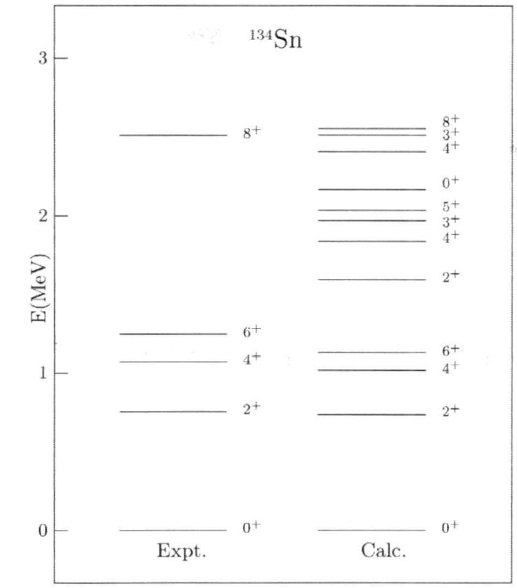

Figure 1. Experimental and calculated spectrum of ^{134}Sn.

states were predicted to lie above the 8^+ state, their downshift in energy being directly related to the new position of the $p_{1/2}$ SP level.

We have also calculated some electromagnetic properties and our predicted values for the first four excited states are reported in table 1, where the two experimental transition rates presently available [7, 11] are also shown. Our calculations have been performed using an effective neutron charge of $0.7e$ [1], which leads to $B(E2)$ values for both the $6^+ \to 4^+$ and $2^+ \to 0^+$ transitions quite close to the experimental ones.

Table 1. Electromagnetic transition rates (in W.u.), quadrupole (in efm^2) and magnetic moments (in nm) in ^{134}Sn.

	Calc	Expt
$B(E2; 2_1^+ \to 0_1^+)$	1.64	1.42 ± 0.20
$B(E2; 4_1^+ \to 2_1^+)$	1.66	
$B(E2; 6_1^+ \to 4_1^+)$	0.82	0.89 ± 0.17
$B(E2; 2_2^+ \to 0_1^+)$	0.35	
$B(E2; 2_2^+ \to 2_1^+)$	2.93	
$B(E2; 2_2^+ \to 4_1^+)$	0.23	
$B(M1; 2_2^+ \to 2_1^+)$	0.02	
$Q(2_1^+)$	-1.6	
$Q(2_2^+)$	-2.8	
$\mu(2_1^+)$	-0.57	
$\mu(2_2^+)$	-0.25	

We hope that the above predictions for ^{134}Sn may provide guidance to future experiments aiming at the identification of the missing states above the yrast 6^+ state as well as at the measurement of other properties, which may certainly give better insight into the structure of this exotic nucleus.

In this context, it should be mentioned that several interesting questions have been posed by the new, although still scarce, experimental data which have become available for some nuclei in the ^{132}Sn region. In particular, special attention has been focused on the properties of the yrast 2^+ states of tin and tellurium isotopes, whose excitation energy shows an asymmetric behavior with respect to $N = 82$. In fact, in ^{134}Sn and ^{136}Te the 2_1^+ energy undergoes a significant drop as compared to the $N = 80$ isotopes, which was traced to a reduction of the neutron pairing above the $N = 82$ shell [14, 15].

With our realistic effective interaction we are able, as shown above, to reproduce with good accuracy the energy of the yrast 2^+ state and its $B(E2)$ transition rate in ^{134}Sn. This was also the case for ^{136}Te (see [16]) whose 2^+ state is described as the corresponding state in ^{134}Sn with the two additional protons coupled to zero angular momentum. As a matter of fact, the 2^+ state, as well as the other three lowest-lying states in ^{134}Sn, are characterized by a very weak configuration mixing. Therefore the small $0^+ - 2^+$ spacing reflects the small difference between the $< (f_{7/2})^2 | V_{\text{eff}} | (f_{7/2})^2 >_{0^+}$ and $< (f_{7/2})^2 | V_{\text{eff}} | (f_{7/2})^2 >_{2^+}$ matrix elements, which turns out to be about 0.4 MeV. This result is due to the negligible contribution [17] of the one particle-one hole excitations in the effective interaction of two neutrons above the $N = 82$ shell, which are instead responsible for the increase of pairing below this shell. The weakness of the cross-shell interaction for ^{132}Sn as compared with other lighter closed cores was also pointed out in Ref. [3], where it was noted that the excitation energy of the 3^- state in ^{132}Sn is only 0.5 MeV smaller than the shell gap for neutrons.

We have investigated the evolution of the yrast 2^+ state and of the other $J \neq 0$ members of the lowest-lying multiplet when adding neutron pairs. In figure 2 we report the excitation energies of the 2_1^+, 4_1^+, and 6_1^+ states as a function of A up to ^{140}Sn. We see that the curves are almost flat, with a slight increase at $N = 90$ for the 2_1^+ and 4_1^+ which becomes more significant for the 6^+ state.

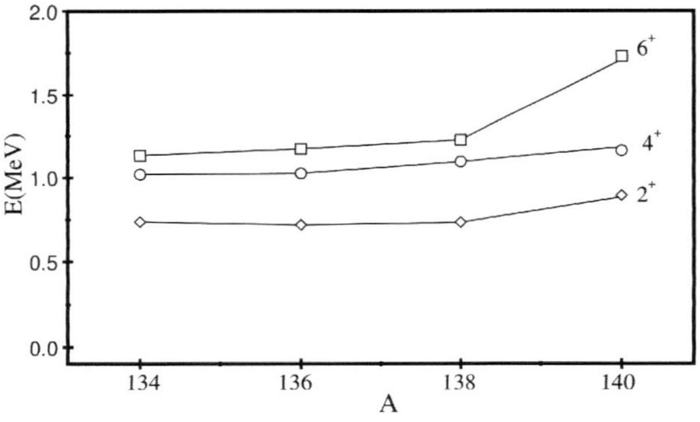

Figure 2. Calculated excitation energies of the yrast 2^+, 4^+, and 6^+ states in tin isotopes with $A = 134$, 136, 138 and 140.

This result is similar to that found in [3], the main difference being that our 4_1^+ and 6_1^+ curves lie about 100-300 MeV below those of [3], while is in contrast to the conclusion of the recent work of [4] claiming for a shell closure at $N = 90$. The results of [4] predict an upshift of the 2^+ state in ^{140}Sn with respect to the lighter isotopes when using an empirical interaction (SMPN) as well as a realistic interaction with three-body monopole corrections (CWG3M). This upshift was related to the filling of the $f_{7/2}$ orbit at $N = 90$ as a consequence of a significant increase

Table 2. Binding energies and one-neutron separation energies (in MeV) in tin isotopes with $A = 134$, 136, 138, and 140.

	^{134}Sn	^{136}Sn	^{138}Sn	^{140}Sn
BE Calc	5.92	11.83	17.68	23.41
BE Expt	5.916 ± 0.150			
S_n Calc	3.55	3.55	3.53	3.50
S_n Expt	3.545 ± 0.152			

in the difference between the effective single-particle energies of the $f_{7/2}$ and $p_{3/2}$ orbits which, starting from 0.854 keV, becomes larger than 2 MeV at $A = 140$. As can be seen in figure 3, our calculations do not predict such an increase. The energy difference between the two orbits remains in fact almost constant and even decreases slightly by about 100 keV when moving from $N = 82$ to $N = 90$. As regards the 6^+ state, we find that its wave function in ^{140}Sn is dominated by the $\nu(f_{7/2})^6(p_{3/2})^2$ configuration, which explains its increasing in energy.

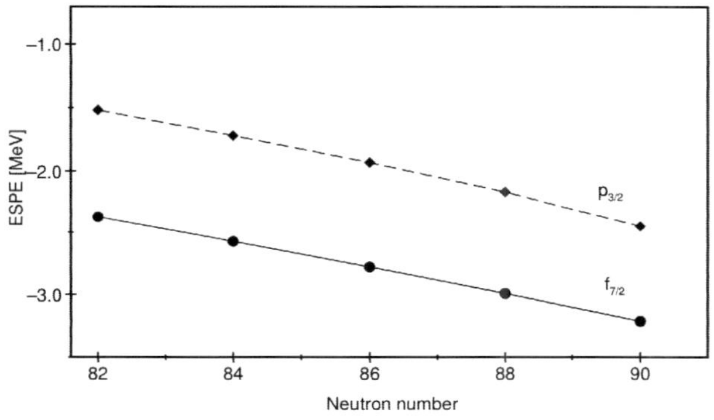

Figure 3. Effective single-particle energies of the $f_{7/2}$ and $p_{3/2}$ levels from $N = 82$ to $N = 90$.

In concluding this section, we discuss binding (relative to ^{132}Sn) and one-neutron separation energies of the even Sn isotopes. The calculated values are reported in table 2, where the available experimental values for ^{134}Sn are also shown. As mentioned in the Introduction, a new measurement of the binding energy of ^{134}Sn has revealed a 0.5 MeV deviation from the previous observed value. With the new mass value the maximum of the neutron-shell gap for N=82 is at $Z = 50$, instead of $Z = 51$ as established by the old measurement. We find an excellent agreement between theory and experiment, which shows the reliability of the $J = 0$ matrix elements of our effective interaction. This confirms the weakening of pairing for two neutrons beyond the $N = 82$ shell. As regards the heavier Sn isotopes, we see that the binding energy keeps increasing with increasing mass number, while the one-neutron separation energy remains practically constant. Our calculations, in line with those of [3] and the estimations of mean field calculations, seem to indicate that in this region we are still quite far from the neutron drip line.

4. Summary

We have presented here the results of a shell-model study of exotic even mass Sn isotopes from A= 134 to A= 140. In our calculations we have employed a realistic low-momentum effective interaction derived from the CD-Bonn NN potential without using any adjustable parameter.

The heaviest Sn isotope for which there is some, albeit scarce, experimental information is ^{134}Sn, so most of our results are predictive in nature. We have shown that the available experimental data are very well reproduced by our calculations. This gives confidence in their predictive power, which we hope can be verified by experiment in the not too distant future.

References

[1] Coraggio L, Covello A, Gargano A and Itaco N 2002 *Phys. Rev.* C **65** 051306(R)
[2] Covello A, Coraggio L, Gargano A and Itaco N 2007 *Eur. Phys. J ST* **150** 93
[3] Kartamyshev M P, Engeland T, Hjorth-Jensen M and Osnes E 2007 *Phys. Rev.* C **76** 024313
[4] Sarkar S and Sarkar M S 2010 *Phys. Rev.* C **81** 064328 and references therein
[5] Machleidt R 2001 *Phys. Rev.* C **63** 024001
[6] Bogner S, Kuo T T S, Coraggio L, Covello A and Itaco N 2002 *Phys. Rev.* C **65** 051301(R)
[7] Beene J R *et al* 2004 *Nucl. Phys.* A **746** 471c
[8] Dworschak M *et al* 2008 *Phys. Rev. Lett.* **100** 072501
[9] Jones K L *et al* 2010 *Nature* **465** 454
[10] Hoff P *et al Phys. Rev. Lett.* **77** 1020
[11] Data extracted using the NNDC On-line Data Service from the ENSDF database, file revised as of October 4, 2010
[12] Coraggio L, Covello A, Gargano A, Itaco N and Kuo T T S 2009 *Prog. Part. Nucl. Phys.* **62** 135 and references therein
[13] Engeland T the Oslo shell-model code 1991-2006, unpublished
[14] Terasaki J, Engel J, Nazarewicz W and Stoitsov M, 2002 *Phys. Rev.* C **66** 054313
[15] Shimizu N, Otsuka T, Mizusaki T and Honma M 2004 *Phys. Rev.* C **70** 054313
[16] Covello A, Coraggio L, Gargano A and Itaco N 2007 *Prog. Part. Nucl. Phys.* **59** 401
[17] Coraggio L, Covello A, Gargano A and Itaco N 2009 *Phys. Rev.* C **80** 021305(R)

Tensor force in effective interaction of nuclear force

2011 J. Phys.: Conf. Ser. 267 012020

(http://iopscience.iop.org/1742-6596/267/1/012020)

Tensor force in effective interaction of nuclear force

Naofumi Tsunoda, Takaharu Otsuka, Koshiroh Tsukiyama
Department of physics, the University of Tokyo, 7-3-1 Hongo, Bunkyo-ku, Tokyo, Japan

Morten Hjorth-Jensen
Department of Physics and Center of Mathematics for Applications, University of Oslo, N-0316 Oslo, Norway

E-mail: `tsunoda@nt.phys.s.u-tokyo.ac.jp`

Abstract. The tensor force in effective interactions is discussed. We show that the tensor force in low-momentum effective interactions $V_{\mathrm{low}k}$ and effective interactions for shell model studies are almost identical to that of the bare nucleon-nucleon interaction. We use realistic nuclear forces and microscopic theories of effective interactions. We calculate $V_{\mathrm{low}k}$ by the technique of similarity transformation. The effective interaction for the nuclear shell model is calculated by folded diagram theory. The monopole contribution from the tensor force is analyzed by a spin-tensor decomposition.

1. Introduction

Radioactive beam accelerator facilities make it possible to perform experiments that test the limits of nuclear stability by studying exotic nuclei. The nuclear shell model has been one of the most successful models for describing nuclei. However, recent studies of neutron rich nuclei indicate that the so-called nuclear magic numbers may only be valid close to stable nuclei. Otsuka *et al.* [1] showed that the tensor force varies the spin-orbit splitting in exotic nuclei, where the proton number and the neutron number are very different. For instance, if neutrons occupy the orbit j', the effective single particle energy of protons in the orbit j is described as follows

$$\Delta\epsilon_p(j) = \frac{1}{2}\left(V_{jj'}^{T=0} + V_{jj'}^{T=1}\right) n_n(j'), \tag{1}$$

where $\Delta\epsilon_p(j)$ represents the change in the effective single particle energy of protons in orbit j and $n_n(j')$ is the occupation number of neutrons in orbit j'. Here $V_{jj'}$ represents the monopole part of the Hamiltonian between two orbits j and j', which is defined as

$$V_{j,j'}^{T} = \frac{\sum_J (2J+1)\langle jj'|V|jj'\rangle_{JT}}{\sum_J (2J+1)}. \tag{2}$$

This is nothing but the angular averaged interaction between two orbits j and j'. The tensor-force contribution always has different sign between a pair of spin-orbit partners. For example, $V_{j_>j'}$ and $V_{j_<j'}$ have opposite sign, where $j_>$ and $j_<$ are the spin-orbit partners, that is,

$j_> = l + 1/2$ and $j_< = l - 1/2$. This effect breaks the conventional magic numbers believed to be universal during the last 50 years.

In Ref. [1], the tensor force in effective interactions for the nuclear shell model is taken as the exchange of $\pi + \rho$ mesons for simplicity. This is qualitatively the same as the tensor force in the bare realistic nuclear force. The tensor force of realistic interactions is fitted to reproduce phase shifts of NN scattering data and ground state properties of the deuteron. However, the effective interaction for the shell model is normally different from the bare realistic nuclear force. First, the free nuclear force exhibits a strong coupling between low-momentum and high-momentum degrees of freedom. This coupling is due to the short range properties of the interaction. For example, the Argonne interactions are defined by using a local operator form in coordinate space and has a strong short range repulsion [2]. Furthermore, the effective interaction for the shell model is defined in a restricted configuration space, normally called the model space. Therefore, it is far from trivial that the tensor force in an effective shell-model interaction can be simply considered to arise from the explicit exchange of $\pi + \rho$ mesons only, as seen in models for the nuclear force. To understand the validity of this assumption, theoretical studies based on realistic nuclear forces and the microscopic theory of effective interactions are needed.

In this work, we construct an effective interaction for the shell model starting from realistic nuclear forces and use perturbative many-body methods for deriving the effective interaction. Then, we analyze the tensor component to see if the tensor force survives under the renormalization.

2. How to construct an effective interaction from realistic nuclear forces

The original eigenvalue problem is

$$H|\Psi_i\rangle = E_i|\Psi_i\rangle. \tag{3}$$

We divide the Hilbert space into a model space, the so-called P-space and an excluded space, the Q-space for short. Here, P and Q are projection operators onto the P-space and the Q-space, respectively, i.e., $P^2 = P$, $Q^2 = Q$, $P + Q = \mathbf{1}$. The basic problem is to find an effective interaction $\tilde{H}$ that reproduces some of the eigenvalues of the original eigenvalue problem of Eq. (3)

$$P\tilde{H}P|\phi_i\rangle = E_i|\phi_i\rangle, \tag{4}$$

where $|\phi_i\rangle = P|\Psi_i\rangle$. This can be accomplished by the following similarity transformation,

$$\tilde{H} = e^{-\omega}He^{\omega} \qquad Q\omega P = \omega. \tag{5}$$

Then it follows immediately that $\omega^2 = \omega^3 = 0\ldots$ and $e^{\omega} = 1 + \omega$. If $P\tilde{H}P$ is an effective interaction, the decoupling condition $Q\tilde{H}P = 0$ must be fulfilled.

Therefore, the central problem is now reduced to how to obtain ω which fulfills the decoupling condition. The physical meaning of ω can be understood as a mapping from the P-space wavefunction to the corresponding Q-space component. In general, there is no unique solution of the decoupling equation and ω depends on the set of original eigenfunctions which are selected to reproduce the corresponding eigenvalues.

If one chooses a set of d eigenfunctions, the decoupling equation has the following formal solution [3]

$$\omega = \sum_{i=1}^{d} Q|\Psi_i\rangle\langle\tilde{\phi}_i|P, \tag{6}$$

where $\langle\tilde{\phi}_i|$ refers to the bi-orthogonal state of $|\phi_i\rangle$, which is defined to be $\langle\tilde{\phi}_i|\phi_j\rangle = \delta_{ij}$. The original eigenfunction $|\Psi\rangle$ is orthogonal because of the hermiticity of the Hamiltonian. However, its projection onto the P-space $|\phi\rangle = P|\Psi\rangle$ is not orthogonal in general.

If we take the P-space as the low-momentum space and the Q-space as the high-momentum space and choose d low-energy eigenstates, we will obtain a low-momentum interaction by the above procedure. We call this potential $V_{\mathrm{low}k}$ and the cutoff parameter Λ is defined as the boundary between the P-space and the Q-space [4, 5]. Using $V_{\mathrm{low}k}$ can be interpreted as rewriting our problem from $H = T - T_{\mathrm{CM}} + \sum V_{\mathrm{bare}}$ to $H = T - T_{\mathrm{CM}} + \sum V_{\mathrm{low}k}$. This means that we change our resolution of the problem and solve the problem not seeing high-momentum details. The problem becomes much easier to solve. In this work, we will neglect induced many-body force.

Nevertheless, if we need to solve a many-body problem, it is still difficult to solve the Schrödinger equation directly even if we use $V_{\mathrm{low}k}$. For example, if one wants to obtain the energy levels of ^{18}O with 18 interacting particles, solving the whole 18-body problem is too large a calculation to perform and we need to calculate a two-body effective interaction defined for a smaller space. A typical example is the sd-shell. However, to determine ω by Eq. (6) one needs the complete knowledge of the d true eigenfunctions $|\Psi_i\rangle$, which we do not know. Hence, what we want to do is to obtain the eigenvalues of the original problem without solving the original problem. In this case, one cannot know ω from Eq. (6) and have to obtain it by an alternative procedure. An efficient way to do this is the so-called $\hat{Q}-$box expansion [6].

If the unperturbed Hamiltonian is degenerate in the model space, that is, $H_0|\phi_i\rangle = E_0|\phi_i\rangle$, where $H = H_0 + H_1$, the effective interaction defined in P-space can be calculated by the following iterative formula

$$V_{\mathrm{eff}}^{(n)} = \hat{Q}(E_0) + \sum_m \hat{Q}_m(E_0)\{V_{\mathrm{eff}}^{(n-1)}\}^m \tag{7}$$

where

$$\hat{Q}(E_0) \equiv PH_1P + PH_1Q\frac{1}{E_0 - QHQ}QH_1P, \tag{8}$$

$$\hat{Q}_m(E_0) \equiv \frac{1}{m!}\frac{d^m\hat{Q}(E_0)}{dE_0^m}. \tag{9}$$

where the derivatives correspond to folded diagrams. By this prescription, we can obtain an effective interaction for the shell model, which gives equivalent results to the original eigenvalue problem within the range of approximation. Again, induced many-body forces are neglected.

3. Tensor force in low-momentum interaction $V_{\mathrm{low}k}$

As mentioned above, the tensor force in the bare realistic nuclear force mainly comes from $\pi + \rho$ meson exchange. In this section, we discuss the effect of the renormalization procedure on the tensor force. For this purpose, we have performed a spin-tensor decomposition of the $V_{\mathrm{low}k}$ interaction with various values of the cutoff Λ [7]. Figure 1 shows the monopole part of the tensor force in $V_{\mathrm{low}k}$ derived using the Argonne V8' (AV8') potential [2]. Except for very low cutoff values like $\Lambda = 1.0$ fm^{-1}, the monopole part of tensor force in $V_{\mathrm{low}k}$ has almost no cutoff dependence. A low value $\Lambda = 1.0$ fm^{-1} in momentum space corresponds to a distance in coordinate space of approximately 1.0 fm. Since the Compton wavelength of a pion is approximately 0.7 fm, a cutoff value $\Lambda = 1.0$ fm^{-1} is too low. With such a cutoff, the renormalization induces strong and non-negligible higher-body forces, which make the problem much more complicated. Therefore, this value is not appropriate for our purpose.

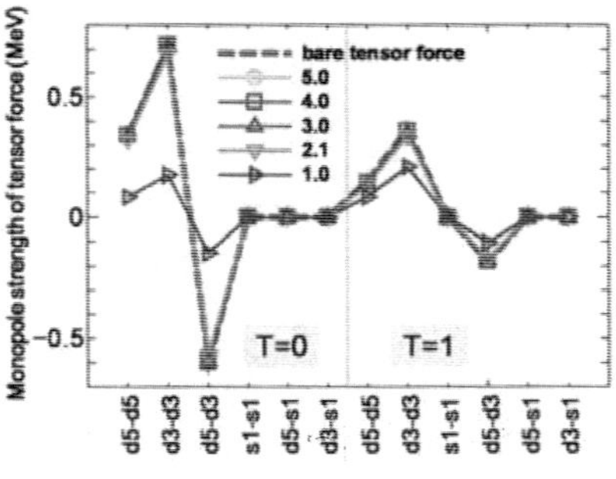

Figure 1. Monopole part of tensor force in low-momentum interaction $V_{\text{low}k}$. Cutoff parameter Λ of $V_{\text{low}k}$ is taken as from 1.0 fm^{-1} to 5.0 fm^{-1}.

We may say that the tensor force survives in a low-momentum interaction $V_{\text{low}k}$ with usual cutoff values, at least for its monopole part. Although we have only showed results for the *sd*-shell, similar conclusions are also reached for the *pf*-shell.

4. Effect of renormalization of short-range tensor force

In this section, we will focus on how the tensor force renormalization of the short range part of the interaction is incorporated into an effective interaction. In our calculation of low-momentum interaction $V_{\text{low}k}$, we consider a two-nucleon system. The tensor force plays a crucial role in the case of the deuteron. The deuteron has isospin $T = 0$ and a non-negligible ${}^3S_1 - {}^3D_1$ mixing. There is no experimental data which indicates an existence of excited states or any other bound state of deuteron. The mixing of ${}^3S_1 - {}^3D_1$ is due to the tensor force.

The Schrödinger equation for the deuteron can be written as the following coupled equations,

$$
\begin{aligned}
&-\frac{\hbar^2}{M}\frac{d^2u(r)}{dr^2} + V_C u(r) + \sqrt{8}V_T w(r) = E_d u(r), \\
&-\frac{\hbar^2}{M}\frac{d^2w(r)}{dr^2} + \left(\frac{6\hbar^2}{Mr^2} + V_C - 2V_T - 3V_{LS}\right) w(r) + \sqrt{8}V_T u(r) = E_d w(r),
\end{aligned} \tag{10}
$$

where $u(r)$ and $w(r)$ are the radial wave functions of the S-wave and the D-wave, respectively. Knowing the solution of Eq. (10), we integrate out the D-wave degrees of freedom to obtain the following effective central force

$$
\begin{aligned}
V_{\text{eff}}(r; {}^3S_1) &= V_C(r; {}^3S_1) + \Delta V_{\text{eff}}(r; {}^3S_1) \\
\Delta V_{\text{eff}}(r; {}^3S_1) &\equiv \sqrt{8}V_T(r)\frac{w(r)}{u(r)}.
\end{aligned} \tag{11}
$$

Here ΔV_{eff} is comparable to V_C in strength. This effective central force makes the deuteron bound. In this sense, the tensor force plays a crucial role in making the deuteron bound. This effect is at least a second order effect in terms of the tensor force, since both the initial and the final state have zero orbital angular momentum. To see how the tensor force is renormalized into a central force, we derive $V_{\text{low}k}$ starting from the full Argonne V8' potential (AV8' full) and tensor subtracted Argonne V8' (AV8' TS) potential. In Fig.2, the red line and the blue line represent the monopole part of $V_{\text{low}k}$ from AV8' full and AV8' TS, respectively, and the black line represents the bare AV8' potential. The difference between $V_{\text{low}k}$ from AV8' full and $V_{\text{low}k}$ from AV8' TS shows the effect of the renormalization of the short-range tensor force. The effect is strongly attractive in the $T = 0$ channel, which is consistent with the considerations from Eq. (11).

To understand this, consider the Schrödinger equation written as

$$
\begin{pmatrix} PHP & PHQ \\ QHP & QHQ \end{pmatrix} \begin{pmatrix} P|\Psi\rangle \\ Q|\Psi\rangle \end{pmatrix} = E \begin{pmatrix} P|\Psi\rangle \\ Q|\Psi\rangle \end{pmatrix}. \tag{12}
$$

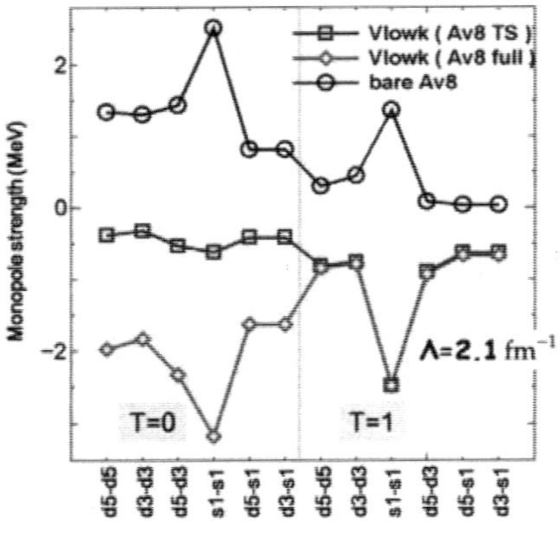

Figure 2. Effect of renormalization of short-range tensor force.

By integrating out the Q-space degrees of freedom, we obtain

$$PHP|\Psi\rangle + PHQ\frac{1}{E - QHQ}QHP|\Psi\rangle = EP|\Psi\rangle. \tag{13}$$

From this equation, we can consider an effective interaction defined in the P-space as

$$H_{\rm eff} = PHP + PHQ\frac{1}{E - QHQ}QHP. \tag{14}$$

Equation (14) is so general that we can choose any definition of the P-space and the Q-space. If we choose the P-space and the Q-space as $^3S_1$ and $^3D_1$ respectively, the resulting $H_{\rm eff}$ is exactly $V_{\rm eff}(r; {}^3S_1)$ of Eq. (11), that is the effective central force in the $^3S_1$ channel. On the other hand, if we choose P-space and Q-space as low-momentum and high-momentum space, we obtain the low-momentum interaction $V_{{\rm low}k}$.

5. Tensor force in effective interaction for the shell model

In this section we discuss the tensor force derived from the effective interaction for the shell model. We have calculated shell model effective interactions for the sd-shell and the pf-shell, by considering the $\hat{Q}-$box to second and third order, with folded diagram correction as explained in Sec. 2, starting from $V_{{\rm low}k}$.

The $\hat{Q}-$box is calculated by considering valence-linked and connected diagrams with unperturbed single particle energy of the harmonic oscillator. Folded diagrams are obtained by numerical derivatives of the $\hat{Q}-$box with respect to the starting energy, the sum of single particle energies of two valence particles, as given in Eq. (7).

Since the Q-space is defined as the complement of the P-space, the intermediate states should be taken up to infinitely high oscillator shells, in principle. As an approximation, we define a sufficiently large Q-space in the evaluation of the $\hat{Q}-$box. Here the $\hat{Q}-$box is calculated by $V_{{\rm low}k}$ with cutoff $\Lambda = 2.1$ fm^{-1}.

Figure 3 shows the monopole part of tensor force of the effective interaction for the sd-shell. We label this effective interaction as $V_{\rm eff}$. One can see again that the monopole part of tensor force of $V_{\rm eff}$ is fairly similar to that of the bare realistic nuclear force. These results are not trivial again, because they mean that the effects of the renormalization of the medium contributions and the truncation of the model space are not predominantly affecting the tensor force. The first-order $\hat{Q}-$box is just the $V_{{\rm low}k}$ interaction, whose tensor force is almost equal to that of the bare nuclear force as discussed already in Sec. 3. Therefore, these results mean that the monopole

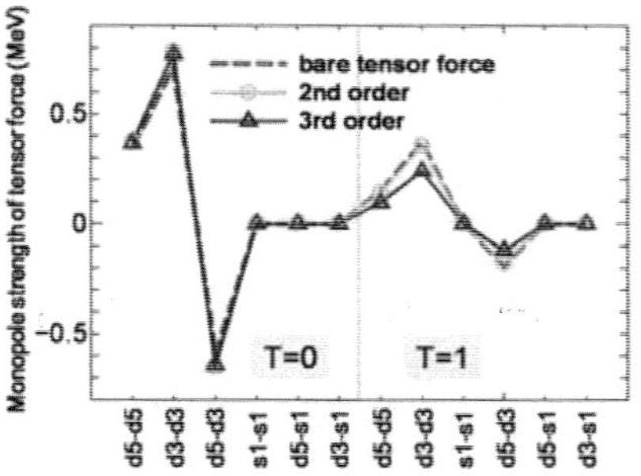

Figure 3. Monopole part of tensor force of the effective interaction for the shell model calculated with the $\hat{Q}-$box expansion.

part of the tensor force is dominated by the first order $\hat{Q}-$box term and the contributions from the other terms are comparatively small. Again, although we have only showed results from calculations in the *sd*-shell, we have obtained similar results for the *pf*-shell as well.

6. Conclusion

We conclude that the tensor force survives under both steps of the renormalization procedure, that is, the renormalization of the high-momentum component of realistic nuclear forces and the medium-renormalization needed to obtain an effective two-body interaction for a specific model space.

7. Acknowledgments

We are very grateful to R. Okamoto and H. Feldmeier for valuable discussions. This work is supported in part by Grant-in-Aid for Scientific Research (A) 20244022 and also by Grant-in-Aid for JSPS Fellows (No. 228635), and by the JSPS Core to Core program "International Research Network for Exotic Femto Systems" (EFES).

References

[1] Otsuka T *et al.* 2005 *Phys. Rev. Lett.* **95** 232502
[2] Wiringa R. B., Stoks V. G. J and Schiavilla R 1995 *Phys. Rev. C* **51** 38
Wiringa R. B and Pieper S. C 2002 *Phys. Rev. C* **89** 182501
[3] Suzuki K 1982 *Prog. Theor. Phys.* **68** 1627
[4] Bogner S. K, Kuo T. T. S, Coraggio L, Covello A and Itaco N 2002 *Phys. Rev. C* **65** 051301(R)
[5] Bogner S. K, Kuo T. T. S, and Schwenk A 2003 *Phys. Rep.* **386** 1
[6] Jensen M. H, Kuo T. T. S and Osnes E 1995 *Phys. Repts.* **261** 125
[7] Brown B. A and Wildenthal B. H, 1988 *Ann. Rev. Nucl. Part. Sci.* **38** 29

Realistic shell-model calculations for neutron-rich calcium isotopes

2011 J. Phys.: Conf. Ser. 267 012021

(http://iopscience.iop.org/1742-6596/267/1/012021)

Realistic shell-model calculations for neutron-rich calcium isotopes

L Coraggio[1], **A Covello**[1,2], **A Gargano**[1] **and N Itaco**[1,2]

[1]Istituto Nazionale di Fisica Nucleare and [2] Dipartimento di Scienze Fisiche, Università di Napoli Federico II,

Complesso Universitario di Monte S. Angelo, Via Cintia - I-80126 Napoli, Italy

E-mail: luigi.coraggio@na.infn.it

Abstract. We study the neutron-rich calcium isotopes performing shell-model calculations with a realistic effective interaction. This is derived from the CD-Bonn nucleon-nucleon potential renormalized by way of the $V_{\rm low-k}$ approach, considering ^{48}Ca as an inert core and including the neutron $0g_{9/2}$ orbital. We compare our results with experiment and with the results of a previous study where ^{40}Ca was assumed as a closed core and the standard $1p0f$ model space was employed. The calculated spectroscopic properties are in both cases in very good agreement with the available experimental data and enable a discussion on the role of the $g_{9/2}$ single-particle state in the heavy-mass Ca isotopes.

1. Introduction

Heavy-mass calcium isotopes ($N > 28$) have been recently the subject of a renewed experimental and theoretical attention, since they lie far from the stability valley, thus allowing to explore the evolution of the shell structure when approaching the neutron drip line.

This has resulted in a large number of experiments in this region [1–10] aiming to study the evolution of the single-particle (SP) orbitals. On the theoretical side, to describe the spectroscopic properties of these nuclei the shell model with various two-body effective interactions has been extensively used in recent years. The most frequently employed interactions are the KB3G [11], FPD6 [12], GXPF1 [13], and a new version of the latter dubbed GXPF1A [14]. The above interactions have all been derived by modifying the two-body matrix elements (TBME) of some realistic effective interactions in order to obtain a better description of the experimental data.

During the last years, however, realistic shell-model effective interactions without any empirical modifications have proved to be able to describe with remarkable accuracy the spectroscopic properties of nuclei in various mass regions [15–18].

On these grounds, we have recently performed a shell-model study of the Ca isotopic chain [19] employing an effective interaction derived from the CD-Bonn potential renormalized by way of the $V_{\rm low-k}$ approach [20, 21]. In that work we have chosen a model space spanned by the four neutron SP levels of the fp shell outside the doubly-closed ^{40}Ca core. This is a standard choice in most of the studies on neutron-rich Ca isotopes, and has therefore allowed a direct comparison between our results and those of other shell-model calculations.

In the present work, starting from the same $V_{\rm low-k}$ as in Ref. [19], we have derived a shell-model effective interaction for a model space spanned by the neutron SP levels $0f_{5/2}$, $1p_{3/2}$,

$1p_{1/2}$, and $0g_{9/2}$ located above the doubly-closed ^{48}Ca core. The aim of the present study is to compare the results of this calculation with those of Ref. [19], so as to try to clarify the role of the $0g_{9/2}$ orbital in heavy-mass calcium isotopes.

The paper is organized as follows. In section 2 we give a brief outline of our calculations and some details about the choice of the SP energies. Section 3 is devoted to the presentation of the results for calcium isotopes beyond ^{48}Ca up to A=56. In section 4 we make some concluding remarks.

2. Outline of calculations

Our shell-model effective interaction has been derived within the framework of perturbation theory [17] starting from the CD-Bonn NN potential [22]. The high-momentum repulsive components of the bare NN potential have been renormalized by employing the so-called $V_{\rm low-k}$ approach [20, 21], so as to obtain a smooth potential that preserves exactly the onshell properties of the original NN potential. Next, the TBME have been derived by way of the $\hat{Q}$-box plus folded-diagram method [17], the $\hat{Q}$-box being a collection of irreducible valence-linked Goldstone diagrams which we have calculated through third order in $V_{\rm low-k}$.

The effective interaction $V_{\rm eff}$ can be written in an operator form as

$$V_{\rm eff} = \hat{Q} - \hat{Q}' \int \hat{Q} + \hat{Q}' \int \hat{Q} \int \hat{Q} - \hat{Q}' \int \hat{Q} \int \hat{Q} \int \hat{Q} + \ \dots \ , \qquad (1)$$

where the integral sign represents a generalized folding operation, and $\hat{Q}'$ is obtained from $\hat{Q}$ by removing terms of first order in $V_{\rm low-k}$. The folded-diagram series is summed up to all orders using the Lee-Suzuki iteration method [23].

We have derived our shell-model effective interaction for the neutron-neutron, proton-proton, and proton-neutron channels using a model space spanned by the four neutron SP levels $0f_{5/2}$, $1p_{3/2}$, $1p_{1/2}$, $0g_{9/2}$, and by the sole proton SP level $0f_{7/2}$, all of them located above the doubly-closed ^{48}Ca core. The complete list of the matrix elements is reported in table 1. As already mentioned in the Introduction, in Ref. [19] the effective hamiltonian has been derived in a model space including the four pf orbitals outside doubly-closed ^{40}Ca, that is the standard choice of the model space in the study of calcium isotopes.

As regards the choice of the neutron SP energies, we refer to [24] for those of the $0f_{5/2}$, $1p_{3/2}$, $1p_{1/2}$ orbitals, obtained calculating the centroids of the observed single-neutron strengths in ^{49}Ca. Since there is no experimental evidence of a SP $J^{\pi} = \frac{9}{2}^{+}$ state in ^{49}Ca, the $0g_{9/2}$ SP energy has been fixed in order to reproduce the energy of the SP $J^{\pi} = \frac{9}{2}^{+}$ state in ^{57}Ni [25]. It is worth mentioning that this choice leads also to a good agreement with experiment for the energies of $J^{\pi} = \frac{9}{2}^{+}$ states in ^{51}Ti and $^{53-57}$Cr. The adopted SP energies are $\epsilon_{3/2} = -4.60$ MeV, $\epsilon_{1/2} = -2.86$ MeV, $\epsilon_{5/2} = -1.20$ MeV, and $\epsilon_{9/2} = -0.55$ MeV.

It should be pointed out that for protons the Coulomb force has been explicitly added to the $V_{\rm low-k}$ before deriving $V_{\rm eff}$.

3. Results

In this section, we report the results of the present calculations for the heavy-mass Ca isotopes from ^{50}Ca to ^{56}Ca (labelled as I) and compare them with those of [19] (labelled as II), where ^{40}Ca was assumed as a closed core. The calculations have been performed by using the Oslo shell-model code [26].

The calculated and experimental [6] energies of ^{50}Ca are reported in table 2. We see that both calculations reproduce quite well the observed energies of positive-parity states. The negative-parity states can be obviously obtained only in calculation I, and the interpretation of their nature is strictly connected with the corresponding states in the ^{48}Ca energy spectrum.

Table 1. Proton-proton, neutron-neutron, and proton-neutron matrix elements (in MeV). They are antisymmetrized and normalized.

$n_a l_a j_a$	$n_b l_b j_b$	$n_c l_c j_c$	$n_d l_d j_d$	J	T_z	TBME	$n_a l_a j_a$	$n_b l_b j_b$	$n_c l_c j_c$	$n_d l_d j_d$	J	T_z	TBME
$0f_{7/2}$	$0f_{7/2}$	$0f_{7/2}$	$0f_{7/2}$	0	1	-1.867	$0f_{5/2}$	$0g_{9/2}$	$0f_{5/2}$	$0g_{9/2}$	2	-1	-0.569
$0f_{7/2}$	$0f_{7/2}$	$0f_{7/2}$	$0f_{7/2}$	2	1	-0.483	$0f_{5/2}$	$0g_{9/2}$	$0f_{5/2}$	$0g_{9/2}$	3	-1	-0.436
$0f_{7/2}$	$0f_{7/2}$	$0f_{7/2}$	$0f_{7/2}$	4	1	0.160	$0f_{5/2}$	$0g_{9/2}$	$1p_{3/2}$	$0g_{9/2}$	3	-1	0.243
$0f_{7/2}$	$0f_{7/2}$	$0f_{7/2}$	$0f_{7/2}$	6	1	0.479	$1p_{3/2}$	$0g_{9/2}$	$1p_{3/2}$	$0g_{9/2}$	3	-1	-0.827
$0f_{5/2}$	$0f_{5/2}$	$0f_{5/2}$	$0f_{5/2}$	0	-1	-0.841	$0f_{5/2}$	$0g_{9/2}$	$0f_{5/2}$	$0g_{9/2}$	4	-1	0.062
$0f_{5/2}$	$0f_{5/2}$	$1p_{3/2}$	$1p_{3/2}$	0	-1	-0.708	$0f_{5/2}$	$0g_{9/2}$	$1p_{3/2}$	$0g_{9/2}$	4	-1	0.170
$0f_{5/2}$	$0f_{5/2}$	$1p_{1/2}$	$1p_{1/2}$	0	-1	-0.396	$0f_{5/2}$	$0g_{9/2}$	$1p_{1/2}$	$0g_{9/2}$	4	-1	0.052
$0f_{5/2}$	$0f_{5/2}$	$0g_{9/2}$	$0g_{9/2}$	0	-1	1.922	$1p_{3/2}$	$0g_{9/2}$	$1p_{3/2}$	$0g_{9/2}$	4	-1	0.060
$1p_{3/2}$	$1p_{3/2}$	$1p_{3/2}$	$1p_{3/2}$	0	-1	-0.802	$1p_{3/2}$	$0g_{9/2}$	$1p_{1/2}$	$0g_{9/2}$	4	-1	-0.059
$1p_{3/2}$	$1p_{3/2}$	$1p_{1/2}$	$1p_{1/2}$	0	-1	-1.028	$1p_{1/2}$	$0g_{9/2}$	$1p_{1/2}$	$0g_{9/2}$	4	-1	0.112
$1p_{3/2}$	$1p_{3/2}$	$0g_{9/2}$	$0g_{9/2}$	0	-1	1.305	$0f_{5/2}$	$0g_{9/2}$	$0f_{5/2}$	$0g_{9/2}$	5	-1	-0.178
$1p_{1/2}$	$1p_{1/2}$	$1p_{1/2}$	$1p_{1/2}$	0	-1	-0.161	$0f_{5/2}$	$0g_{9/2}$	$1p_{3/2}$	$0g_{9/2}$	5	-1	0.196
$1p_{1/2}$	$1p_{1/2}$	$0g_{9/2}$	$0g_{9/2}$	0	-1	0.980	$0f_{5/2}$	$0g_{9/2}$	$1p_{1/2}$	$0g_{9/2}$	5	-1	-0.275
$0g_{9/2}$	$0g_{9/2}$	$0g_{9/2}$	$0g_{9/2}$	0	-1	-1.067	$1p_{3/2}$	$0g_{9/2}$	$1p_{3/2}$	$0g_{9/2}$	5	-1	-0.191
$0f_{5/2}$	$1p_{3/2}$	$0f_{5/2}$	$1p_{3/2}$	1	-1	0.138	$1p_{3/2}$	$0g_{9/2}$	$1p_{1/2}$	$0g_{9/2}$	5	-1	0.487
$0f_{5/2}$	$1p_{3/2}$	$1p_{3/2}$	$1p_{1/2}$	1	-1	-0.061	$1p_{1/2}$	$0g_{9/2}$	$1p_{1/2}$	$0g_{9/2}$	5	-1	-0.401
$1p_{3/2}$	$1p_{1/2}$	$1p_{3/2}$	$1p_{1/2}$	1	-1	0.244	$0f_{5/2}$	$0g_{9/2}$	$0f_{5/2}$	$0g_{9/2}$	6	-1	0.153
$0f_{5/2}$	$0f_{5/2}$	$0f_{5/2}$	$0f_{5/2}$	2	-1	-0.207	$0f_{5/2}$	$0g_{9/2}$	$1p_{3/2}$	$0g_{9/2}$	6	-1	0.194
$0f_{5/2}$	$0f_{5/2}$	$0f_{5/2}$	$1p_{3/2}$	2	-1	0.089	$1p_{3/2}$	$0g_{9/2}$	$1p_{3/2}$	$0g_{9/2}$	6	-1	0.400
$0f_{5/2}$	$0f_{5/2}$	$0f_{5/2}$	$1p_{1/2}$	2	-1	-0.232	$0f_{5/2}$	$0g_{9/2}$	$0f_{5/2}$	$0g_{9/2}$	7	-1	-0.945
$0f_{5/2}$	$0f_{5/2}$	$1p_{3/2}$	$1p_{3/2}$	2	-1	-0.126	$0f_{7/2}$	$0f_{5/2}$	$0f_{7/2}$	$0f_{5/2}$	1	0	-2.430
$0f_{5/2}$	$0f_{5/2}$	$1p_{3/2}$	$1p_{1/2}$	2	-1	-0.285	$0f_{7/2}$	$0f_{5/2}$	$0f_{7/2}$	$0f_{5/2}$	2	0	-1.562
$0f_{5/2}$	$0f_{5/2}$	$0g_{9/2}$	$0g_{9/2}$	2	-1	0.266	$0f_{7/2}$	$0f_{5/2}$	$0f_{7/2}$	$1p_{3/2}$	2	0	-0.638
$0f_{5/2}$	$1p_{3/2}$	$0f_{5/2}$	$1p_{3/2}$	2	-1	0.219	$0f_{7/2}$	$1p_{3/2}$	$0f_{7/2}$	$1p_{3/2}$	2	0	-0.830
$0f_{5/2}$	$1p_{3/2}$	$0f_{5/2}$	$1p_{1/2}$	2	-1	0.353	$0f_{7/2}$	$0f_{5/2}$	$0f_{7/2}$	$0f_{5/2}$	3	0	-0.422
$0f_{5/2}$	$1p_{3/2}$	$1p_{3/2}$	$1p_{3/2}$	2	-1	0.133	$0f_{7/2}$	$0f_{5/2}$	$0f_{7/2}$	$1p_{3/2}$	3	0	0.169
$0f_{5/2}$	$1p_{3/2}$	$1p_{3/2}$	$1p_{1/2}$	2	-1	0.221	$0f_{7/2}$	$0f_{5/2}$	$0f_{7/2}$	$1p_{1/2}$	3	0	-0.295
$0f_{5/2}$	$1p_{3/2}$	$0g_{9/2}$	$0g_{9/2}$	2	-1	-0.554	$0f_{7/2}$	$1p_{3/2}$	$0f_{7/2}$	$1p_{3/2}$	3	0	-0.369
$0f_{5/2}$	$1p_{1/2}$	$0f_{5/2}$	$1p_{1/2}$	2	-1	-0.197	$0f_{7/2}$	$1p_{3/2}$	$0f_{7/2}$	$1p_{1/2}$	3	0	0.790
$0f_{5/2}$	$1p_{1/2}$	$1p_{3/2}$	$1p_{3/2}$	2	-1	-0.127	$0f_{7/2}$	$1p_{1/2}$	$0f_{7/2}$	$1p_{1/2}$	3	0	-0.678
$0f_{5/2}$	$1p_{1/2}$	$1p_{3/2}$	$1p_{1/2}$	2	-1	-0.386	$0f_{7/2}$	$0f_{5/2}$	$0f_{7/2}$	$0f_{5/2}$	4	0	-0.891
$0f_{5/2}$	$1p_{1/2}$	$0g_{9/2}$	$0g_{9/2}$	2	-1	0.684	$0f_{7/2}$	$0f_{5/2}$	$0f_{7/2}$	$1p_{3/2}$	4	0	-0.189
$1p_{3/2}$	$1p_{3/2}$	$1p_{3/2}$	$1p_{3/2}$	2	-1	-0.308	$0f_{7/2}$	$0f_{5/2}$	$0f_{7/2}$	$1p_{1/2}$	4	0	-0.546
$1p_{3/2}$	$1p_{3/2}$	$1p_{3/2}$	$1p_{1/2}$	2	-1	-0.534	$0f_{7/2}$	$1p_{3/2}$	$0f_{7/2}$	$1p_{3/2}$	4	0	-0.085
$1p_{3/2}$	$1p_{3/2}$	$0g_{9/2}$	$0g_{9/2}$	2	-1	0.477	$0f_{7/2}$	$1p_{3/2}$	$0f_{7/2}$	$1p_{1/2}$	4	0	-0.344
$1p_{3/2}$	$1p_{1/2}$	$1p_{3/2}$	$1p_{1/2}$	2	-1	-0.452	$0f_{7/2}$	$1p_{1/2}$	$0f_{7/2}$	$1p_{1/2}$	4	0	-0.573
$1p_{3/2}$	$1p_{1/2}$	$0g_{9/2}$	$0g_{9/2}$	2	-1	0.397	$0f_{7/2}$	$0f_{5/2}$	$0f_{7/2}$	$0f_{5/2}$	5	0	0.157
$0g_{9/2}$	$0g_{9/2}$	$0g_{9/2}$	$0g_{9/2}$	2	-1	-0.771	$0f_{7/2}$	$0f_{5/2}$	$0f_{7/2}$	$1p_{3/2}$	5	0	0.246
$0f_{5/2}$	$1p_{3/2}$	$0f_{5/2}$	$1p_{3/2}$	3	-1	0.163	$0f_{7/2}$	$1p_{3/2}$	$0f_{7/2}$	$1p_{3/2}$	5	0	-1.027
$0f_{5/2}$	$1p_{3/2}$	$0f_{5/2}$	$1p_{1/2}$	3	-1	0.044	$0f_{7/2}$	$0f_{5/2}$	$0f_{7/2}$	$0f_{5/2}$	6	0	-1.796
$0f_{5/2}$	$1p_{1/2}$	$0f_{5/2}$	$1p_{1/2}$	3	-1	0.307	$0f_{7/2}$	$0g_{9/2}$	$0f_{7/2}$	$0g_{9/2}$	1	0	-1.138
$0f_{5/2}$	$0f_{5/2}$	$0f_{5/2}$	$0f_{5/2}$	4	-1	0.198	$0f_{7/2}$	$0g_{9/2}$	$0f_{7/2}$	$0g_{9/2}$	2	0	-0.846
$0f_{5/2}$	$0f_{5/2}$	$0f_{5/2}$	$1p_{3/2}$	4	-1	0.207	$0f_{7/2}$	$0g_{9/2}$	$0f_{7/2}$	$0g_{9/2}$	3	0	-0.288
$0f_{5/2}$	$0f_{5/2}$	$0g_{9/2}$	$0g_{9/2}$	4	-1	0.228	$0f_{7/2}$	$0g_{9/2}$	$0f_{7/2}$	$0g_{9/2}$	4	0	-0.331
$0f_{5/2}$	$1p_{3/2}$	$0f_{5/2}$	$1p_{3/2}$	4	-1	-0.243	$0f_{7/2}$	$0g_{9/2}$	$0f_{7/2}$	$0g_{9/2}$	5	0	-0.066
$0f_{5/2}$	$1p_{3/2}$	$0g_{9/2}$	$0g_{9/2}$	4	-1	-0.613	$0f_{7/2}$	$0g_{9/2}$	$0f_{7/2}$	$0g_{9/2}$	6	0	-0.499
$0g_{9/2}$	$0g_{9/2}$	$0g_{9/2}$	$0g_{9/2}$	4	-1	-0.277	$0f_{7/2}$	$0g_{9/2}$	$0f_{7/2}$	$0g_{9/2}$	7	0	0.016
$0g_{9/2}$	$0g_{9/2}$	$0g_{9/2}$	$0g_{9/2}$	6	-1	-0.116	$0f_{7/2}$	$0g_{9/2}$	$0f_{7/2}$	$0g_{9/2}$	8	0	-1.795
$0g_{9/2}$	$0g_{9/2}$	$0g_{9/2}$	$0g_{9/2}$	8	-1	-0.004							

As a matter of fact, in ^{48}Ca the $J^\pi = 3_1^-$, 4_1^-, 5_1^- states, which can be interpreted as $1p-1h$ $Z = 20$ cross-shell excitations, are located ~ 0.5 MeV higher in energy than the corresponding ones in ^{50}Ca. This may be seen as an indication that, when increasing the number of neutrons, the structure of the above negative-parity states changes because of configuration mixing with states built on the neutron $0g_{9/2}$ orbital. Our results seem to indicate that the overlap with our model space is larger for $J^\pi = 5_1^-$ than for $J^\pi = 3_1^-$, 4_1^-. The experimental $J^\pi = 2_3^+$, 4_1^+ states have no theoretical counterpart in calculation I since their structure is dominated by the $\nu(0f_{7/2})^7(1p_{3/2})^3$ configuration [6, 19]. Note that the results for positive-parity states with calculations I and II are close to each other, thus confirming the soundness of the perturbation theory of the shell-model effective interaction. We have also calculated the $B(E2; 2_1^+ \rightarrow 0_1^+)$ transition rate in ^{50}Ca employing effective operators obtained at third order in perturbation theory, consistently with the derivation of H_{eff} as in Ref. [19]. The calculated value is 7.1 e^{2}fm^4 to be compared with the experimental one 7.5 ± 0.2 e^{2}fm^4 [27], while with calculation II we have obtained 10.9 e^{2}fm^4 [19].

Table 2. Calculated energy levels (in MeV) of ^{50}Ca and ^{52}Ca compared with the corresponding experimental ones [6, 28] up to 5.2 MeV excitation energy. The energies of the states whose parity and/or angular momentum are uncertain are reported in parenthesis.

		^{50}Ca				^{52}Ca	
J^π	Expt.	Calc. I	Calc. II	J^π	Expt.	Calc. I	Calc. II
0^+	0.000	0.000	0.000	0^+	0.000	0.000	0.000
2^+	1.026	0.939	0.953	2^+	2.562	2.358	2.396
2^+	(3.004)	2.856	2.941	1^+	(3.150)	3.205	3.135
1^+	(3.532)	3.472	3.547	3^-	(3.992)	4.702	
3^-	(3.998)	4.698					
2^+	(4.036)		3.582				
0^+	(4.470)	4.839	4.926				
4^+	4.515		4.567				
4^-	(4.832)	5.590					
2^+	(4.870)	4.957	5.231				
5^-	(5.110)	5.188					

In table 3 we have reported the experimental and calculated energy spectra of ^{51}Ca. We see that also for this nucleus the comparison between theory and experiment is remarkably good, apart from the $J^\pi = \frac{7}{2}{}_1^+$ state. The structure of this state is mainly obtained coupling the $J^\pi = 3_1^-$ state in ^{50}Ca to a neutron in the $1p_{3/2}$ orbital, which explains the overestimation of the calculated energy by ~ 450 keV. In calculation I we find that the first excited $J^\pi = \frac{9}{2}^+$ state is essentially a seniority $v = 1$ state. This lends further support to our choice of the $0g_{9/2}$ SP energy. It should be mentioned that in Ref. [8] both $J^\pi = \frac{7}{2}{}_1^+$ and $J^\pi = \frac{9}{2}{}_1^+$ are interpreted as $1p-1h$ $Z = 20$ cross-shell excitations.

As regards the negative-parity states, the agreement between calculations I and II is very good. Note that the $J^\pi = \frac{7}{2}{}_1^-$ state is outside the model space of calculation I since its structure is mainly made up by one neutron-hole in ^{48}Ca core and 4 neutrons filling the $1p_{3/2}$ orbital.

The nucleus ^{52}Ca is the last one of this isotopic chain for which some experimental excited states have been observed [4–6]. The experimental energies of these states are reported in table 2 together with the corresponding calculated ones up to 5.2 MeV. Both calculations I and II reproduce quite well the subshell closure of the $1p_{3/2}$ orbital.

Table 3. Energy levels (in MeV) of ^{51}Ca and ^{53}Ca. The calculated ^{51}Ca energy levels are compared with the corresponding experimental ones [6, 8]. The energies of the states whose parity and/or angular momentum are uncertain are reported in parenthesis. The ^{53}Ca energy levels are reported up to 4.0 MeV excitation energy.

	^{51}Ca				^{53}Ca	
J^π	Expt.	Calc. I	Calc. II	J^π	Calc. I	Calc. II
$\frac{3}{2}^-$	(0.000)	0.000	0.000	$\frac{1}{2}^-$	0.000	0.000
$\frac{1}{2}^-$	(1.721)	1.753	1.586	$\frac{5}{2}^-$	2.112	2.039
$\frac{5}{2}^-$	(2.379)	2.190	2.269	$\frac{3}{2}^-$	2.169	2.402
$\frac{3}{2}^-$	(2.937)	2.829	2.932	$\frac{9}{2}^+$	2.604	
$\frac{7}{2}^-$	(3.437)		3.429			
$\frac{5}{2}^-$	(3.479)	3.564	3.552			
$\frac{7}{2}^+$	(3.845)	4.309				
$\frac{9}{2}^+$	(4.155)	4.126				
$\frac{9}{2}^-$	(4.322)	4.281	4.435			

For the heavier Ca isotopes, ^{53,54,55}Ca, only the ground states have been identified with spin and parity assignments [7]. In tables 3 and 4 we have reported our predicted spectra for these three nuclei and for ^{56}Ca up to 4.0, 4.5, 2.4, and 3.0 MeV excitation energy, respectively. From the inspection of the above tables, it is clear that the disagreement between calculations I and II increases with mass number, since the role of $0g_{9/2}$ orbital becomes more and more relevant when adding valence neutrons. This has also repercussions on the two-neutron separation energy S_{2n} for ^{56}Ca which is about 2 MeV larger in calculation I, as can be inferred from the inspection of figure 1 where the calculated ground-state energies of the even-mass isotopes relative to ^{48}Ca are compared with the measured ($A = 50,\ 52$) and estimated ones ($A = 54,\ 56$) [29].

Table 4. Theoretical energy levels (in MeV) of ^{54}Ca, ^{55}Ca, and ^{56}Ca. See text for details.

	^{54}Ca			^{55}Ca			^{56}Ca	
J^π	Calc. I	Calc. II	J^π	Calc. I	Calc. II	J^π	Calc. I	Calc. II
0^+	0.000	0.000	$\frac{5}{2}^-$	0.000	0.000	0^+	0.000	0.000
2^+	2.587	2.061	$\frac{9}{2}^+$	0.208		2^+	2.268	0.963
0^+	2.783	3.138	$\frac{1}{2}^-$	0.237	1.104	7^-	2.514	
5^-	3.116		$\frac{3}{2}^-$	1.924	1.541	2^-	2.745	
3^+	3.157	2.748	$\frac{5}{2}^+$	2.131		3^-	2.825	
4^-	3.430		$\frac{13}{2}^+$	2.204		5^-	2.960	
2^+	4.517	4.172	$\frac{7}{2}^-$	2.325	1.931	4^+	2.961	1.487
			$\frac{7}{2}^+$	2.368		2^+	2.982	2.224
			$\frac{5}{2}^-$	2.381	2.390			

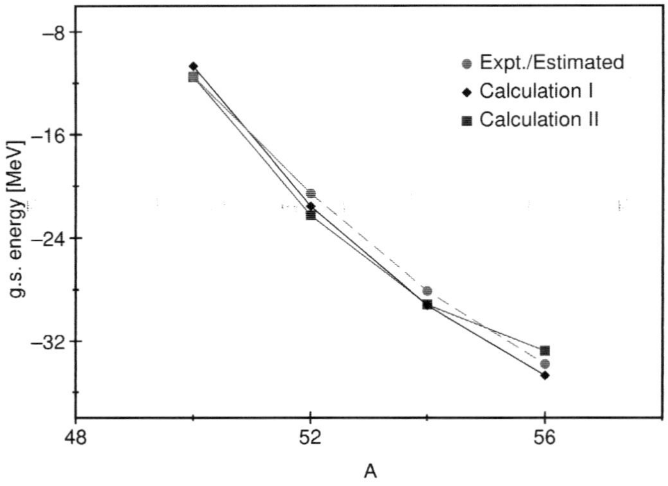

Figure 1. Ground-state energies for calcium isotopes from $A = 50$ to 56. See text for details.

4. Concluding remarks

In summary, we have given here a shell-model description of heavy calcium isotopes assuming ^{48}Ca as a closed core and including the neutron $0g_{9/2}$ orbital. This has been done by renormalizing the NN potential CD-Bonn by way of the $V_{\rm low-k}$ approach and then deriving the shell-model hamiltonian within the framework of perturbation theory. We have compared the results of the present study with those of the previous one [19] which started from doubly-closed ^{40}Ca and employed the $1p0f$ model space.

The results of both calculations are in good agreement with the available experimental data, and do not differ significantly up to ^{53}Ca. Actually, from ^{54}Ca on the differences between the energy spectra calculated within the two different model spaces increase with the number of valence neutrons. Clearly, new experimental data are needed to establish the real role of the neutron $0g_{9/2}$ orbital in the description of the structure of nuclei north-east of ^{48}Ca core.

References

[1] Liddick S N *et al* 2004 *Phys. Rev. Lett.* **92** 072502
[2] Liddick S N *et al* 2004 *Phys. Rev. C* **70** 064303
[3] Dinca D -C *et al* 2005 *Phys. Rev. C* **71** 041302(R)
[4] Gade A *et al* 2006 *Phys. Rev. C* **74** 021302(R)
[5] Perrot F *et al* 2006 *Phys. Rev. C* **74** 014313
[6] Rejmund M, Bhattacharyya S, Navin A, Mittig W, Gaudefroy L, Gelin M, Mukherjee M, Rejmund F, Roussel-Chomaz P and Theisen C 2007 *Phys. Rev. C* **76** 021304(R)
[7] Mantica P F *et al* 2008 *Phys. Rev. C* **77** 014313
[8] Fornal B *et al* 2008 *Phys. Rev. C* **77** 014304
[9] Maierbeck P *et al* 2009 *Phys. Lett. B* **675** 22
[10] Bhattacharyya S *et al* 2009 *Phys. Rev. Lett.* **101** 032501
[11] Poves A, Sánchez-Solano J, Caurier E and Nowacki F 2001 *Nucl. Phys. A* **694** 157
[12] Richter W A, der Merwe M G V, Julies R E and Brown B A 1991 *Nucl. Phys. A* **523** 325
[13] Honma M, Otsuka T, Brown B A and Mizusaki T 2002 *Phys. Rev. C* **65** 061301(R)
[14] Honma M, Otsuka T, Brown B A and Mizusaki T 2005 *Eur. Phys. J. A* **25, s01** 499
[15] Coraggio L, Covello A, Gargano A and Itaco N 2006 *Phys. Rev. C* **73** 031302(R)
[16] Coraggio L, Covello A, Gargano A, Itaco N, Kuo T T S, Entem D R and Machleidt R 2007 *Phys. Rev. C* **75** 024311
[17] Coraggio L, Covello A, Gargano A, Itaco N and Kuo T T S 2009 *Prog. Part. Nucl. Phys.* **62** 135
[18] Coraggio L, Covello A, Gargano A and Itaco N 2009 *Phys. Rev. C* **80** 021305

[19] Coraggio L, Covello A, Gargano A and Itaco N 2009 *Phys. Rev. C* **80** 044311
[20] Bogner S, Kuo T T S and Coraggio L 2001 *Nucl. Phys. A* **684** 432c
[21] Bogner S, Kuo T T S, Coraggio L, Covello A and Itaco N 2002 *Phys. Rev. C* **65** 051301(R)
[22] Machleidt R 2001 *Phys. Rev. C* **63** 024001
[23] Suzuki K and Lee S Y 1980 *Prog. Theor. Phys.* **64** 2091
[24] Cottle P D and Kemper K W 2008 *Phys. Rev. C* **78** 037304
[25] Rudolph D, Weisshaar D, Cristancho F, Eberth J, Fahlander C, Iordanov O, Skoda S, Teich C, Thelen O and Thomas H 1999 *Eur. Phys. J. A* **6** 377–380 ISSN 1434-6001
[26] Engeland T the Oslo shell-model code 1991-2006, unpublished
[27] Valiente-Dobón J J *et al* 2009 *Phys. Rev. Lett.* **102** 242502
[28] Data extracted using the NNDC On-line Data Service from the ENSDF database, file revised as of June 4, 2010.
[29] Audi G, Wapstra A H and Thibault C 2003 *Nucl. Phys. A* **729** 337

Effective tensor forces and neutron rich nuclei

Giampaolo Co'[1,2], **Viviana De Donno**[1,2], **Marta Anguiano**[3] **and Antonio M Lallena**[3]

[1] Dipartimento di Fisica, Università del Salento, Via Arnesano, I-73100 Lecce, ITALY
[2] INFN, Sezione di Lecce, Via Arnesano, I-73100 Lecce, ITALY
[3] Departamento de Física Atómica, Molecular y Nuclear, Universidad de Granada, E-18071 Granada, SPAIN

E-mail: gpco@le.infn.it

Abstract. We study the effects of the tensor term of the effective nucleon-nucleon interaction on nuclear excited states. Our investigation has been conducted by using a self-consistent Random Phase Approximation approach. We investigate various nuclei in different regions of the isotopes chart. Results for a set of calcium isotopes are shown.

The tensor parts of the nucleon-nucleon interaction are usually neglected in Hartree-Fock (HF) and Random Phase Approximation (RPA) calculations. In these theories, the effects of the tensor force are taken into account by modifying the values of the parameters of the other terms of the effective force. We have investigated the validity of this approach, and we present here some results of our study.

In our work we have considered only the tensor-isospin term of the nuclear interaction, which we have described as

$$v_b^{t\tau}(r) = v_{AV8'}^{t\tau}(r)(1 - e^{-br^2}) \ , \qquad (1)$$

where we have indicated with $v_{AV8'}^{t\tau}(r)$ the tensor-isospin term of the Argonne V8' microscopic potential [1]. In the above expression, this term is multiplied by a function which considers the effects of the short-range correlations. The only free parameter is b, whose value rules the strength of the tensor part of the force. Small values of b increase the active range of the correlation reducing the strength of the force.

As example of the effects produced by the tensor force on the nuclear excited states, we show in Fig. 1 the evolution of the RPA eigenenergy of the first 0^- excited state in ^{16}O , obtained with a phenomenological RPA calculation. Similar sensitivity of the 0^- excitation energies to the presence of the tensor term of the force has been found in all the nuclei we have investigated, ^{12}C, ^{16}O, ^{40}Ca, ^{48}Ca and ^{208}Pb. We present here only the ^{16}O case for sake of simplicity. The results of Fig. 1 have been obtained by using single particle energies and wave functions calculated by diagonalizing a Woods-Saxon potential in a harmonic oscillator basis. The residual interaction, used in the RPA calculation, is a zero-range Landau-Migdal force whose parameters values are those given in Ref. [2]. The value obtained with this interaction is shown in the figure by the dashed line. When the full Argonne V8' tensor term is added, we obtain the value represented by the dotted line. The full line shows the experimental value of 10.96 MeV [3]. The squares indicate the values of the 0^- eigenenergies as a function of b. These values become smaller with the increase of the tensor force strength. The evolution of the 0^- eigenenergy, from that obtained with the pure Landau-Migdal force up to that obtained when the full tensor term is active, is very smooth.

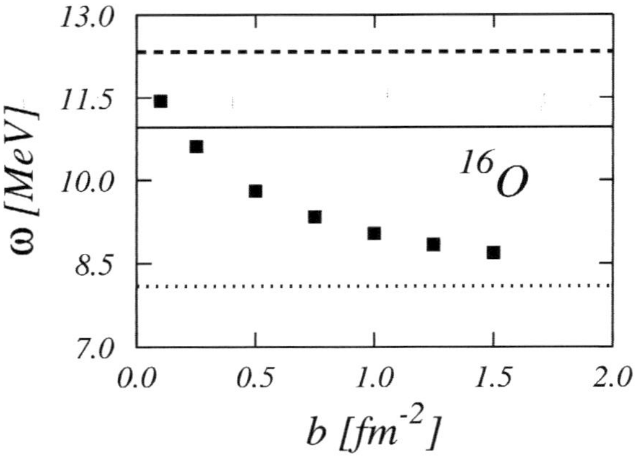

Figure 1. Energy of the first 0^- excited state in ^{16}O calculated by using a phenomenological RPA approach. The dashed line indicates the values obtained without tensor force. The dotted line shows the value obtained when the full Argonne V8' term is active. The squares indicate the values obtained for various values of the parameter b of Eq. 1. The full line shows the experimental value.

The phenomenological RPA calculations described above have been used to identify an observable very sensitive to the tensor force. We have considered this quantity in order to select the values of b.

The tensor force produces effects on the single particle basis and also on the excited states of the nucleus. For this reason we have conducted our study by doing self-consistent RPA calculations. This means that we used in the RPA calculations the single particle wave functions and energies obtained by solving the HF equations. The effective interaction in both type of calculations is the same.

The effect of the tensor force on the single particle energies has been well clarified by Otsuka and collaborators [4]. For example, the energies of two proton spin-obit partner single particle levels are modified by the tensor-isospin interaction acting with a neutron single particle level. If the total angular momentum of the occupied neutron single particle level is $j = l+1/2$ the energy of the $j = l - 1/2$ proton level is lowered and that of the $j = l + 1/2$ proton level is increased. This effect is reversed when the occupied neutron level has $j = l - 1/2$. For this reason, the nuclei where all the neutron spin-orbit partners single particle levels are fully occupied do not show any tensor force effect.

We have conducted our study by comparing results obtained with and without tensor force. We have considered two different parametrization of the Gogny interaction. They are the traditional D1S [5] force and the more modern D1M parametrization [6] which describes better the behaviour of the the microscopic equation of state of neutron matter at high densities.

The D1S and D1M interactions do not have tensor terms. We add to these interactions a tensor-isospin term of the form given by Eq. 1. The procedure we have adopted to construct new parametrization of the force is iterative. We first made a HF calculation without tensor force to produce single particle energies and wave functions. Then we made RPA calculations with the tensor force, selecting the value of b which reproduces the experimental energy of the 0^- in ^{16}O. With this new interaction we have recalculated the HF single particle energies and wave functions, and we changed the spin-orbit term of the interaction to reproduce the splitting between the neutron $1p_{3/2}$ and $1p_{1/2}$ levels. With this single particle basis we performed a new set of RPA calculations and the procedure has been repeated up to convergence. In this way, starting from the D1S and D1M forces, we have constructed two new parametrizations which we call respectively D1SV8 and D1MV8. The differences with the original parametrizations are only in the spin-orbit term and, obviously, in the presence of the tensor term. We did not make a complete refit of the forces.

In our studies, we have also used the GT2 interaction [7] which contains a tensor term, and we made calculations by switching off this term. Since the results obtained in this way are very similar to those obtained with the other two interactions, for sake of simplicity, we do not present them here.

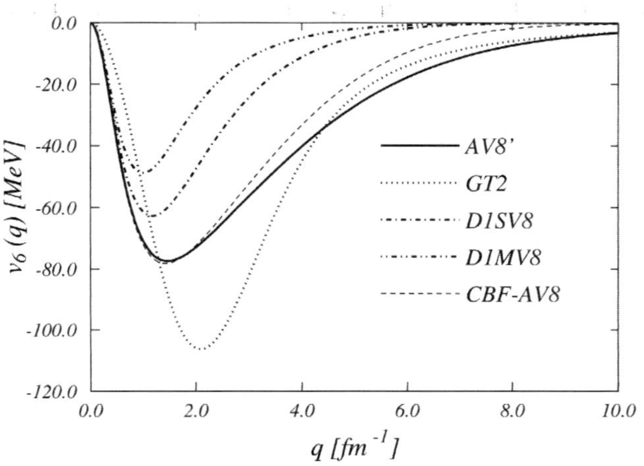

Figure 2. Tensor-isospin terms of the interactions used in our study as a function of the momentum of the interacting pair.

In Fig. 2 we show the momentum dependent part of the tensor forces, defined as

$$v_6(q)\, S_{12}(\mathbf{q}) = \int d^3r\, e^{i\mathbf{q}\cdot\mathbf{r}}\, v^{t\tau}(r)\, S_{12}(\mathbf{r}) = -4\pi \int dr\, r^2\, j_2(qr)\, v^{t\tau}(r) S_{12}(\mathbf{r}) \quad , \qquad (2)$$

where we have indicated with S_{12} the usual expression of the tensor operator [8].

In this figure, the full line shows the the tensor term of the Argonne V8' interaction used to build the D1SV8 and D1MV8 forces, indicated by the dashed-dotted and dashed-doubly dotted lines. With the dotted line we show the tensor term of the GT2 interaction [7]. The dashed line has been obtained by multiplying the bare Argonne V8' term with the scalar part of the short range correlation calculated in Correlated Basis Function theory [8]. This line indicates that the effect of the short range correlations obtained by a microscopic calculations are much smaller than those required by our procedure.

With the interactions presented above, we have studied a set of isotopes of doubly magic nuclei. These nuclei have been chosen in such a way that the single particle levels below the Fermi energy are fully occupied. In this way we avoided deformation problems and we minimized pairing effects. We have studied ground states and excitation spectra of nuclei in the oxygen, calcium, zirconium and tin regions and also ^{208}Pb, for a total number of 16 different isotopes.

We have seen that the effects of the tensor force are not important on the binding energies, as already pointed out in Ref. [9]. We have obtained variations of less than 1% of the values obtained without tensor terms.

We have also studied the protons and neutrons density distributions for all the nuclei mentioned above. Also in this case the effects of the tensor force are irrelevant. The largest differences between calculations with and without tensor force are of the order of few parts on a thousand.

On the opposite, we have observed that the effect of the tensor force on the single particle energies in not negligible. As example we show in Fig. 3 the single particle spectrum of the three calcium isotopes considered in our study. For simplicity we present only the results obtained with D1S and D1M interactions, and their corresponding interactions containing the tensor term. In the figure, the thin lines show the results obtained without tensor force and the thick

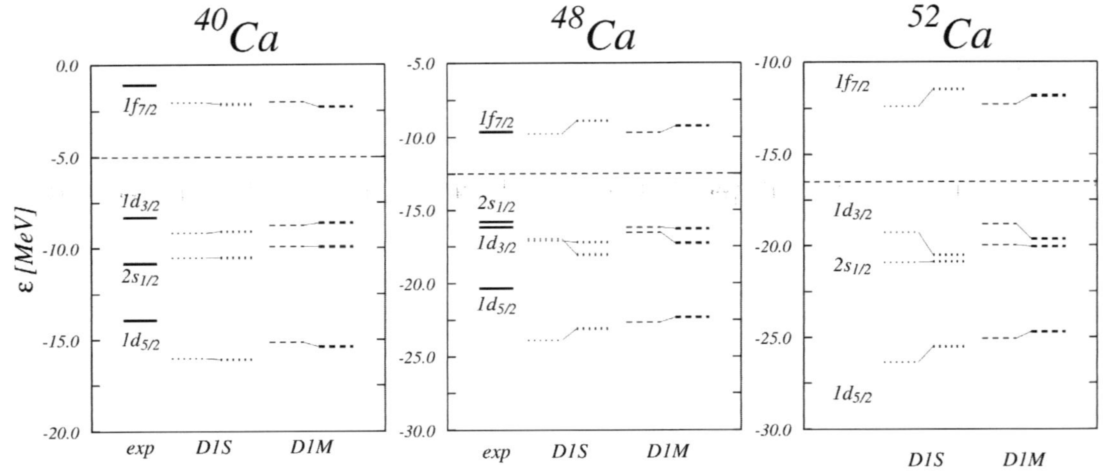

Figure 3. Single particle energies of the calcium isotopes calculated with the interactions D1S and D1M, thin lines, and with their corresponding interactions containing the tensor term, the D1SV8 and D1MV8, thick lines.

lines those obtained with the tensor force. The full lines show the experimental single particle energies taken from Ref. [10].

Table 1. Energies, in MeV, of the 4^- excited state for various Ca isotopes.

	exp	D1S	D1SV8	D1SV8*	D1M	D1MV8	D1MV8*
^{40}Ca	7.66	7.50	7.41	7.32	7.04	6.70	6.64
^{48}Ca	6.11	7.91	8.61	8.75	6.81	7.30	7.40
^{52}Ca		7.04	9.03	9.15	6.77	7.84	7.97

As expected, in ^{40}Ca the tensor term has no effect, since all the neutron spin-orbit partner levels are occupied. The small differences between the energies are due to the different spin-orbit terms between the D1S and D1SV8 forces and the D1M and D1MV8 forces.

The effect is present in the single particle spectrum of the other two isotopes since in these nuclei the neutron $j = l + 1/2$ levels are occupied, while their spin-orbit partners are empty. The figure shows that with the inclusion of the tensor term the energy of the $1d_{5/2}$ levels increases, and that of the $1d_{3/2}$ decreases. For the ^{48}Ca nucleus, this effect helps in reproducing the empirical sequence of single particle states. For the D1S interaction the tensor force lowers the energy of the $1d_{3/2}$ level below that of the $2s_{1/2}$ as it is experimentally observed. The effect in ^{52}Ca is not so dramatic. From the experimental point of view it is not yet clear if the spin and parity of the ^{51}K ground state is $3/2^+$ or $1/2^+$. All our calculations indicate $3/2^+$.

The action of the tensor force affects also the energy of the first proton level above the Fermi surface, the $1f_{7/2}$ level, by increasing its value. This means that the gap between the occupied

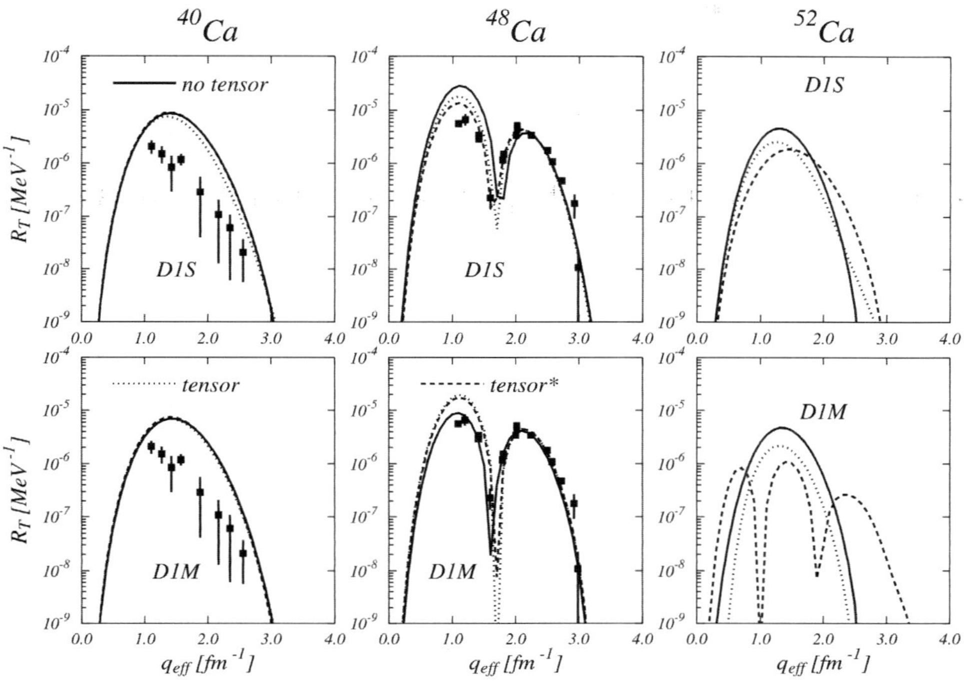

Figure 4. Electron scattering transverse responses calculated by using interactions without tensor, full lines, with tensor, dotted lines and by switching off the tensor force only in the RPA calculations, dashed lines. The experimental data are taken form Refs. [11] and [12].

$1d_{3/2}$ proton level and the $1f_{7/2}$ empty level increases if the tensor force is included. For this reason the energy of excited states dominated by the proton $[1f_{7/2}, 1d^{-1}_{3/2}]$ particle-hole pair should increase. This is what we observe from the results shown in Tab. 1 where we present the RPA eigenergies for the 4^- states in the three calcium isotopes obtained from calculations done with forces with (D1SV8 and D1MV8) and without (D1S and D1M) tensor force. We also show results of calculations where the tensor force has been switched off only in the RPA calculation. They are indicated by the D1SV8* and D1MV8* labels.

The results behave exactly as expected. The tensor force has practically no effect on the ^{40}Ca results, while there is a clear increase in ^{48}Ca and ^{52}Ca. On the other hand, the experimental values do not seem to have this trend. It would be interesting to know the experimental value of the 4^- energy in ^{52}Ca.

To have information about the wave function of the excited states, we have calculated the inelastic electron scattering transverse responses. In Fig. 4, we compare our results with the experimental data from Ref. [11] and [12].

The effects of the tensor force are irrelevant in the ^{40}Ca case. We observe some sensitivity to the tensor force in the case of ^{48}Ca . In the D1S case the tensor force lowers the first peak, while the effect is reversed in the case of the D1M interaction. The main effect is that on the single particle wave functions, on the HF calculation. The situation is different for the ^{52}Ca, isotope. In this case the tensor force effects are relevant on both the HF results, and also on the results obtained by the RPA calculation. It is interesting the situation of the D1M calculation in ^{52}Ca, where the presence of the tensor force in the RPA eliminates the contamination of the neutron $[2d_{5/2},2p^{-1}_{1/2}]$ particle-hole pair, and it changes the shape of the response.

In general, the tensor force does not produce relevant effects on the bulk properties of the nuclear systems, such as binding energies, density distributions and excitation energies. Concerning the excited states, however, there are situations where the inclusion of the tensor form is important. Certainly the energies of the 0^- states are extremely sensitive to the presence of the tensor term of the force. Furthermore, the effects on the single particle energies can modify excitations dominated by specific particle-hole pairs. We have presented the case of the 4^- excitations in calcium isotopes which clearly shows tensor force effects when the neutrons do not occupy all the spin-orbit partners levels.

The accuracy required by modern nuclear structure calculations requires the inclusion of the tensor term in the effective nucleon-nucleon interaction, especially for the study of nuclei with neutron excess.

References

[1] Pudliner B S, Pandharipande V R, Carlson J, Pieper S C and Wiringa R B 1997 *Phys. Rev. C* **56** 1720

[2] De Donno V, Co' G, Maieron C, Anguiano M, Lallena A M and Moreno-Torres M 2009 *Phys. Rev. C* **79** 044311

[3] Lederer C M and Shirley V S 1978 *Table of isotopes, 7th ed.* (John Wiley and sons, New York)

[4] Otsuka T, Suzuki T, Fujimoro R, Grawe H and Akaishi Y 2005 *Phys. Rev. Lett.* **95** 232502

[5] Berger J F, Girod M and Gogny D 1991 *Comp. Phys. Comm.* **63** 365

[6] Goriely S, Girod M, Hilaire S and Péru S 2009 *Phys. Rev. Lett.* **102** 252501

[7] Otsuka T, Matsuo T and Abe D 2006 *Phys. Rev. Lett.* **97** 162501

[8] Arias de Saavedra F, Bisconti C, Co' G and Fabrocini A 2007 *Phys. Rep.* **450** 1

[9] Co' G and Lallena A M 1998 *Nuov. Cim. A* **111** 527

[10] Doll P, Wagner G J, Knöpfle K T and Mairle G 1976 *Nucl. Phys. A* **263** 210

[11] Williamson C F and et al 1987 unpublished

[12] Wise J and et al 1985 *Phys. Rev. C* **31** 1699

The past and future of magnetic moment measurements of short-lived nuclear states

2011 J. Phys.: Conf. Ser. 267 012023

(http://iopscience.iop.org/1742-6596/267/1/012023)

The past and future of magnetic moment measurements of short-lived nuclear states

Noémie Benczer-Koller and Gerfried J. Kumbartzki

Department of Physics and Astronomy, Rutgers University, New Brunswick, NJ 08903, USA

E-mail: `nkoller@physics.rutgers.edu`

Abstract. Over the last few years there has been considerable progress in both experimental facilities and theoretical calculations of magnetic moments of low-lying, short-lived nuclear states. In particular, the competition between collective and single particle excitations in the region around $70 \leq A \leq 100$ has been examined. Recent data on Pd isotopes are presented. Newly developed techniques are applicable to work with radioactive beams. However, experimental limitations have emerged and will be discussed especially in their relations to the determination of absolute magnetic moments as opposed to relative measurements of moments in chains of isotopes.

1. Introduction

Magnetic moments of excited states provide a specific signature for defining the structure of the states. In particular, the relative importance of neutron and proton configurations can be approximately inferred from even a single determination of the *sign* of the magnetic moment. Precision magnetic moment measurements, coupled with B(E2) values, enable a definite description of the microscopic configuration.

In general, while calculations based on standard nuclear models are in fair agreement with the measurements, some surprises have sprung in many cases. The superdeformed bands in ^{194}Hg seem to have $g \approx Z/A$ while the normal deformation bands in the same isotope exhibit clear aspects of single particle configurations. The g factors of the Nd isotopes vary from Z/A for the fully deformed collective states (with spins and parities $2^+, 4^+, 6^+, 8^+$, or 10^+) in ^{150}Nd to single-particle values for the lighter isotopes, even reaching negative values for the 6_1^+ state in ^{144}Nd. The $g(^{42,44}\mathrm{Ca};2_1^+)$ factors are positive while other Ca nuclei with neutrons in the $f_{7/2}$ shell have negative g factors. The ^{92,94}Zr have both negative and positive g factors for the first two 2^+ states which have been identified as consisting, respectively, of symmetric and antisymmetric proton-neutron configurations.

Many techniques have been applied for these measurements. In the past, the technique yielding the largest magnetic fields involved implanting the isotope of interest in a ferromagnetic foil (IMPAC or IPAC techniques). Recently, the magnetic moments of short-lived ($\approx$ ps lifetimes), low-lying excited states in nuclei across the periodic table have been measured by the "transient field" method, TF [1, 2]. This procedure requires the production of an excited state in an aligned spin state, followed by traversal of a ferromagnetic foil in which the magnetic moment will precess about the induced hyperfine field. In most recent experiments the excitation of the state of interest has been carried out by Coulomb excitation, in inverse kinematics, of a beam consisting of the isotope of interest by a target (such as ^{12}C, ^{28}Si, ^{24}Mg, or ^{58}Ni), lighter than the beam isotope. The ferromagnetic foils of choice have been iron or gadolinium. The induced hyperfine fields can not be derived from basic principles. However, the strength of the interaction has been parametrized from the effective hyperfine fields observed in many experiments on nuclei with known magnetic moments.

In this work, the experimental techniques have been pushed to the limits because of the interest in higher excited states such as the 2_2^+, 4_1^+ and 0_2^+ states. The possibility of reaching states in radioactive nuclei not yet available at existing accelerators by capturing α particles from a ^{12}C target nuclei was explored. This latter technique, fusion evaporation reactions [3] or production by beam fragmentation at intermediate energies [4], result in excitations of nuclei to higher energy levels. These nuclei traverse the ferromagnetic foils while in these precursor states and the precessions observed are often related to the g factors of these higher energy states rather than to the g factor of the state of interest.

The details of the procedures used recently are described in Section 2. The resulting magnetic moment measurements of states in ^{106}Pd and ^{100}Pd are described in Section 3.

2. Experimental developments

Almost all current experiments involve the production of the state of interest in a spin aligned mode. The excited ion, moving at moderate velocities [1, 2], traverses a ferromagnetic foil where the nuclear spin (magnetic moment) precesses under the influence of an effective hyperfine field. Coincidences between a target particle, which provided the excitation reaction, and the γ rays from decaying excited states of the ion are recorded. The angular correlation and the corresponding slope, evaluated at the angle at which the γ detectors are placed with respect to the particle detector, need to be measured quite accurately

2.1. Targets

The targets are similar to those described in previous publications [5, 6, 7]. They consist essentially of a gadolinium foil of appropriate thickness (4-6 mg/cm^2). A light scattering material is deposited on the front of this foil. A copper foil in which the probe ions and beam will stop is evaporated on the down-beam side of the target. The target is kept at around 50 - 100K by a closed cycle Displex refrigerator. However, the exact temperature at the target spot under beam conditions cannot be measured. The magnetization as a function of temperature was measured off line in an AC magnetometer. It is noted that for most targets the magnetization remains constant between 50K and 100K, drops by 10 - 20 % at low temperatures and follows a Brillouin curve above 100K.

2.2. Excitation of the state of interest

The most favorable mode of excitation is Coulomb excitation in normal or inverse kinematics. The latter are particularly propitious for work with beams of radioactive isotopes for which targets cannot be prepared.

In the current experiments, Coulomb excitation at energies above the Coulomb barrier was used to excite the 2_2^+ and 4_1^+ states in ^{106}Pd, while the α capture from ^{12}C target reaction was used to excite the 2_1^+ state in ^{100}Pd. At higher beam energies, more energy is deposited in the target than in most previous lower-energy experiments using the transient field. Thus, the temperature of the gadolinium foil at the beam spot position might be higher than expected, leading to a locally reduced magnetization.

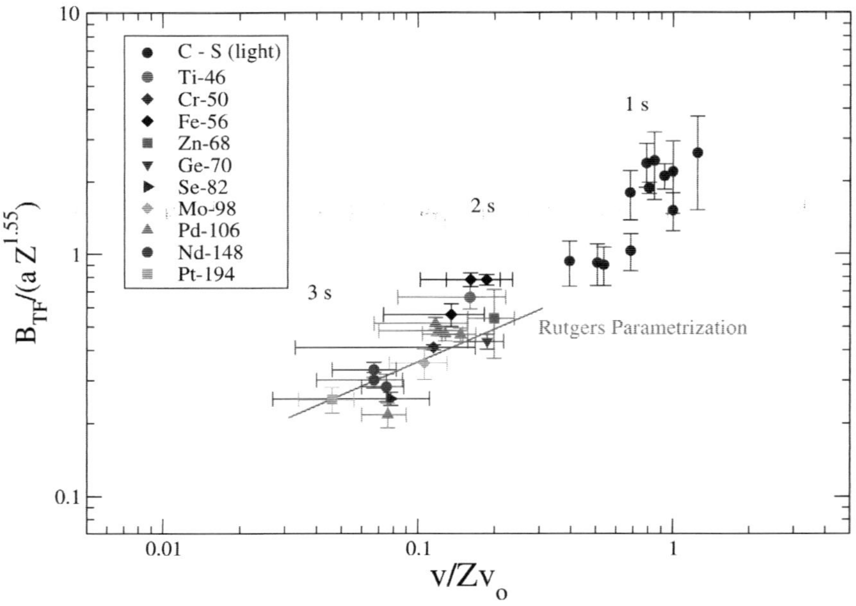

Figure 1. Transient field in gadolinium for nuclei with known g factors

2.3. Detectors and acquisition

The γ rays were detected usually in four Ge detectors or, if possible, in Clover Ge detectors. The light target particles from which the beam was scattered were detected in a forward-situated particle detector (Canberra PIPS, solar cells, or multidetector array).

2.4. Measurement of the slope of the angular correlation

In general, the angular correlation of the γ rays with respect to the recoiling particles is determined by taking coincidence data for various γ-ray detector angles. However, in order to save beam time, the angular correlation can be derived from the precession data using the segmentation of the Clover detectors. For this purpose, the data in Clover segments located at the same angle with respect to the beam direction are analyzed separately. Each Clover detector has a slightly different geometry inside its housing. Hence the angular separation of the detector segments and their relative efficiencies have to be determined for the specific distance between the detectors and the target that is used in the experiment. The angular correlation parameters are then determined as described in Ref. [6].

2.5. Calibration of the transient field

The transient field $\mathbf{B_{TF}}$ was calibrated [8] by measuring the precession of the angular correlation for states for which the g factor was known from independent measurements in nuclei ranging from Ne to Pt [8]. The number of possible calibration points is rather limited and their accuracy varies.

Fig. 1 represents a new summary of calibration points. Only recent measurements for states for which g is known from previous data together with light-ion data compiled by Stuchbery [4] are shown. The velocity of the ions is given in terms of v/Zv_0, a measure of the electronic occupation of the ion in a given charge state. Here v is the ion's velocity, Z is the ion atomic number and $v_0 = e^2/\hbar c$ is the Bohr velocity for the K electron. It is generally believed that

Table 1. g factors in ^{106}Pd.

I^π	τ(ps)	Previous work	Theory Kim [10]	Theory Robinson [13]	This work
2_1^+	17.3(11)	+0.398(21)[11]	+0.392	+0.5	+0.48(1)
4_1^+	1.9			+0.453	+0.47(10)
2_2^+	4.5	+0.30(6) [12]		+0.664	+0.45(10)

the transient field arises mainly from vacancies in either the $1s$, $2s$ or $3s$ electronic shells of the ions traversing the ferromagnetic foil. The hyperfine fields generated by these vacancies vary by orders of magnitude as the major quantum number n of the shell increases. The three regions corresponding to the $1s$ ($v/Zv_0 \approx 1$), $2s$ ($v/Zv_0 \approx 0.1$), and $3s$ ($v/Zv_0 \leq 0.1$) electronic configurations are shown in Fig. 1. The Rutgers parametrization is drawn for the range in which it was originally derived.

Above $v/Zv_0 \approx 1$ the field strength is expected to decrease as more and more ions are fully stripped. New transient field measurements at higher probe ion velocities and with higher statistics suggest a more complex structure than hitherto used in the various parametrization formulas. A better understanding of the transient field requires more precision calibration data and presents a challenge for future experiments.

3. Results and discussion

3.1. The 2_2^+ and 4_1^+ states in ^{106}Pd: is ^{106}Pd a vibrational nucleus?

^{106}Pd lies in the region of the Cd isotopes that have, until recently [9, 10], been good examples of vibrational nuclei. For such nuclei definite predictions have been made for the g factors and the ratios of their transition probabilities from the two-phonon excitations, namely the 4_1^+, 2_2^+, and 0_2^+ states. Specifically, if this triplet of states has a vibrational structure, all three states should have the same g factor, namely $g \approx Z/A$.

The g factor of the 2_1^+ state has been measured using the IMPAC technique by Johansson *et al.*, $g(^{106}\text{Pd}; 2_1^+) = 0.398(21)$ [11] and has been one of the components in the calculation of the Rutgers parametrization. The g factor of the 2_2^+ state has also been measured before [12] but $g(4_1^+)$ has not been measured previously.

The current experiments were run at 230, 280 and 290 MeV and on two different targets. The results are displayed in Table 1. The g factors quoted for this work were calculated assuming a transient field strength given by the Rutgers parametrization. The new value of $g(2_1^+)$ is larger than that measured by Johansson. The reason for this discrepancy is not known.

However, the structure information is fully contained in the *relative* g factors which are independent of any parametrization or calibration. The precision of the g factor measurements for the two-phonon states is relatively low because of the weak excitation of the states even at higher beam energies. Nevertheless, the data show that the g factors for the one-phonon and two-phonon states are indeed comparable.

Two theoretical calculations have been carried out for the $2_1^+, 2_2^+$, and 4_1^+ states in ^{106}Pd (Table 1). K.-H. Kim *et al.* followed an IBM-2 approach [10]. S. J. Q. Robinson [13] used the interaction JJ45PN of Brown and Hjorth-Jensen in shell model calculations which involved 4 proton holes in the $f_{7/2}, p_{3/2}, p_{1/2}$ and $g_{9/2}$ configuration space and 4 neutron holes in the $g_{9/2}$ and $d_{5/2}$ space [14].

The determination of the excitation energies of Kim *et al.* is close to the experimental values, while Robinson's are somewhat high and the wave functions are very fractionated, suggesting

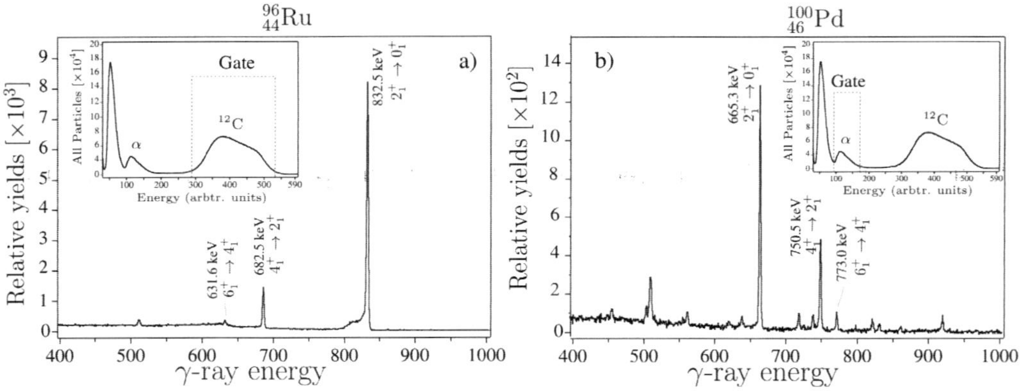

Figure 2. ^{96}Ru and ^{100}Pd γ-ray spectra in coincidence with gates selecting either a)^{12}C recoils or b) α particles from the decay of the residual ^{8}Be nuclei.

collectivity. The B(E2)'s predicted by Robinson are much smaller than the experimental values, while Kim's results are in fair agreement with experiment. Robinson's g factors, are in agreement with the measured values. Kim only calculated $g(2_1^+)$. The quadrupole moment, $Q(2_1^+)$= -0.51(7)b [15], is in disagreement with the vibrational model prediction of $Q = 0$.

In view of the totality of the information it is not possible to assign a definite structure to ^{106}Pd.

3.2. The α capture reaction and ^{100}Pd: $^{12}C(^{96}Ru,^{8}Be)^{100}Pd$

The α-capture from the ^{12}C target by the projectile opens for investigations regions of proton-rich radioactive nuclei not available as beams at existing accelerators. The Bonn group has reported similar measurements of g factors in a few isotopes. The method has the advantage of providing a built-in calibrator because the Coulomb excitation of the beam itself, obtained simultaneously with the products of the α-transfer reaction, serves as a check of the analysis.

The two reactions are clearly separated experimentally when γ spectra are observed in coincidence with either the ^{12}C or the α-particles from the ^{8}Be breakup (Fig. 2).

However, the method has downsides as well: the cross sections are relatively small, the angular correlation is nearly isotropic, higher excited states are populated, and the kinematics are not well defined. Nevertheless, assuming small side feeding, and with reasonable running time, the g factor $g(^{100}\text{Pd};2_1^+) = +0.35(15)$ was obtained.

The resulting g factor, $g(^{100}\text{Pd};2_1^+)$, has a large error that arises from the weak population and small spin alignment of the state of interest (Fig. 3). The option of increasing the beam intensity or energy has to be exercised with caution and applied only if the target temperature can be held low, a challenge for future measurements.

4. Conclusions

- The g factors of the two-phonon states in ^{106}Pd are the same as the g factor of the 2_1^+, one-phonon state, in agreement with the vibrational model predictions. However, the ratios of B(E2) values and the quadrupole moments are not.
- The α-transfer reaction produces a small spin alignment, hence a small sensitivity for

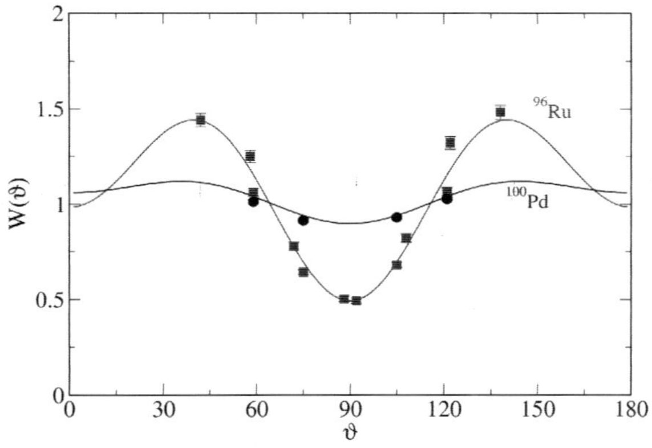

Figure 3. Angular correlations in Coulomb-excited ^{96}Ru and in ^{100}Pd produced in α-transfer reactions.

determination of the precession of the angular correlation.

- New techniques lead to possibly high-precision results but the systematic uncertainties inherent in the parametrization of the transient field need to be reduced. New independent magnetic moment measurements are required as calibration points.

4.1. Acknowledgments

The authors wish to acknowledge the help of their collaborators, D. A. Torres, G. Gürdal, Y. Y. Sharon, L. Zamick, S. J. Q. Robinson, K.-H. Speidel and the Yale WNSL faculty, students, postdocs and staff. The authors also thank the US National Science Foundation for its support.

5. References

[1] Benczer-Koller N and Kumbartzki G J 2007 *J. Phys. G:Nucl.Part.Phys.* **34** R321
[2] Speidel K H, Kenn O and Nowacki F 2003 *Prog. Part. Nucl. Phys.* **49** 91
[3] Mayer R, Kumbartzki G, Weissman L, Benczer-Koller N, Broude C, Cizewski J A, Hass M, Holden J, Janssens R V F, Lauritsen T, Lee I, Macchiavelli A O, McNabb D P and Satteson M 1998 *Phys. Rev.* **C 58** R2640
[4] Stuchbery A E 2004 *Phys. Rev.* **C 69** 064311
[5] Boutachkov P, Benczer-Koller N, Kumbartzki G J, Escuderos A, Sharon Y Y, Zamick L, Robinson S J Q, Ai H, Chamberlain M, Gürdal G, Heinz A, McCutchan E A, Qian J, Werner V, Williams E, Aleksandrova K, Copos C A, Kovacheva D A and Manchev P 2007 *Phys. Rev.* **C 75** 021302(R)
[6] Boutachkov P, Robinson S J Q, Escuderos A, Kumbartzki G, Benczer-Koller N, Stefanova E, Sharon Y Y, Zamick L, McCutchan E A, Werner V, Ai H, Garnsworthy A B, Gürdal G, Heinz A, Qian J, Thompson N J, Williams E and Winkler R 2007 *Phys. Rev.* **C 76** 054311
[7] Mücher D, Gürdal G, Speidel K H, Kumbartzki G J, Benczer-Koller N, Robinson S J Q, Sharon Y Y, Zamick L, Lisetskiy A F, Casperson R J, Heinz A, Krieger B, Leske J, Maier-Komor P, Werner V, Williams E and Winkler R 2009 *Phys. Rev.* **C 79** 054310
[8] Shu N K B, Melnik D, Brennan J M, Semmler W and Benczer-Koller N 1980 *Phys. Rev.* **C 21** 1828
[9] Garrett P E, Green K L and Wood J L 2008 *Phys. Rev.* **C 78** 044307
[10] Kim K H, Gelberg A, Mizusaki T, Otsuka T and von Brentano P 1996 *Nucl. Phys.* **A 604** 163
[11] Johansson K, Karlsson E, Norlin L O, Windahl R Å and Ahmed M R 1972 *Nucl. Phys.* **A 188** 600
[12] Bowman J D, Kaufmann E N, Bhattacherjee S K and Levanoni M 1968 *Phys. Rev. Lett.* **20** 1176
[13] Robinson S J Q 2010 private communication
[14] Hjorth-Jensen M, Kuo T and Osnes E 1995 *Phys. Rep.* **261** 125
[15] Hosoyama K, Torizuka Y, Kawazoe Y and Ui H 1973 *Phys. Rev. Lett.* **30** 388

Assignment of single particle configurations in odd-A nuclei near A~100 with angular correlation measurements

2011 J. Phys.: Conf. Ser. 267 012024

(http://iopscience.iop.org/1742-6596/267/1/012024)

Assignment of single particle configurations in odd-A nuclei near A~100 with angular correlation measurements

A.V. Ramayya*[1], J.H. Hamilton[1], J.K. Hwang[1], S.H. Liu[1], A.V. Daniel[2], C. Goodin[1], Y.X. Luo[3], J.O. Rasmussen[3], N.J. Stone[4,5], S.J. Zhu[6] and K. Li[1]

[1]Physics Department, Vanderbilt University, Nashville, Tennessee 37235, USA
[2]Flerov Laboratory of Nuclear Reactions, JINR, Dubna, Russia
[3]Lawrence Berkeley National Laboratory, Berkeley, California 94720, USA
[4]Department of Physics, Oxford University, Oxford OX1 3PU, United Kingdom
[5]Department of Physics, University of Tennessee, Knoxville, Tennessee 37996, USA
[6]Department of Physics, Tsinghua University, Beijing 100084, People's Republic of China

E-mail: a.v.ramayya@vanderbilt.edu

Abstract. The multipole mixing ratios of $\Delta I = 1$ transitions between levels in rotational bands built on single-particle states in odd neutron nuclei are dependent on the configurations of the states. In particular, the mixing ratio can be used to distinguish between several possible single-particle configurations if interpreted with the particle plus axial-rotor model (PRM). This work features the first determination of the ground-state configurations of ^{109,111}Ru. The single-particle structures of the ground states of ^{101}Zr and ^{103,105,107}Mo as well as excited states in ^{103,107}Mo are also investigated, with a new result found in ^{107}Mo.

1. Introduction

The single particle configurations for odd-A deformed nuclei are based on either Nilsson shell model calculations or by comparing the states with the states in neighboring nuclei where shell model calculations are performed. The multipole mixing ratios of $\Delta I = 1$ transitions between levels in rotational bands built on single-particle states in odd neutron nuclei are dependent on the configurations of the states. In particular, the mixing ratio can be used to distinguish between several possible single-particle configurations if interpreted with PRM. The neutron single-particle configurations of ^{101}Zr, ^{103,105,107}Mo, and ^{109,111}Ru are determined by measuring the mixing ratios of $\Delta I = 1$ transitions in their rotational bands. The mixing ratios are interpreted in terms of PRM. These nuclei are known to be highly deformed [1], with the $h_{11/2}$ neutron orbital playing a role in driving the deformation although the proton-neutron interaction may be also important. These nuclei are produced in spontaneous fission of ^{252}Cf. In this paper, we described the technique of measuring angular correlations with large gamma detector arrays and discussed the assignment of single particle configurations. The details on angular correlation techniques and the measurements can be found in references [2,3].

The advantages of doing angular correlation experiments with large detector arrays such as Gammasphere as opposed to traditional setups with 2–4 detectors are major increase in angular resolution and detection efficiency. However, there are many difficulties for measuring angular correlations with such large detector arrays. Specifically, the number of pairs increases as the square of the number of detectors. Since the detector

---

* To whom any correspondence should be addressed.

efficiency and the solid angle correction factors are different for each pair, handling a problem of this magnitude requires careful analysis techniques, and we describe these below. Following the description of our method, we show results for unattenuated angular correlations.

2. Experimental technique and analysis procedure

The data for this analysis were taken by using the Gammasphere detector array, which was located at Lawrence Berkeley National Laboratory. A ^{252}Cf spontaneous fission source with an α activity of 62 μCi was sandwiched between two iron foils. The foils were thick enough to stop the fission fragments, eliminating the need for a Doppler correction. Approximately 5.7×10^{11} three- and higher-fold γ coincidence events were recorded.

The coincidence intensity between a cascade of two γ rays as a function of angle is fitted to Eq. (1).

$$W(\theta) = 1 + G_2A_2(\delta)P_2(\cos\theta) + G_4A_4(\delta)P_4(\cos\theta) \quad (1)$$

In this equation the A_k s are calculable coefficients depending on the spin sequence of the levels involved and the multipolarities and mixing ratios of the transitions between them. The G_k coefficients relate to attenuations of the correlations and are equal to unity unless some interaction is present during the lifetime of the intermediate state of the cascade that can cause precession of the nuclear spin. For the cascades involved in this work, the nuclear hyperfine field in iron is less than about 50 T. Unless the life-time of the intermediate state is greater than ~0.7 ns, the G_ks are close to unity.

The mixing ratios of $\Delta I = 1$ transitions within a rotational band are given by

$$\delta\left(\frac{E2}{M1}\right) = \sqrt{\frac{5}{12}\frac{1}{K}\frac{E_\gamma}{1.2\times10^3}\frac{Q_0}{(g_K-g_R)}\frac{<I_fK20|I_iK>}{<I_fK10|I_iK>}}, \quad (2)$$

where Q_0 is the quadrupole moment; g_K and g_R are the intrinsic and rotational g factors, respectively; and E_γ is the $\Delta I=1$ transition energy in keV. For these calculations, $2g_R$ is taken to be equal to Z/A and the quadrupole moment Q_0 is calculated from the deformation parameter, β, using the relation

$$\beta = 91.7\frac{Q_0}{ZA^{2/3}} \quad (3)$$

with β assumed to take a common value, β = 0.3. The single-particle g_K factor depends on the Nilsson configuration of the particle and can be calculated by using the $M1$ matrix elements tabulated in Ref. [4] for various single-particle configurations by

$$g_K = g_l + \frac{g_s-g_l}{2K}\Sigma(\alpha^2_{l,K-\frac{1}{2}} + \alpha^2_{l,K+\frac{1}{2}}) \quad (4)$$

where the α's are Nilsson coefficients tabulated in Ref. [4] as a function of the deformation parameter β, and g_l, g_s are the orbital and spin g factors. In these calculations, which involve neutron states, the values $g_l=0$ and $g_s^{eff} = 0.6g_s^{free} = -2.296$ are used.

Gammasphere consists of 110 germanium detectors, corresponding to 5995 unique detector pairs. The detectors are placed at 17 different azimuthal angles and 60 various polar angles. For each azimuthal angle, there are 5–10 detectors placed symmetrically with respect to the polar angle. Due to the symmetries of Gammasphere, the angle between any two detectors will be one of only 64 possible values, and each of these 64 angle bins has many pairs of detectors. For our experiment, only 101 detectors were present, so the number of detectors in each bin is slightly less than if Gammasphere were at full capacity. The angular properties of Gammasphere and the process of determining angle bins are given in greater detail in [5]. The angle bin information for our experiment is given in Table 1. For the angular correlation, we sorted our triple coincidence data into 64 two dimensional histograms corresponding to the 64 angle bins. The histograms were produced by reading the full data set event by event. For each γ–γ coincidence, the angle between the detectors which detected γ_1 and γ_2 was calculated and the event was added to the appropriate histogram. For each bin, n, the histogram was then fit to find the number of coincidences of the cascade of interest, N_n. This N_n must be corrected for relative detector efficiency, the number of detectors in the particular bin, and the detector response functions, all discussed below. The value for the corrected N_n is given by

$$N_n = \varepsilon_n(E_1, E_2) \int_0^{\pi} W(\theta) R_n(\theta, E_1, E_2) \sin(\theta)\, d\theta, \qquad (5)$$

where $\varepsilon_n(E1,E2)$ is the function describing the relative efficiency of the pair and $R_n(\theta,E1,E2)$ is the response function of the pair. The 64 histograms can also be made with additional gates when added selectivity is needed. For example, when considering the 4^+ - 2^+ - 0^+ cascade in ^{110}Ru, the events in each histogram might be required to be in coincidence with the 2^+ - 0^+ transitions in its fission partners ^{138,140}Xe.

Table 1. Angle bins in Gammasphere with the 101 detectors used in our experiment.

Bin	cos(θ)	Num. Pairs	Bin	cos(θ)	Num. Pairs	Bin	cos(θ)	Num. Pairs
1	0.939	203	23	0.309	56	45	-0.391	90
2	0.934	58	24	0.298	100	46	-0.471	52
3	0.840	47	25	0.282	101	47	-0.495	89
4	0.823	45	26	0.269	104	48	-0.500	56
5	0.815	104	27	0.175	100	49	-0.577	116
6	0.809	56	28	0.168	44	50	-0.580	101
7	0.764	100	29	0.139	23	51	-0.594	90
8	0.762	45	30	0.064	92	52	-0.613	104
9	0.755	50	31	0.057	104	53	-0.656	50
10	0.745	30	32	0.000	273	54	-0.745	30
11	0.656	50	33	-0.057	104	55	-0.755	50
12	0.613	104	34	-0.064	90	56	-0.762	45
13	0.594	88	35	-0.139	23	57	-0.764	100
14	0.580	100	36	-0.168	44	58	-0.809	56
15	0.577	116	37	-0.175	101	59	-0.815	104
16	0.500	56	38	-0.269	104	60	-0.823	44
17	0.495	88	39	-0.282	101	61	-0.840	44
18	0.471	52	40	-0.298	100	62	-0.934	58
19	0.391	91	41	-0.309	56	63	-0.939	196
20	0.357	58	42	-0.327	46	64	-1.000	46
21	0.333	60	43	-0.333	60			
22	0.327	44	44	-0.357	58			

The number of coincidences in each angle bin is corrected for the relative efficiencies of the detector pairs. The efficiencies of the individual detectors of Gammasphere may vary greatly compared to the mean efficiency. If the detector pairs i,j belong to bin n, the efficiency of the bin is given by

$$\epsilon_n(E_1, E_2) = \tfrac{1}{2} \textstyle\sum_{i,j} [\epsilon_i(E_1)\epsilon_j(E_2) + \epsilon_i(E_2)\epsilon_j(E_1)]. \qquad (6)$$

To calculate the $\varepsilon_{i,j}(E)$'s, a linear γ spectrum was constructed for each detector. A separate linear spectrum was constructed for the sum of all detectors. The summed spectrum was divided by the number of working detectors and then the individual detector spectra were divided by the normalized sum spectrum. The resulting spectra give the relative efficiency curve for each detector.

For a typical angular correlation measurement, it is necessary to calculate a solid angle correction factor Q_k for each parameter A_k. However, for very low intensity transitions, the sensitivity of the angular correlation measurement can be improved by determining the detector response function $R_n(\theta,E1,E2)$ for each pair of detectors. For a given detector pair, the response function describes the distribution of possible angles about the central angle of the pair as a function of energy. The response functions for each pair can then be summed to find the response function of each angle bin.

We calculated the response function using a simple Monte Carlo simulation, with the γ-ray transport simulated up to the first collision. This is equivalent to the traditional calculation of Q_k. The mean free path, $\lambda(E)$, of γ-rays was calculated using the known Gammasphere detector properties. The energy dependence of $R_n(\theta,E1,E2)$ is negligible, and so only $R_n(\theta)$ was calculated.

To determine the angular correlation, each of the 64 histograms was fit to find the intensity of the peak of interest, N_n. Our fitting method is based on the analysis of two dimensional γ–γ coincidence spectra. To fit a given histogram, a window within the histogram is selected. Within this window, the positions of the peaks are defined by using projections of the histogram on the two axes. The surface is then approximated by three types of background and the sum of the two dimensional γ peaks found in the projections. The three types of background consist of a smooth background and two series of ridges parallel to the axes corresponding to γ lines in the x and y direction. The last part of the fitting procedure is to solve a well known NNLS (non-negative least square) problem [6].

Using the N_n^{exp} found from fitting the histograms, we can estimate the two parameters A_2 and A_4 using the minimization expression

$$min \sum_n \frac{(N^{exp}(\theta_n)-N(\theta_n))^2}{\delta^2_{n,exp}+\delta^2_{n.model}}. \qquad (7)$$

The experimental uncertainty in the γ peak intensity is always much greater than the model error, so δ_{model} was neglected. The 64 angle bins are further grouped into 17 larger bins, thereby increasing the statistics of each bin. Since we have to determine only three parameters, 17 data points are more than adequate.

3. Experimental results

One example of angular correlation measurements is shown in Fig. 1. In Fig 1, ^{138}Ba has the 2^+ level with a very short lifetime of 0.202(8) ps which indicates that the correlation should be unattenuated. Then, we obtained the expected value of $G_2 \cong 1$. The consistency of this result with the expected value shows the viability of our method for measuring angular correlations with Gammasphere.

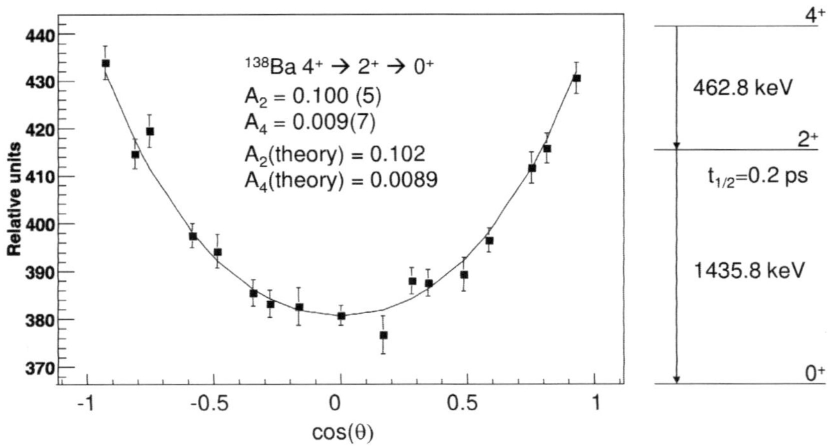

Fig. 1. Angular correlation of the $4^+\rightarrow2^+\rightarrow0^+$ cascade in ^{138}Ba.
The levels are shown on the right.

The relevant partial level schemes of odd-A Mo and Ru nuclei are shown in Fig. 2. In Table 2 are shown the A_2 and A_4 coefficients and also the extracted δ values for the ΔI=1 transitions within the rotational bands. To interpret the mixing ratios of the ΔI=1 transitions within rotational bands, the experimental values were compared to calculations made with a particle plus axial rotor model for various single-particle states. In the fifth and sixth columns in Table 2 are shown the calculated δ values for various assumed configurations.

Table 2. Angular correlation Angular correlations measured in this work. In most cases, there were two values for the experimental mixing ratio (δ(exp)) of the γ transition in parentheses. The calculated mixing ratios (δ (cal)) are compared with the experimental mixing ratios to determine the configuration of the γ transition.

Nucleus	Cascade	A_2^{exp}, A_4^{exp}	δ (exp)	δ (cal)	Configuration
^{101}Zr	9/2$^+$→ 5/2$^+$ → 3/2$^+$	−0.12(2),−0.00(4)	−0.16(9)	−0.13	3/2[411]
	310.0-(98.2)		−2.1(4)		
	7/2$^-$ → 5/2$^+$ → 3/2$^+$	0.08(2), 0.02(3)	−0.13(10)		
	222.9-(98.2)		−2.2(6)		
^{103}Mo	7/2$^-$ → 5/2$^+$ → 3/2$^+$	0.09(1), 0.01(2)	−0.19(5)	−0.15	3/2[411]
	251.4-(102.8)		−1.9(2)		
	11/2$^-$ → 7/2$^-$ → 5/2$^+$	−0.05(1), 0.01(2)	*g* factor		
	144.5-251.4		No mixing		
	13/2$^-$ → 9/2$^-$ → 7/2$^-$	−0.23(3), − 0.04(5)	−0.29 to −1.94	−0.15	5/2[532]
	372.3-(124.9)		Single range		
	9/2$^-$ → 7/2$^-$ → 5/2$^+$	+0.25(4), − 0.07(6)	$-0.49^{+0.14}_{-0.22}$		
	(124.9)-251.4		$-2.66^{+0.92}_{-1.40}$		
^{105}Mo	11/2$^-$ → 7/2$^-$ → 5/2$^-$	−0.12(1), 0.02(2)	−0.12(3)	−0.15	5/2[532]
	283.2-(95.3)		−2.9(5)		
	13/2$^-$ → 9/2$^-$ → 7/2$^-$	−0.17(1),−0.01(2)	−0.25(4)	−0.17	5/2[532]
	390.6-(138.3)		−2.3(2)		
^{107}Mo	9/2$^-$ → 7/2$^+$ → 5/2$^+$	0.17(1),−0.00(3)	−0.44 to −1.44	−0.20	5/2[402]
	306.4-(152.1)		Single range		
	11/2$^+$ → 7/2$^+$ → 5/2$^+$	−0.21(3),−0.01(4)	−0.31 to −1.76		
	414.5-(152.1)		Single range		
	9/2$^-$ → 7/2$^-$ → 5/2$^+$	0.14(3), 0.06(4)	−0.18(9)	−0.16	7/2[523]
	(110.2)-348.3		$-0.0012^{+0.057}_{-0.0001}$		
^{109}Ru	11/2$^+$ → 7/2$^+$ → 5/2$^+$	−0.16(2), 0.01(3)	−0.25(6)	−0.26	5/2[402]
	472.8-(185.1)		−2.0(3)		
	13/2$^+$ → 9/2$^+$ → 7/2$^+$	−0.20(3),−0.03(4)	$-0.35^{+0.09}_{-0.12}$	−0.24	5/2[402]
	540.7-(222.7)		−1.8(4)		
^{111}Ru	9/2$^-$ → 7/2$^+$ → 5/2$^+$	0.15(2), 0.03(2)	$-1.29^{+0.94}_{-0.34}$		
	166.6-(150.2)		−0.48(5)	−0.21	5/2[402]
	7/2$^-$ → 7/2$^+$ → 5/2$^+$	−0.37(1),−0.00(2)	−0.32(2)		
	103.8-(150.2)		−1.73(9)		
	11/2$^+$ → 7/2$^+$ → 5/2$^+$	−0.22(1),−0.03(4)	−1.23(16)		
	431.3-(150.2)		−0.51(7)		

Orlandi *et al.* [7] established the 3/2[411] assignment of the ground band of ^{101}Zr, which is also supported in this work. They found the ground band in ^{103}Mo to have the same configuration, which is also supported by this work. They further showed the negative parity 5/2$^-$ bandhead at 347 keV in ^{103}Mo to have the 5/2[532] configuration and identified this configuration as the ground-state band in ^{105}Mo. The present results in ^{105}Mo support this assignment. In ^{107}Mo, the configuration of the ground state was assigned by Urban *et al.* [8] to be 5/2[413] based on the measurement of *E2/M1* branching ratios in the rotational band. Two independent values of the mixing ratio of the 152.1-keV transition between the lowest levels of this band in ^{107}Mo, are reported in this work. They are both negative and, taking the stronger $M1$ limit from the experimental fit ranges, give a δ around −0.5, showing strong preference for the 5/2[402] assignment over the 5/2[413] for which the predicted δ value is positive and close to unity. It would seem that Urban *et al.* [8], who obtained δ, without sign, chose δ positive contrary to the present direct sign measurement. For the negative-parity band-head at 348.4 keV, the configuration was assigned 7/2[523] [8], which is supported by the present result on mixing in the 110.2-keV transition. In ^{109,111}Ru, Wu *et al.* [9] attempted to determine the single-particle configurations of the ground-state bands based on $E2/M1$ branching ratios but were not able to distinguish between the 5/2[413] and 5/2[402] configurations. In this work angular correlation data gives clear evidence that the ground states of ^{109,111}Ru both have a 5/2[402] configuration. Calculations for a 5/2[402] state yield δ = −0.26 and δ = −0.21, respectively, in excellent agreement with the measured values of −0.25(6) and −0.32(2), whereas calculations for the 5/2[413] configuration yield large positive values.

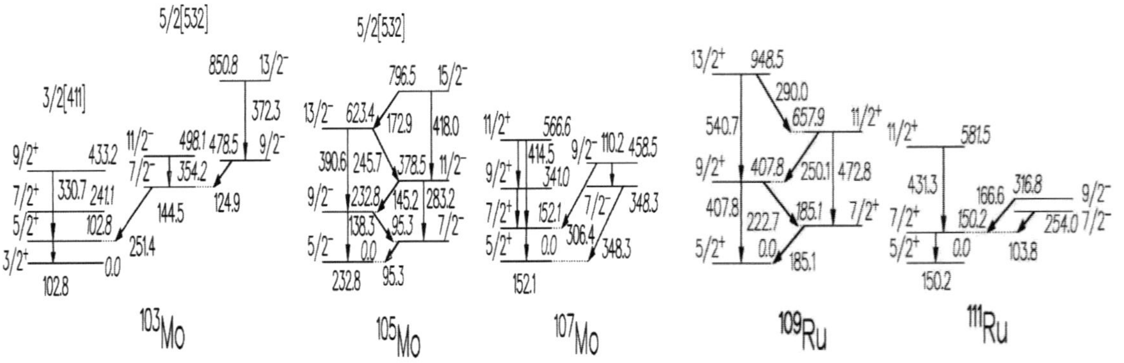

Fig. 2.

References

[1] Skalski J, Mizutori S, and Nazarewicz W 1997 *Nucl. Phys.* A **617** 282.
[2] Daniel A V et al. 2007 *Nucl. Inst. And Meth. in Phys. Res.* B **262** 399.
[3] Goodin C et al. 2009 *Phys. Rev.* C **80** 014318.
[4] Browne E and Femenia F 1971 *Nucl. Data Tables* **10** 81 and Nilsson S G 1955 *Nat. Fys. Medd. Dan. Vis. Selsk.* **29**.
[5] Patel D et al. 2002 *J. Phys.* G **28** 649.
[6] Lawson C L and Hanson R J 1974 *Solving Least Square Problems* (NJ: Prentice Hall, Englewood Cliffs).
[7] Orlandi R *et al.* 2006 *Phys. Rev.* C **73** 054310.
[8] Urban W *et al.* 2005 *Phys. Rev.* C **72** 027302.
[9] Wu C Y *et al.* 2006 *Phys. Rev.* C **73** 034312.

A separable approximation for Skyrme interactions and charge-exchange excitations

2011 J. Phys.: Conf. Ser. 267 012025

(http://iopscience.iop.org/1742-6596/267/1/012025)

A separable approximation for Skyrme interactions and charge-exchange excitations

A P Severyukhin[1], **V V Voronov**[1] **and Nguyen Van Giai**[2]

[1]Bogoliubov Laboratory of Theoretical Physics, Joint Institute for Nuclear Research, 141980 Dubna, Moscow region, Russia

[2]Institut de Physique Nucléaire, CNRS-IN2P3, Université Paris-Sud, F-91406 Orsay Cedex, France

E-mail: `voronov@theor.jinr.ru`

Abstract. A finite rank separable approximation for the quasiparticle random phase approximation with Skyrme interactions is applied to study properties of the Gamow-Teller resonances and the spin-dipole resonances in some spherical nuclei. It is shown that characteristics calculated within the suggested approach are in a good agreement with available experimental data.

1. Introduction

A study of the charge-exchange nuclear modes is an interesting problem not only from the nuclear structure point of view but it is very important for the nuclear astrophysics applications. Many versions of the self-consistent approach based on the Random Phase Approximation (RPA) or the quasiparticle RPA (QRPA) were developed during last years, see for example [1-10]. A comparison of such calculations with recent experimental data demonstrates that model mentioned above can not reproduced correctly the strength distribution of the spin-isospin resonances. It is necessary to take into account a coupling with more complex configurations that results in shifting some strength up [11, 12, 13]. Moreover the tensor forces can give an additional shift [10, 11]. It is much simpler to include a coupling in QRPA calculations if one uses separable forces [12, 13, 14]. This idea stimulated us to develop the finite rank approximation for the Skyrme interactions [15, 16] that enables one to perform calculations in the large configuration space. Applications of our method to study the low-lying states and giant resonances within the QRPA and beyond are given in [15, 16, 17, 18]. Recently, we have proposed an extension of our approach for the charge-exchange nuclear excitations [19]. Before to investigate effects of the coupling we need to be sure that the finite rank approximation for the Skyrme interactions is enough good to reproduce main characteristics of such nuclear modes.

In this report we describe briefly our method for the charge-exchange excitations and present our investigations of properties of the Gamow-Teller resonances (GTR) and the spin-dipole resonances (SDR) in some spherical nuclei.

2. Method of calculation

The starting point of the method is the HF-BCS calculation [23] of the parent ground state, where spherical symmetry is imposed on the quasiparticle wave functions. The continuous part of the single-particle spectrum is discretized by diagonalizing the Skyrme HF hamiltonian on a harmonic oscillator basis. We work in the quasiparticle representation defined by the canonical Bogoliubov transformation:

$$a_{jm}^{+} = u_j \alpha_{jm}^{+} + (-1)^{j-m} v_j \alpha_{j-m} \tag{1}$$

where jm denote the quantum numbers $nljm$. We use the Skyrme interaction [20] in the p-h channel, while the pairing correlations are generated by the surface peaked density-dependent zero-range force

$$V_{pair}(\mathbf{r}_1, \mathbf{r}_2) = V_0 \left(1 - \frac{\rho(r_1)}{\rho_c}\right) \delta(\mathbf{r}_1 - \mathbf{r}_2) \tag{2}$$

Here $\rho(r_1)$ is the particle density in the coordinate space, ρ_c is equal to the nuclear saturation density, the strength V_0 is the parameter fixed to reproduce the odd-even mass difference of nuclei in the study region. In order to avoid divergences, it is necessary to introduce a cutoff in the single-particle space. This cutoff limits the active pairing space above the Fermi level. We used a procedure that is proposed in Ref. [16, 21].

The residual interaction in the p-h channel V_{res} can be obtained as the second derivative of the energy density functional with respect to the particle density ρ. Following our previous paper [15] we simplify V_{res} by approximating it by its Landau-Migdal form. For Skyrme interactions all Landau parameters with $l > 1$ are zero and they are functions of the coordinate $\mathbf{r}$. We keep only the $l = 0$ terms in V_{res} and the expressions for F_0', G_0' in terms of the Skyrme force parameters can be found in Ref. [22]. The Coulomb and spin-orbit residual interactions are dropped. Therefore we can write the residual interaction in the isovector channel in the following form:

$$V_{res}(\mathbf{r}_1, \mathbf{r}_2) = N_0^{-1}[F_0'(r_1) + G_0'(r_1)\sigma^{(1)} \cdot \sigma^{(2)}]\tau^{(1)}\tau^{(2)}\delta(\mathbf{r}_1 - \mathbf{r}_2) \tag{3}$$

where $\sigma^{(i)}$ and $\tau^{(i)}$ are the spin and isospin operators, and $N_0 = 2k_F m^*/\pi^2\hbar^2$ with k_F and m^* standing for the Fermi momentum and nucleon effective mass in nuclear matter.

In what follows we use the second quantized representation and $\hat{V}_{res}$ can be written as:

$$\hat{V}_{res} = \frac{1}{4}\sum_{1234} (V_{1234} - V_{1243}) : a_1^{+} a_2^{+} a_4 a_3 := \hat{V}_M + \hat{V}_{SM} \tag{4}$$

where a_1^{+} (a_1) is the particle creation (annihilation) operator and 1 denotes quantum numbers (n_1, l_1, j_1, m_1). Following our method [15, 16, 17, 18] the residual interactions can be reduced to a finite rank separable form:

$$\hat{V}_M = -2\sum_{JM}\sum_{k=1}^{N} \kappa_F^{(k)} : \hat{M}_{JM}^{(k)+} \hat{M}_{JM}^{(k)} :, \tag{5}$$

$$\hat{V}_{SM} = -2\sum_{JM}\sum_{L=J;J\pm1}\sum_{k=1}^{N} \kappa_G^{(k)} : \hat{S}_{LJM}^{(k)+} \hat{S}_{LJM}^{(k)} :, \tag{6}$$

$$\begin{pmatrix} \kappa_F^{(k)} \\ \kappa_G^{(k)} \end{pmatrix} = -N_0^{-1}\frac{Rw_k}{2r_k^2}\begin{pmatrix} F_0'(r_k) \\ G_0'(r_k) \end{pmatrix}, \tag{7}$$

where R is a large enough cutoff radius for a N-point integration Gauss formula with abscissas r_k and weights w_k[15].

The operators entering the normal products in Eqs.(5,6) are defined as follows:

$$\hat{M}^{(k)+}_{JM} = (-1)^{J-M}\,\hat{J}^{-1} \sum_{j_n j_p m_n m_p} \langle j_n m_n j_p - m_p \mid JM\rangle \times f^{(Jk)}_{j_n j_p}(-1)^{j_p - m_p} a^{+}_{j_n m_n} a_{j_p m_p}, \tag{8}$$

$$\hat{S}^{(k)+}_{LJM} = (-1)^{J-M}\,\hat{J}^{-1} \sum_{j_n j_p m_n m_p} \langle j_n m_n j_p - m_p \mid JM\rangle \times g^{(LJk)}_{j_n j_p}(-1)^{j_p - m_p} a^{+}_{j_n m_n} a_{j_p m_p}. \tag{9}$$

$f^{(Jk)}_{j_n j_p}$ are the single-particle matrix elements of the multipole operators,

$$f^{(Jk)}_{j_n j_p} = u_{j_n}(r_k) u_{j_p}(r_k) \langle j_n || i^J Y_J || j_p\rangle, \tag{10}$$

and $g^{(LJk)}_{j_n j_p}$ are the single-particle matrix elements of the spin-multipole operators,

$$g^{(LJk)}_{j_n j_p} = u_{j_n}(r_k) u_{j_p}(r_k) \langle j_n || i^L T_{LJ} || j_p\rangle. \tag{11}$$

In the above equations, $\langle j_n || i^J Y_J || j_p\rangle$ is the reduced matrix element of the spherical harmonics Y_{JM}, $\hat{J} = \sqrt{2J+1}$, $T_{LJM}(\hat{r},\sigma) = [Y_L \times \sigma]^M_J$. The radial wave functions $u_j(r)$ are related to the HF single-particle wave functions,

$$\phi_{i,m}(1) = \frac{u_i(r_1)}{r_1} \mathcal{Y}^m_{l_i,j_i}(\hat{r_1},\sigma_1). \tag{12}$$

We introduce the phonon creation operators

$$Q^{+}_{JMi} = \sum_{j_n j_p} \left(X^{Ji}_{j_n j_p} A^{+}(j_n j_p; JM) - (-1)^{J-M} Y^{Ji}_{j_n j_p} A(j_n j_p; J-M) \right), \tag{13}$$

$$A^{+}(j_n j_p; JM) = \sum_{m_n m_p} \langle j_n m_n j_p m_p \mid JM\rangle \alpha^{+}_{j_n m_n} \alpha^{+}_{j_p m_p}, \tag{14}$$

where the index J denotes total angular momentum and M is its z-projection in the laboratory system. One assumes that the ground state is the QRPA phonon vacuum $\mid 0\rangle$ and the excited states are defined as $Q^{+}_{JMi} \mid 0\rangle$. Making use of the linearized equation-of-motion approach one can get the QRPA equations [23]:

$$\begin{pmatrix} \mathcal{A} & \mathcal{B} \\ -\mathcal{B} & -\mathcal{A} \end{pmatrix} \begin{pmatrix} X \\ Y \end{pmatrix} = \omega \begin{pmatrix} X \\ Y \end{pmatrix}. \tag{15}$$

$\mathcal{A}^{(\lambda)}_{(j_1 j_1')(j_2 j_2')}$ matrix is related to forward-going graphs and the $\mathcal{B}^{(\lambda)}_{(j_1 j_1')(j_2 j_2')}$ matrix is related to backward-going graphs. The dimension of the matrices $\mathcal{A}, \mathcal{B}$ is the space size of the two-quasiparticle configurations. The explicit solution of the corresponding QRPA equations is described in detail elsewhere [19]. Our approach enables one to reduce remarkably the dimensions of the matrices that must be inverted to perform nuclear structure calculations in very large configuration spaces. It is shown that the matrix dimensions never exceed $4N \times 4N$ independently of the configuration space size.

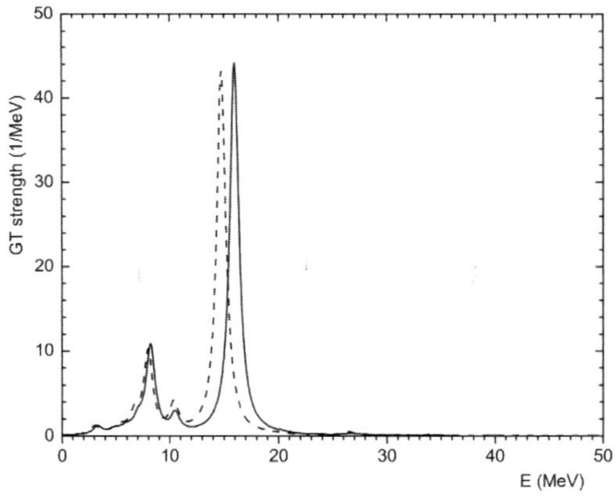

Figure 1. Gamow-Teller strength distribution for ^{132}Sn. The calculation within the LM treatment for the p-h interaction (dashed lines), and the calculation with the full velocity-dependent terms of the p-h interaction (solid lines) are shown.

Table 1. Ikeda sum rule values. S_-, S_+ are the sum rule values of the T_- and T_+ channels, respectively.

		S_-	S_+	$S_- - S_+$
^{90}Zr	full	30.5	0.5	30.0
	LM	30.5	0.5	30.0
	Expt.	28.0±1.6	1.0±0.3	27.0±1.6
^{132}Sn	full	96.8	0.8	96.0
	LM	96.8	0.8	96.0

3. Results

In this work, we use the SGII parameterization [22] of the Skyrme force. As the first example we consider the Gamow-Teller (GT) excitations in ^{90}Zr and ^{132}Sn. We calculate the Ikeda sum rule for the T_- and T_+ channels with the LM forces and compare with results of calculations with the full Skyrme interactions. Results of our calculations for ^{90}Zr and ^{132}Sn are shown in table 1. To compare with experimental data in ^{90}Zr we use the same energy interval $E_x \leq 50MeV$ for the T_- channel and $E_x \leq 32MeV$ for the T_+ channel [24, 25]. One can see that the LM approximation for the Skyrme forces reproduces experimental data for the Ikeda sum rules and gives the same results as calculations with the full Skyrme interactions. This conclusion is valid for the ^{132}Sn case also.

The GT strength distributions in ^{132}Sn that were calculated within different approaches are shown in the Fig. 1. To present our results we use the energy averaging procedure with the Lorentz weight function [14]. The Lorentzian smearing parameter is taken equal to 1 MeV. It seen from this figure that the GT strength distributions calculated with LM and Skyrme forces are similar but the GT resonace for the LM case is shifted down by 1 MeV. This is due to an approximation for the velocity-depended terms for the LM forces. It is worthwhile to mention that, an inclusion of the tensor correlations can shift up to 10 % of the GT strength to the energy region above 30 MeV [10, 26, 27]. Moreover one needs to go beyond the QRPA[12] and such calculations are in progress now.

Let us consider the spin-dipole (SD) excitations. Results of our calculations for the SD states

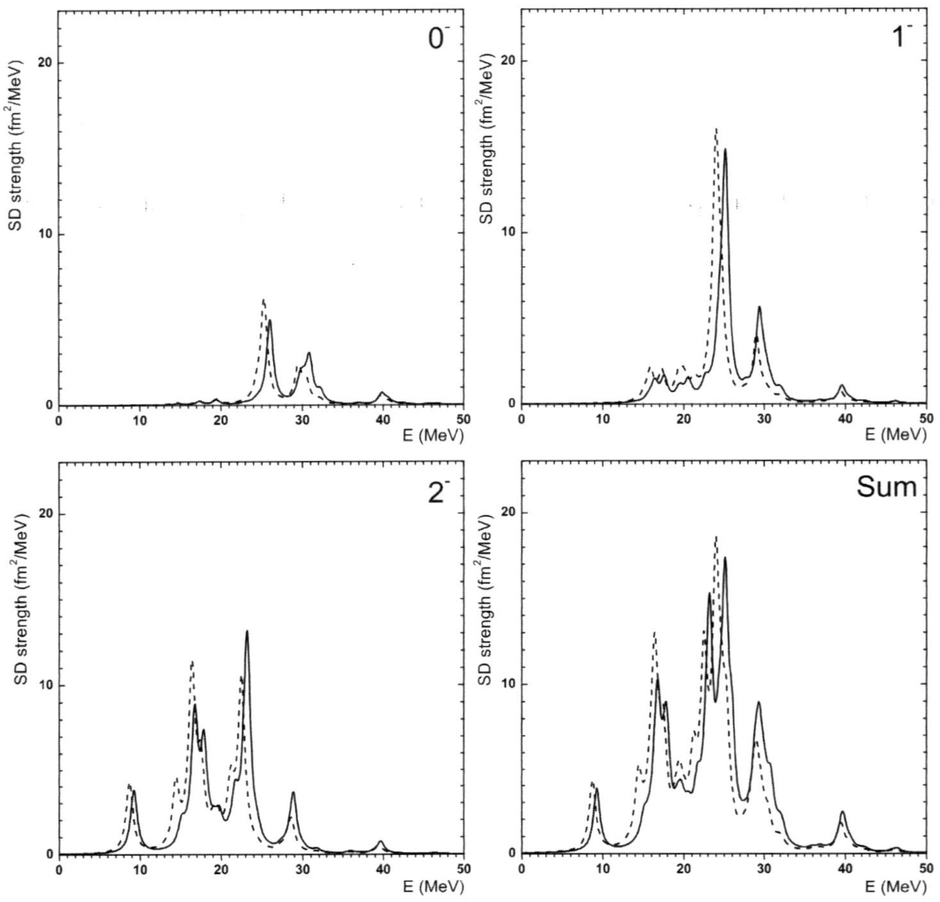

Figure 2. Strength distributions in the T_- channel of the spin-dipole excitations in ^{48}Ca. See the legend to Fig. 1 for details.

with different values of the spin and parity for the T_- channel in ^{48}Ca are shown in Fig.2. One can conclude that our approach reproduces the SD strength distribution for different states that are alike the full Skyrme force.

4. Conclusions

A finite rank separable approximation for the QRPA calculations with Skyrme interactions, which was proposed in our previous work, is applied to study the charge-exchange nuclear modes. The suggested approach enables one to reduce remarkably the dimensions of the matrices that must be inverted to perform nuclear structure calculations in very large configuration spaces. As an illustration we present results of our investigations for the Gamow-Teller and spin-dipole resonances in some spherical nuclei. A comparison of our results with experimental data and other theoretical calculations demonstrates that our method gives a good description of properties of the charge-exchange excitations. An inclusion of the tensor terms of the Skyrme interaction and taking into account the coupling of the QRPA states with more complex components of the wave functions are in progress now.

Acknowledgments

We are grateful to Prof. H. Sagawa for valuable discussions. A.P.S. and V.V.V. thank the hospitality of IPN-Orsay where a part of this work was done. This work is partly supported by the IN2P3-JINR agreement.

References

[1] Engel J, Bender M, Dobaczewski J, Nazarewicz W and Surman R 1999 *Phys. Rev.* C **60** 014302
[2] Khan E, Sandulescu N, Grasso M and Nguyen Van Giai 2002 *Phys. Rev.* C **66** 024309
[3] Péru S, Berger J F and Bortignon P F 2005 *Eur. Phys. J.* A **26** 25
[4] Vretenar D, Afanasjev A V, Lalazissis G A and Ring P 2005 *Phys. Rep.* **409** 101
[5] Terasaki J and Engel J 2006 *Phys. Rev.* C **74** 044301
[6] Fracasso S and Coló G 2007 *Phys. Rev.* C **76** 044307
[7] Sagawa H, Yoshida S, Zhou X-R, Yako K and Sakai H 2007 *Phys. Rev.* C **76** 024301
[8] Paar N, Vretenar D, Khan E and Coló G 2007 *Rep. Prog. Phys.* **70** 691
[9] Haozhao Liang, Nguyen Van Giai and Jie Meng 2008 *Phys. Rev. Lett.* **101** 122502
[10] Bai C L, Sagawa H, Zhang H Q, Zhang X Z, Coló G and Xu F R 2009 *Phys. Lett.* B **675** 28
[11] Bertsch G F and Hamamoto I 1982 *Phys. Rev.* C **26** 1323
[12] Kuzmin V A and Soloviev V G 1984 *J. Phys.* G **10** 1507
[13] Kuzmin V A and Soloviev V G 1985 *J. Phys.* G **11** 603
[14] Soloviev V G 1992 *Theory of Atomic Nuclei: Quasiparticles and Phonons* (Bristol and Philadelphia: Institute of Physics)
[15] Nguyen Van Giai, Stoyanov Ch and Voronov V V 1998 *Phys. Rev.* C **57** 1204
[16] Severyukhin A P, Voronov V V and Nguyen Van Giai 2008 *Phys. Rev.* C **77** 024322
[17] Severyukhin A P, Stoyanov Ch, Voronov V V and Nguyen Van Giai 2002 *Phys. Rev.* C **66** 034304
[18] Severyukhin A P, Voronov V V and Nguyen Van Giai 2004 *Eur. Phys. J.* A **22** 397
[19] Severyukhin A P, Voronov V V and Nguyen Van Giai submitted to *Phys. Rev.* C
[20] Vautherin D and Brink D M 1972 *Phys. Rev.* C **5** 626
[21] Krieger S J *et al* 1990 *Nucl. Phys.* A **517** 275
[22] Nguyen Van Giai and Sagawa H 1981 *Phys. Lett.* B **106** 379
[23] Ring P and Schuck P 1980 *The Nuclear Many Body Problem* (Berlin: Springer)
[24] Wakasa T *et al* 1997 *Phys. Rev.* C **55** 2909
[25] Raywood K J *et al* 1990 *Phys. Rev.* C **41** 2836
[26] Bai C L, Zhang H Q, Zhang X Z, Xu F R, Sagawa H and Coló G (2009) *Phys. Rev.* C **79** 041301(R)
[27] Severyukhin A P and Sagawa H in preparation.

An upgraded algorithm for shell model calculations and its implementation in medium-heavy nuclei

2011 J. Phys.: Conf. Ser. 267 012026

(http://iopscience.iop.org/1742-6596/267/1/012026)

An upgraded algorithm for shell model calculations and its implementation in medium-heavy nuclei

Bianco D[1], Andreozzi F[1], Lo Iudice N[1], Porrino A[1], Knapp F[2, 3]

[1]Dipartimento di Scienze Fisiche, Universitá di Napoli Federico II, and INFN, Sezione di Napoli, Napoli, Italy

[2] INFN, Sezione di Napoli, Napoli, Italy

E-mail: `loiudice@na.infn.it`

Abstract. A matrix diagonalization algorithm we developed in the past years has been revised and implemented for large scale shell model calculations using a spin uncoupled basis so as to obtain a sparse Hamiltonian matrix. The dimensions of such a matrix are reduced drastically by a procedure that combines an effective projection of the good angular momentum with an importance sampling specific of the method. Its numerical implementation on ^{116}Sn illustrates the simplicity of the method and the fast convergence of the iterative procedure.

1. Introduction

The standard methods [1] for diagonalizing matrices of very large dimensions are unmanageable even with the powerful computing facilities available now. On the other hand, one is mostly interested at generating the lowest eigenvalues and eigenstates.

Lanczos algorithm is the ideal tool for this purpose [2]. This algorithm has been implemented in several ways for performing large scale nuclear shell model calculations. *Antoine* [3, 4] is maybe the most widely adopted code in this field. It uses an uncoupled basis (m-scheme), which simplifies tremendously the construction of the Hamiltonian matrix and reduces effectively its dimensions by taking full advantage of its sparse nature.

Considerable effort has been devoted also at the attempt of reducing further the sizes of the Hamiltonian matrix. The quantum Monte Carlo diagonalization (QMCD) method [5, 6], adopted with success for systematics in the (s, d) and (f, p) nuclear regions, generates stochastically a truncated basis for the diagonalization of the many-body Hamiltonian. Specific procedures have been developed to restore the total spin and other symmetries, broken by the stochastic sampling procedure.

The Monte Carlo technique applied to shell model (SMMC) was first developed as a tool alternative to the direct diagonalization of the Hamiltonian [7, 8, 9]. The SMMC, indeed, evaluates expectation values and strength functions, providing important information on ground state properties and electromagnetic responses of medium-mass nuclei. The drawback of the SMMC approach is represented by the famous *sign problem*, not completely solved yet.

A method for achieving a drastic truncation of the shell model space is based on the density matrix renormalization group (DMRG) [10] borrowed from condensed matter [11]. In nuclear

[3] On leave of absence from the Faculty of Mathematics and Physics, Charles University in Prague, Czech Republic

structure, this approach came out to be more effective in the coupled scheme and, in this form, was used to compute the low energy spectra of nuclei in the (f,p) shell, like ^{48}Cr and ^{56}Ni [12, 13].

As pointed out already, all the above methods, apart from the SMMC, exploit the Lanczos algorithm. Although extremely powerful, this method presents some critical points. In fact, the states generated by the Lanczos iterative procedure, though mathematically orthogonal, might loose this property because of numerical errors that propagate rapidly with the number of iterations. It might happen, therefore, that the same state occurs more than once or states of unwanted angular momenta appear.

Few years ago, we developed a method [14, 15] that faces directly the diagonalization of the Hamiltonian. It is an iterative algorithm which generates a selected set of eigenvectors of a large matrix and is extremely simple to be implemented. It is also endowed with an importance sampling that yields a drastic reduction of the space and allows to extrapolate to the exact eigenvalues.

The method was implemented in the coupled j-scheme only, which is too time consuming in the construction of the Hamiltonian matrix. We have now implemented the algorithm in the the m-scheme and developed a more efficient iterative procedure which allows to face matrices of much larger dimensions, especially when combined with an importance sampling [16]. A brief outline of the method together with illustrative results are presented here.

2. The algorithm

Let us consider an operator $\hat{A}$, assumed for simplicity to be self-adjoint, and the matrix $A = \{a_{ij}\}$ representing $\hat{A}$ in an orthonormal basis $\{| 1\rangle, | 2\rangle, \ldots, |i >, \ldots, | N\rangle\}$. The diagonalization procedure goes through several iteration loops. The first loop is composed of the following steps:

1a) Consider the lowest n_0 basis states $|i >$ ($n_0 << N$), construct and diagonalize the $n_0 \times n_0$ submatrix $A_0 = (a_{ij})$,

1b) select the lowest v eigenvalues $\lambda_1^{(0)}, \ldots, \lambda_k^{(0)}, \ldots, \lambda_v^{(0)}$ and the corresponding eigenvectors

$$| \varphi_k^{(0)}\rangle = \sum_{i=1}^{n_0} c_i^{(0)} \mid i\rangle\rangle, \tag{1}$$

1c) consider the subspace spanned by the v eigenvectors $| \varphi_k^{(0)}\rangle$ plus the basis states $|j> = |n_0+1>, \ldots |n_1>$, compute $b_{kj}^{(0)} = \langle \varphi_k^{(0)} \mid \hat{A} \mid j\rangle$ and construct the new submatrix

$$A_1 = \begin{pmatrix} \lambda_1^{(0)} & 0 \;\; 0 & 0 & b_{1(n_0+1)}^{(0)} & \cdot\;\cdot & b_{1n_1}^{(0)} \\ 0 & . \;\; 0 & 0 & . & \cdot\;\cdot & . \\ 0 & 0 \;\; . & 0 & . & \cdot\;\cdot & . \\ 0 & 0 \;\; 0 & \lambda_v^{(0)} & b_{v(n_0+1)}^{(0)} & \cdot\;\cdot & b_{vn_1}^{(0)} \\ b_{(n_0+1)1}^{(0)} & \cdot\;\;\cdot & b_{(n_0+1)v}^{(0)} & a_{(n_0+1)(n_0+1)} & \cdot\;\cdot & a_{(n_0+1)n_1} \\ . & \cdot\;\;\cdot & . & . & \cdot\;\cdot & . \\ . & \cdot\;\;\cdot & . & . & \cdot\;\cdot & . \\ b_{n_1 1}^{(0)} & \cdot\;\;\cdot & b_{n_1 v}^{(0)} & a_{n_1(n_0+1)} & \cdot\;\cdot & a_{n_1 n_1} \end{pmatrix}, \tag{2}$$

1d) diagonalize A_1 and extract the new lowest v eigenvalues $\lambda_k^{(1)}$ $(k = 1, v)$ and the corresponding eigenvectors $| \varphi_1^{(1)}\rangle, \ldots | \varphi_v^{(1)}\rangle$,

1e) consider now the new subspace spanned by $| \varphi_1^{(1)}\rangle, \ldots \; | \varphi_v^{(1)}\rangle$ plus the basis states $|j >= |n_1+1 >, \ldots |n_2 >$, construct the new submatrix A_2 just as done in point 1c) for A_1 and, after its diagonalization, extract the new lowest v eigenvalues $\lambda_1^{(2)}, \dots \lambda_v^{(2)}$ and the corresponding eigenvectors $\mid \varphi_1^{(2)}\rangle, \dots \mid \varphi_v^{(2)}\rangle$.

The just outlined procedure is iterated until the full basis is exhausted. This zero approximation loop yields the approximate v eigenvalues and eigenvectors

$$E_k^{(1)} \equiv \lambda_k^{(N)}, \qquad \mid \psi_k^{(1)}\rangle \equiv \mid \varphi_k^{(N)}\rangle = \sum_{i=1}^{N} c_k^{(N)}(i) \mid i\rangle. \tag{3}$$

These eigensolutions are the new entries for a new iteration. More specifically, we consider the basis composed of the eigenvectors $\psi_k^{(1)}$ plus the original basis states $\{\mid j\rangle\}$. Since the vectors $\psi_k^{(1)}$ are linear combinations of the $\mid j\rangle$ states, this new basis is no longer orthonormal and may be even redundant. We have therefore to solve an eigenvalue problem of the general form and resort to the Choleski decomposition method. With this modifications, the subsequent iteration loops proceed as the first one and generate a sequence of v vectors $\psi_1^{(i)}, \dots, \psi_v^{(i)}$ that converge to the exact eigensolutions [14].

3. Implementation of the algorithm in the m-scheme

We start with the modified Hamiltonian

$$H_J = H + c[\mathbf{J}^2 - J(J+1)]^2, \tag{4}$$

where H is a two-body Hamiltonian of general form, $\mathbf{J}$ the total spin operator, and c a positive constant. The Hamiltonian acts in a model space spanned by the states $|i> = |\alpha_1, \alpha_2, \dots, \alpha_p>$, where p denotes the number of valence nucleons, $\alpha_i = \{a_i m_i\}$ the single particle (s.p.) quantum numbers and $a_i = \{n_i l_i j_i\}$ the s.p. shells. The states $|i>$ have a good magnetic quantum number $M = m_1 + \dots + m_i + \dots + m_p$.

We define a subspace M_0 spanned by a set of configurations $\{a_i^{n_i}\} = (a_1^{n1}, a_2^{n2}, \dots a_i^{n_i} \dots)$, where n_i is the occupation number of the s.p. shell a_i. Under the action of $\mathbf{J}^2$, the states $|i>$ of a given configuration $\{a_i^{n_i}\}$ transform among themselves in virtue of the commutation relations $[\mathbf{J}, j_i^2] = 0$.

For fixed J, we diagonalize the Hamiltonian matrix H_J in M_0 by using, for instance, the algorithm outlined in Sect. 2. If the constant c is chosen so as to push up in energy the eigenvalues with total spin $J' \neq J$, the algorithm generates v lowest eigenvalues $E_1^{(0)}, \dots, E_v^{(0)}$ and eigenvectors $\psi_1^{(0)}, \dots, \psi_v^{(0)}$. These have a good J in virtue of the rotational invariant Hamiltonian and the mentioned invariance of each configuration $\{a_i^{n_i}\}$ with respect to $\mathbf{J}^2$.

The eigenvectors $\psi_1^{(0)}, \dots, \psi_v^{(0)}$ are coupled by H only to those basis states of the complementary subspace $M_0^{(C)}$ that differ from the ones in M_0 by at most two s.p. states or, in the second quantization language, two particle-two holes $(2p-2h)$. Let us denote them by $|j>_2$. We now consider the subspace spanned by the eigenvectors $\psi_1^{(0)}, \dots, \psi_v^{(0)}$ plus the basis states $|j>_2$ and adopt the algorithm to diagonalize H_J in this subspace. A new set of lowest eigenvalues $E_1^{(1)}, \dots, E_v^{(1)}$ and eigenvectors $\psi_1^{(1)}, \dots, \psi_v^{(1)}$ is generated.

These new eigenvectors are coupled only to the basis states, to be denoted by $|j>_4$, differing from the original $|j>_0$ by at most $4p - 4h$. The diagonalization of H_J in the new subspace, spanned by $\psi_1^{(1)}, \dots, \psi_v^{(1)}$ plus $\{|j>_4\}$, generates a new set of v lowest eigenvectors of good J.

By iterating the procedure just outlined, we cover eventually the full space and the set of v eigensolutions tend to the exact ones.

3.1. The importance sampling revisited

Even after having exploited the sparsity of the Hamiltonian matrix, the diagonalization procedure might become prohibitively lengthy as the dimensions of the space increase. It is therefore useful, if not necessary, to resort to the sampling procedure. This has been simplified as follows.

Let us fix a sequence of positive small numbers ϵ_i of decreasing values $\epsilon_0 > \epsilon_1 > \ldots > \epsilon_i > \ldots > \epsilon_l$. Once the diagonalization of H_J in the initial subspace M_0 yields the lowest v eigenvalues $E_1^{(0)}, \ldots, E_v^{(0)}$ with the corresponding eigenvectors $\psi_1^{(0)}, \ldots, \psi_v^{(0)}$, we pick only the $(2p-2h)$ basis states $|j>$ that fulfill the condition

$$\frac{b_0^2(kj)}{a_{jj} - E_k^{(0)}} > \epsilon_0 \tag{5}$$

where $k = 1, v$ and a_{jj} ($j = n_0 + 1, N$) is the diagonal matrix element corresponding to the state $|j>$ coupled to $\varphi_k^{(0)}$ through the matrix element $b_0(kj) =< j|H_J|\psi_k^{(0)} >$.

We then adopt the diagonalization algorithm to diagonalize H_J in the subspace spanned by the v states $\{\psi_k^{(0)}\}$ plus the basis states $\{|j>\}$ selected according to the sampling condition (5). New eigenvalues $E_1^{(1)}, \ldots, E_v^{(1)}$ and eigenvectors $\psi_1^{(1)}, \ldots, \psi_v^{(1)}$ are generated. These eigenfunctions are now linear combinations of the original $(0p-0h)$ states of the subspace M_0 plus the $(2p-2h)$ states selected according to the sampling condition (5).

We now explore again the subspace complementary to M_0 and select all the states $|j>$ that fulfill the condition (5) with the following replacements $E_k^{(0)} \rightarrow E_k^{(1)}$, $b_0(kj) \rightarrow b_1(kj) =< j|H_J|\psi_k^{(1)} >$, $\epsilon_0 \rightarrow \epsilon_1$. The diagonalization algorithm applied to the new subspace spanned by the v states $\{\psi_k^{(1)}\}$ plus the set of states $\{|j>\}$ yields new eigensolutions $E_1^{(2)}, \ldots, E_v^{(2)}$ and eigenvectors $\psi_1^{(2)}, \ldots, \psi_v^{(2)}$. These updated eigenvectors include $2p-2h$ states not admitted in the first run (since $\epsilon_1 < \epsilon_0$) plus $(4p-4h)$ states. We now iterate the sampling procedure applying at the i_{th} iteration the sampling condition

$$\frac{b_i^2(kj)}{a_{jj} - E_k^{(i)}} > \epsilon_i, \tag{6}$$

where $b_i(kj) =< j|H_J|\psi_k^{(i)} >$ ($k = 1, v,\ j = n_0 + 1, N$). Once the full space is covered, the process yields a set of eigenvalues and eigenfunctions that approach the exact values in the limit $\epsilon_l \longrightarrow 0$. In practice, as we shall see, the exact eigensolutions are reached for a sufficiently small $\epsilon_l \neq 0$.

4. Numerical implementation

The sampling algorithm was applied to ^{116}Sn. The model space is defined by the neutron shells $\{2d_{5/2}, 1g_{7/2}, 2d_{3/2}, 3s_{1/2}, 1h_{11/2}\}$. The single particle energies were borrowed from Ref. [17], while a renormalized G matrix [18] derived from the CD-Bonn potential [19] was taken as a two-body interaction. The $E2$ transition strengths were computed using the proton and neutron effective charges $e_p = 1.5$ and $e_n = 1$.

An initial subspace M_0 of dimensions $n_0 \simeq 50$ was sufficient to generate ten lowest eigenstates with good J. Once chosen M_0, we applied the iterative sampling procedure using a set of decreasing values of ϵ_i. At each iteration, the sampling condition picks up new states that differ from the $0p-0h$ states of the original subspace M_0 by an increasing number of $p-h$ states.

Fig. 1 plots the lowest ten $J^\pi = 0^+$ eigenvalues of ^{116}Sn versus the dimensions n of the sampled matrix, normalized to the full matrix dimensions N. The value n/N is uniquely

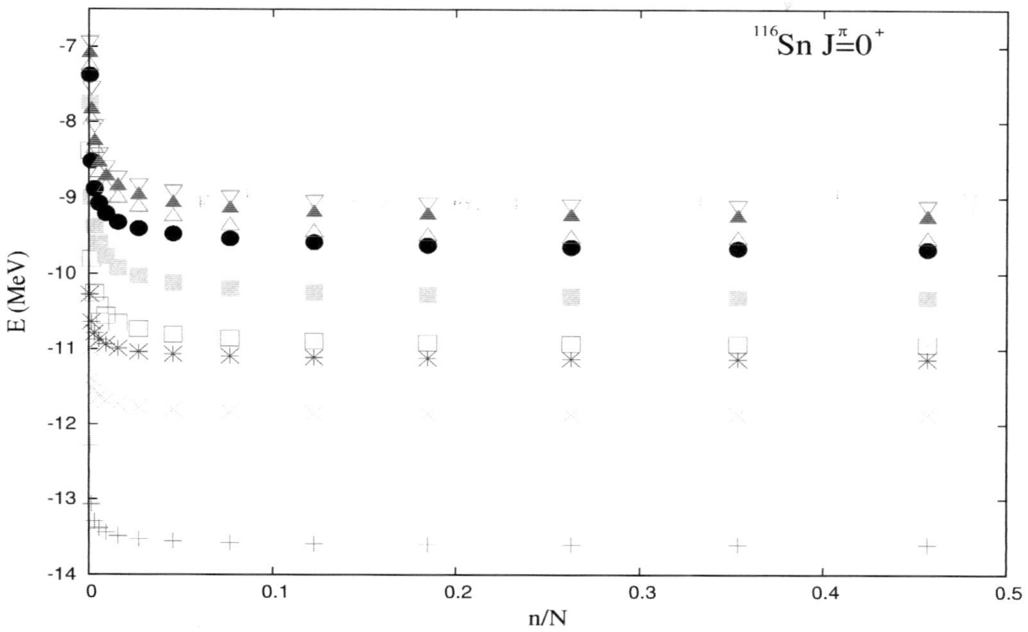

Figure 1. Convergence rate of the lowest $J^\pi = 0^+$ eigenvalues in ^{116}Sn

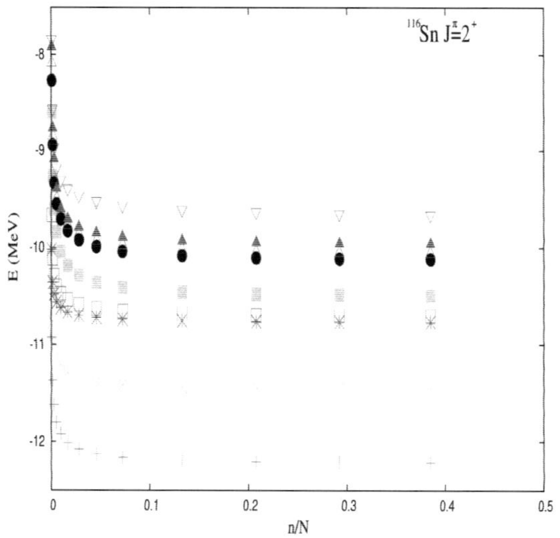

Figure 2. Convergence rate of the lowest $J^\pi = 2^+$ levels of ^{116}Sn

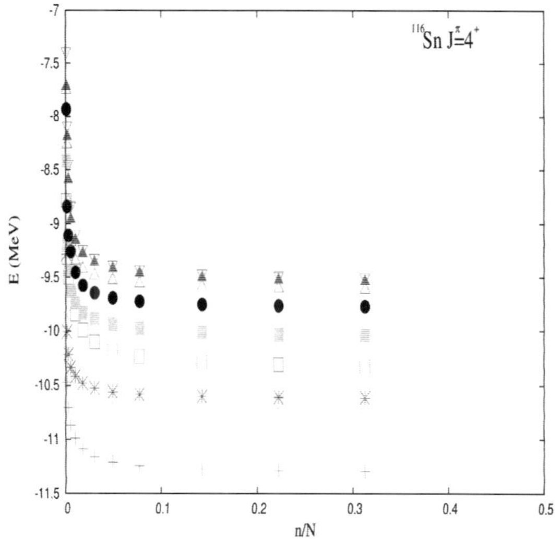

Figure 3. Convergence rate of the lowest $J^\pi = 4^+$ levels of ^{116}Sn

determined by the running sampling parameter ϵ. As shown in the plots, all eigenvalues approach closely the exact values for less than 10% of the basis states. This can be seen also in Figs 2 and 3 for the $J^\pi = 2^+$ and $J^\pi = 4^+$ lowest levels, respectively. A rapid convergence is achieved also for the $E2$ strengths (Figs. 4 and 5). The smooth behavior of both energies and strengths versus n/N is to be noticed. This allows to make an extrapolation to the exact quantities.

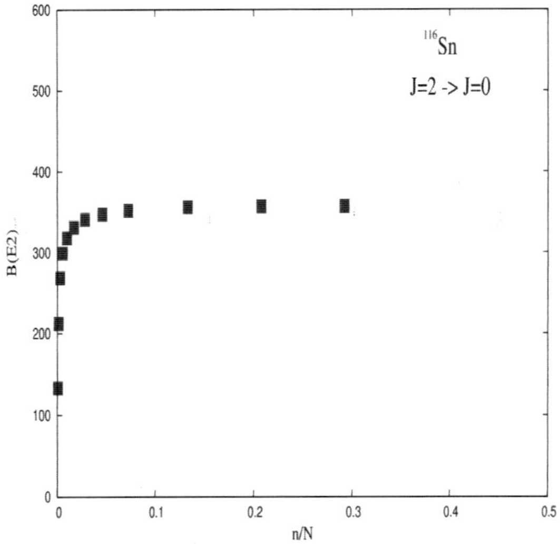

Figure 4. Convergence properties of $B(E2;2^+ \rightarrow 0^+)$ in ^{116}Sn

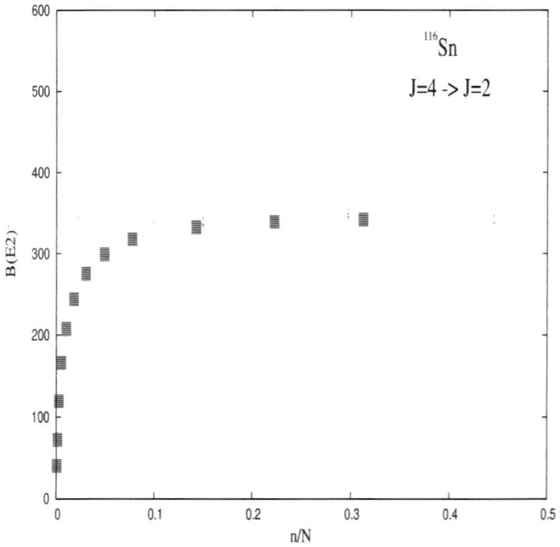

Figure 5. Convergence properties of $B(E2;4^+ \rightarrow 2^+)$ in ^{116}Sn

5. Concluding remarks

The case of ^{116}Sn, illustrated here, shows that the algorithm, implemented in the m-scheme, enables us to generate quite a large number of low-lying eigenstates for each J value.

The importance sampling described above allows to achieve a rapid convergence of the iterative procedure for both energies and wavefunctions, as illustrated for the levels of few J values and for some $E2$ transition strengths.

This rapid convergence induces effectively a cut of the dimensions of the matrix Hamiltonian by at least an order of magnitude. In virtue of this property, the method should be able to treat Hamiltonian matrices of very large dimensions. Very recent calculations [16] confirm this expectation.

References

[1] J.H. Wilkinson, J H 1965 *The Algebraic Eigenvalue Problem* (Oxford: Clarendom Press)
[2] Golub G H and Van Loan C F 1996 *Matrix Computations* (Baltimore:John Hopkins University Press)
[3] Caurier E and Nowacki F 1999 *Acta Physica Polonica* **30** 705
[4] Caurier E *et al* 2005 *Rev. Mod. Phys.* **77** 427
[5] Otsuka T, Honma M and Mizusaki T 1998 *Phys. Rev. Lett.* **81** 1588
[6] Otsuka T *et al* 2001 *Prog. Part. Nucl. Phys.* **47** 319
[7] White J A , Koonin S E and Dean D J 2000 *Phys. Rev. C* **61** 034303
[8] Koonin S E Dean D J and Langanke K 1997 *Phys. Rep.* **278** 1
[9] Koonin S E Dean D J and Langanke K 1997 *Annu. Rev. Nucl. Part. Sc.* **47** 463
[10] Dukesky J and Pittel S 2001 *Phys. Rev. C* **63** 061303
[11] White S R 1993 *Phys. Rev. B* **48** 10345
[12] Pittel S and Sandulesku N 2006 *Phys. Rev. C* **73** 014301
[13] Thakur B, Pittel S, Sandulescu N 2008 *Phys.Rev.* C **78** 041303
[14] Andreozzi Porrino A and Lo Iudice N 2002 *J. Phys. A: Math. Gen.* **35** L61
[15] Andreozzi F, Lo Iudice N and Porrino A 2003 *J. Phys. : Nucl. Part. Phys.* **29** 2319
[16] Andreozzi F, Bianco D, Lo Iudice N, Porrino A and Knapp F to be submitted for publication.
[17] Dikmen E, Öztürk O, Valliere M 2009 *J. Phys. : Nucl. Part. Phys.* **36** 045102
[18] Hjorth-Jensen M *et al* 1995 *Physics Reports* **261** 125
[19] Machleidt R 1989 *Adv. Nucl. Phys.* **19** 189

Finite-range separable pairing interaction within the new N^3LO DFT approach

2011 J. Phys.: Conf. Ser. 267 012027

(http://iopscience.iop.org/1742-6596/267/1/012027)

Finite-range separable pairing interaction within the new N^{3}LO DFT approach

Veselý P.[1], Dobaczewski J.[1,2], Michel N.[1], and Toivanen J.[1]

[1] Department of Physics, P.O. Box 35 (YFL), University of Jyväskylä, FI-40014, Finland

[2] Institute of Theoretical Physics, University of Warsaw, ul. Hoża 69, 00-681 Warsaw, Poland

E-mail: `vesely@ipnp.troja.mff.cuni.cz`

Abstract. For over four decades, the Skyrme functional within various parametrizations has been used to calculate nuclear properties. In the last few years there was a number of attempts to improve its performance and introduce generalized forms. In particular, the most general phenomenological quasi-local energy density functional, which contains all combinations of density, spin-density, and their derivatives up to the sixth order (N^{3}LO), was proposed in [1]. Since in the phenomenological-functional approaches, the particle-particle (pp) channel is treated independently of the particle-hole (ph) channel, there remains a question of what pairing interaction is suitable to use within the N^{3}LO energy functional. In the present study, we use the separable, finite-range, translationally invariant interaction given in [2] and we present the first test-case application of this pairing interaction within the code HOSPHE that is able to treat the N^{3}LO energy density functional.

1. Construction of the N^{3}LO functional

Following [1], the construction of the N^{3}LO functional can be outlined in three steps. First we define basic building blocks: (i) the general non-local densities, which include both the scalar and vector (spin) terms,

$$\rho(\vec{r}\sigma, \vec{r}'\sigma') = \frac{1}{2}\rho_{00}(\vec{r}, \vec{r}')\delta_{\sigma,\sigma'} + \frac{1}{2}\sum_{\mu=-1,0,1} \langle\sigma|\sigma_\mu|\sigma'\rangle \rho_{1\mu}(\vec{r}, \vec{r}'), \tag{1}$$

(ii) the derivative operator,

$$\nabla_{1,\mu=\{-1,0,1\}} = -i\left\{\frac{1}{\sqrt{2}}(\nabla_x - i\nabla_y), \nabla_z, -\frac{1}{\sqrt{2}}(\nabla_x + i\nabla_y)\right\}, \tag{2}$$

and (iii) the relative momentum operator,

$$k_{1\mu} = \frac{1}{2i}(\nabla_{1\mu} - \nabla'_{1\mu}). \tag{3}$$

In the second step, we construct more general building blocks: (iv) the tensor derivative operator of order m and rank I,

$$D_{mI} = \underbrace{[\nabla ... [\nabla[\nabla\nabla]_{I_2}]_{I_3} ...]_I}_{m} \ , \tag{4}$$

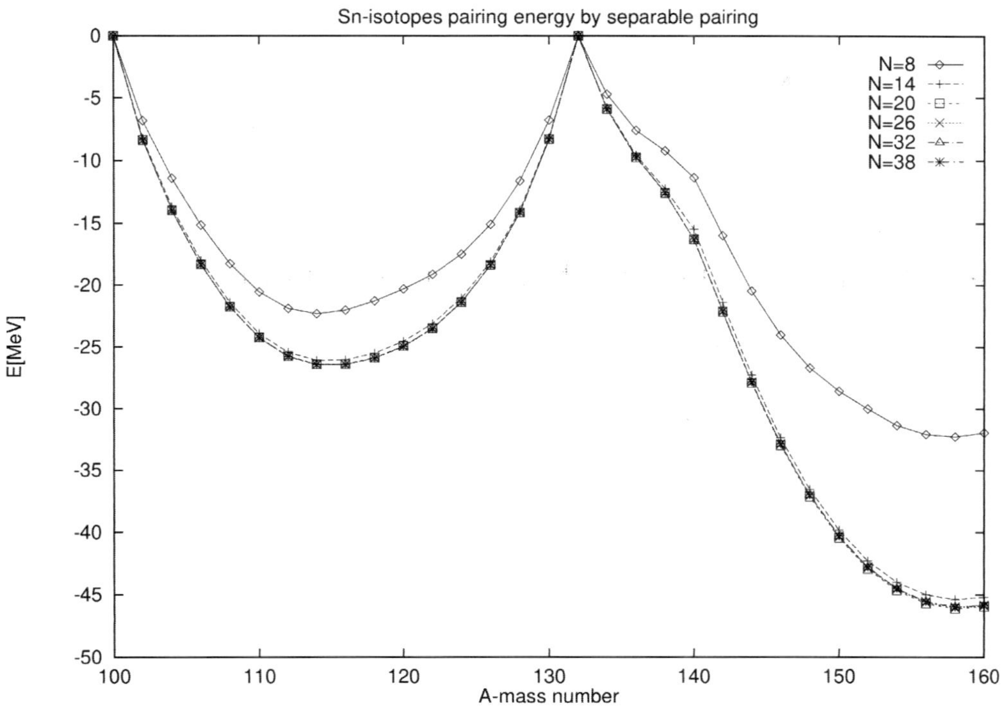

Figure 1. Pairing energies for the chain of even-even tin isotopes, obtained in the HO bases of N=8, 14, 20, 26, 32, and 38 shells.

and in analogous way (v) the tensor operator of relative momentum,

$$K_{mI} = \underbrace{[k...[k[kk}_{m}]_{I_2}]_{I_3}...]_{I} \; , \tag{5}$$

and finally (vi) the generalized local (primary) density,

$$\rho_{nLvJ}(\vec{r}) = \left\{ \left[K_{nL} \otimes \rho_v(\vec{r}, \vec{r}') \right]_J \right\}_{\vec{r}=\vec{r}'} \; . \tag{6}$$

In the third step, we use the building blocks to construct the energy density,

$$\mathcal{H}(\vec{r}) = \sum_{mI,nLvJ} \sum_{n'L'v'J'} C^{n'L'v'J'}_{mI,nLvJ} \left[\rho_{n'L'v'J'}(\vec{r}) [D_{mI} \rho_{nLvJ}(\vec{r})]_{J'} \right]_0 \; , \tag{7}$$

and the energy density functional (EDF) is defined as the integral of $\mathcal{H}(\vec{r})$. Each term in the summation in Eq. (7) can be characterized by its order, which is defined as:

$$\text{order} = m + n + n' \; . \tag{8}$$

This general EDF contains a large number of coupling constants $C^{n'L'v'J'}_{mI,nLvJ}$; however, there are several symmetries (spherical, Galilean, gauge) that significantly reduce the number of terms in the energy density (7) (see [1] for details).

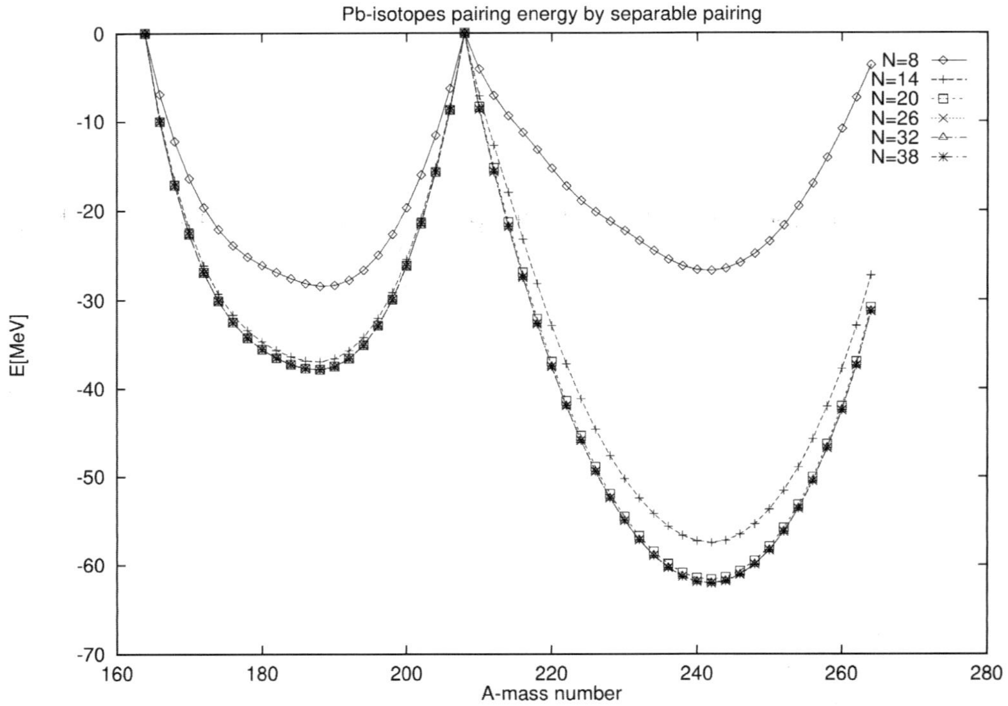

Figure 2. Same as in Fig. 1 but for the lead isotopes.

2. Finite-Range Separable Pairing Interaction

In the phenomenological-functional approaches, the particle-particle (pp) channel is treated independently of the particle-hole (ph) channel. Traditionally, for the Skyrme functional, the delta pairing interaction,

$$\hat{V}^{pp} = V_0 \delta(\vec{r} - \vec{r}'), \tag{9}$$

or its density-dependent variant have been used. The main disadvantage of this pairing interaction is, however, its cut-off dependence, as discussed, e.g., in [3]. This requires using renormalization or regularization procedures (see [4] and references cited therein), which may depend on the order of derivatives appearing in the mean-field Hamiltonian. Because of this reason, we decided to apply in the pp channel of our phenomenological functional the finite-range separable pairing interaction, as it was introduced in [2], which does not lead to any cut-off dependence.

The finite-range separable pairing interaction is defined in the momentum representation as:

$$\langle \vec{k}_1 \vec{k}_2 | V^{pp} | \vec{k}_1' \vec{k}_2' \rangle = (2\pi)^3 \delta(\vec{K} - \vec{K}') \langle \vec{k} | V^{pp} | \vec{k}' \rangle \ . \tag{10}$$

From the beginning, the total momentum, $\vec{K} = \vec{k}_1 + \vec{k}_2$, is separated from the relative momentum, $\vec{k} = \frac{1}{2}(\vec{k}_1 - \vec{k}_2)$. The term $\delta(\vec{K} - \vec{K}')$ in Eq. (10) ensures the conservation of the total momentum of the interacting pair. In terms of relative momenta, the pairing interaction for neutrons ($\nu = n$) and protons ($\nu = p$) is then defined by

$$\langle \vec{k} | V_\nu^{pp} | \vec{k}' \rangle = -G_\nu p(k) p(k') \tag{11}$$

where functions $p(k)$ have the form of sum of m Gaussian terms

$$p(k) = \frac{1}{m}\sum_{i=1}^{m} \mathrm{e}^{-a_i^2 k^2} \ . \tag{12}$$

After the Fourier transform to the coordinate representation, one obtains

$$\langle \vec{r}|V_\nu^{pp}|\vec{r}'\rangle = -G_\nu P(r)P(r') \tag{13}$$

with

$$P(r) = \frac{1}{m}\sum_{i=1}^{m} \frac{1}{(4\pi a_i^2)^{3/2}} \mathrm{e}^{-\frac{r^2}{4a_i^2}} \ . \tag{14}$$

One can now express this pairing interaction by using its matrix elements in the standard harmonic oscillator (HO) basis. To this end, one can use two more transformations – the Talmi-Moshinsky transformation to separate the total and relative coordinates and coupling of angular momenta of the interacting pair [5]. The final, fully separable, form of the pairing energy reads:

$$\begin{aligned} E^{pp} = &-\frac{1}{4}G_+ \sum_J \sum_N \sum_{\mu\nu\mu'\nu'} V_{\mu\nu}^{J,NJ} V_{\mu'\nu'}^{J,NJ} \Big(\langle \Psi_{\alpha_\mu j_\mu}||\kappa_0^{\prime +J}||\Psi_{\alpha_\nu j_\nu}\rangle \langle \Psi_{\alpha'_\mu j'_\mu}||\kappa_0^{J}||\Psi_{\alpha'_\nu j'_\nu}\rangle + \\ &\qquad + \langle \Psi_{\alpha_\mu j_\mu}||\kappa_1^{\prime +J}||\Psi_{\alpha_\nu j_\nu}\rangle \langle \Psi_{\alpha'_\mu j'_\mu}||\kappa_1^{J}||\Psi_{\alpha'_\nu j'_\nu}\rangle \Big) \\ &-\frac{1}{4}G_- \sum_J \sum_N \sum_{\mu\nu\mu'\nu'} V_{\mu\nu}^{J,NJ} V_{\mu'\nu'}^{J,NJ} \Big(\langle \Psi_{\alpha_\mu j_\mu}||\kappa_0^{\prime +J}||\Psi_{\alpha_\nu j_\nu}\rangle \langle \Psi_{\alpha'_\mu j'_\mu}||\kappa_1^{J}||\Psi_{\alpha'_\nu j'_\nu}\rangle + \\ &\qquad + \langle \Psi_{\alpha_\mu j_\mu}||\kappa_1^{\prime +J}||\Psi_{\alpha_\nu j_\nu}\rangle \langle \Psi_{\alpha'_\mu j'_\mu}||\kappa_0^{J}||\Psi_{\alpha'_\nu j'_\nu}\rangle \Big), \end{aligned} \tag{15}$$

where $G_\pm = (G_n \pm G_p)/2$ and $V_{\mu\nu}^{J,NJ}$ are the separable interaction matrix elements:

$$V_{\mu\nu}^{J,NL} = \sqrt{(2j_\mu+1)(2j_\nu+1)} \left\{ \begin{array}{ccc} l_\mu & l_\nu & J \\ \frac{1}{2} & \frac{1}{2} & 0 \\ j_\mu & j_\nu & J \end{array} \right\} \sum_{nl} M_{n_\mu l_\mu n_\nu l_\nu}^{NLnl} \sqrt{4\pi} \left(\int_0^\infty r^2 \mathrm{d}r P(r) R_{nl}(r, b_r) \right), \tag{16}$$

which include the Talmi-Moshinsky brackets $M_{n_\mu l_\mu n_\nu l_\nu}^{NLnl}$, a 9j-symbol coming from the coupling of the two nucleons to total angular momentum J, and a radial integral which can be calculated analytically:

$$\int_0^\infty r^2 \mathrm{d}r P(r) R_{nl}(r, b_r) = \frac{1}{b_r^{3/2}} \sqrt{\frac{(n+\frac{1}{2})!}{2n!}} \frac{1}{m} \sum_{i=1}^{m} \frac{1}{(4\pi a_i^2)^{3/2}} \left(\frac{4a_i^2 b_r^2}{1+2a_i^2 b_r^2} \right)^{3/2} \left(\frac{1-2a_i^2 b_r^2}{1+2a_i^2 b_r^2} \right)^n \delta_{l0} \ . \tag{17}$$

From relations (15)–(17), it follows that the free parameters of this pairing-interaction model are the coupling constants G_+ and G_-, and m parameters a_i. These parameters should be fitted to the experimental data.

3. Calculations and preliminary results

The finite-range separable pairing interaction was implemented into the recently developed code for calculation of spherical nuclei within the N^{3}LO energy functional – HOSPHE [6].

For the first set of test calculations we chose two chains of even-even isotopes for tin ($A = 100 - 160$) and lead ($A = 164 - 264$). Since our aim here is just to check qualitative results of the method, we use the following set for $m = 1$:

$$G_+ = 738\,\mathrm{MeV\,fm}^3, \qquad G_- = 0, \qquad a_1 = 0.636\,\mathrm{fm}.$$

These values were fixed [2] to reproduce the gap equation of nuclear matter calculated by the Gogny force.

The obtained pairing energies are shown in figure 1 (Sn isotopes) and figure 2 (Pb isotopes). They were calculated within the Hartree-Fock-Bogolyubov method with the finite-range separable pairing force acting in the pp channel and the Skyrme SLy4 interaction acting in the ph channel. It can be seen that the pairing energies rapidly converge already for $N = 20$ HO shells. Further refinement of the nuclear pairing properties can probably be obtained by adding more Gaussian terms, and hence by fitting more parameters in the phenomenological pairing force, i.e. with $m \geq 2$.

Acknowledgments
This work is supported by the Academy of Finland and the University of Jyväskylä within the FIDIPRO programme and by the Polish Ministry of Science and Higher Education under Contract No. N N 202 328234.

References

[1] Carlsson B G, Dobaczewski J, and Kortelainen M, 2008 *Phys. Rev.* C **78**, 044326; 2010 *Phys. Rev.* C **81**, 029904(E)
[2] Tian Y, Ma Z Y, and Ring P, 2009 *Phys. Lett.* B **676**, 44
[3] Dobaczewski J, Nazarewicz W, Werner T R, Berger J-F, Chinn C R, and Dechargé J, 1996 *Phys. Rev.* C **53**, 2809
[4] Borycki P J, Dobaczewski J, Nazarewicz W, and Stoitsov M V, 2006 *Phys. Rev.* C **73**, 044319
[5] Tian Y, Ma Z Y, and Ring P, 2009 *Phys. Rev.* C **79**, 064301
[6] Carlsson B G, Dobaczewski J, Toivanen J, and Veselý P, 2010 *Comp. Phys. Commun.* **181**, 1641

Configuration interactions constrained by energy density functionals

2011 J. Phys.: Conf. Ser. 267 012028

(http://iopscience.iop.org/1742-6596/267/1/012028)

Configuration interactions constrained by energy density functionals

B Alex Brown[1], Angelo Signoracci[1] and Morten Hjorth-Jensen[2]

[1]Department of Physics and Astronomy, and National Superconducting Cyclotron Laboratory, Michigan State University, East Lansing, Michigan 48824-1321, USA

[2]Department of Physics and Center for Mathematical Applications, University of Oslo, N-0316, Oslo, Norway

E-mail: brown@nscl.msu.edu

Abstract. A new method for constructing a Hamiltonian for configuration interaction calculations with constraints to energies of spherical configurations obtained with energy-density-functional (EDF) methods is presented. This results in a unified model that reproduces the EDF binding-energy results in the limit of single Slater determinants, but can also be used for obtaining energy spectra and correlation energies with renormalized nucleon-nucleon interactions. The three-body and/or density-dependent terms that are necessary for good nuclear saturation properties are contained in the monopole corrections obtained from the EDF. Applications to binding energies and spectra in the region of ^{132}Sn are given.

In nuclear structure theory the two main computational methods for heavy nuclei based upon the nucleon fermionic degrees of freedom are the Hartree-Fock or energy-density-functional (EDF) method and the configuration interaction (CI) method. The EDF method is often limited to the configuration for a single Slater determinant. The EDF has parameters that are fitted to global properties of nuclei such as binding-energies and rms charge radii [1], [2] [3].

The CI method takes into account many Slater determinants. CI often uses a Hamiltonian derived from experimental single-particle energies and a microscopic nucleon-nucleon interaction [4]. A given CI Hamiltonian is applied to a limited mass region that is related to the configurations of a few valence orbitals outside of a closed shell and the associated renormalized nucleon-nucleon interaction that is specific to that mass region [4], [5]. Spectra and binding energies (relative to the closed core) obtained from such calculations for two to four valence particles are in good agreement with experiment [4], [5]. As many valence nucleons are added the agreement with experimental spectra [6] and binding energies deteriorates (see Fig. 36 in [4]). An important part that is missing from these CI calculations is the effective two-body interaction that comes from the three-body interaction of two valence nucleons interacting with one nucleon in the core [7]. To improve agreement with experimental spectra one often adjusts some of the valence two-body matrix elements. The most important part of this adjustment can be traced to the monopole component of the two-body matrix elements that controls how the effective single-particle energies evolve as a function of proton and neutron number [7].

Fig. (1) shows Wick's theorem applied to a closed shell for the one-body kinetic energy, the two-body interaction and the three-body interaction. The part contained in the dashed box represents the closed-shell and effective one-body parts of the Hamiltonian that might be contained in an EDF approach. Up to now this has been treated phenomenologically in the

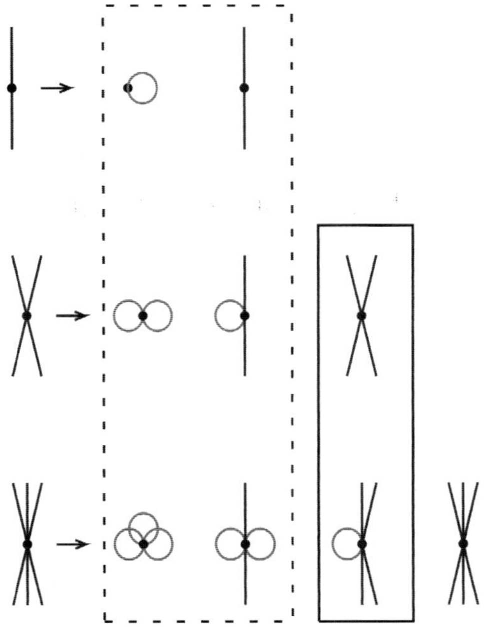

Figure 1. Schematic diagram for the terms in the Hamiltonian obtained from Wick's theorem for a closed shell. The red lines represent the summation over the orbitals in the closed shell. The black lines represent the valence particles and/or holes.

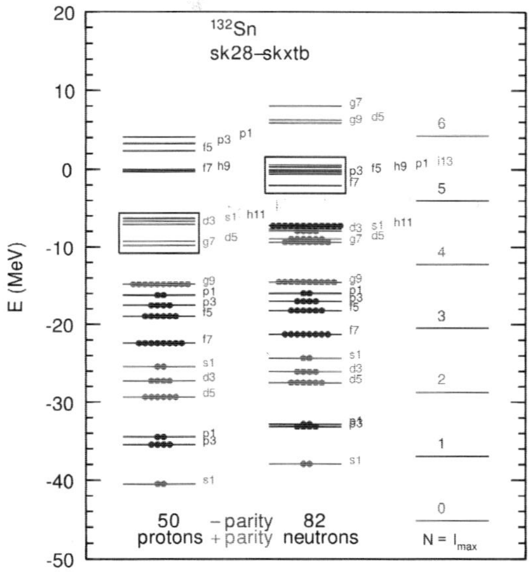

Figure 2. Single-particle potential energies for ^{132}Sn obtained with the Skxtb Skyrme Hamiltonian. The orbitals in the box are those in the model space for the CI calculations. The lines on the right-hand side show a typical energy spacing for the harmonic oscillator major shells.

framework of the Skyrme Hartree-Fock or relativistic Hartree methods with some parameters (typically 6-10) fitted to global experimental data. There are efforts underway to relate the parameters of these phenomenological approaches to the underlying two and three body forces between nucleons, and also to extend the functional forms to obtain improved agreement with experiment [3]. The part contained in the solid-line box is the residual interaction used for CI calculations. The remaining term is a valence three-body interaction.

In this paper we consider results for a ^{132}Sn closed shell, but the method can be applied to other closed-shell nuclei. The single-particle energy for orbital a is defined as

$$e_a = E(^{132}\mathrm{Sn} + a) - E(^{132}\mathrm{Sn}), \tag{1}$$

where $E(^{132}\mathrm{Sn})$ is the energy of the closed-shell configuration for ^{132}Sn, and $E(^{132}\mathrm{Sn} + a)$ is the energy of the closed-shell configuration plus one nucleon constrained to be in orbital a. Eq. (1) defines the one-body part of the CI calculations. Often experimental data are used for the energies in Eq. (1). In this paper we will use the results of EDF calculations for these energies.

For this paper we will use the EDF results based on the Skxtb Skyrme interaction [1], [8]. This is the Skx interaction from [1] with the addition of a tensor part fitted to observed single-particle energies [8]. The single-particle eigenvalues of the potential obtained with Skxtb are shown in Fig. (2). The results for the single-particle energies as defined by Eq. (1) for ^{133}Sb are shown in Fig. (3). The energy of the ^{133}Sb ground state relative to that of ^{132}Sn is -9.56 MeV with Skxtb compared to the experimental value of -9.68 MeV. An important property of Skxtb is that the experimental single-particle energies for the low-lying single-particle states around ^{132}Sn are reproduced with an rms deviation of about 300 keV [1], [8]. We are not aware of any other class of Skyrme interactions that can do better for the single-particle energies as defined by Eq. (1).

The major new aspect of our method is to take the monopole part of the effective two-body interaction from

$$\bar{V}_{ab} = E(^{132}\text{Sn} + a + b) - E(^{132}\text{Sn}) - e_a - e_b, \tag{2}$$

where $E(^{132}\text{Sn}+a+b)$ is the symmetry-constrained spherical EDF energy of the configuration for a closed shell plus two nucleons in orbitals a and b. This monopole interaction contains both the two and three body terms shown by the solid-line box in Fig. (1) as well as higher-order terms to the extent that they are contained in the EDF phenomenology. We modify the monopole part of the microscopic valence interaction to reproduce the results of Eq. (2). With this modification, the CI calculations closely reproduce the EDF calculations for single-Slater determinants, even when relatively many valence nucleons are added. Thus, the CI calculations are constrained to reproduce the trends of closed-shell energies and effective single-particle energies obtained with the EDF. For our model space orbitals, Eq. (2) involves about one hundred configurations for two nucleons (proton-proton, neutron-neutron and proton-neutron), but these calculations in a spherical basis are computationally fast.

The two-body part of the CI Hamiltonian is obtained with the usual renormalization procedure [5]. For our examples, the active valence orbitals are $(0g_{7/2}, 1d_{5/2}, 1d_{3/2}, 2s_{1/2}, 0h_{11/2})$ for protons $(0h_{9/2}, 1f_{7/2}, 1f_{5/2}, 2p_{3/2}, 2p_{1/2}, 0i_{13/2})$ for neutrons. The first-order matrix elements are obtained with the V_{lowk} method [9] with a cut off of $\Lambda = 2.2$ fm^{-1} for the N^{3}LO nucleon-nucleon interaction [10]. Core-polarization corrections are calculated in second-order up to $6\hbar\omega$ in the excitations energy. For the harmonic-oscillator basis we use $\hbar\omega = 7.874$ MeV, obtained from the global parametrization $\hbar\omega = 45\ \text{A}^{-2/3}\text{-}25\text{A}^{-1/3}$ MeV.

Two-body matrix elements $< abJ \mid V \mid abJ >$ for two protons in $a = b = 0g_{7/2}$, one proton in $a = 0g_{7/2}$ and one neutron in $b = 1f_{7/2}$, and two neutrons in $a = b = 1f_{7/2}$ are shown in Fig. (4) for the effective interaction obtained with harmonic-oscillator (black dots) and EDF (red dots) radial wavefunctions. Overall, the EDF results are close to the oscillator values, with the largest exception for some low-spin matrix elements that differ by up to about 100 keV.

For the lowest proton orbital with $a = b = (0\text{g}_{7/2})$ the renomalized N^{3}LO monopole interaction is $\bar{V}_{\text{N}^3\text{LO}}$=0.191 MeV (it is repulsive due to the Coulomb interaction). The result obtained from Eq. (2) with Skxtb is $\bar{V}_{\text{Skxtb}}$=0.374 MeV. Therefore, the contribution from the effective two-body interaction in the valence space due to the three-body interaction of two-valence nucleons with one nucleon in the core is $0.374 - 0.191 = 0.183$ MeV. This correction is included in CI by modifying all of the valence TBME for the $0g_{7/2}$ orbital by

$$< abJ \mid V \mid abJ >_{\text{eff}} = < abJ \mid V \mid abJ >_{\text{N}^3\text{LO}} - \bar{V}_{\text{N}^3\text{LO}} + \bar{V}_{\text{Skxtb}}. \tag{3}$$

Similar corrections are made for all other orbital pairs in our model space. The EDF corrections from Eq. (3) for the lowest orbitals are shown in Fig. (5).

The results for ^{134}Te are shown in Fig. (6). The agreement between experiment and theory is reasonable. It can be improved (as discussed below) if the Skxtb single-particle energies were replaced by the experimental values. The purpose of this comparison is to show what can be obtained with Skxtb+N^{3}LO without any adjustment of the model parameters.

The EDF monopole corrections are on the order of 0.2 to 0.3 MeV. When many nucleons are added, the monopole contribution goes as

$$\Delta E = n(n-1)\bar{V}/2, \tag{4}$$

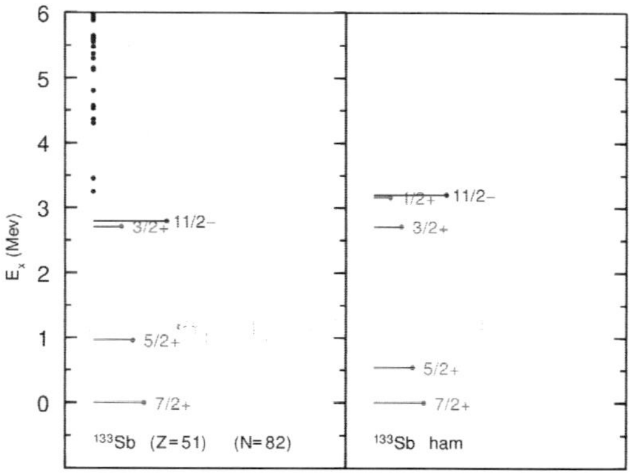

Figure 3. Comparison of experiment and theory (ham) for ^{133}Sb. Theory (ham) is based on the Skxtb EDF for the single-particle energies obtained from Eq. (1). The length of the line indicates the spin with positive parity (red) and negative parity (blue). Experimental levels that are unknown or uncertain are shown by the black dots.

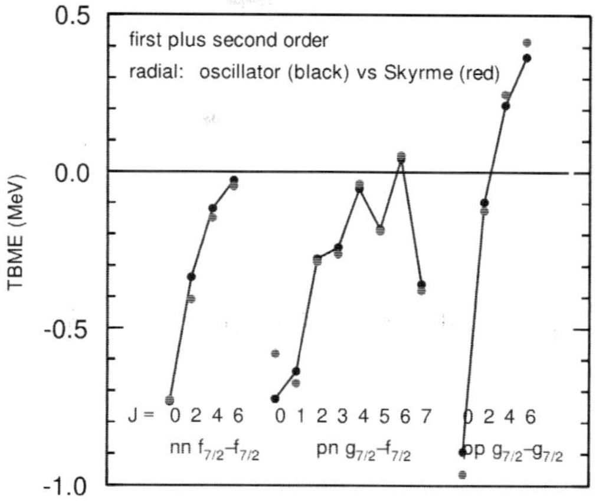

Figure 4. Two-body matrix elements $< abJ \mid V \mid abJ >$ for two protons in $a = b = 0g_{7/2}$, one proton in $a = 0g_{7/2}$ and one neutron in $b = 1f_{7/2}$, and two neutrons in $a = b = 1f_{7/2}$. Comparison between results with harmonic-oscillator (solid line) and EDF (red dots) radial wavefunctions is made.

where n is the number of valence nucleons. When we constrain the configuration of ^{146}Gd ($N = 82$ and $Z = 64$) to be $(0g_{7/2})^8(1d_{5/2})^6$ for the valence protons, the CI calculation gives a binding energy increase of 97.92 MeV (relative to ^{132}Sn). The EDF calculation (with the same assumption for the configuration) gives 98.57 MeV. These are close to each other due to our EDF monopole correction to the valence matrix elements. If the EDF monopole correction were not included in CI the results would differ by 19.3 MeV. The microscopic valence interaction on its own is too strong and gives an "over-saturation" (see Fig. 36 in [4]).

The difference $98.57 - 97.92 = 0.65$ MeV for ^{146}Gd could be considered as the contribution from the valence three-body monopole interaction on the right-hand side of Fig. (1) together with higher-order corrections implicit in the EDF. With $\Delta E_3 = n(n-1)(n-2)\bar{V}_3/6$ and $n = 14$ we obtain $\bar{V}_3 = 0.0018$ MeV. Thus, valence three-body interaction is very small. But one may need to take it into account for the binding energies of nuclei with ten or more valence particles due to its n^3 dependence.

Next we would start to improve upon the CI+EDF calculation for ^{146}Gd by going beyond the closed-shell configuration. Allowing for the addition of two-particle two-hole configurations we obtain a binding energy difference with ^{132}Sn of 100.45 MeV compared to the experimental value of 101.58 MeV. The full space calculation for ^{146}Gd should be close to experiment.

When experimental energies are known, the traditional shell-model approach is to use these for the CI calculations [4]. If we replace the single-particle energies with those that reproduce the experimental results for ^{133}Sb as shown on the left-hand side of Fig. (3) (the $2s_{1/2}$ single-particle energy is placed 328 keV below that of the $1d_{3/2}$), the results for ^{134}Te [Fig. (7)] are in better agreement with experiment [compared to Fig. (6)] and consistent with the results obtained by Coragio et al. [4]. But for our calculations that connect directly to EDF methods, it is important to use both single-particle energies and monopole corrections from the same EDF. It is expected

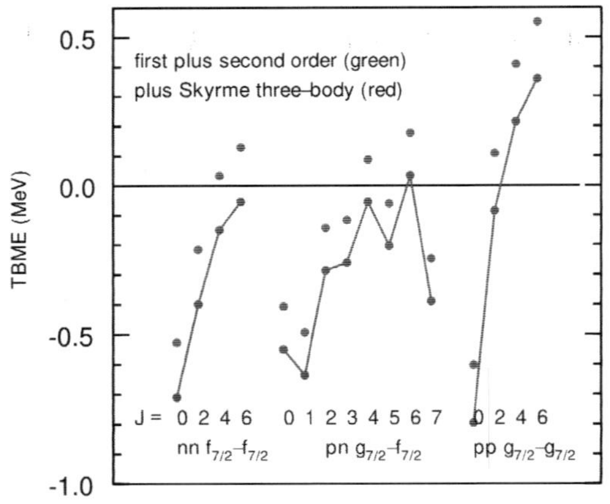

Figure 5. Comparison between results with the EDF radial wavefunctions and with the addition of the EDF monopole correction (red dots).

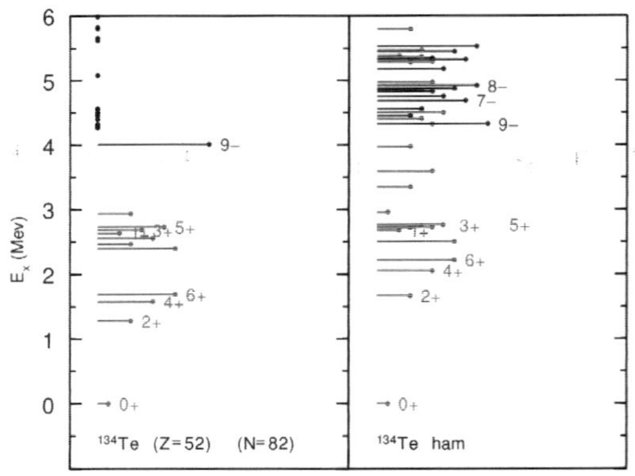

Figure 6. Comparison of experiment and theory (ham) for ^{134}Te. Theory (ham) is based on the Skxtb EDF for the single-particle energies and two-body monopoles, and the N^{3}LO (V_{lowk}) interaction in an oscillator basis renormalized to second order for the valence two-body matrix elements. The length of the line indicates the spin with positive parity (red) and negative parity (blue). Experimental levels that are unknown or uncertain are shown by the black dots.

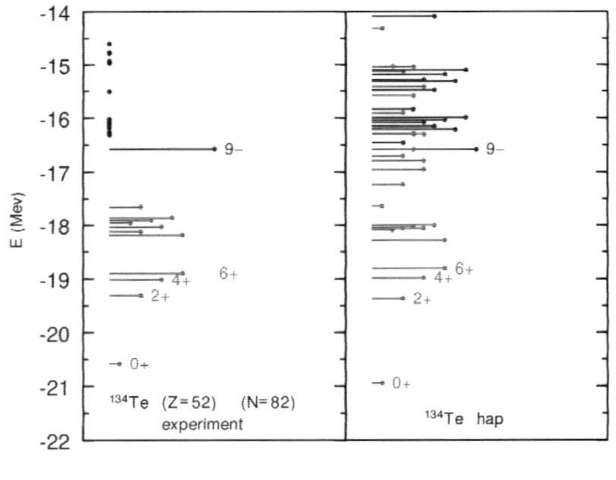

Figure 7. Comparison of experiment and theory (hap) for ^{134}Te. Theory (hap) is the same as (ham) in Fig. (6) except that the single-particle energies have been replaced by their experimental values. The length of the line indicates the spin with positive parity (red) and negative parity (blue). Experimental levels that are unknown or uncertain are shown by the black dots.

that the single-particle spectra from EDF will be improved with better parametrization and extended functional forms [3].

In conclusion, we have provided a new method that is able to constrain the monopole part of CI calculations to the EDF results in the limit of single-Slater determinants. This constrained CI contains all interactions implicit in EDF including the three-body and/or density dependent contribution. Results for the ^{132}Sn region show that the three-body contribution to the monopole interaction is generally repulsive and is similar in magnitude to the monopole part of the two-body interaction obtained from renomalized realistic interactions such as N^{3}LO. The application of the CI approach is limited to about eight valence nucleons due to matrix dimensions. But there are approximate methods that can be used to apply our new Hamiltonian to nuclei with more valence nucleons, such as Monte-Carlo methods, exact pairing (for protons or neutrons) [11], or deformed HF+GCM (within the valence model space) for protons and neutrons.

Acknowledgments This work is partly supported by NSF Grant PHY-0758099 and the DOE UNEDF-SciDAC grant DE-FC02-09ER41585.

References

[1] Brown B A 1998 *Phys. Rev. C* **58** 220
[2] Klupfel P, Reinhard P G, Burvenich T J and Maruhn J A 2009 *Phys. Rev. C* **79** 034310
[3] Kortelainen M, Lesinski T, More J, Nazarewicz W, Sarich J, Schnuck N, Stoitsov M V and Wild S 2010 *Phys. Rev. C* **82** 024313
[4] Coraggio L, Covello A, Gargano A, Itaco N and Kuo T T S 2009 *Prog. in Part. and Nucl. Phys.* **62** 135
[5] Hjorth-Jensen M, Kuo T T S and Osnes E 1995 *Phys. Rep.* **261** 125
[6] Brown B A and Wildenthal B H 1988 *Ann. Rev. of Nucl. Part. Sci.* **38** 29
[7] Zuker A P 2003 *Phys. Rev. Lett.* **90** 042502
[8] Brown B A, Duguet T, Otsuka T, Abe D and Suzuki T 2006 *Phys. Rev. C* **74** 061303(R)
[9] Bogner S K, Furnstahl R J and Schwenk A 2010 *Prog. in Part. and Nucl. Phys.* **65** 94
[10] Entem D R and Machleidt R 2003 *Phys. Rev. C* **68** 041001
[11] Volya A, Brown B A and Zelevinsky V 2001 *Phys. Lett.* **B509** 37

Three-body interactions in the $1f_{7/2}$ shell?

2011 J. Phys.: Conf. Ser. 267 012029

(http://iopscience.iop.org/1742-6596/267/1/012029)

Three-body interactions in the $1f_{7/2}$ shell?

P Van Isacker[1] **and I Talmi**[2]
[1]GANIL, CEA/DSM–CNRS/IN2P3, BP 55027, F-14076 Caen Cedex 5, France
[2]Weizmann Institute of Science, Rehovot, Israel

Abstract. It is shown that the three-body forces in the $1f_{7/2}$ shell, for which recently evidence was claimed on the basis of spectroscopic properties of the Ca isotopes and $N = 28$ isotones, can be most naturally explained as an effective interaction due to excluded higher-lying shells, in particular the $2p_{3/2}$ orbit.

Three-body interactions play an important role in nuclear many-body theories. They may be part of the "bare" interaction. They may also arise from the renormalization of the "bare" two-body interaction, yielding the effective interaction which may be used in the shell model. This is obtained by replacing the effects of admixtures of highly excited configurations by a modification of the "bare" interaction.

On the other hand, shell-model calculations, during the last 60 years, have been successfully using almost exclusively two-body effective interactions. It should be realized, however, that three-body interactions may contribute to single-nucleon energies which are usually taken from experiment. They may well include contributions of interactions of two core nucleons and a valence nucleon.

Also matrix elements of two-body interactions which are used in shell-model calculations are usually determined from experiment. They may include contributions due to the interaction of a core nucleon with two valence nucleons. Even if three-body interactions are weak, the large number of core nucleons makes their possible contribution significant. Still, no explicit evidence for effective three-body interactions between valence nucleons has been observed.

This feature is clearly seen in spectra of several j^n configurations of identical nucleons. Experimentally observed level spacings in these proton or neutron configurations are well reproduced by the use of two-body effective interactions with rather pure shell-model wave functions. Only a few levels seem to be perturbed.

In lighter nuclei, the $1f_{7/2}$ shell of neutrons is expected in calcium nuclei with $Z = 20$ and neutron number N from 21 to 28. The $1f_{7/2}$ proton shell should be found in nuclei with $N = 28$ and $Z = 21$–28. Shell-model calculations assuming $1f_{7/2}$ configurations were carried out [1, 2, 3, 4]. It turned out that the situation is rather complicated. All levels of these configurations are observed in experiment. Ground states of nuclei (binding energies) seem to agree with predictions based on two-body effective interactions. Spacings of many excited states, however, strongly deviate from such predictions.

These deviations were interpreted many years ago as due to mixing of $(1f_{7/2})^n$ states with states of configurations which are rather close in energy, in particular, states in which one $1f_{7/2}$ nucleon is raised into the $2p_{3/2}$ orbit. The position of the neutron $2p_{3/2}$ in ^{41}Ca is only 2 MeV above the $7/2^-$ ground state. The position of the proton $2p_{3/2}$ state is about 3 MeV above the

^{49}Sc ground state. Configuration mixings of this kind were considered for calcium isotopes [5, 6] and for $N = 28$ isotones [7, 8]. The results turned out to be in good agreement with experiment.

By their definition, effective interactions are determined by the model space for which they are intended. If pure $(1f_{7/2})^n$ states are adopted, two-body interactions do not yield good agreement with experiment. Many years ago, three-body interactions were introduced to obtain better agreement of calculated energies with experimental data [9, 10]. Recently, a detailed discussion of the effects of three-body interactions was presented by Zelevinsky [11].

In the 1960s atomic spectroscopists considered second-order perturbations, due to two-body interactions, in which one particle is excited to a higher orbit. They showed that under certain conditions, the effect may assume the form of additional two-body interactions as well as three-body ones [12, 13].

In a long and very detailed paper [14], Poves and Zuker considered not only the $1f_{7/2}$ states of identical nucleons. They analyzed the complete $1f_{7/2}$ shell of protons and neutrons, from ^{40}Ca to ^{56}Ni. They checked the agreement obtained by using various interactions suggested by many-body theory and modified them to obtain better agreement with experiment. They considered perturbations from nearby configurations due to two-body interactions. This way, they obtained the effective interactions in the $1f_{7/2}$ shell which includes two-body as well as three-body interactions. The agreement of results obtained by using the effective interaction which they derived for the $1f_{7/2}$ shell, with experiment was good.

The results of ref. [14] clearly indicated that effective three-body interactions in $1f_{7/2}$ shell arise from interactions with close configurations. In other words, they are due to a choice of shell-model space which is too restricted. Recently, however, a paper was published [15], in which the author claims to show the "manifestation of three-body forces in $f_{7/2}$-nuclei". Although the report by Poves and Zuker is quoted in ref. [15], no mention is made of the fact that in ref. [14] it was shown how configuration mixing leads to three-body interactions. In ref. [14] the latter were actually used for calculating successfully energy levels. The author of ref. [15] contrasts the two approaches and states that "the three-body effective interaction has a somewhat different manifestation which can be experimentally identified."

We do not intend to consider the $1f_{7/2}$ shell in detail. Our aim was to reiterate the fact that configuration mixing, due to two-body interactions leads to three-body effective interactions [16]. We were unaware of ref. [14] and hence, we used earlier calculations, in which such mixings were considered [5, 6, 7, 8]. We took non-diagonal matrix elements of the two-body interactions, obtained in these references and computed the resulting three-body interactions. The results show that the three-body interaction obtained by Volya [15] is rather similar to those following from configuration mixing.

Using the non-diagonal matrix elements of the two-body interaction from refs. [5, 6, 7, 8], we calculate matrix elements of the three-body effective interaction. These matrix elements completely define the three-body interaction. We make use of the expansion of $(1f_{7/2})^n$ wave functions in terms of coefficients of fractional parentage (c.f.p.) [17]. Every antisymmetric wave function in the j^n configuration may be expanded as

$$\Psi(j^n aJ) = \sum_{a_0 J_0} [j^{n-1}(a_0J_0)jJ|\}j^n aJ]\Psi(j^{n-1}(a_0J_0)j_nJ), \tag{1}$$

where $[j^{n-1}(a_0J_0)jJ|\}j^naJ]$ are coefficients of fractional parentage (c.f.p.) and a, a_0 are additional quantum numbers which may be necessary to characterize the states. The j^{n-1} wave functions on the r.h.s. of (1) are antisymmetric in $n-1$ particles to which j_n of the nth particle is coupled to yield the total J. Each of them may not be fully antisymmetric but their linear combination (1) is.

The c.f.p. are very convenient for calculating matrix elements of various operators. To calculate matrix elements of two-body or three-body interactions, use is made of the fact that both states are fully antisymmetric and the interaction is symmetric in all n particles. Hence, the contribution of each of the $n!/k!(n-k)!$, $k = 1, 2, \ldots < n$ terms of the k-body interaction to the matrix element is the same. Thus, we obtain

$$\begin{aligned}
\langle j^n aJ|\hat{V}|j^n bJ\rangle &= \frac{n!}{k!(n-k)!}\langle j^n aJ|\hat{V}_{12\ldots k}|j^n bJ\rangle \\
&= \frac{n!}{k!(n-k)!}\sum_{a_0 b_0 J_0}\langle j^{n-1}a_0 J_0|\hat{V}_{12\ldots k}|j^{n-1}b_0 J_0\rangle \\
&\quad \times [j^{n-1}(a_0 J_0)jJ|\}j^n aJ][j^{n-1}(b_0 J_0)jJ|\}j^n bJ] \\
&= \frac{n}{n-k}\sum_{a_0 b_0 J_0}\langle j^{n-1}a_0 J_0|\hat{V}|j^{n-1}b_0 J_0\rangle \\
&\quad \times [j^{n-1}(a_0 J_0)jJ|\}j^n aJ][j^{n-1}(b_0 J_0)jJ|\}j^n bJ]. \qquad (2)
\end{aligned}$$

This procedure may be further applied until the j^3 or j^2 configuration is reached.

We first consider the j^2 configuration and the effect of mixing with an excited jj' configuration. The contribution to the energy in second-order perturbation theory is given by

$$\sum_{j'}\frac{|\langle j^2 J_0|\hat{V}|jj'J_0\rangle|^2}{E(j^2 J_0) - E(jj'J_0)}, \qquad (3)$$

where the matrix elements of the two-body interaction $\hat{V}$ are taken between antisymmetric and normalized states $\langle j^2 J_0|$ and $|jj'J_0\rangle$. The expression (3) is the matrix element of an additional two-body interaction in the $|j^2 J_0\rangle$ state.

To obtain matrix elements of the three-body interaction, the j^3 configuration should be considered. Antisymmetrized and normalized states of the excited configuration may be obtained as

$$\Psi(j^2(J_1)j'J) = \frac{1}{\sqrt{3}}(\Psi(j^2(J_1)j_3'J) - \Psi(j^2(J_1)j_1'J) - \Psi(j^2(J_1)j_2'J)). \qquad (4)$$

In calculating the matrix element between the state (4) and the state $\Psi(j^3 J)$, the two-body interaction $\hat{V}_{12} + \hat{V}_{13} + \hat{V}_{23}$ may be replaced by $3\hat{V}_{12}$. Using c.f.p. we obtain for it the expression

$$\begin{aligned}
&-\sqrt{3}\sum_{J_0}[j^2(J_0)jJ|\}j^3 aJ]\langle j^2(J_0)j_3 J|\hat{V}_{12}|j^2(J_1)j_1'J\rangle \\
&-\sqrt{3}\sum_{J_0}[j^2(J_0)jJ|\}j^3 aJ]\langle j^2(J_0)j_3 J|\hat{V}_{12}|j^2(J_1)j_2'J\rangle. \qquad (5)
\end{aligned}$$

To evaluate the second term in (5), we use a change of coupling transformation ((10.11) in ref. [17]) and obtain for the excited state the expression

$$|j_1 j_3(J_1)j_2'J\rangle = \sum_{J_2}(-1)^{j+j'+J_1+J_2}\hat{J}_1\hat{J}_2\left\{\begin{matrix} j & J_2 & J \\ j' & J_1 & j\end{matrix}\right\}|j_1 j_2'(J_2)j_3 J\rangle, \qquad (6)$$

with $\hat{J} = \sqrt{2J+1}$. Using (6), the integration over j_3 may be carried out. It yields the only term with non-vanishing contribution which has $J_2 = J_0$. The second term in (5) is thus equal to

$$-\sqrt{3}(-1)^{j+j'+J_1}\sum_{J_0}\hat{J}_0\hat{J}_1[j^2(J_0)jJ|\}j^3 aJ]\left\{\begin{matrix} j & J_0 & J \\ j' & J_1 & j\end{matrix}\right\}\langle j_1 j_2 J_0|\hat{V}_{12}|j_1 j_2' J_0\rangle. \qquad (7)$$

The first term in (5) yields a similar result

$$\sqrt{3}(-1)^{j+j'+J_1}\sum_{J_0}\hat{J}_0\hat{J}_1[j^2(J_0)jJ|\}j^3aJ]\left\{\begin{matrix} j & J_0 & J \\ j' & J_1 & j \end{matrix}\right\}\langle j_1j_2J_0|\hat{V}_{12}|j_2j_1'J_0\rangle. \tag{8}$$

The two expressions may be combined yielding the result

$$-\sqrt{6}(-1)^{j+j'+J_1}\sum_{J_0}\hat{J}_0\hat{J}_1[j^2(J_0)jJ|\}j^3aJ]\left\{\begin{matrix} j & J_0 & J \\ j' & J_1 & j \end{matrix}\right\}\langle j^2J_0|\hat{V}_{12}|jj'J_0\rangle, \tag{9}$$

where the state $|jj'J_0\rangle$, as well as $|j^2J_0\rangle$, is antisymmetric and normalized.

To obtain the contribution of second-order perturbation, (9) should be multiplied by a similar summation in which J_0 is replaced by J_0'. The result is thus equal to

$$\begin{aligned} & 6\sum_{j'J_1}(2J_1+1)\sum_{J_0J_0'}\hat{J}_0\hat{J}_0'[j^2(J_0)jJ|\}j^3aJ][j^2(J_0')jJ|\}j^3aJ] \\ & \times\left\{\begin{matrix} j & J_0' & J \\ j' & J_1 & j \end{matrix}\right\}\left\{\begin{matrix} j & J_0 & J \\ j' & J_1 & j \end{matrix}\right\}\frac{\langle j^2J_0|\hat{V}_{12}|jj'J_0\rangle\langle jj'J_0'|\hat{V}_{12}|j^2J_0'\rangle}{E(j^3aJ)-E(j^2(J_1)j'J)}. \end{aligned} \tag{10}$$

In order to obtain an effective three-body interaction which is valid for all j^n configurations, we make an assumption on the energy denominator in (10). We make the approximation that it is the same for all states of the j^n and $j^{n-1}j'$ configurations. This assumption, which is probably more justified in atomic spectroscopy, was made in refs. [12, 13]. The energy denominator in (10) is thus replaced by $E(j^3)-E(j^2j')=\Delta E$ which is independent of a, J_1 and J. It may be taken to be, according to perturbation theory, equal to the difference of single-nucleon energies $\epsilon_j-\epsilon_{j'}$.

In this approximation, the summation over even values of J_1 in (10) may be carried out by using identities of Racah coefficients. We obtain, due to J_0 and J_0' having even values,

$$\begin{aligned} & 2\sum_{J_1\text{ even}}(2J_1+1)\left\{\begin{matrix} j & J_0 & J \\ j' & J_1 & j \end{matrix}\right\}\left\{\begin{matrix} j & J_0' & J \\ j' & J_1 & j \end{matrix}\right\} \\ & = \sum_{J_1}(2J_1+1)\left\{\begin{matrix} j & J_0 & J \\ j' & J_1 & j \end{matrix}\right\}\left\{\begin{matrix} j & J_0' & J \\ j' & J_1 & j \end{matrix}\right\} \\ & \quad +\sum_{J_1}(-1)^{J_1}(2J_1+1)\left\{\begin{matrix} j & J_0 & J \\ j' & J_1 & j \end{matrix}\right\}\left\{\begin{matrix} j & J_0' & J \\ j' & J_1 & j \end{matrix}\right\} \\ & = \frac{\delta_{J_0J_0'}}{2J_0+1}+\left\{\begin{matrix} j & J & J_0 \\ j & j' & J_0' \end{matrix}\right\}. \end{aligned} \tag{11}$$

Using this result in (10) we obtain the second-order contribution given by

$$\begin{aligned} & 3\sum_{j'}\sum_{J_0J_0'}\hat{J}_0\hat{J}_0'[j^2(J_0)jJ|\}j^3aJ][j^2(J_0')jJ|\}j^3aJ] \\ & \times\left[\frac{\delta_{J_0J_0'}}{2J_0+1}+\left\{\begin{matrix} j & J & J_0 \\ j & j' & J_0' \end{matrix}\right\}\right]\frac{\langle j^2J_0|\hat{V}_{12}|jj'J_0\rangle\langle jj'J_0'|\hat{V}_{12}|j^2J_0'\rangle}{\Delta E}. \end{aligned} \tag{12}$$

As explained above, (12) includes also matrix elements of an effective two-body interaction. Their contribution may be obtained by using (2) with $n=3$ and $k=2$. The matrix element in the j^2 configuration is given by (3) in which, according to our approximation, the energy

Table 1. The $T = 3/2$ three-body interaction matrix elements in the $1f_{7/2}$ shell obtained by Volya and the effective three-body interaction matrix elements derived with the formula (14) for various p3f7 interactions. Energies are in keV.

	$Z = 20$			$N = 28$		
J	Volya [15]	p3f7 [5]	p3f7 [6]	Volya [15]	p3f7 [7]	p3f7 [8]
3/2	−559 (273)	−412	−173 (89)	−128 (88)	−53 (29)	−115
5/2	2 (185)	−207	−104 (48)	−18 (70)	−54 (26)	−57
7/2	53 (70)	138	76 (58)	55 (28)	42 (24)	37
9/2	272 (98)	392	157 (96)	122 (41)	36 (34)	110
11/2	51 (130)	29	46 (91)	102 (43)	72 (77)	6
15/2	−24 (73)	77	28 (27)	−53 (29)	5 (5)	23

denominator is the same as in (12). Thus, the matrix elements of the additional effective two-body interaction are given by

$$3[j^2(J_0)jJ|\}j^3aJ]^2 \sum_{j'} \frac{|\langle j^2 J_0|\hat{V}_{12}|jj'J_0\rangle|^2}{\Delta E}. \tag{13}$$

Subtracting (13) from (12), we obtain the matrix elements of the effective three-body interaction in the state $|j^3aJ\rangle$ given by

$$3\sum_{j'}\sum_{J_0 J_0'} \hat{J}_0\hat{J}_0'[j^2(J_0)jJ|\}j^3aJ][j^2(J_0')jJ|\}j^3aJ]$$
$$\times\left\{\begin{matrix} j & J & J_0 \\ j & j' & J_0' \end{matrix}\right\}\frac{\langle j^2J_0|\hat{V}_{12}|jj'J_0\rangle\langle jj'J_0'|\hat{V}_{12}|j^2J_0'\rangle}{\Delta E}. \tag{14}$$

The contribution in second-order perturbation is contained in both (13) and (14). This contribution is always attractive (negative). This need not be the case for the three-body interaction (14). The reason is that in (11), terms with odd values of J_1 are introduced. In terms with $J_0 \neq J_0'$, it follows from (11) that $\sum_{J_1 \mathrm{even}} = -\sum_{J_1 \mathrm{odd}}$ and hence, the contribution of terms with odd-J_1 values to the three-body terms vanishes. If, however, $J_0 = J_0'$, these odd-J_1 terms contribute to both (13) and (14). They must be cancelled when both interactions are considered. The additional two-body terms (13) are usually, however, absorbed into the other effective two-body interaction.

The above formalism can be applied to the $1f_{7/2}$ shell. Several interactions are available that include effects from the $2p_{3/2}$ shell. Two of them were derived on the basis of spectroscopic properties of the calcium isotopes [5, 6] and two more from those of the $N = 28$ isotones [7, 8]. In particular, the authors of these references give numerical values for the matrix elements involving the $2p_{3/2}$ shell which enter the expression (14) and for the difference in single-particle energies, $\epsilon_{2p_{3/2}} - \epsilon_{1f_{7/2}}$. We refer to these interactions as p3f7, followed by the relevant reference.

We have used these interactions, without any modification, to calculate the effective three-body interaction which is induced when the model space is restricted to $1f_{7/2}$. The results are shown in Table 1. For $j = 7/2$ a three-nucleon state is completely specified by its total angular momentum J and there is no need for the additional label a. There are six three-nucleon states with $J = 3/2$, $5/2$, $7/2$, $9/2$, $11/2$ and $15/2$, and each of these states defines a component of the three-body interaction. The three-body matrix elements of Volya [15] result from separate fits to the calcium isotopes and to the $N = 28$ isotones, and they should thus be compared with corresponding matrix elements obtained with the p3f7 interactions from refs. [5, 6] and [7, 8], respectively. The numbers in parentheses in Table 1 are the variances of the parameters in the fits, either given by Volya or computed from the errors on the matrix elements which are known for two of the interactions [6, 7].

There are substantial variations in the calculated effective three-body matrix elements. In particular, those derived from the $Z = 20$ p3f7 interactions are generally larger than those obtained with the $N = 28$ p3f7 interactions. This results from a combination of a larger difference $\epsilon_{2p_{3/2}} - \epsilon_{1f_{7/2}}$ and smaller two-body matrix elements $\langle (1f_{7/2})^2 J_0 | \hat{V} | 1f_{7/2} 2p_{3/2} J_0 \rangle$ in the latter interactions.

There are certainly differences between these calculations and the results of Volya. However, in spite of these differences and uncertainties, it is clear that the matrix elements are correlated: with one exception, attractive (repulsive) matrix elements in the analysis of Volya turn out to be attractive (repulsive) in our analysis. The exception concerns the $J = 15/2$ interaction which is attractive in Volya's analysis while it is repulsive for the p3f7 interactions.

We considered two approaches to the $1f_{7/2}$ shell. A dogmatic one in which states of nucleons outside closed shells are taken to be pure $1f_{7/2}$ states. In this approach, to obtain the observed energies, it is necessary to introduce three-body interactions. In the other approach, it is recognized that there are rather close configurations whose states may mix with $(1f_{7/2})^n$ states by two-body interactions. It was shown by Poves and Zuker [14] and repeated here, that the effect of such mixing in second-order perturbation theory leads to three-body effective interactions in the $(1f_{7/2})^n$ configuration.

It is true that the first requirement of any calculation is to reproduce well the energies of a system. This is due to the fact that the energy is stationary at the correct wave functions. Small deviations from the exact wave functions may have a large effect on other observables like electromagnetic moments and transitions. Looking at the magnetic moments, it is clear that states of actual nuclei are not well described by pure $(1f_{7/2})^n$ states. In pure j^n configurations of identical nucleons, the g-factors of all states should be equal. In contrast to other shells, this is not the case for the $1f_{7/2}$ shell. Introducing three-body interactions does not change any state with the exception of the two $J = 2$ and two $J = 4$ for $n = 4$ configurations.

Magnetic moments are very sensitive to certain small admixtures in the wave functions. Many years ago, Arima and Horie [18] showed that changes in them are proportional to the amplitude of admixtures similar to those discussed above. In such excitations, one $j = l + 1/2$ nucleon is raised into the $j' = l - 1/2$. In the present case, excitation of a $1f_{7/2}$ nucleon into the $1f_{5/2}$ orbit are important. Even an admixture of 1–2% is sufficient to obtain the magnetic moment of ^{43}Ca which is rather different from that of ^{41}Ca. Examples of using configuration mixing to obtain the correct magnetic moments can be found for calcium isotopes in ref. [19] and for $N = 28$ isotones in ref. [20].

Other evidence for the impurity of $(1f_{7/2})^n$ states in actual nuclei follows from the results of various direct reactions. They exhibit definite admixtures of $l = 1$ components in wave functions. Admixtures of configurations affect energies of states as well as other observables. Large quadrupole moments and strong E2 transitions in light calcium isotopes may indicate appreciable admixtures of core-excited collective states. If energies of states are correctly obtained by using three-body forces with rather pure $(1f_{7/2})^n$ wave functions, no place for the necessary admixtures is available.

In ref. [15] it is stated that "while going beyond the single j-shell is imperative to account for the observed g-factors, the agreement between the experiment and the common large-scale shell model results is not perfect". From this fact the author concludes that this "may be a hint for the presence of a three-body force." We would like to remind the reader that such arguments may be valid in QED which is a very exact theory. They are certainly irrelevant in looking for deviations from calculations based on the shell model which is a drastic approximation to the exact theory of nuclear structure.

References

[1] Lawson R D and Uretsky J L 1957 *Phys Rev* **106** 1369
[2] Talmi I 1957 *Phys Rev* **107** 326
[3] Ginocchio J N and French J B 1963 *Phys Lett* **7** 137
[4] McCullen J D, Bayman B F and Zamick L 1964 *Phys Rev* **134** B515
[5] Engeland T and Osnes E 1966 *Phys Lett* **20** 424
[6] Federman P and Talmi I 1966 *Phys Lett* **22** 469
[7] Auerbach N 1967 *Phys Lett B* **24** 260
[8] Lips K and McEllistrem M T 1970 *Phys Rev C* **1** 1009
[9] Eisenstein I and Kirson M 1973 *Phys Lett B* **47** 315
[10] Quesne C 1970 *Phys Lett B* **31** 7
[11] Zelevinsky V 2009 *Yadernaya Fizika [Phys At Nucl]* **72** 1107
[12] Rajnak K and Wybourne B G 1963 *Phys Rev* **132** 280
[13] Racah G and Stein J 1967 *Phys Rev* **156** 58
[14] Poves A and Zuker A 1981 *Phys Reports* **70** 235
[15] Volya A 2009 *Eur Phys Lett* **86** 52001
[16] Van Isacker P and Talmi I 2010 *Eur Phys Lett* **90** 32001
[17] See *eg* Talmi I 1993 *Simple Models of Complex Nuclei. The Shell Model and Interacting Boson Model* (Harwood, Academic, Chur)
[18] Arima A and Horie H 1954 *Prog Theor Phys* **12** 623
[19] Taylor M J *et al* 2003 *Phys Lett B* **559** 187
[20] Speidel K-H *et al* 2000 *Phys Rev C* **62** 031301

Particle-hole intruder levels in ^{67}Cu, collectivity, monopole shifts, and the hockey-stick behaviour of $\ell - 1/2$ $5/2^-$ levels in neutron-rich odd-mass Cu nuclei

2011 J. Phys.: Conf. Ser. 267 012030

(http://iopscience.iop.org/1742-6596/267/1/012030)

Particle-hole intruder levels in ^{67}Cu, collectivity, monopole shifts, and the hockey-stick behaviour of ℓ – ½ 5/2$^-$ levels in neutron-rich odd-mass Cu nuclei

W B Walters [1] and C J Chiara [1,2]

[1]University of Maryland, College Park, MD 20742 USA

[2]Argonne National Laboratory, Argonne, IL 60439 USA

E-mail: wwalters@umd.edu

Abstract. A new sequence of gamma rays with energies of 572, 499, 585, and 674 keV has been identified in ^{67}Cu populating the 7/2$^-$ level at 2362 keV. Owing to the strong population of the 2362-keV level via an ℓ = 3 proton pickup reaction, that level is assigned to be an $f_{7/2}^{-1}$ 2-particle-1-hole "intruder" proton configuration, and the new levels are found to form a sequence consistent with intruder sequences in the adjacent odd-mass Cu isotopes and in the odd-mass Sb isotopes. The changing position of the intruder sequence in the odd-mass Cu isotopes is discussed and related to the onset of collectivity associated with the presence of $g_{9/2}$ neutrons beyond N = 40. The increase in collectivity is also discussed for a number of isotonic and isotopic chains as more protons or neutrons, respectively, are added beyond an oscillator shell boundary. For most of these systems, the ℓ – ½ levels show a systematic "hockey-stick-like" behaviour with a sharp decrease in energy with the addition of the first protons or neutrons, owing to both the added collectivity and the tensor interaction, and then a lower slope when collectivity changes are diminished and only the tensor interaction is influencing the changes in level positions.

1. Introduction

Following the identification of the lowered energies of the 5/2$^-$ levels in ^{71,73}Cu [1], many other studies have been undertaken to obtain additional information for the levels of other odd-mass Cu levels [2,3,4,5,6,7]. Moreover, a number of theoretical studies have also been undertaken that pertain to the systematic changes in these levels, and, in particular, the rapid downward movement of the ℓ - ½ 5/2$^-$ levels as ℓ + ½ $g_{9/2}$ neutrons are added [8,9,10]. In this work, new level structure has been determined for ^{67}Cu from which level sequences associated with coupling to the ^{66}Ni core are readily identified [11]. The recent identification of an unexpected K = ½ 1-particle-hole intruder level at 493 keV in ^{67}Co$_{40}$ has led to keen interest in intruder levels in Cu and Co for the even-even closed shell Ni isotopes [12,13].

2. Experimental measurements

These new data for the levels of ^{67}Cu were obtained by determining the gamma rays following the reaction of a 430-MeV ^{64}Ni beam with a ^{238}U target using the ATLAS accelerator and Gammasphere at Argonne National Laboratory. The data sorting has been described in recent papers by Hoteling *et al.* [14,15] No isomers with half lives in the range from 20 ns to ~10 μs were observed. An $f_{7/2}$ 2-particle-1-hole state was identified at 2362(30) keV by Zeidman *et al.* in the (d,^{3}He) reaction with an ℓ value of 3 and a spectroscopic factor of 3.1 [16]. A number of additional levels in ^{67}Cu were identified by Nybø *et al.* in the ^{64}Ni(α,p)^{67}Cu reaction [17]. The systematic behaviour of particle-hole intruder levels was surveyed by Heyde *et al.* [18] Such levels are of interest as, although their behaviour is sensitive to the overall collectivity changes in a particular isotopic sequence, they appear to reflect, in particular for protons, the occupancy of the valence high-j neutrons.

3. Results and discussion

A partial level scheme for ^{67}Cu showing the four new transitions and four new proposed levels that populate the $7/2^-$ 2-particle-1-hole intruder level at 2362 keV is shown in Fig. 1. This figure includes the collective levels populated in Coulomb excitation by Stefanescu et al. [3] The approximately equally spaced levels connected by strong dipole and weak quadrupole crossover transitions are quite similar to those identified for the odd-mass Sb nuclei whose low-energy levels are shown in Fig. 2a [19]. In particular, all of the levels shown for the odd-mass Sb nuclei drop in energy, "in parallel" beyond the N = 4 oscillator-shell boundary at N = 64 up through N = 70. Beyond N = 70, the 2-particle-1-hole $9/2^+$ intruder level rises rapidly in energy, while the "tensor-interaction-induced" energy reduction for the $g_{7/2}$ proton state continues to drop in energy, but at a somewhat lesser rate.

The behaviour of "tensor-interaction-induced" energy changes can be seen further in Figs. 2b, 2c, and 2d where the odd-neutron levels are plotted for N = 29, 51, and 53 isotones, respectively. For all three of these sets of levels, protons are being added beyond an oscillator-shell boundary, either Z = 20 or Z = 40: in each case, there is a sharp drop with the addition of the first and/or second pair of protons, and then a gentle drop for additional protons. In Fig. 2b for the N = 29 isotones, the familiar "hockey-stick" shape has been added. In all three cases, the energies of the adjacent 2^+ levels are plotted, along with other levels of interest.

A box has been placed around the region of Sb isotopes comparable to the levels observed in the odd-mass Cu isotopes, namely the region beyond [for Sb] N = 64 where $h_{11/2}$ neutrons are added to the nucleus. The levels shown within that boxed region — the "$9/2^+$ and $1/2^+$ collective" levels, the "$9/2^+$ intruder" levels, and the $\ell - ½$ $7/2^+$ levels— all drop in energy. Similar behaviour is observed for all of the levels in the odd-mass Cu isotopes with N = 38, 40, 42, and 44. For Sb, however, as more neutrons are added, the collective levels and the intruder rise rapidly, and the $\ell - ½$ $7/2^+$ level descends at a lower rate. This comparison supports the notion that beyond N = 46, the collective levels and intruder levels in Cu will rise in energy and the $\ell - ½$ $5/2^-$ level will descend slowly.

A summary of the low-energy levels in the odd-mass Cu isotopes is plotted in Fig. 3. These levels show behaviour similar to those in the odd-mass Sb isotopes. That is, the observed $1/2^-$, $5/2^-$, "$7/2^-$ intruder", and "$7/2^-$ collective" levels drop in energy as N increases from 40 to 42 driven largely by collectivity, as reflected by the sharp increases in B(E2) values reported by Stefanescu *et al.* for the $1/2^-$ and $7/2^-$ collective levels [3]. As with the Sb nuclei, the intruders level off at N = 42 and N = 44 and are likely to rise sharply as N approaches 50. The $5/2^-$ levels which exhibit much smaller B(E2) values than the adjacent $1/2^-$ and $7/2^-$ collective levels also drop rapidly for N = 40 and N = 42, and then more slowly for N = 44 and N = 46, exhibiting the "hockey-stick" shaped curve similar to that shown in Fig. 2b for the N = 29 isotones. In contrast, the $9/2^+$ level that has relatively large spectroscopic factors in the lighter Cu isotopes exhibits little change in position as N increases from 40 to 42. As the position of this level is not likely to be changed by quadrupole collectivity, the insensitivity to N is not surprising. The positions of the levels of $^{79}Cu_{50}$ calculated by three different approaches are also shown in Fig. 3, along with the experimental data for isotonic $^{81}Ga_{50}$, whose ground-state spin has recently been determined to be 5/2 [20,21,22].

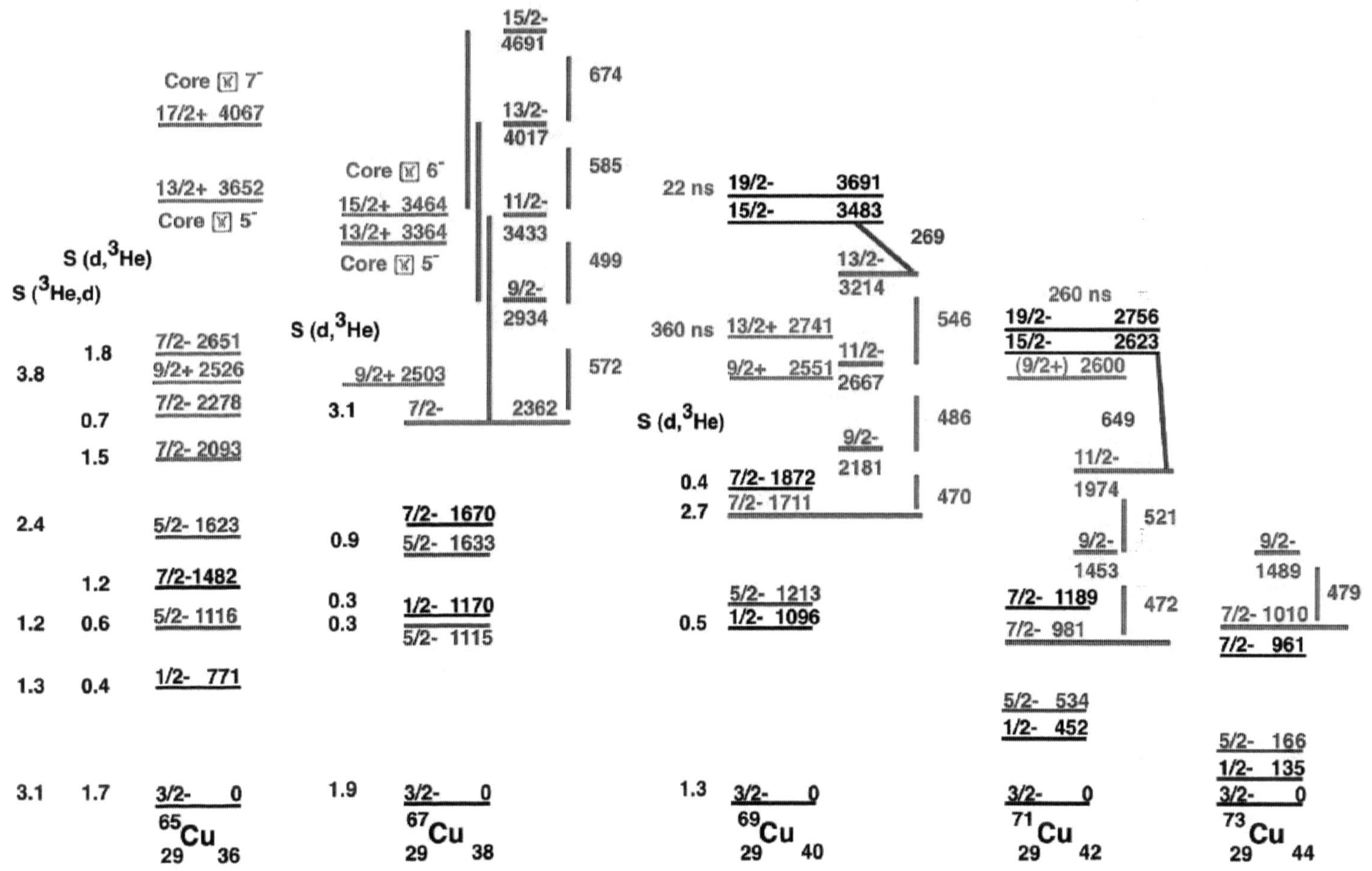

Figure 1. Energy levels in odd-mass ^{65,67,69,71,73}Cu isotopes. New levels are shown in ^{67}Cu at 2934, 3433, 4017, and 4691 keV that form a level sequence of dipole and quadrupole transitions which populate the 7/2$^-$ intruder level at 2362 keV, shown with a longer line. Spectroscopic factors are shown for pickup and stripping reactions. Intruder levels and excited states that decay by dipole and quadrupole sequences are also shown for ^{69,71,73}Cu. All of the levels shown in this plot, except the 1/2$^-$ levels, have been observed in this experiment and are consistent with earlier data from Ishii et al. [7], Stefanescu et al. [3], and Franchoo et al. [1].

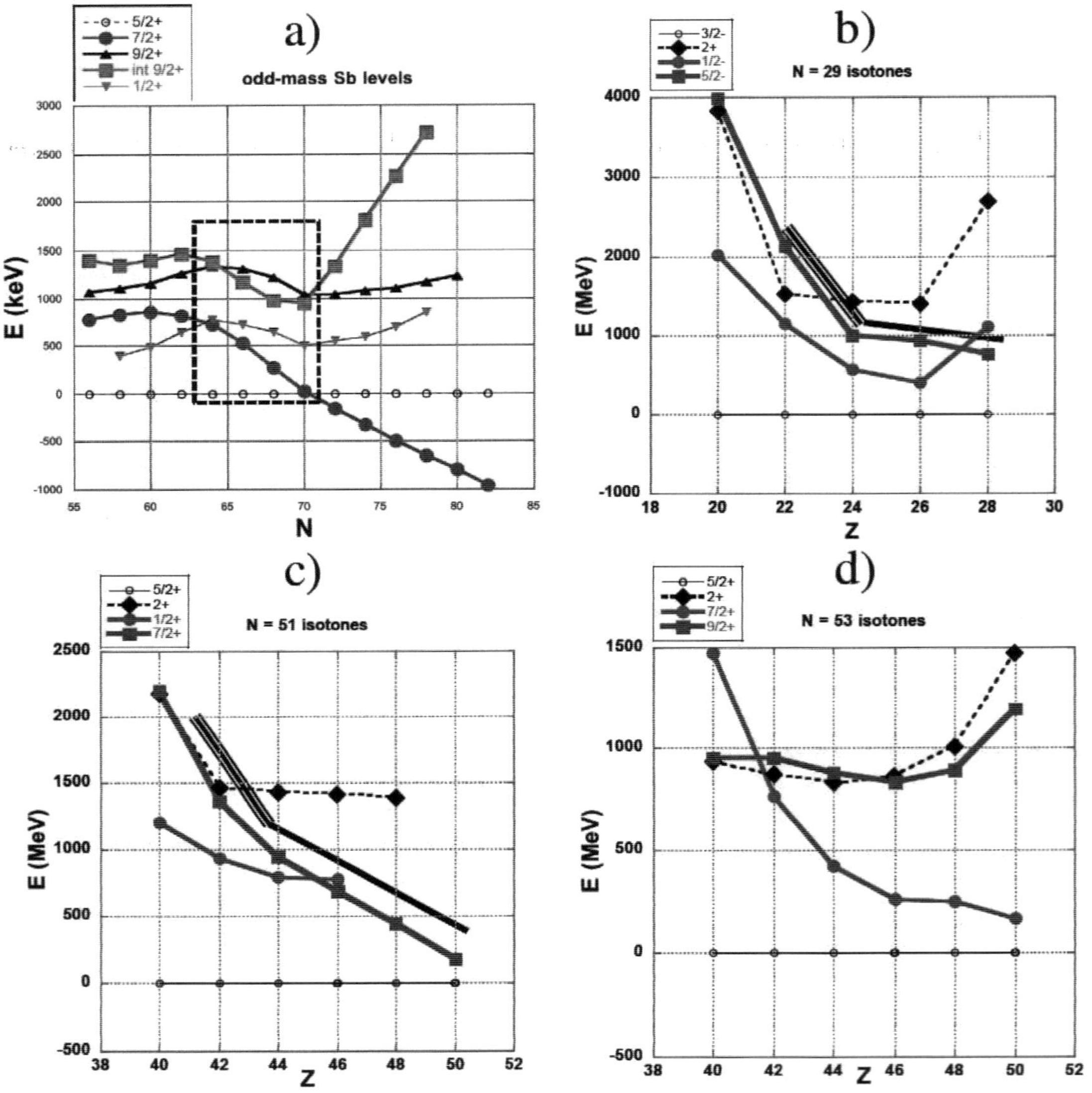

Figure 2a. Plot of the positions of selected low-energy levels in the odd-mass Sb isotopes, including the 2-particle-1-hole $9/2^+$ intruder levels as described by Heyde et al. [18] The box described by the dashed lines surrounds those levels that are "equivalent" to the odd-mass Cu levels shown in Fig. 1 where the intruder levels come down in energy.

Figure 2b. Plot of the positions of the lowest $1/2^-$ and $5/2^-$ levels in the odd-mass N = 29 isotones, along with the position of the lowest 2^+ level in the adjacent N = 28 isotones. The boxes associated with the positions of the $5/2^-$ levels are shown to exhibit the "hockey-stick" shape of $5/2^-$ levels as described in the text.

Figure 2c. Plot of the positions of the lowest $1/2^+$ and $7/2^+$ levels in the N = 51 isotones, along with the position of the lowest 2+ level in the adjacent N = 50 isotones.

Figure 2d. Plot of the positions of the lowest $9/2^+$ and $7/2^+$ levels in the N = 53 isotones, along with the position of the lowest 2^+ level in the adjacent N = 52 isotones.

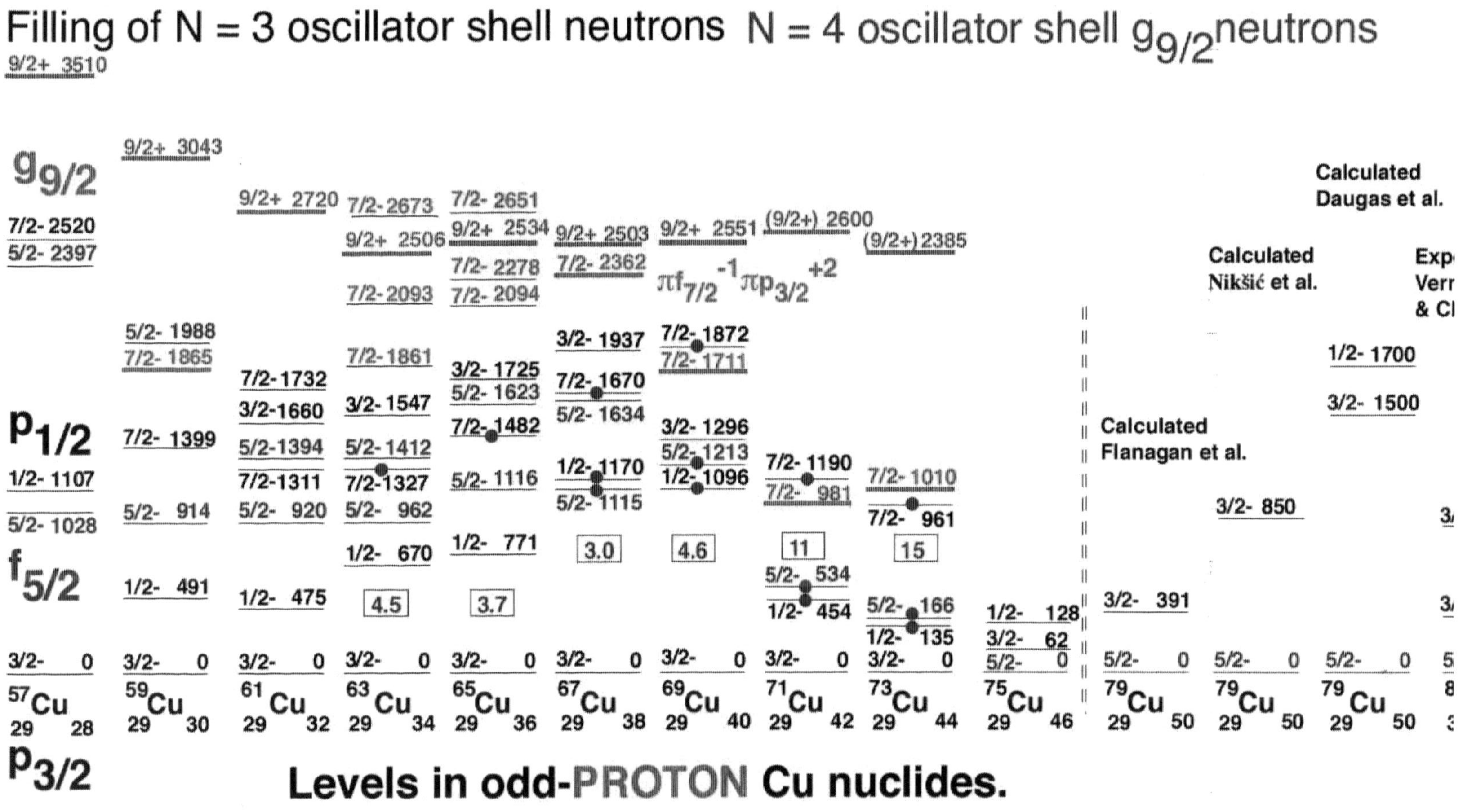

Figure 3. Summary plot of the energy levels of the odd-mass Cu isotopes, including calculated positions for the levels of $^{79}Cu_{50}$ and observed levels i
The filled circles show levels populated in Coulomb excitation reactions. The numbers in the boxes show the B(E2) values for the $3/2^-$ to $7/2^-$ transiti
including recent values taken from Stefanescu et al., that illustrate the rapid rise in B(E2) as N increases beyond 40. [3]

In conclusion, new levels have been indentified in ^{67}Cu that populate the 2-particle-1-hole intruder 7/2$^-$ level at 2362 keV. These levels now make it possible to observe the behaviour of this intruder sequence for $^{67,69,71,73}Cu_{38,40,42,44}$ and identify the similarities with analogous intruder levels in odd-mass Sb nuclides. This behaviour highlights the importance of collective effects as neutrons or protons are added beyond oscillator shell boundaries. These results indicate the need to view full systems in developing models to describe nuclei for which few or no results are available.

Acknowledgements

This work has been supported by the US Department of Energy under grant DE-FG02-94-ER40834, and Contract DE-AC02-06CH11357. The authors wish to thank Drs. M P Carpenter, R V F Janssens, D Seweryniak, and S Zhu for helpful discussions.

References

[1] Franchoo S *et al.* 1998 *Phys. Rev. Lett.* **81** 3100
[2] Franchoo S *et al.* 2001 *Phys. Rev. C* **64** 054308
[3] Stefanescu I *et al.* 2008 *Phys. Rev. Lett.* **100** 112502
[4] Stefanescu I *et al* 2009 *Phys. Rev. C* **79** 034319
[5] Daugas J M *et al.* 2010 *Phys. Rev. C* **81** 034304
[6] Flanagan K T *et al.* 2009 *Phys. Rev. Lett.* **103** 142501
[7] Ishii T *et al.* 2000 *Phys. Rev. Lett.* **84** 39
[8] Oros-Peusquens A M and Mantica P F 2000 *Nucl. Phys. A* **669** 81
[9] Otsuka T *et al.* 2001 *Phys. Rev. Lett.* **87** 082502
[10] Otsuka T *et al.* 2005 *Phys. Rev. Lett.* **95** 232502
[11] Chiara C J, Stefanescu I, Hoteling N, Broda R, Fornal B, Janssens R V F, Walters W B , Carpenter M P , Gürdal G, Hecht A A , Hoffman C R, Kay B P, Kondev F G, Królas W, Lauritsen T, Lister C J, McCutchan E A, Pawlat T, Seweryniak D, Sharp N, Stone N J, Stone J R, Wang X, Wöhr A, Wrzesinski J, Zhu S 2010 to be submitted for publication in *Phys. Rev. C.*
[12] Pauwels D *et al.* 2008 *Phys. Rev. C* **78** 041307R
[13] Pauwels D *et al.* 2009 *Phys. Rev. C* **79** 044309
[14] Hoteling N *et al.* 2006 *Phys. Rev. C* **74** 064313
[15] Hoteling N *et al.* 2007 *Phys. Rev. C* **77** 044314
[16] Zeidman B and Nolan J A 1978 *Phys. Rev. C* **18** 2122
[17] Nybø N *et al.* 2001 *Phys. Scr.* **63** 181
[18] Heyde K *et al.* 1983 *Phys. Rep.* **102** 291
[19] Shroy R E *et al.* 1979 *Phys. Rev. C* **19** 1324
[20] Verney D *et al.* 2007 *Phys. Rev. C* **76** 054312
[21] T. Nikšić *et al.* 2005 *Phys. Rev. C* **71** 014308
[22] Cheal B *et al.* 2010 *Phys. Rev. Lett.* **104** 252502

$19/2^+$ isomer of ^{125}Cd and realistic shell-model calculations of neutron-rich Cd isotopes

2011 J. Phys.: Conf. Ser. 267 012031

(http://iopscience.iop.org/1742-6596/267/1/012031)

$19/2^+$ isomer of ^{125}Cd and realistic shell-model calculations of neutron-rich Cd isotopes

G S Simpson[1,2], **A Scherillo**[2,3], **J Genevey**[1], **R Orlandi**[2,4], **J A Pinston**[1], **I S Tsekhanovich**[2], **N Warr**[3], **A Covello**[4,5], **A Gargano**[6]

[1] LPSC, Université Joseph Fourier Grenoble 1, CNRS/IN2P3, Institut National Polytechnique de Grenoble, F-38026 Grenoble Cedex, France

[2] Institut Laue-Langevin, B.P. 156, F-38042 Grenoble Cedex 9, France

[3] Institut für Kernphysik, Universität zu Köln, Zülpicherstr. 77, D-50937 Köln, Germany.

[4] Department of Physics and Astronomy, The University of Manchester, Brunswick Street, M13 9PL, United Kingdom

[5] Dipartimento di Scienze Fisiche, Università degli Studi di Napoli, Federico II, Complesso Universitario di Monte Sant'Angelo, Via Cintia, I-80126 Napoli, Italy

[6] INFN, Complesso Universitario di Monte Sant'Angelo, Via Cintia, I-80126 Napoli, Italy

E-mail: `simpson@lpsc.in2p3.fr`

Abstract.
The decay of a $T_{1/2} = 19(3)$ μs isomeric state from ^{125}Cd has been observed using γ-ray spectroscopy at the Lohengrin mass spectrometer of the Institut Laue-Langevin, Grenoble. Two coincident γ rays were observed to be emitted from the decay of this isomeric state, contradicting recently published data on this nucleus. Realistic shell-model calculations have been performed to interpret the decay scheme, allowing a spin and parity of $19/2^+$ to be assigned to the isomeric state. All configurations contributing to this isomeric state have amplitudes less than 7 %. Experimental data have recently been published on the decays of μs isomeric states in ^{127,128,130}Cd. Realistic shell-model calculations for these nuclei are presented which reproduce the known experimental decay schemes reasonably well.

1. Introduction

One of the main aims of nuclear structure studies is to test effective interactions used in shell-model calculations in regions far from stability. The most sensitive examinations of these interactions are performed close to doubly magic nuclei, such as ^{132}Sn. This nucleus is one of only two heavy, neutron-rich, doubly magic nuclei currently experimentally accessible, the other being ^{78}Ni. Nuclei with proton particles outside ^{132}Sn have been reasonably well studied in the past [1], as they are easier to produce. In this article we present the results of a study on neutron-rich Cd nuclei, which have just a few proton and neutron holes with respect to ^{132}Sn. This work is a continuation of our studies of the neutron-rich indium nuclei [2].

Recently, new experimental information has become available on the decay of intermediate spin states in the cadmium isotopes with $A = 125 - 130$, populated via the decays of μs isomers [3, 4, 5], which supplement previous β-decay data on ^{126,128}Cd [6]. Hole-hole nuclei, with respect to ^{132}Sn, are particularly interesting to study as Dillmann *et al.* [7] suggested that a weakened neutron-proton interaction may be responsible for the high energy of the 1^+ state in ^{130}In, as a consequence of a weakening of the $N = 82$ shell gap. In the same article it was also postulated that the weakening of this gap may be the cause of a divergence between experimental mass measurements of Cd isotopes, with $N \geq 77$, and systematic trends. Based on a comparison with shell-model calculations, Hoteling *et al.* [5] also postulated that the appearance of a μs isomeric state in ^{125}Cd, and more neutron-rich Cd isotopes, may also be a consequence of the weakened shell gap. The high energy of the 1^+ state in ^{130}In has been correctly calculated in a previous article by our group without the need to modify the shell-model interaction [2] and recent measurements of the masses of cadmium isotopes are now in line with systematic trends [8]. In the present work we compare predictions of state-of-the-art realistic shell-model calculations to existing data for the neutron-rich cadmium isotopes, with $N \geq 77$, and present new data for the decay of a $19/2^+$ isomeric state in ^{125}Cd.

Little experimental data exist on the $N = 77$ nucleus ^{125}Cd. A β-decaying $11/2^-$ isomeric state was reported some time ago [9, 10]. More recent works concern the decay of μs isomeric states [5, 11] in ^{125}Cd, however these data are inconsistent. Hoteling *et al.* reported a cascade of seven γ rays (409, 490, 720, 743, 868, 923 keV) originating from the decay of a single μs isomeric state, with spin $(33/2^-)$. This differs from an earlier work by Hellström *et al.* [11] where four delayed γ rays (486, 665, 720, 743 keV) were assigned to this nucleus. These transitions were thought to originate from the decays of two isomeric states as the 720- and 743-keV transitions had stronger intensities than the other two transitions. The 720–743-keV cascade was thought to arise from the decay of the $19/2^+_1$ state to the $15/2^-_1$ level and then to the $11/2^-$ state, via an $M2 - E2$ cascade, from a comparison with level systematics of neighboring nuclei. The second isomeric state was unassigned. The inconsistencies between the level schemes reported in Refs. [5, 11] have lead us to re-examine the decay of this nucleus. In our previous work on the neutron-rich indium nuclei [2] we have mentioned that two γ rays of energy 719.1 and 742.7 keV, were observed in coincidence with $A = 125$ ions at the Lohengrin mass spectrometer. These data will be shown in more detail in a forthcoming article and are partially presented here and are compared to modern large-scale shell-model calculations for ^{125}Cd, performed with a realistic effective interaction.

2. Experimental technique

Neutron-rich $A = 125$ ions were produced by the thermal-neutron induced fission of a thin 7×1 cm^2, ^{239}Pu target, with a total mass of 3 mg, at the high-flux reactor. These ions were then selected and mass separated, according to their mass-to-ionic charge ratios A/q, by the Lohengrin mass spectrometer. Ions were detected by an ionisation chamber filled with 35 mbar of isobutane, allowing identification of the A/q ratio. The flight time of the ions through the spectrometer was around 2.1 μs. Delayed γ rays were observed observed at the spectrometer's focal point by a Clover Ge detector [12] and a Miniball triple-cluster Ge detector [13], both placed perpendicular to the ion beam in a compact geometry, either side of the 6-cm thick ionization chamber. Any γ rays detected in the Ge detectors up to 40 μs after the arrival of an ion were recorded by the the data acquisition system. The time between any γ–γ coincidences, within the ion–γ window, was limited to be less than 250 ns in the data analysis software. More details on the experiment can be found in [2].

3. Results

Two delayed γ rays of energy 719.1(3) and 742.7(3) keV, have been observed in coincidence with $A = 125$ ions. These were previously assigned to ^{125}Cd in [5, 11] where the higher reaction energies used at GSI and MSU also allow clear isotopic identification. These two γ rays are the only ones consistently reported in both references [5, 11]. Proof that only the 719.1- and 742.7-keV γ rays are present in the same delayed cascade is obtained from $\gamma - \gamma$ coincidences recorded up 20 μs after the arrival of an A=125 ion. A gate set on the 719.1-keV γ ray, shown in Fig. 1, reveals a coincidence with a line at 742.7 keV and *vice versa*. Compton background from

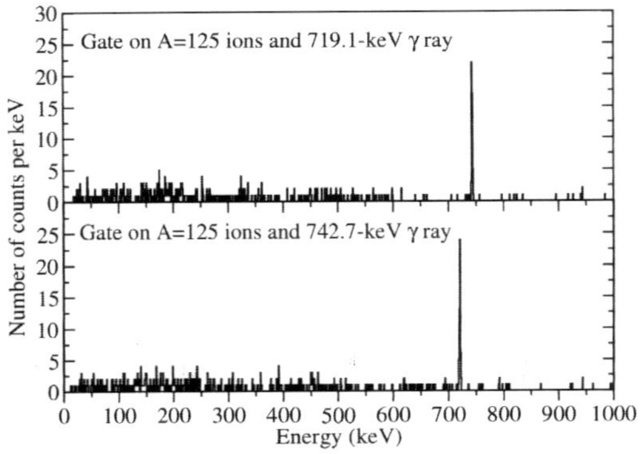

Figure 1. A γ-ray coincidence spectra obtained with gates on the 719.1 and 742.7 keV γ rays up to 20 μs after the arrival of an A=125 ion.

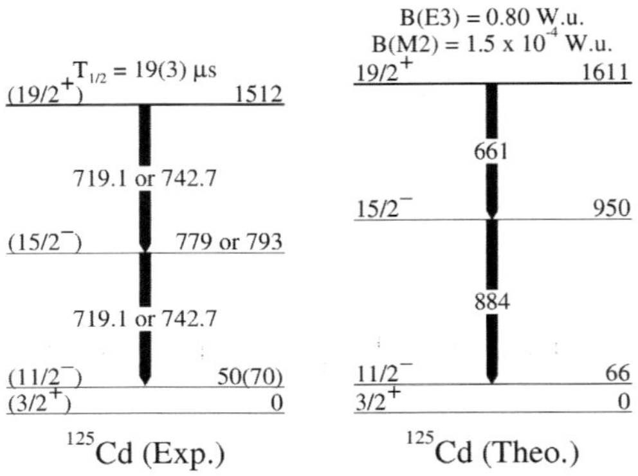

Figure 2. Experimental and calculated partial decay schemes of ^{125}Cd as obtained in the present work.

the unsuppressed detectors is also present. No clear traces of any other transitions are present, despite the good statistics for the 719.1- and 742.7-keV lines; ~2000 γ-singles photopeak counts and ~50 photopeak $\gamma - \gamma$ coincidence counts in a time window of up to 20 μs after the arrival of an A=125 ion. At this point one may conclude that a delayed cascade containing only two γ rays of energy 719.1 and 742.7 keV has been observed in the present work.

A 19(3) μs half-life has been found from a summed time spectrum of the 719.1- and 742.7-keV γ rays. It is therefore likely that the spin assignment of the $(19/2^-)$ for the 1463-keV state in [5] is incorrect, as a 719.1- or 742.7-keV $E2$ transition, to the $15/2^-$ state, would not have a 19-μs half-life.

In the odd-A Sn isotopes close to ^{132}Sn, the $19/2^+_1$ state forms an yrast spin trap with a μs half-life. For the $N = 77$ isotone of ^{125}Cd, ^{127}Sn, the isomeric state decays via two competing branches, a strongly converted, low-energy $E2$ transition and an $M2/E3$ decay [2, 14]. The $M2/E3$ branch populates the $15/2^-$ level, which then decays to the $11/2^-$ ground state. The isomeric $E2$ transition populates the $15/2^+$ state, which then feeds the $13/2^-$ and $15/2^-$ states, which then both decay to the ground state. Therefore the $E2$ and $M2/E3$ decay branches out of the isomer in ^{127}Sn consist of cascades of 5 and 2 transitions, respectively. As only 2 delayed γ rays have been observed from ^{125}Cd in the present work, we assign the measured 719.1–742.7-keV cascade to a $19/2^+ \rightarrow 15/2^- \rightarrow 11/2^-$ decay sequence. The $11/2^-$ state is a known β-decaying isomer [9, 10].

The additional unobserved γ rays reported in [5, 11] probably come from a high-spin, short-lived isomeric state, or states. The flight time of 2.1 μs through the Lohengrin spectrometer is about an order of magnitude longer than that of the MSU and GSI separators. The different reactions used to produce ^{125}Cd, thermal-neutron induced fission at the ILL, fragmentation of ^{136}Xe at MSU and relativistic projectile fission at GSI, may also have some influence on the states populated.

4. Shell-model calculations

In order to confirm the spins and parities assigned to the delayed cascade in ^{125}Cd and to further study the neighboring neutron-rich Cd nuclei, shell-model calculations have been performed using a two-body effective interaction derived from the CD-Bonn nucleon-nucleon potential [16]. The study is performed along the same lines as those for the In and even-Cd nuclei studied in [2] and [17]. The nucleus ^{132}Sn was considered as a closed core, with proton and

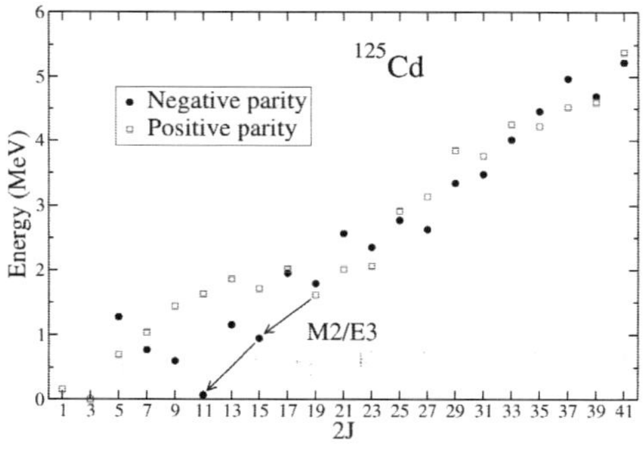

Figure 3. Calculated yrast states, for each parity, of ^{125}Cd.

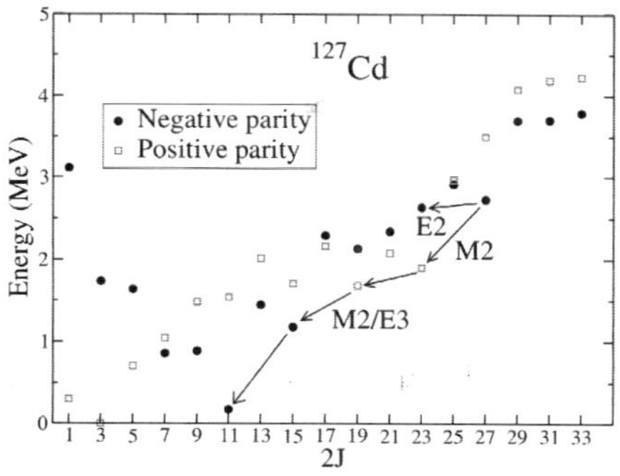

Figure 4. Calculated yrast states, for each parity, of ^{127}Cd.

neutron holes occupying levels in the 28–50 and 50–82 shell, respectively. This corresponds to the orbits $\pi(0f_{5/2}, 1p_{3/2}, 1p_{1/2}, 0g_{9/2})$, $\nu(0g_{7/2}, 1d_{5/2}, 1d_{3/2}, 2s_{1/2}, 0h_{11/2})$ for which we take the single-particle energies 2.750, 1.650. 0.302, 0.000 and 2.443, 1.655, 0.000, 0.332, 0.069 MeV, respectively. No truncation was used for the calculations within this valence space, which have been performed using the NuShellX code [15]. Effective proton and neutron charges of $e(\pi) = 1.35e$ and $e(\nu) = 0.78e$, and free-nucleon g factors of $g_l(\pi) = 1.000$, $g_l(\nu) = 0.000$ $g_s(\pi) = 5.586$ and $g_s(\nu) = -3.826$ were assumed, as justified in [17], when calculating transition rates. Calculations were also performed for ^{127,128,130}Cd, using identical interactions, single-particle energies and valence space, and are reported below.

4.1. ^{125}Cd

The theoretical calculations for ^{125}Cd predict that the $19/2^+_1$ state forms an yrast spin trap, as shown in Fig. 3, decaying via a $19/2^+ \to 15/2^- \to 11/2^-$ sequence. The predicted γ-decay energies, shown in Fig. 2, are in reasonable agreement with those measured. The $19/2^+$ state has a mixed configuration and its strongest component, arising from the $\pi g_{9/2}^{-2} \otimes \nu d_{3/2}^{-2} h_{11/2}^{-2} s_{1/2}^{-1}$ configuration, has an amplitude of 6.9 %. The calculated reduced transition probabilities of decays from the $19/2^+$ to the $15/2^-$ state are $B(M2) = 1.47 \times 10^{-4}$ and $B(E3) = 0.796$ Weisskopf units (W.u.), suggesting a mixed $M2/E3$ transition. When these transition rates are applied to the experimentally observed γ-ray energies, half-lives of 12.0 and 9.7 μs are obtained for the 719.1- and 742.7-keV γ rays, in reasonable agreement with the experimentally measured value of 19(3) μs. The calculations do not however, allow us to identify which of the two γ rays is the isomeric transition.

The results of the calculations further strengthen the assignment of the observed cascade to a $19/2^+ \to 15/2^- \to 11/2^-$ decay sequence as the $15/2^+$ state is calculated to be 104 keV above the $19/2^+$ level. If one assumes that the $15/2^+$ state lies below the $19/2^+$ level, then its calculated reduced transition probability is $BE(2) = 5.35$ W.u. For a hypothetical transition in the energy range of 10–27 keV a calculated half-life of 1.2 μs is obtained. This transition would be strongly converted and not detectable in the present experiment, however this predicted half-life is a factor of 16 shorter than the one experimentally measured. In contrast, the calculated half-life of the $M2/E3$ transition is in much better agreement with the experimental measurement and is a factor of only 1.6 to 2.0 too short. If an $E2$ transition, with an energy greater than 28 keV, between the $19/2^+$ and $15/2^+$ levels was responsible for the isomerism then Cd K X-rays would be expected to be present in the spectra of Fig. 1. As the detection efficiency at this energy is similar to that at ~750 keV, then such a scenario can be rejected.

The additional delayed γ rays from ^{125}Cd observed in [5, 11], originate from a higher lying, short-lived isomer, however there is presently insufficient data to allow us to identify this second isomer. The lowest $33/2^-$ state in ^{125}Cd, identified as a μs isomeric state in [5], is calculated to decay to the $31/2^-_1$, via a 537-keV, $M1$ transition, with an half-life of only 39 ps.

4.2. ^{127}Cd

The calculations for ^{127}Cd similarly predict an isomeric character for the $19/2^+_1$ state, which decays by a mixed $M2/E3$ transition to the $15/2^-$ state, as in ^{125}Cd. Figure 4 shows yrast levels for each parity. Values of $B(M2) = 8.21 \times 10^{-4}$ and $B(E3) = 0.571$ W.u. are calculated for this transition, which are very similar to those of the same transition in ^{125}Cd. In addition, a second isomer is predicted with spin $27/2^-$. This state can decay by two branches; a 45-keV $E2$ transition to the $23/2^-_1$ state ($B(E2) = 0.85$ W.u.), which will be strongly converted, and a 674 keV $M2$ transition to the $23/2^+_1$ level ($B(M2) = 6.58 \times 10^{-4}$ W.u.). Such an isomeric decay would be analogous to that of the $27/2^-$ isomers recently measured in ^{125,127,129}Sn [18]. The calculated $27/2^-_1$ state in ^{127}Cd contains strong components arising from the $\pi g^{-2}_{9/2} \otimes \nu h^{-3}_{11/2}$ and $\pi g^{-2}_{9/2} \otimes \nu d^{-2}_{3/2} h^{-1}_{11/2}$ configurations, with amplitudes of 29.3 and 28.7 % respectively.

4.3. ^{128}Cd

We have previously reported partial calculations of ^{128}Cd in [2]. Since these calculations were reported, additional experimental information on this nucleus has become available [3]. The calculated levels for this nucleus are presented in Fig. 5. The only difference between the predictions previously reported in [2] and those shown in Fig. 5 is the single-particle energy of the $\nu h_{11/2}$ orbit which has been changed from 100 to 69 keV, as experimentally determined in [19]. The calculations are in reasonable agreement with the experimental data, and correctly predict the isomeric nature of the 10^+ state. However, the calculated $B(E2)$ value, 1.0 W.u., is a factor of $\sim$2 too high [$B(E2)_{exp} = 0.39(1)$]. The isomeric nature of the 5^- state, is also correctly predicted, though as a consequence of the inversion of the 4^+ and 5^- states a half-life about an order of magnitude longer is predicted. The problem of the calculated energy of the 2^+ state being too high was already discussed in [2] and an explanation for the surprisingly low energy of the first 2^+ state is given in [20], from a comparison with Hartree-Fock calculations.

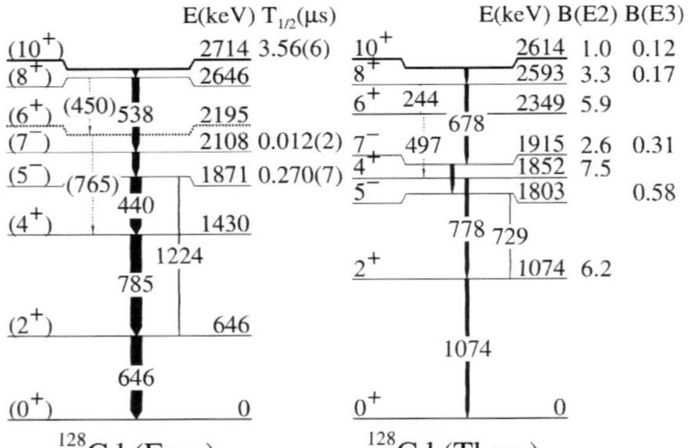

Figure 5. Experimental decay scheme of ^{128}Cd from [3] and that calculated in this work. The $B(E2)$ and $B(E3)$ reduced transition rates are given in Weisskopf units and refer to stretched transitions only.

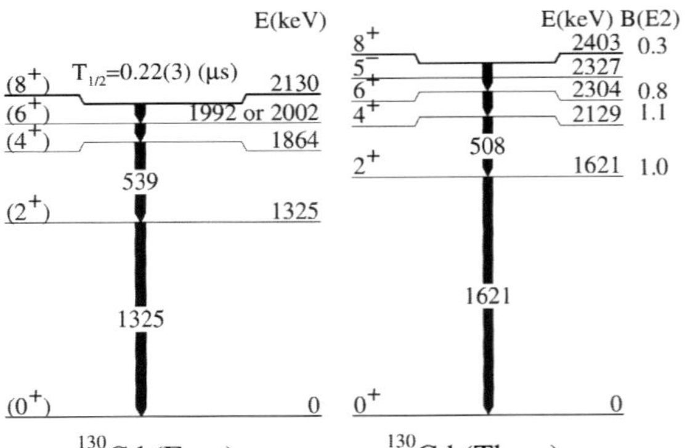

Figure 6. Experimental decay scheme of ^{130}Cd from [4] and that calculated in this work. The $B(E2)$ reduced transition rates are given in Weisskopf units and refer to stretched transitions only.

4.4. ^{130}Cd

The calculations are again able to correctly predict the isomeric nature of the 8^+ state of ^{130}Cd, whose decay has recently been experimentally measured [4]. A comparison between the predictions and the experimental level scheme is shown in Fig. 6 and good agreement is obtained. The calculated $B(E2)$ value of 0.3 W.u. is, however, much lower than the experimentally determined values of 1.7(2) or 1.3(2) W.u., depending on the energy of the isomeric transition. The predicted position of the 5_1^- state is in agreement with the second set of shell-model calculations reported in [4], which do not include excitations across the $Z = 50$ and $N = 82$ closures. A different set of shell-model calculations also appear in [4], which include excitations across the $Z = 50$ shell gap. These report the 5_1^- state lying above the 8^+ isomer.

5. Conclusion

A 19(3)-μs isomeric state has been observed in ^{125}Cd decaying by two γ rays. The energies of these two γ rays are consistent with those observed in recent works [5, 11], however other γ rays previously assigned to this cascade are not present. The origin of the isomerism has been assigned, by a comparison with a realistic shell-model calculation, to the decay of the $19/2_1^+$ state to the $15/2_1^-$ state, via a mixed $M2/E3$ transition.

Level schemes for the nuclei ^{127,128,130}Cd have been calculated, using the same effective interaction, and have been presented. These predictions are also in reasonable agreement with recent experimental data [3, 4]. The good agreement between our calculations and reliable experimental data seems to confirm that modern shell-model interactions are an adequate framework to describe nuclear structure properties of far-from-stability nuclei in the ^{132}Sn region [21].

References

[1] Pinston J A and Genevey J 2004 *J. Phys. G* **30** R57.
[2] Scherillo A *et al.* 2004 *Phys. Rev. C* **70** 054318.
[3] Cáceres L*et al.* 2009 *Phys. Rev. C* **79** 011301.
[4] Jungclaus A *et al.* 2007 *Phys. Rev. Lett.* **99** 132501.
[5] Hoteling N *et al.* 2007 *Phys. Rev. C* **76** 044324.
[6] Kautzsch T *et al.* 2000 *Eur. Phys. J. A* **9** 201.
[7] Dillmann I *et al.* 2003 *Phys. Rev. Lett.* **91** 162503.
[8] Breitenfeldt M *et al.* 2010 *Phys. Rev. C* **81** 034313.
[9] Mach H *et al.* 1986 *Phys. Rev. C* **34** 1117.
[10] Huck H *et al.* 1989 *Phys. Rev. C* **39** 997.
[11] Hellström M *et al. Proceedings of the Third International Conference on Fission and Properties of Neutron-Rich Nuclei* edited by Hamilton J H, Ramayya A V and Carter H K, World Scientific, Singapore, 22 (2003).
[12] Duchêne G *et al.* 1999 *Nucl. Instr. and Meth.* **A432** 90.
[13] Eberth J *et al.* 2001 *Prog. Part. Nucl. Phys.* **46** 389.
[14] Pinston J A *et al.* 2000 *Phys. Rev. C* **61** 024312.
[15] Rae W D M NuShellX for Windows 2008 *http://knollhouse.eu*, unpublished.
[16] Machleidt R 2001 *Phys. Rev. C* **63** 024001.
[17] Genevey J *et al.* 2003 *Phys. Rev. C* **67** 054312.
[18] Lozeva R *et al.* 2008 *Phys. Rev. C* **77** 064313.
[19] Fogelberg B *et al.* 2004 *Phys. Rev. C* **70** 034312.
[20] Rodriguez T R, Egido J L and Jungclaus A 2008 *Phys. Lett.* **B 668** 410.
[21] Coraggio L, Covello A, Gargano A, and Itaco N, 2009 *Phys. Rev. C* **80** 021305(R).

Symmetry and supersymmetry in nuclear pairing: exact solutions

Symmetry and supersymmetry in nuclear pairing: exact solutions

A.B. Balantekin[1], **Y. Pehlivan**[2]

[1] Physics Department, University of Wisconsin, Madison, WI 53706 USA
[2] Mimar Sinan Fine Arts University, Besiktas, Istanbul 34349, Turkey

E-mail: baha@physics.wisc.edu,yamac@physics.wisc.edu

Abstract. Pairing plays a crucial role in nuclear spectra and attempts to describe it has a long history in nuclear physics. The limiting case in which all single particle states are degenerate, but with different s-wave pairing strengths was only recently solved. In this strong coupling limit the nuclear pairing Hamiltonian also exhibits a supersymmetry. Another solution away from those limits, namely two non-degenerate single particle states with different pairing strengths, was also given. In this contribution these developments are summarized and difficulties with possible generalizations to more single particle states and d-wave pairing are discussed.

1. Introduction

Pairing plays a very prominent role in nuclear structure physics. The s-wave pairing problem can be conveniently formulated using the quasi-spin operators:

$$\hat{S}_j^+ = \sum_{m>0} (-1)^{(j-m)} a^\dagger_{j\,m} a^\dagger_{j\,-m}, \quad \hat{S}_j^- = \sum_{m>0} (-1)^{(j-m)} a_{j\,-m} a_{j\,m}, \tag{1}$$

and

$$\hat{S}_j^0 = \frac{1}{2} \sum_{m>0} \left(a^\dagger_{j\,m} a_{j\,m} + a^\dagger_{j\,-m} a_{j\,-m} - 1 \right) = \hat{N}_j - \frac{1}{2}\Omega_j, \tag{2}$$

where $\Omega_j = j + \frac{1}{2}$ is the maximum number of pairs that can occupy the level j and the number operator is

$$\hat{N}_j = \frac{1}{2} \sum_{m>0} \left(a^\dagger_{j\,m} a_{j\,m} + a^\dagger_{j\,-m} a_{j\,-m} \right). \tag{3}$$

Quasi-spin operators generate mutually commuting SU(2) algebras:

$$[\hat{S}_i^+, \hat{S}_j^-] = 2\delta_{ij}\hat{S}_j^0, \qquad [\hat{S}_i^0, \hat{S}_j^\pm] = \pm\delta_{ij}\hat{S}_j^\pm \tag{4}$$

which are realized in the $s_j = \frac{1}{2}\Omega_j$ representations. (The Casimir operator of these SU(2) algebras for each j is $s_j(s_j+1)$). As we describe later, it is possible to define similar operators for d-wave pairing situations.

2. Exact Solutions

Over the years considerable attention was paid to exactly solvable pairing Hamiltonians with one- and two-body interactions. The pairing interaction was first presented by Racah in LS-coupling scheme [1] and was generalized to the jj-coupling scheme [2]. Exactly solvable cases include

- The exact quasi-spin limit [3]:

$$\hat{H} = -|G| \sum_{jj'} \hat{S}_j^+ \hat{S}_{j'}^- . \tag{5}$$

- Richardson's solution for the case when the single particle energies are added to the Hamiltonian in Eq.(5) [4]

$$\hat{H} = \sum_{jm} \epsilon_j a_{j\,m}^\dagger a_{j\,m} - |G| \sum_{jj'} \hat{S}_j^+ \hat{S}_{j'}^- . \tag{6}$$

- A model of Gaudin's [5], which is closely related to the Richardson's solution.
- The limit with separable pairing in which the energy levels are degenerate (the one-body term becomes a constant for a given number of pairs) [6, 7, 8]:

$$\hat{H} = -|G| \sum_{jj'} c_j^* c_{j'} \hat{S}_j^+ \hat{S}_{j'}^- . \tag{7}$$

- Most general separable case with two orbitals [9].

Let us first examine the degenerate case. It is convenient to define the operators

$$\hat{S}^+(0) = \sum_j c_j^* \hat{S}_j^+ \quad \text{and} \quad \hat{S}^-(0) = \sum_j c_j \hat{S}_j^- . \tag{8}$$

Then the Hamiltonian in Eq. (7) can be written as

$$\hat{H} = -|G| \hat{S}^+(0) \hat{S}^-(0). \tag{9}$$

In the 1970's Talmi [10] showed that under certain assumptions, a state of the form

$$\hat{S}^+(0)|0\rangle = \sum_j c_j^* \hat{S}_j^+ |0\rangle \tag{10}$$

where $|0\rangle$ is the particle vacuum, is an eigenstate of a class of Hamiltonians including the one above. Indeed

$$\hat{H} \hat{S}^+(0)|0\rangle = \left(-|G| \sum_j \Omega_j |c_j|^2 \right) \hat{S}^+(0)|0\rangle . \tag{11}$$

How can we identify other one-pair states, which are eigenstates of the Hamiltonian in Eq. (7)? For example for two levels j_1 and j_2, the state orthogonal to the one in Eq. (10)

$$\left(\frac{c_{j2}}{\Omega_{j_1}} \hat{S}_{j1}^+ - \frac{c_{j1}}{\Omega_{j_2}} \hat{S}_{j2}^+ \right) |0\rangle , \tag{12}$$

is also an eigenstate with E=0. It turns out that there is a systematic way to find such states. First one introduces the operators

$$\hat{S}^+(x) = \sum_j \frac{c_j^*}{1 - |c_j|^2 x} \hat{S}_j^+ , \quad \hat{S}^-(x) = \sum_j \frac{c_j}{1 - |c_j|^2 x} \hat{S}_j^- . \tag{13}$$

(Note that the operators in Eq. (8) are the same as the operators in Eq. (13) calculated at $x = 0$). Then one can show that [6, 7]

$$\hat{S}^+(0)\hat{S}^+(z_1^{(N)})\ldots\hat{S}^+(z_{N-1}^{(N)})|0\rangle \tag{14}$$

is an eigenstate of the Hamiltonian in Eq. (7) with energy

$$E_N = -|G|\left(\sum_j \Omega_j |c_j|^2 - \sum_{k=1}^{N-1} \frac{2}{z_k^{(N)}}\right) \tag{15}$$

if the following Bethe ansatz equations are satisfied:

$$\sum_j \frac{-\Omega_j/2}{1/|c_j|^2 - z_m^{(N)}} = \frac{1}{z_m^{(N)}} + \sum_{k=1(k\neq m)}^{N-1} \frac{1}{z_m^{(N)} - z_k^{(N)}} \qquad m = 1, 2, \ldots N-1. \tag{16}$$

Similarly

$$\hat{S}^+(x_1^{(N)})\hat{S}^+(x_2^{(N)})\ldots\hat{S}^+(x_N^{(N)})|0\rangle \tag{17}$$

is an eigenstate with zero energy if the following Bethe ansatz equations are satisfied:

$$\sum_j \frac{-\Omega_j/2}{1/|c_j|^2 - x_m^{(N)}} = \sum_{k=1(k\neq m)}^{N} \frac{1}{x_m^{(N)} - x_k^{(N)}} \qquad \text{for every} \quad m = 1, 2, \ldots, N. \tag{18}$$

It should be emphasized that the states in Eqs. (14) and (17) are eigenstates of the Hamiltonian in Eq. (7) if available single-particle levels are at most half full. One can show that, if the single-particle levels are more than half full, the state

$$\hat{S}^-(z_1^{(N)})\hat{S}^-(z_2^{(N)})\ldots\hat{S}^-(z_{N-1}^{(N)})|\bar{0}\rangle \tag{19}$$

is an eigenstate with the same energy as in Eq. (15) if the Bethe ansatz equations given in Eq.(16) are satisfied [7]. In Eq. (19) $|\bar{0}\rangle$ designates the state where all single-particle levels are completely filled.

One should emphasize that although the solutions presented here are exact solutions, to obtain explicit expressions for the energies and eigenstates, one still needs to obtain solutions of the Bethe ansatz equations. So far methods to solve the Bethe ansatz equations were developed only in a limited number of cases [7, 11].

Note that it is also possible to calculate the quantum invariants of these pairing Hamiltonians [12].

3. Supersymmetry

The above discussion indicates that the states of the Hamiltonian in Eq. (7) with N pairs (Eq. 14) and with $N_{max} + 1 - N$ pairs (Eq. 17) have the same eigenenergy. Furthermore the zero energy states are missing if the single-particle states are more than half full. This situation is reminiscent of the spectral conditions of the supersymmetric quantum mechanics [13]. Indeed the operator [8]

$$\hat{T} = \exp\left(-i\frac{\pi}{2}\sum_i(\hat{S}_i^+ + \hat{S}_i^-)\right), \tag{20}$$

transforms the empty single-particle state, $|0\rangle$, to the fully occupied state, $|\bar{0}\rangle$:

$$\hat{T}|0\rangle = |\bar{0}\rangle. \tag{21}$$

It is easy to prove the equation above by noting

$$\hat{T}^\dagger \hat{S}^\pm \hat{T} = \hat{S}^\mp. \tag{22}$$

One can define new operators

$$\hat{B}^- = \hat{T}^\dagger \hat{S}^-(0), \qquad \hat{B}^+ = \hat{S}^+(0)\hat{T}. \tag{23}$$

Supersymmetric quantum mechanics tells us that the partner Hamiltonians $\hat{H}_1 = \hat{B}^+\hat{B}^-$ and $\hat{H}_2 = \hat{B}^-\hat{B}^+$ have identical spectra except for the extremal (usually ground) state of $\hat{H}_1$. Here two Hamiltonians $\hat{H}_1$ and $\hat{H}_2$ are actually identical and equal to the pairing Hamiltonian, Eq. (7). Hence the role of the supersymmetry is to connect the states in Eqs. (14) and (19), i.e. this supersymmetry connects particle and hole states.

4. Exact solution for two single-particle states

It turns out that one can find an exact solution for the case where there are only two single-particle levels [9]:

$$\frac{\hat{H}}{|G|} = \sum_j 2\varepsilon_j \hat{S}_j^0 - \sum_{jj'} c_j^* c_{j'} \hat{S}_j^+ \hat{S}_{j'}^- + \sum_j \varepsilon_j \Omega_j, \tag{24}$$

where ε_j and c_j's are dimensionless and the sums are performed over only two single-particle states.

The eigenstates of the Hamiltonian in Eq. (24) can be written using the step operators:

$$\mathcal{J}^+(x) = \sum_j \frac{c_j^*}{2\varepsilon_j - |c_j|^2 x} S_j^+ \tag{25}$$

as

$$\mathcal{J}^+(x_1)\mathcal{J}^+(x_2)\ldots\mathcal{J}^+(x_N)|0\rangle. \tag{26}$$

Defining the auxiliary quantities

$$\beta = 2\frac{\varepsilon_{j_1} - \varepsilon_{j_2}}{|c_{j_1}|^2 - |c_{j_2}|^2} \qquad \delta = 2\frac{\varepsilon_{j_2}|c_{j_1}|^2 - \varepsilon_{j_1}|c_{j_2}|^2}{|c_{j_1}|^2 - |c_{j_2}|^2}, \tag{27}$$

one obtains the energy eigenvalues as

$$E_N = -\sum_{n=1}^{N} \frac{\delta x_n}{\beta - x_n}. \tag{28}$$

In the above equations, the parameters x_k are to be found by solving the Bethe ansatz equations

$$\sum_j \frac{\Omega_j |c_j|^2}{2\varepsilon_j - |c_j|^2 x_k} = \frac{\beta}{\beta - x_k} + \sum_{n=1(\neq k)}^{N} \frac{2}{x_n - x_k}. \tag{29}$$

5. Prospects for other exact solutions

We have shown that there is an exact solution of the s-wave pairing problem with two non-degenerate orbitals in terms of the solutions of the Bethe ansatz equations. (This problem is of course numerically diagonalizable in an SU(2) × SU(2) basis). It seems to be very difficult to generalize the Bethe ansatz method to the case of three non-degenerate orbitals. (Of course the three single-particle level problem is diagonalizable in an SU(2) × SU(2) × SU(2) basis).

However, the Bethe ansatz method seems to be generalizable to at least some d-wave pairing situations. In his work, mentioned earlier Talmi showed that, for a Hamiltonian with only one- and two-body interactions, if the doubly-magic ground state energy is normalized to zero,

$$H|0\rangle = 0, \tag{30}$$

then the conditions

$$[H, S^\dagger(0)]|0\rangle = VS^\dagger(0)|0\rangle \tag{31}$$

and

$$[[H, S^\dagger(0)], S^\dagger(0)] = W\left(S^\dagger(0)\right)^2 \tag{32}$$

are sufficient to determine the entire energy spectra: the eigenstates are given by $(S^\dagger(0))^n|0\rangle$ with eigenvalues $nV + n(n-1)W/2$. In addition, if one considers d-wave pairing operator

$$D_M^\dagger = \sum_{jj'} \alpha_{jj'} \left(a_j^\dagger \times a_{j'}^\dagger\right)_M^{(2)}, \tag{33}$$

then the additional condition

$$[[H, S^\dagger], D_M^\dagger] = WS^\dagger D_M^\dagger \tag{34}$$

makes eigenstates with d-wave pairs possible

$$HD_M^\dagger|0\rangle \propto D_M^\dagger|0\rangle. \tag{35}$$

Note that these conditions suggest existence of an algebraic structure. In fact, Ginocchio model [14] satisfies the double commutators given above.

One should explore if the Bethe ansatz method can be generalized in other cases where we also have an algebraic framework. A possible answer lies in Gaudin's method mentioned earlier. It can be shown that the infinite-dimensional algebra

$$[J^+(\lambda), J^-(\mu)] = 2\frac{J^0(\lambda) - J^0(\mu)}{\lambda - \mu}, \tag{36}$$

$$[J^0(\lambda), J^\pm(\mu)] = \pm\frac{J^\pm(\lambda) - J^\pm(\mu)}{\lambda - \mu}, \tag{37}$$

$$[J^0(\lambda), J^0(\mu)] = [J^\pm(\lambda), J^\pm(\mu)] = 0 \tag{38}$$

can be used to find eigenvalues and the eigenvectors of the Hamiltonian

$$H(\lambda) = J^0(\lambda)J^0(\lambda) + \frac{1}{2}J^+(\lambda)J^-(\lambda) + \frac{1}{2}J^-(\lambda)J^+(\lambda). \tag{39}$$

For details the reader is referred to, e.g. Refs. [15], [16], and [17]. (It should be emphasized that the Hamiltonian in Eq. (39) is *not* the Casimir operator of the infinite-dimensional algebra given above). The connection between the Hamiltonian in Eq. (39) and the Hamiltonian originally solved by Richardson (Eq. 6) is uncovered by using the realization

$$J^0(\lambda) = \sum_{i=1}^{N} \frac{\hat{S}_i^0}{\epsilon_i - \lambda} \quad \text{and} \quad J^\pm(\lambda) = \sum_{i=1}^{N} \frac{\hat{S}_i^\pm}{\epsilon_i - \lambda}, \tag{40}$$

where $S_i^\pm$ and S_i^0 are the quasi-spin operators. A possible procedure to include d-wave pairing would first identify the underlying algebraic structure (i.e. the SO(8) symmetry of the Ginocchio

Model), then introduce the analogs of the Eqs. (36) through (39) using a realization analogous to that in Eq. (40). In fact a boson pairing model, using the SU(1,1) analog of the quasi-spin in this fashion, was already studied in some detail [16, 18]. However, such a program with arbitrary fermion pairings remains to be largely unexplored.

One should finally remark that some of the techniques discussed here are more generally applicable to other Hamiltonians with special one- and two-body interactions. One such example is the Hamiltonian describing a dense electron gas near the center of a core-collapse supernova [19].

Acknowledgments
This work was supported in part by the U.S. National Science Foundation Grant No. PHY-0855082 and in part by the University of Wisconsin Research Committee with funds granted by the Wisconsin Alumni Research Foundation.

References

[1] Racah G 1943 *Phys. Rev.* **63** 367.
[2] Racah G and Talmi I 1953 *Physica* **18** 1097; *Phys. Rev.* **89**, 913; Flowers BH 1952 *Proc. Royal Soc. London Ser. A-Math. and Phys. Sci.* **212**, 248.
[3] Kerman AK 1961 *Ann. Phys.* (NY) **12**, 300.
[4] Richardson RW 1967 *Phys. Rev.* **159**, 792.
[5] Gaudin M 1976 *J. Physique* **37**, 1087; *La Fonction d'onde de Bethe*, Collection du Commissariat a l'énergie atomique, Masson, Paris, 1983.
[6] Pan F, Draayer JP and Ormand WE 1998 *Phys. Lett.* B **422**, 1 [arXiv:nucl-th/9709036].
[7] Balantekin A B, de Jesus J H and Pehlivan Y 2007 *Phys. Rev.* C **75** 064304 [arXiv:nucl-th/0702059].
[8] Balantekin A B and Pehlivan Y 2007 *J. Phys.* G **34** 1783 [arXiv:0705.1318 [nucl-th]].
[9] Balantekin A B and Pehlivan Y 2007 *Phys. Rev.* C **76** 051001 [arXiv:0710.3941 [nucl-th]].
[10] Talmi I 1971 *Nucl. Phys.* **A172**, 1.
[11] Balantekin A B , Guven N and Pehlivan Y 2008 *AIP Conf. Proc.* **1072** 82 [arXiv:0808.0405 [nucl-th]].
[12] Pehlivan Y [arXiv:0806.1810].
[13] Witten E 1981 *Nucl. Phys.* B **188**, 513.
[14] Ginocchio J N 1978 *Phys. Lett.* B **79**, 173; 1980 *Ann. Phys. (NY)*, **126** 234.
[15] Balantekin A B , Dereli T and Pehlivan Y 2005 *J. Phys. A* **38**, 5697 [arXiv:math-ph/0505071].
[16] Balantekin A B, Dereli T and Pehlivan Y 2005 *Int. J. Mod. Phys.* E **14**. 47 [arXiv:nucl-th/0505023].
[17] Balantekin A B 2008 *Proc. Int. School of Physics "Enrico Fermi": Nuclear Structure far from Stability: New Physics and New Technology, Volume 169* ed Covello A , Iachello F, Ricci R A and Maino G (Amsterdam: IOP Press) p 423 [arXiv:0711.0710 [nucl-th]].
[18] Balantekin A B, Dereli T and Pehlivan Y 2004 *J. Phys. G* **30**, 1225 [arXiv:nucl-th/0407006].
[19] Balantekin A B and Pehlivan Y 2007 *J. Phys. G* **34**, 47 [arXiv:astro-ph/0607527].

Shell model Monte Carlo approach: the heavy nuclei

2011 J. Phys.: Conf. Ser. 267 012033

(http://iopscience.iop.org/1742-6596/267/1/012033)

Shell model Monte Carlo approach: the heavy nuclei

Y. Alhassid[1,*], **L. Fang**[1], **and H. Nakada**[2]

[1]Center for Theoretical Physics, Sloane Physics Laboratory
Yale University, New Haven, CT 06520, USA
[2]Department of Physics, Graduate School of Science
Chiba University, Inage, Chiba 263-8522, Japan

*E-mail: **yoram.alhassid@yale.edu**

Abstract. The auxiliary-field Monte Carlo (AFMC) method, also known as the shell model Monte Carlo (SMMC), enables us to calculate microscopically statistical and collective properties of nuclei in the presence of strong correlations. These calculations can be carried out in shell model spaces that are many orders of magnitude larger than spaces that can be treated with conventional diagonalization methods. A recent major development has been the extension of the AFMC approach to heavy nuclei. Such applications to heavy nuclei have been a major challenge. On the conceptual level, a crucial question is whether a truncated spherical shell model Hamiltonian can describe the proper collectivity observed in heavy nuclei and, in particular, the rotational character of strongly deformed nuclei. On the technical level, the low excitation energies make it necessary to perform calculations down to much lower temperatures. At such low temperatures the propagator becomes ill-conditioned and requires the introduction of stabilization methods.

1. Introduction

The microscopic derivation of nuclear properties from the underlying nuclear forces is a major challenge in nuclear theory. These nuclear properties are required in the modeling of astrophysical processes such as nucleosynthesis and supernovas. Microscopic theories are also necessary for the interpretation of data from new radioactive beam facilities.

Much progress has been made in applications of ab initio methods such as the Green function Monte Carlo (GFMC) method [1] and the no core shell model (NCSM) [2]. However, both GFMC and NCSM are limited to light nuclei. Density functional theory (DFT) has been successful in the global treatment of nuclei [3, 4], but the inclusion of correlations beyond the mean field and the description of excitations require extensions of DFT [5, 6].

A promising approach is the configuration-interaction (CI) shell model method, in which the correlated state is expressed as a coherent superposition of a large number of uncorrelated configurations. The CI approach is widely used in atomic, molecular and nuclear physics, as well as in finite-size condensed matter systems (e.g., quantum dots). It accounts for both shell effects and correlations, and has been very successful in describing the low-energy spectroscopy of nuclei up to $A \sim 52$ using a single major valence shell [7]. However, applications of the CI method in medium-mass and heavy nuclei are limited as the dimensionality of the model space becomes prohibitively large for conventional diagonalization methods.

This difficulty was overcome in part through the use of the auxiliary-field Monte Carlo (AFMC) method, also known in nuclear physics as the shell model Monte Carlo (SMMC) approach [8, 9, 10]. Fermionic Monte Carlo methods are often limited by the so-called sign problem. However, the dominating collective components of effective nuclear interactions [11] have a good sign. Consequently, certain nuclear properties can be calculated with good-sign interactions [12, 13]. Furthermore, interactions containing small terms that have a bad sign can be treated by an extrapolation method [9]. AFMC enables us to carry out calculations in model spaces that are many orders of magnitude larger than those that can be solved by conventional methods.

The AFMC method has proven to be a very powerful tool in calculating the statistical properties of nuclei [14, 15, 16, 17]. However, most of the AFMC calculations carried out to date were limited to medium-mass nuclei whose deformation is not large. In such even-even nuclei the gap to the first excited state is of the order of 1-2 MeV and the ground state can be reached with modest values of the inverse temperature β. The situation is very different in heavy nuclei (e.g., the rare-earth region) where the deformation can be large and the first excitation energy is of the order of 100 keV. This leads to major technical difficulties in extending AFMC to heavy deformed nuclei. Furthermore, well-deformed nuclei are characterized by rotational bands. A central issue is whether it is possible to describe this rotational behavior in the framework of a truncated spherical shell model.

We have recently extended the AFMC approach to such heavy deformed nuclei [18]. In Sec. 2 we discuss the AFMC and its extension to heavy nuclei and in Sec. 3 we demonstrate its applications to a typical heavy deformed rare-earth nucleus, ^{162}Dy.

2. Auxiliary-field Monte Carlo (AFMC) approach

We briefly outline the AFMC method and then discuss technical aspects that arise in heavy nuclei.

2.1. *AFMC method*

The AFMC method is based on a Hubbard-Stratonovich (HS) transformation [19], a representation of the imaginary-time many-particle propagator as a functional integral over propagators of non-interacting particles moving in external time-dependent auxiliary fields.

In the HS representation, the Gibbs operator $e^{-\beta\hat{H}}$ of a nucleus describes by the Hamiltonian $\hat{H}$ at inverse temperature $\beta = 1/T$ is decomposed as

$$e^{-\beta H} = \int \mathcal{D}[\sigma] G_\sigma U_\sigma \, , \tag{1}$$

where G_σ is a Gaussian weight and U_σ is the propagator of non-interacting nucleons moving in time-dependent auxiliary fields $\sigma(\tau)$.

The integrand in Eq. (1) describes a one-body propagator and is relatively easy to calculate using matrix algebra in the single-particle space. However, the number of auxiliary fields to be integrated over is very large. In AFMC, this integration is carried out by Monte Carlo methods that select the most important configurations.

2.2. *The heavy nuclei*

The extension of AFMC to heavy nuclei has required to overcome several major challenges that are summarized below.

- *AFMC in proton-neutron formalism.* In heavy nuclei, protons and neutrons occupy different shells. This has required the major task of rewriting the AFMC code in proton-neutron formalism. A different formulation based on isospin T_z projection was used in Ref. [20]. Our formulation, however, is much more efficient since the matrices representing the auxiliary-field have the dimension of either the proton or neutron single-particle space rather than the combined dimension.

- *Stabilization at low temperatures.* The first excitation in even-even rare-earth nuclei is about an order of magnitude smaller than in medium-mass nuclei. To calculate ground-state observables it is then necessary to propagate to much longer imaginary time $\beta \sim 20$. At large β, the propagator becomes ill-conditioned, i.e., the ratio between its largest and smallest eigenvalues becomes very large.
In strongly correlated electron systems a similar problem was resolved by stabilization methods in the grand-canonical ensemble [21], keeping the small and large scales separated in the propagation. The method is based on a singular value decomposition (SVD) or a modified Gram-Schmidt (MGS) decomposition of the propagation matrix. We have used the MGS decomposition of a matrix $M = LDV$ or $M = UDR$ where L (R) is a lower (upper) triangular matrix with diagonal elements 1 and D is a diagonal matrix. The MGS decomposition can be up to ~ 20 times faster than SVD [22].
We extended the stabilization methods to the canonical ensemble by stabilizing separately each term in the Fourier series of particle-projected observables.
- *Choice of model space.* It is necessary to use a sufficiently large model space to describe the rotational character of deformed nuclei. To determine the required spherical orbitals we consider a deformed Woods-Saxon plus spin-orbit mean-field potential and determine the average occupation of the spherical orbitals. We include in our model space all spherical orbitals whose occupation probabilities are between 0.1 and 0.9. The effect of other orbitals is taken into account by the renormalization of the effective interaction.
We find that the proton $1f_{7/2}$ orbital with an occupation probability of 0.15 is the only orbital above the $Z = 82$ gap that has an occupation probability larger than 0.1. For example, the occupation probability of the proton $0h_{9/2}$ orbital is only 0.03. We therefore choose the proton model space to include the $50-82$ shell plus the $1f_{7/2}$ orbital. Similarly we determine the neutron model space to be the $82-126$ shell plus the $0h_{11/2}$ and $1g_{9/2}$ orbitals.
- *Discretization effects.* In practice, the HS transformation is calculated by dividing β into a finite number of small time slices $\Delta\beta$. This results in a systematic discretization error in observables, such as the thermal energy. We calculate observables at several values of $\Delta\beta$ and then extrapolate linearly to $\Delta\beta = 0$. Above a certain value of β we find weak dependence on $\Delta\beta$ and take instead an average.
- *Particle-number projection.* The stabilized calculations, required at larger values of β, are time-consuming. We have implemented an approximate particle-number projection that uses a smaller number of quadrature points but is sufficiently accurate for the AFMC calculations. Time-reversal symmetry of the propagators for good-sign interactions can also be used to halve the computational time.

The single-particle Hamiltonian is taken from a Woods-Saxon plus spin-orbit potential. For the interaction we use the dominant components of realistic effective nuclear interactions: multipole-multipole interactions (quadrupole, octupole and hexadecupole) and a monopole pairing interaction [18]. These interaction components are attractive and have a good Monte Carlo sign.

3. Results

In this section we discuss AFMC results for a typically heavy deformed nucleus ^{162}Dy.

3.1. Thermal signatures of rotational collectivity

While AFMC enables us to overcome the model's size problem, it has its own limitations. AFMC is a method for calculating thermal and ground-state observables but it is not suitable for extracting a detailed spectrum. Since the particular collective character of a nucleus is often identified by characteristic patterns of its spectrum, one has to rethink how various collectivity types manifest in statistical observables.

In AFMC we measure observables at a given temperature (canonical ensemble), while experiments in nuclei provide information on nuclear properties at fixed energy (microcanonical ensemble). However, it is relatively easy to transform the microcanonical results to the canonical ensemble. This requires the experimental knowledge of a complete set of levels and their spins, and is therefore limited to low excitation energies.

A thermal observable that can be used to distinguish between different types of collectivity is $\langle \mathbf{J}^2 \rangle$, where $\mathbf{J}$ is the total angular momentum of the nucleus. Assuming a nucleus characterized by a ground-state rotational band with moment of inertia I_g, we have at sufficiently low temperatures T

$$\langle \mathbf{J}^2 \rangle \approx 2 I_g T \ . \tag{2}$$

This result can also be derived by applying the equipartition theorem for the nuclear rotation.

In Fig. 1 we show $\langle J^2 \rangle$ versus T for the well-deformed rare-earth nucleus ^{162}Dy. The solid circles with error bars are the stabilized AFMC results at low temperature while the solid straight line is a fit to Eq. (2). We find a moment of inertia of $I_g = 35.8 \pm 1.5$ MeV^{-1}, in agreement with the experimental value of $I_g = 37.2$ MeV^{-1} (determined by the excitation energy, 80.7 keV, of the first excited 2^+ state). The dashed line is calculated using the experimentally known lowest five experimental rotational bands. The dashed-dotted line is a fit assuming a vibrational nucleus. The good agreement of the AFMC results with Eq. (2) confirms the rotational character of ^{162}Dy within the truncated spherical model space discussed in Sec. 2.2.

The excitation energy of a rotational nucleus at low temperatures is given by

$$E(T) = E_0 + T \ , \tag{3}$$

where E_0 is a ground-state energy. Fig. 2 shows the AFMC thermal energy of ^{162}Dy as a function of T. The solid line is a fit to Eq. (3) with $E_0 = -375.387 \pm 0.019$ MeV. The good agreement with the AFMC results is consistent with the rotational character of this nucleus.

3.2. Level density

The level density is related to the partition function by an inverse Laplace transform. The average level density is obtained in the saddle-point approximation [23]

$$\rho(E) \approx \frac{1}{\sqrt{2\pi T^2 C}} e^{S(E)} \ , \tag{4}$$

where $S(E)$ is the canonical entropy and C is the canonical heat capacity.

In AFMC we calculate the thermal energy $E(\beta)$ versus inverse temperature β and integrate the thermodynamic relation $-\partial \ln Z / \partial \beta = E(\beta)$ to find the canonical partition function $Z(\beta)$. The entropy and heat capacity are then calculated using $S(E) = \ln Z + \beta E$ and $C = -\beta^2 \partial E / \partial \beta$.

Fig. 3 shows the AFMC thermal energy as a function of the inverse temperature β up to $\beta = 20$ MeV^{-1}. The dot-dashed line is the thermal Hartree-Fock-Bogoliubov (HFB) energy. The AFMC results give a correlation energy of $E_{\rm HFB} - E_0$ of about 3.12 MeV ($E_{\rm HFB}$ is the ground-state energy in the HFB approximation).

The state density of ^{162}Dy is shown in Fig. 4. The solid blue circles are the AFMC results using Eq. (4) while the dot-dashed line is the HFB state density. The latter describes the intrinsic level density, while the AFMC state density includes the contribution from rotational bands. Thus, the ratio between the AFMC and HFB densities provides a microscopic estimate of the rotational enhancement factor $K_{\rm rot}$.

We used several experimental data sets to determine the experimental level density. At low energies, i.e., below ~ 2 MeV, there is a complete set of experimental levels [24, 25] (solid

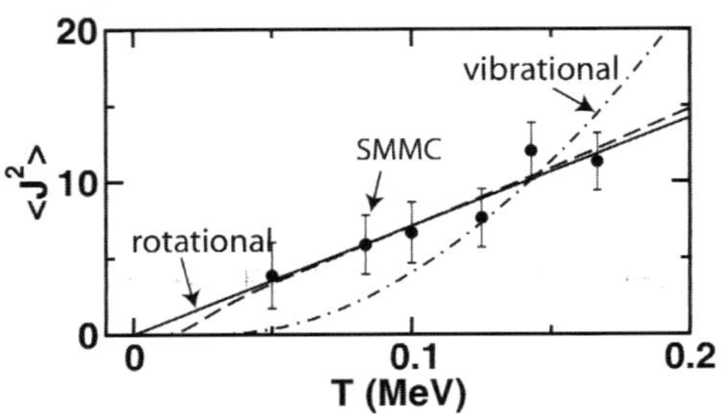

Figure 1. $\langle J^2 \rangle$ versus T for ^{162}Dy. The solid circles with error bars are the AFMC results. The solid line is a fit to Eq. (2), while dot-dashed line is a fit to a vibrational model. The dashed line is obtained from the lowest experimentally known five rotational bands. From Ref. [18].

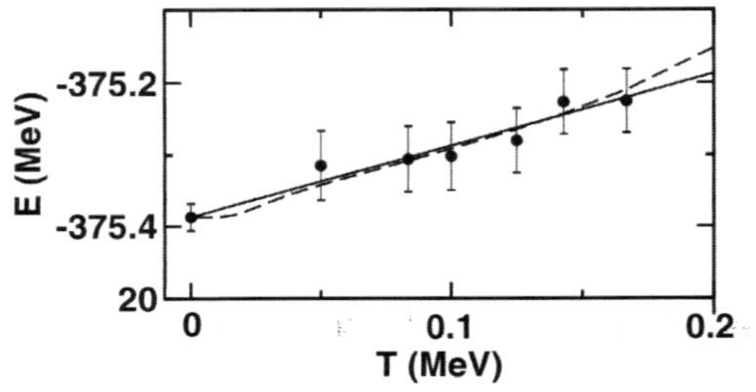

Figure 2. Thermal energy $E(T)$ versus T for ^{162}Dy. The solid circles are the AFMC results and the solid line is a fit to Eq. (3). The dashed line is calculated from the lowest five rotational bands of ^{162}Dy. From Ref. [18].

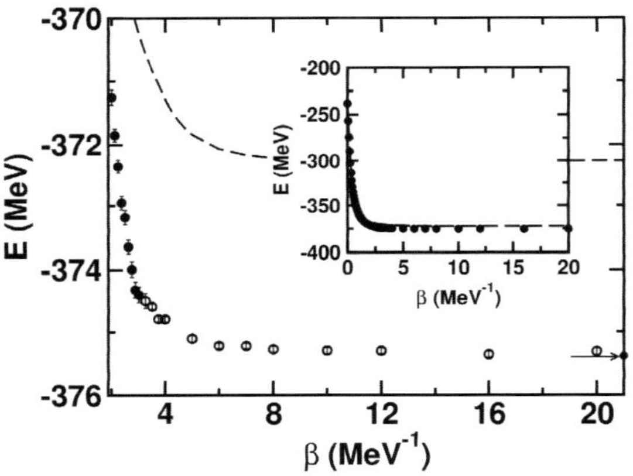

Figure 3. Thermal energy $E(\beta)$ versus inverse temperature β for ^{162}Dy and $\beta > 2$ MeV^{-1}. The dashed line is the HFB energy. The inset shows the complete set of AFMC energies versus β.

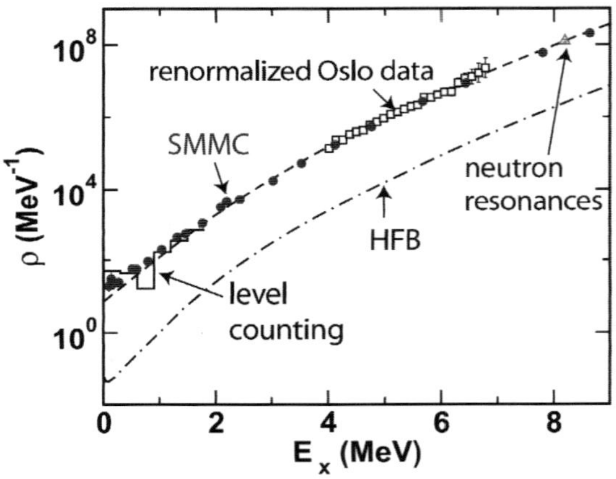

Figure 4. State density of ^{162}Dy. The solid blue circles are the AFMC results and the dot-dashed line is the HFB state density. Also shown are three data sets (level counting, Oslo data and neutron resonance data) and the experimentally determined composite formula (dashed line). After Ref. [18].

histograms in Fig. 4). At intermediate energies below $\sim$ 7 MeV there is the (renormalized) data set obtained by the Oslo method [26, 27] (open squares), and at $E_x = 8.196$ MeV we have the neutron resonance data [28] (red triangle). The dashed line in Fig. 4 is a composite level density formula [29] fitted to all three experimental data sets. This composite formula describes a constant temperature level density formula below a matching energy of $E_M = 1.752$ MeV and a backshifted Fermi gas formula above E_M. We observe that the AFMC level density is in very good agreement with the experimentally determined composite formula.

4. Conclusion

The AFMC method enables us to calculate the statistical and collective properties of nuclei in very large model spaces. We discussed an extension of the method to heavy deformed nuclei. In particular, we find that the rotational character of such nuclei can be reproduced in a truncated spherical shell model approach.

Acknowledgments

This work is supported in part by the U.S. DOE grant No. DE-FG-0291-ER-40608 and as Grant-in-Aid for Scientific Research (C), No. 22540266 by the JSPS. Computational cycles were provided by the NERSC high performance computing facility at LBL.

References

[1] Carlson J and Schiavilla R 1998 *Rev. Mod. Phys.* **70** 743; Pieper S C and Wiringa R B 2001 *Ann. Rev. Nucl. Part. Sci.* **51** 53. For a recent review, see Wiringa R B 2009 *AIP Conf. Proc.* **1128**, p. 1
[2] For a recent review, see Navrtil P, Quanglioni S, Stetcu I and Barrett B R 2009 *J. Phys.* G **36** 083101
[3] Bender M, Heenen P -H and Reinhard P -G 2003 *Rev. Mod. Phys.* **75** 121
[4] Bertsch G F, Bertulani C A, Nazarewicz W, Schunck N and Stoitsov M V 2009 *Phys. Rev.* C **79** 034306; Stoitsov M, Nazarewicz W and Schunck N 2009 *J. Mod. Phys.* E **18** 816
[5] Bender M, Bertsch G F and Heenen P -H 2005 *Phys. Rev. Lett.* **94** 102503; Bender M , Bertsch G F and Heenen P H 2006 *Phys. Rev.* C **73** 034322
[6] Bertsch G F, Girod M, Hilaire S, Delaroche J -P, Goutte H and Peru S 2007 *Phys. Rev. Lett.* **99** 032502; Terasaki J, Engel J and Bertsch G F 2008 *Phys. Rev.* C 78 044311
[7] For a recent review, see Caurier E, Martinez-Pinedo G, Nowacki F, Poves A and Zuker A P 2005 *Rev. Mod. Phys.* **77** 427
[8] Lang G, Johnson C W, Koonin S E and Ormand W E 1993 *Phys. Rev.* C **48** 1518
[9] Alhassid Y, Dean D J, Koonin S E, Lang G and Ormand W E 1994 *Phys. Rev. Lett.* **72** 613
[10] Alhassid Y 2001 *Int. J. Mod. Phys.* B **15** 1447
[11] Dufour M and Zuker A P 1996 *Phys. Rev.* C **54** 1641
[12] Alhassid Y, Bertsch G F, Dean D J and Koonin S E 1996 *Phys. Rev. Lett.* **77** 1444
[13] Nakada H and Alhassid Y 1997 *Phys. Rev. Lett.* **79** 2939; 1998 *Phys. Lett.* B **436** 231
[14] Alhassid Y, Liu S and Nakada H 1999 *Phys. Rev. Lett.* **83** 4265
[15] Alhassid Y, Bertsch G F and Fang L 2003 *Phys. Rev.* C **68** 044322
[16] Y. Alhassid Y, Bertsch G F, Fang L and Liu S 2005 *Phys. Rev.* C **72** 064326
[17] Alhassid Y, Liu S and Nakada H 2007 *Phys. Rev. Lett.* **99** 162504
[18] Alhassid Y, Fang L and Nakada H 2008 *Phys. Rev. Lett.* **101** 082501
[19] Hubbard J 1959 *Phys. Rev. Lett.* **3** 77; R. L. Stratonovich R L 1957 *Dokl. Akad. Nauk. S.S.S.R.* **115** 1097
[20] Özen C and Dean D J 2006 *Phys. Rev.* C **73** 014302
[21] Loh Jr E Y and Gubernatis J E 1992 in *Electronic Phase Transitions*, ed. Hanke W and Kopaev Yu V, p. 177
[22] Golub G H and Van Loan C F 1996 *Matrix Computations* (John Hopkins University Press)
[23] Bohr A and Mottelson B 1969 *Nuclear Structure* Vol. 1 New York: Benjamin)
[24] Firestone B R and Shirley V S 1996 *Table of Isotopes* (New York: Wiley); Helmer R G and Reich C W 1999 *Nucl. Data Sheets* **87** 317
[25] Aprahamian A *et al.* 2006 *Nucl. Phys.* A **764** 42
[26] Schiller A *et al.* 2000 *Nucl. Instrum. Methods Phys. Res.*, Sect. A **447** 498
[27] Guttormsen M *et al.* 2003 *Phys. Rev.* C **68** 064306; Guttormsen M, private communication.
[28] *Handbook for Calculations of Nuclear Reaction Data* 1998 (IAEA, Vienna).
[29] Gilbert A and Cameron AGW 1965 *Can. J. Phys.* **43** 1446

Improved method of analysis for Recoil Distance Measurements of nuclear lifetimes

Improved method of analysis for Recoil Distance Measurements of nuclear lifetimes

P.G. Bizzeti

Dipartimento di Fisica e Astronomia, Università di Firenze and I.N.F.N., Sezione di Firenze, Via G. Sansone 1, 50059 Sesto Fiorentino (Firenze), Italy

E-mail: `bizzeti@fi.infn.it`

Abstract. The Recoil Distance Method is known as a very powerful tool for measuring the lifetimes of excited nuclear levels in the picosecond region. Usually, for each level, coincidences with the shifted part of a feeding transition are used to avoid the bias from the accumulated delay in the higher lying cascade, by means of the Differential Decay Curve Method. Here, an integral version of the Decay Curve Method is proposed and discussed. This technique has several noticeable advantages, particularly when the overall statistics of the results is limited owing to the weak beam current, as expected with radioactive ion beams.

1. Introduction

Present and future facilities for radioactive ion beams have been reported in a number of talks in this conference. They will provide excellent opportunities for nuclear spectroscopy, but to exploit them at best we must be ready to refine our current methods of measurement and analysis. Here, I will consider the measurements of nuclear lifetimes in the picosecond region, with the Recoil Distance Method (RDM).

In this case, the excited nuclei, usually produced by fusion-evaporation, have enough momentum to exit from the target and fly in the vacuum, until they reach a stopper at a variable distance x from the target. Decays in flight can be distinguished from those at rest in the stopper, due to their Doppler shift $\Delta E = E_0 \beta \cos\theta$, where $\beta = v/c$ and θ is the angle between the direction of the emitted γ and the velocity of the decaying nucleus.

If there was only one significant lifetime, the fraction of decays at rest would show the typical exponential decay with respect to the distance x. This is not the case when the accumulated delay in the feeding cascade is comparable with the lifetime to be measured, as the unshifted intensity is given by the convolution of the exponential decay with the feeding-time distribution $f(t)$. In this case, the shifted and unshifted components of the relevant γ ray are measured in coincidence with the Doppler shifted part of a feeding transition, and the results are analyzed with the Differential Decay Curve Method (DDCM) [1, 2].

If all nuclei move with the same velocity in the direction of the beam, a simple relation holds between the unshifted intensity, the time derivative of the shifted intensity and the lifetime.

In fact,

$$I_u^{(s)}(t) = \int_0^t f(t') \exp\left(\frac{t-t'}{\tau}\right) \mathrm{d}t' \quad (1)$$

$$I_s^{(s)}(t) = \int_0^t f(t') \left[1 - \exp\left(\frac{t-t'}{\tau}\right)\right] \mathrm{d}t' \quad (2)$$

By deriving Eq. 2 with respect to t one obtains

$$I_u^{(s)}(t) = \tau \frac{\mathrm{d}I_s^{(s)}(t)}{\mathrm{d}t} = v\tau \frac{\mathrm{d}I_s^{(s)}(x)}{\mathrm{d}x} \quad (3)$$

where $x = vt$ and the velocity v is assumed to be constant. A typical example of data analysis with DDCM is shown in Fig. 1. The curve of shifted intensity $I_s(x)$ is fitted with a quadratic spline, whose derivative is used to deduce the lifetime from the unshifted intensity I_u at each measurement point. As we can see, in this case the result is remarkably stable in a rather wide region of confidence. However, when the errors on the individual points increase (as would be the case with lower statistical accuracy) or the slope of the shifted curve has a less regular trend, the error in the evaluation of the derivative can become rather large, especially at the two borders of the explored region. Moreover, working with very weak currents, one will be obliged to use a rather thick target. As a consequence, nuclei produced at different depths in the target will experience different energy loss before exiting in the vacuum, and a rather wide distribution of velocities of nuclei flying in the vacuum implies that their *average velocity* $\bar{v}$ will be different for the different values of the target-to-stopper distance x: faster nuclei, in fact, arrive to the stopper earlier, and contribute less to the average velocity. In this situation (that could be the rule for measurements with the weak beams of radioactive ions) the simple strategy of the DDCM (which assumes a constant value of $\bar{v}$) needs to be revised.

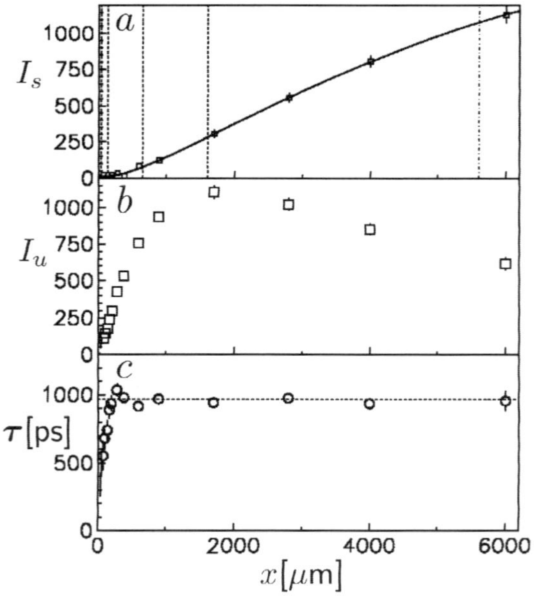

Figure 1. Example of analysis of RDM results with the DDCM. ($J^\pi = 9^-$ level of ^{146}Sm). The trend of the Doppler shifted intensities I_s (part a) is fitted with a quadratic spline (whose nodes are indicated by vertical lines). The unshifted intensities I_u at the different distances (part b) are divided by the corresponding derivatives of the spline and by the average velocity v, to obtain the values of the mean life depicted in the part c. The resulting value of the mean life is $\tau = 0.97(5)$ns. From [3], Il Nuovo Cimento A, **111**, 797 (1998), with kind permission of Società Italiana di Fisica.

2. The integral form of the Decay Curve Method

It is preferable, in such a situation, to use in the place of the DDCM Eq. 3, the one obtained by integration of it with respect to the time:

$$\int_{t_1}^{t_2} I_u(t)\mathrm{d}t \quad = \quad \tau\left[I_s(t_2) - I_s(t_1)\right] \tag{4}$$

or its more general extension, which takes into account the possible dependence on x of the average velocity $\bar{v}_x$ (along the beam direction) of nuclei decaying in flight:

$$\int_{x_1}^{x_2} I_u(x)\mathrm{d}x \quad = \quad \tau\left[\bar{v}_x(x_2)\, I_s(x_2) - \bar{v}_x(x_1) I_s(x_1)\right] \tag{5}$$

It is important to note that Eq. 5 remains valid independently of the velocity distribution of nuclei moving in the vacuum, provided that $\bar{v}(x)$ be deduced from the observed Doppler Shifts at the distances x_1 and x_2. In fact, let us assume that $g(x, v_x)\mathrm{d}v_x\mathrm{d}x$ gives the number of nuclei with velocity component (along the beam direction) in the interval $v_x, v_x{+}\mathrm{d}v_x$, whose decay feeds the relevant level while they lie in the space interval $x, x{+}\mathrm{d}x$. The Doppler shifted intensity I_s and the unshifted intensity I_u are given by

$$I_u(x) \quad = \quad \int_0^x \mathrm{d}x' \int_0^\infty \mathrm{d}v_x\; g(x', v_x) \exp\left(-\frac{x - x'}{v_x \tau}\right) \tag{6}$$

$$I_s(x) \quad = \quad \int_0^x \mathrm{d}x' \int_0^\infty \mathrm{d}v_x\; g(x', v_x) \left[1 - \exp\left(-\frac{x - x'}{v_x \tau}\right)\right] \tag{7}$$

We integrate I_u from 0 to a generic distance X_0:

$$\begin{aligned}
\int_0^{X_0} I_u(x)\mathrm{d}x \quad &= \quad \int_0^{X_0} \mathrm{d}x' \int_0^\infty \mathrm{d}v_x g(x', v_x) \int_{x'}^{X_0} \mathrm{d}x\; \exp\left(-\frac{x - x'}{v_x \tau}\right) \\
&= \quad \int_0^x \mathrm{d}x' \int_0^\infty \mathrm{d}v_x\; g(x', v_)\; v_x \tau\; \left[1 - \exp\left(-\frac{X_0 - x'}{v_x \tau}\right)\right] \\
&= \quad \tau\; I_s(X_0)\; \bar{v}_x(X_0)
\end{aligned} \tag{8}$$

where

$$\bar{v}_x(X_0) \quad = \quad \frac{\int_0^{X_0} \mathrm{d}x' \int_0^\infty \mathrm{d}v_x\; v_x\; g(x', v_x) \left[1 - \exp(-\frac{x - x'}{v_x \tau})\right]}{\int_0^{X_0} \mathrm{d}x' \int_0^\infty \mathrm{d}v_x\; g(x', v_x) \left[1 - \exp(-\frac{x - x'}{v_x \tau})\right]} \tag{9}$$

is exactly the one which can be deduced from the measured Doppler shift for a γ transition *in coincidence* with the same coincidence gate used to determine I_s and I_u. To obtain Eq. 5, it is sufficient to take the difference of the results of Eq. 8 for $X_0 = x_2$ and $X_0 = x_1$.

In principle, one could use also the differential form

$$I_u(x) \quad = \quad \tau\, \frac{\mathrm{d}\left[\bar{v}_x(x)\; I_s(x)\right]}{\mathrm{d}x} \tag{10}$$

which can be obtained from Eq. 8 by derivation of its second line with respect to X_0. Also in this form, the result is independent of the particular shape of the velocity distribution. The differential form can be less useful than the integral one in many practical cases, particularly when the overall statistics is limited, but it can be interesting to compare the results one would obtain from it to those obtained from the DDCM formula 3, assuming (erroneously) a constant value of $\bar{v}_x$. A simple example is discussed in Section 3.

3. Effect of the velocity distribution

As a consequence of the finite target thickness, the velocities v_x of nuclei emerging from the target are statistically distributed in a finite interval around their average value v_0. The consequences for the DDCM analysis have been evaluated with a simple numerical simulation, whose results are shown in Fig. 2. The delayed feeding of the relevant level, assumed to have mean life τ, is attributed to two cascading transitions having mean lives $\tau_1 = \frac{2}{3}\tau$ and $\tau_2 = \frac{5}{6}\tau$. The assumed statistical distribution of the velocity v_x is rather broad, extending from $\frac{1}{2}v_0$ to $\frac{3}{2}v_0$. The average value $\bar{v}$ of velocities v_x for nuclei decaying in flight (dashed curve in Fig. 2*a*) approaches v_0 at large values of the distance x and decreases appreciably when this distance decreases. As we can see, the difference between the two curves of part *a*, showing the values of $\bar{v}I_s$ with the calculated values of $\bar{v}(x)$ or assuming a constant $\bar{v} = v_0$, is rather small, but it is significantly enhanced in the derivatives (Fig. 2*c*) and therefore also in the deduced mean lives (Fig. 2*d*). Neglecting the dependence of $\bar{v}$ on the distance x would therefore result in a systematic error on the deduced mean life. To overcome this problem, computer codes have been developed to take into account, by a Monte-Carlo procedure [4], the process of energy loss and scattering in the target material of nuclei produced at different depths. In this way, one can exploit at best the information contained in the experimental results, but at the expense of systematic uncertainties related to the model used to describe the slowing-down process.

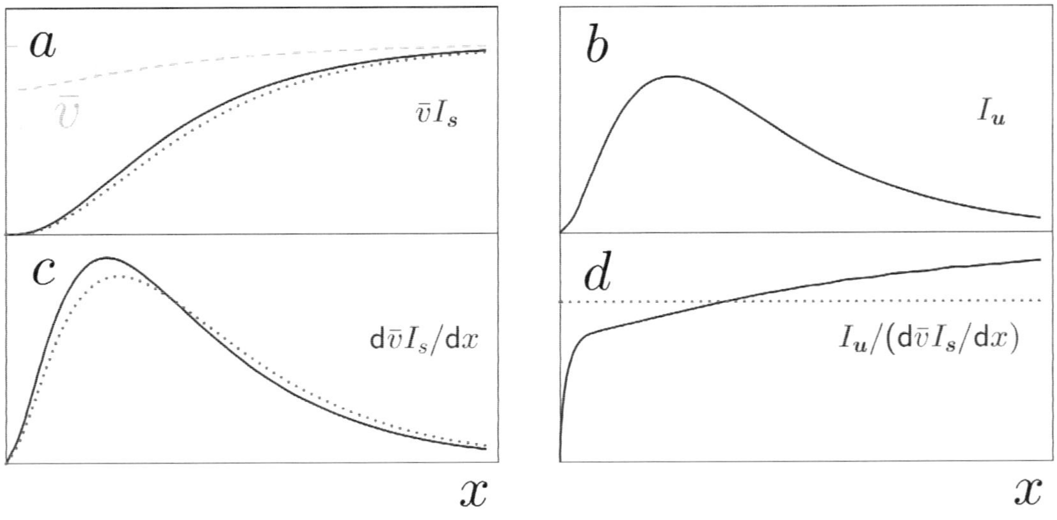

Figure 2. Numerical simulation of the analysis of RDM data with a relatively thick target (arbitrary units). The velocity distribution of the excited nuclei emerging from the target is assumed to be constant in the interval $\frac{1}{2}v_0 < v_x < \frac{3}{2}v_0$ and zero outside. The assumed distribution of the feeding time is $f(t) = [\exp(-t/\tau_1) - \exp(-t/\tau_2)]/(\tau_1 - \tau_2)$, with $\tau_1 = \frac{2}{3}\tau$, $\tau_2 = \frac{5}{6}\tau$. Solid lines show the results obtained assuming a constant $\bar{v}_x = v_0$, dotted lines take into account the correct dependence of $\bar{v}_x$ on x, which is depicted with a dashed line in part *a*. The product $\bar{v}_x I_s$ is shown in part *a*, and its derivative with respect to x in part *c*. Part *b* shows the unshifted intensity I_u, and the estimated values of the mean life τ (ratios of values of part *b* and part *c*) are given in part *d*.

4. The optimal measurement

The proposed method, in its integral form (Eq. 8), provides a simple and model independent alternative, particularly useful when one has to deal with limited statistics. In such a case, it is unlikely that lifetime measurements can be performed for a large number of nuclear levels. Most probably, they will concern one or few levels, and the measurement conditions can be optimized for each of them.

The ideal experiment will consist of a long measurement at the largest distance x_2, to obtain the values of I_s and $\bar{v}$ with sufficient accuracy, and of a series of shorter measurements at many points in the interval from x_1 to x_2, whose sum (or weighted sum) will give the integral of I_u with comparable accuracy. The value of I_s at the lower distance x_1 is usually rather small, and its measurement is not very significant for the result.

In this way, there is no need for a detailed Monte-Carlo simulation of the slowing-down process *in the target.* A detailed investigation of the slowing-down process *in the stopper* would still be necessary only for very short lifetimes approaching the stopping time in the stopper material.

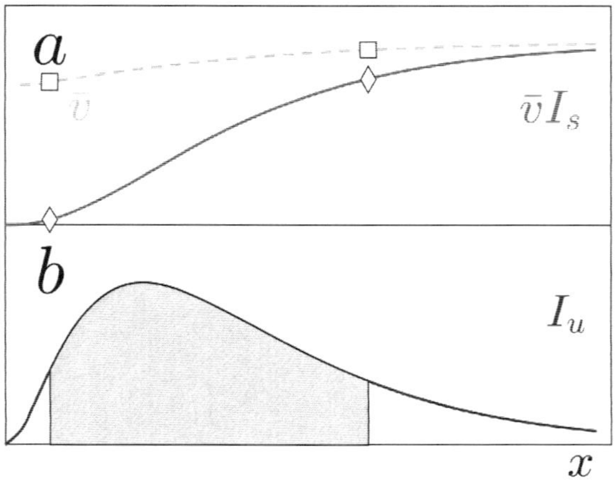

Figure 3. Relevant quantities for the evaluation of the mean life by means of the Integral Decay Curve Method. The integral of the I_u curve (part *b*), indicated by the shadowed area, must be compared with the difference of products $\bar{v}I_s$ at the two extremes of the interval (part *a*), $\bar{v}$ being the average velocity of nuclei decaying in flight (depicted as a dashed line).

References

[1] Dewald A, Harissopulos S and von Brentano P 1989 *Z. Phys.* A **334**, 163–175

[2] Böhm G,, Dewald A, Petkov P and, von Brentano P 1999 *Nuclear Instruments and Methods in Physics Research* A **329**, 248–261

[3] Bizzeti PG, Bizzeti-Sona AM, Bazzacco D, Lunardi S, Rossi Alvarez C, de Angelis G, De Poli M, Maron G, Napoli DR, Dewald A, Klemme T and Tiesler H 1998 *Nuovo Cimento* A **111**, 697–703

[4] Petkov P, Tonev D, Gableske J, Dewald A, Klemme T and von Brentano P 1999 *Nuclear Instruments and Methods in Physics Research* A **431**, 208–223

Neutron-particle and proton-hole excitations in the N=128 isotones ^{208}Hg and ^{209}Tl from spectroscopy following ^{208}Pb+^{238}U deep-inelastic reactions

2011 J. Phys.: Conf. Ser. 267 012035

(http://iopscience.iop.org/1742-6596/267/1/012035)

Neutron-particle and proton-hole excitations in the N=128 isotones ^{208}Hg and ^{209}Tl from spectroscopy following ^{208}Pb+^{238}U deep-inelastic reactions

B Fornal[1], **B Szpak**[1], **R V F Janssens**[2], **R Broda**[1], **M P Carpenter**[2], **G D Dracoulis**[3], **K H Maier**[1], **G J Lane**[3], **J Wrzesiński**[1], **S Zhu**[2]

[1]Institute of Nuclear Physics, Polish Academy of Sciences, Kraków, Poland,

[2]Argonne National Laboratory, Argonne, IL, USA,

[3]Department of Nuclear Physics, Australian National University, Canberra, Australia

Abstract. Gamma rays in the $\pi^{-2}\nu^2$ nucleus ^{208}Hg and $\pi^{-1}\nu^2$ nucleus ^{209}Tl have been studied at Gammasphere using deep-inelastic reactions induced by a 1360 MeV ^{208}Pb beam on a thick ^{238}U target. Previously unknown yrast γ-ray cascades above the 8^+ and $17/2^+$ nanosecond isomers in ^{208}Hg and ^{209}Tl, respectively, were identified in coincidence with known γ rays deexciting the isomers. Yrast levels up to spin 13^- in ^{208}Hg have been located, and they are interpreted in light of the structure of the ^{210}Pb isotone and with the help of shell model calculations. Shell model calculations by using the V_{low-k} realistic interaction have been performed for ^{210}Pb and ^{208}Hg, and compared with experiment. The V_{low-k} effective Hamiltonian seems to account well for the properties of these neutron-rich shell model nuclei.

The nuclei ^{208}Hg and ^{209}Tl are N=128 isotones, with two neutron valence particles and two and one valence proton holes, respectively, outside doubly-magic ^{208}Pb. The structure of these nuclei is well worth studying, since it has the potential of providing information about empirical proton-neutron interactions in the as yet unexplored sector of the nuclidic chart with Z<82 and N>126. What little is known about the properties of nuclei in this region comes mainly from a recent study of ^{238}U fragmentation at relativistic energies [1], where the lowest yrast states in ^{208}Hg and ^{209}Tl were identified up to isomeric excitations with spin and parity 8^+ and $17/2^+$. In ^{208}Hg, an isomeric excitation with a half-life of 99 ns was located at 1297+x keV. It decays via a highly converted, undetected low energy transition followed by a cascade of three γ rays 203, 425 and 669 keV, which establish three other levels at 1297, 1094 and 669 keV. All those states were proposed to be members of the $\pi^{-2}(0^+)\nu g_{9/2}^2$ multiplet, with the 8^+ isomer arising from the coupling of maximally aligned spin of the two $g_{9/2}$ neutrons. Similarly, in ^{209}Tl, a cascade of three transitions 138, 324, 661 keV was found to deexcite a metastable state with a half-life of 95 ns that was placed at 1123+y keV. This isomeric state was interpreted as the highest spin member, $J^\pi = 17/2^+$, of the $\pi s_{1/2}^{-1}\nu g_{9/2}^2$ multiplet.

Another opportunity to access yrast excitations in these neutron-rich species is offered by deep-inelastic processes that occur in collisions of a ^{208}Pb beam with a ^{238}U target [2]. As shown in our earlier works, the production of neutron-rich nuclei in such processes is made

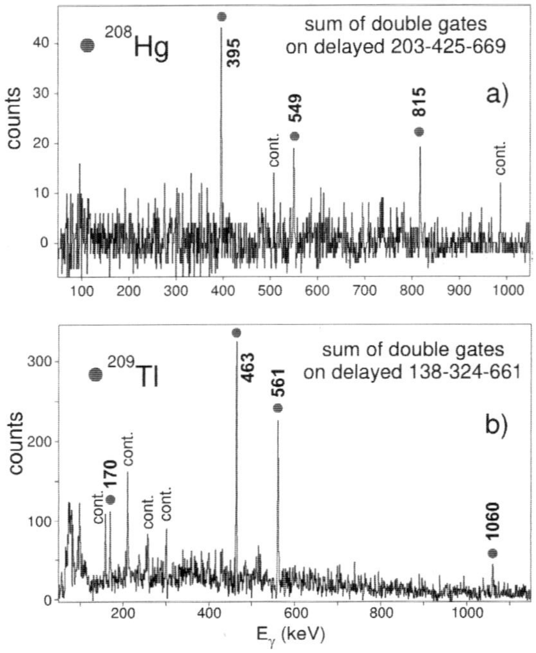

Figure 1. γ-ray coincidence spectra for (a) ^{208}Hg and (b) ^{209}Tl, showing γ rays in prompt-delayed coincidence with double gates on the transitions deexciting the known isomers. Contaminant lines of known origin are also indicated.

possible by a tendency towards N/Z equilibration of the di-nuclear system formed during the collisions [3, 4, 5, 6]. In the reactions under investigation, the lighter colliding partner, ^{208}Pb, had lower N/Z ratio than the target nucleus and, as a result, the production of nuclei with larger neutron excess than the projectile was favored. In the case of Pb products, yields sufficient for spectroscopic studies with the high detection sensitivity Gammasphere spectrometer [7] extended in neutron number up to ^{211}Pb [8].

From the observed production yields for the $^{202-211}$Pb isotopes, it became clear that Tl nuclei with masses up to at least 209 and Hg isotopes with masses up to possibly 208 had to be present in the data set with cross sections sufficient to examine in detail the coincidence relationships required to establish significant level schemes. Having all this in mind, we analyzed data from the experiment that was performed earlier at the Argonne National Laboratory using Gammasphere and a 1360 MeV ^{208}Pb beam from the ATLAS accelerator focused on a thick, 50 mg/cm^2, ^{238}U target. $\gamma - \gamma-$time coincidence data were collected with a composite trigger, requiring three or more Compton-suppressed γ rays to be in coincidence for the in-beam events, and two or more such transitions in coincidence for the out-of-beam events. A total of 2.3×10^9 events were recorded. The beam, coming in bursts with an approx. 0.3 ns time width, was pulsed with an approx. 1.6 μs repetition time, providing clean separation of prompt and isomeric events, which simplified observation of $\gamma - \gamma$ correlations across isomers.

At the beginning of data analysis, we noticed that a sizeable amount of coincidence events was present between known γ rays emitted in the decay of isomeric states located earlier in ^{209}Tl and ^{208}Hg [1]. Based on this observation, a search of γ transitions occurring above the metastable states in those two reaction products was performed by gating on γ rays deexciting the isomers. For this purpose a three-dimensional histogram which contained triple coincidence events with one prompt and two delayed γ rays was used. In the case of ^{208}Hg, all combinations of double gates set on the delayed 203, 425 and 669 keV lines [Fig. 1a] displayed new transitions at energies 395, 549 and 815 keV that definitely occur in this nucleus above the 99 ns isomer. A similar gating procedure applied to the known γ rays in ^{209}Tl revealed a sequence of transitions with energies 170, 463, 561 and 1060 keV [Fig. 1b] that undoubtedly belong to ^{209}Tl and precede in time the 95 ns isomer. By examining coincidence relationships and relative intensities of the newly identified lines, it was found that in ^{209}Tl they form a 463-561-170-1060 keV cascade feeding the isomeric state. For ^{208}Hg, such an analysis suggested that the 395 keV transition populates the 99 ns isomer from a level at 1692+x, whereas the 815 and 549 keV γ rays occur higher in the level scheme. Due to low statistics, however, it was not possible to establish firmly whether the 395, 815 and 549 keV γ rays are in mutual coincidence.

Further insight into the structure of ^{208}Hg could be gained by analyzing the new findings from the perspective of level arrangement in the ^{210}Pb isotone. There should be a close resemblance between the spectroscopy of yrast states in the ^{208}Hg and ^{210}Pb nuclei as in both cases the low lying structure should be dominated by excitations arising from spin recouplings of the two valence neutrons occupying the $g_{9/2}$, $i_{11/2}$ and $j_{15/2}$ single particle states. The lowest orbitals in ^{206}Hg are, in contrast, $s_{1/2}$ and $d_{3/2}$. In this case, the two-proton holes can couple to at most $I=2^+$ on the energy expense of approx. 1 MeV and, therefore, proton excitations are not expected to play a significant role in the low-lying yrast structure of ^{208}Hg. The ^{210}Pb level scheme, of which the high spin part was established in [9], is shown in Fig. 2. The 2^+, 4^+, 6^+ and 8^+ states are members of the $\nu g^2_{9/2}$ multiplet. Above the 8^+ isomer, three distinct yrast excitations are present at 1806, 2511 and 3152 keV with spin-parity 10^+, 11^- and 13^-, respectively. The 10^+ state arises mostly from the $\nu g_{9/2} i_{11/2}$ configuration, whereas the 11^- and 13^- excitations involve a neutron on the $j_{15/2}$ orbital and have $\nu g_{9/2} j_{15/2}$ and $\nu i_{11/2} j_{15/2}$ character, respectively.

A similar sequence of levels is expected in ^{208}Hg assuming that, to the first approximation, the two proton holes coupled to spin $J^\pi=0^+$ play a role of spectators. Indeed, as sugested in [1] the first four excited states are the 2^+, 4^+, 6^+ and 8^+ excitations having mostly a $\pi^{-2}(0^+)\nu g^2_{9/2}$ character. At higher excitation energies, a level located 395 keV above the 8^+ isomer, i.e., at 1692+x keV, and populated with a relatively high intensity is a candidate for the 10^+ state with the $\pi^{-2}(0^+)\nu g_{9/2} i_{11/2}$ dominant configuration. The decay pattern of the 10^+, 11^- and 13^- levels in ^{210}Pb suggests also that the 815 and 549 keV γ rays observed in ^{208}Hg might correspond to $11^- \rightarrow 10^+$ and $13^- \rightarrow 11^-$ transitions.

Our further efforts to interpret the high-spin levels identified in ^{208}Hg were heavily influenced by the results of shell model calculations. These calculations, in a space involving $50<Z<82$ protons and $126<N<184$ neutrons with respect to the ^{208}Pb core, have been performed with the OXBASH code [10]. The single particle energies were taken from experiment. The basis for the residual interaction was the realistic interaction calculated by Kuo and Herling, using the Kuo-Brown method [11], from the interaction between free nucleons inferred from experiments [12]. We have used it, but with the modifications applied in [13, 14] to better fit experimental energies. This covered the interactions between neutron particles, meaning orbitals above the N=126 shell gap, and those between proton holes, meaning orbitals below the Z=82 gap. But for the proton-hole – neutron-particle nucleus ^{208}Hg, the interaction between orbitals above and below the gap is of primary importance. This interaction has been calculated [15] from the H7B interaction [16].

The calculated energies of the yrast states in ^{210}Pb, shown in the left portion of Fig. 2, agree well with experiment as one may expect, since the Kuo-Herling interaction was already proven

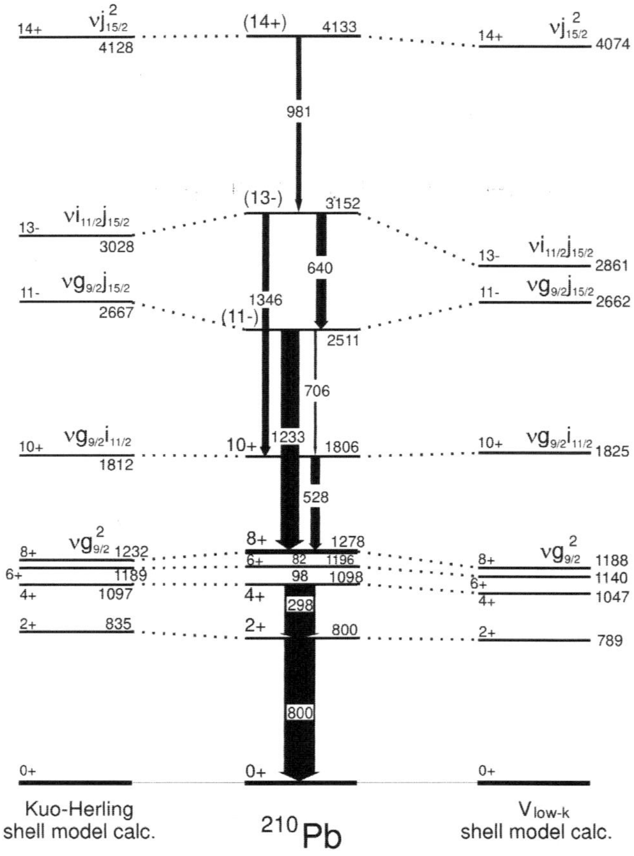

Figure 2. The ^{210}Pb level scheme from [9] with the results of shell model calculations employing the Kuo-Herling interaction (left-hand side) and V_{low-k} realistic effective interaction (right-hand side).

to describe particle-particle excitations. What is more interesting, the results of Kuo-Herling calculations for ^{208}Hg, shown in Fig. 3, account well for the structure proposed from experiment, supporting the assignments of 10^+, 11^- and 13^- quantum numbers to the levels located in the present work above the 99 ns isomer.

It has occurred to us, that the new findings in nuclei belonging to the yet unexplored region of the nuclidic chart with Z<82 and N>126, could serve as a testing ground for the so called V_{low-k} realistic shell model interactions that became recently available. These interactions are derived from realistic nucleon-nucleon (NN) potentials without any adjustable parameters.

We have obtained V_{low-k} effective interaction for the 50<Z<82 and 126<N<184 model space by using the Computational Environment for Nuclear Structure (CENS) by T. Engeland, M. Hjorth-Jensen and G.R. Jansen [17], based on [18]. First, the high-momentum repulsive components of the CD-Bonn NN potential were renormalized by way of the so-called V_{low-k} approach [19, 20], with a momentum cut-off Λ=2.2 fm^{-1}. Next, the V_{low-k} potential was used,

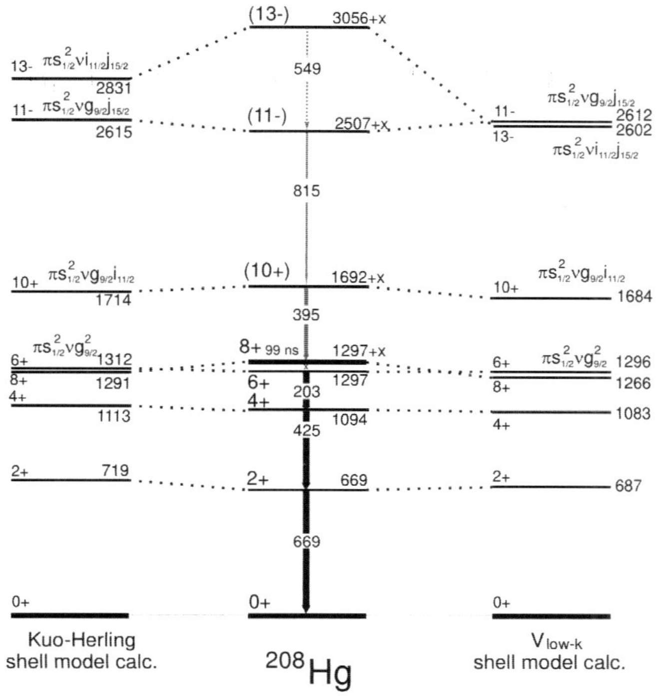

Figure 3. The proposed level scheme for ^{208}Hg. Also shown are the results of shell model calculations performed with the Kuo-Herling interaction (lef-hand side) and V_{low-k} realistic effective interaction (right-hand side).

with the addition of the Coulomb force for protons, to calculate the effective interaction V_{eff} within the framework of the Q-box folded-diagram expansion [21]. The computation of the Q-box was performed at second order in V_{low-k}. It included four two-body terms: the V_{low-k}, the two core polarization diagrams V_{1p1h} and V_{2p2h}, corresponding to one particle-one hole and two particle-two hole excitations, and a ladder diagrams accounting for excluded configurations above the chosen model space. The shell model calculations have been performed using the OXBASH code [10]. We let the valence protons occupy the five levels $0g_{7/2}$, $1d_{5/2}$, $1d_{3/2}$, $2s_{1/2}$, and $0h_{11/2}$ of the Z=50-82 shell, while for valence neutrons the model space included the $2g_{9/2}$, $1i_{11/2}$, $1j_{15/2}$, $3d_{5/2}$, $4s_{1/2}$, $2g_{7/2}$, and $3d_{3/2}$ orbitals of the N=126-184 shell. We used the experimental single particle energies defined with respect to the ^{208}Pb core. The results for ^{210}Pb are displayed on the right hand side in Fig. 2. Considering that no adjustable parameters have been involved in the procedure of deriving the shell model two-body matrix elements from the NN potential, the agreement between experimental and calculated level energies is noteworthy. The only major discrepancy, of the order of 300 keV, occurs for the 13^- excitation which involves two neutrons on high spin orbitals, $\nu i_{11/2} j_{15/2}$, coupled to the maximum spin: 13^-.

A similar comparison done for ^{208}Hg and displayed in Fig. 3, shows that energies of the 2^+,

4^+, 6^+, 8^+, 10^+, and 11^- states are also reproduced well by the V_{low-k} effective interaction. Significant deviation is observed only for the 13^- state originating mostly from the $\pi^{-2}(0^+)\nu$ $i_{11/2}j_{15/2}$ configuration, as it was in the case of the corresponding state in ^{210}Pb.

In summary, detailed γ-ray coincidence measurements using Gammasphere for the reaction system ^{208}Pb+^{238}U have provided much information about γ-ray transitions and yrast excitations in ^{208}Hg and ^{209}Tl above the known nanosecond isomers. The new findings include high-spin states in ^{208}Hg arising from two-proton-hole, two-neutron-particle couplings. Considerable similarities between the yrast structure of the ^{210}Pb isotone, as well as comparisons with results of shell model calculations with the Kuo-Herling interaction were important in the interpretation of the ^{208}Hg levels. The newly identified structure was confronted with results of shell model calculations employing realistic V_{low-k} interaction. The satisfactory agreement that was observed between experiment and theory demonstrates noticeable predictive power of shell model calculations based on V_{low-k} realistic Hamiltonian in neutron-rich nuclei.

Acknowledgments

This work is supported by the Polish Ministry of Science and Higher Education under contract N-N202-263238 and by the US Department of Energy, Office of Nuclear Physics, under contract DE-AC02-06CH11357.

References

[1] Al-Dahan N et al. 2009 *Phys. Rev.* C **80** 061302(R)
[2] Fornal B et al. 2001 *Phys. Rev. Lett* **87** 212501
[3] Freiesleben H, Kratz J V 1984 *Phys. Rep.* **106** 1
[4] Broda R 2006 *J. Phys. G* **32** R151
[5] Krolas W et al. 2003 *Nucl. Phys.* A **724** 289
[6] Fornal B et al. 1995 *Acta Phys. Pol.* B **26** 357
[7] Lee I Y 1990 *Nucl. Phys.* A **520** 641c
[8] Lane G J et al. 2005 *Phys.Lett.* B **606** 34
[9] Rejmund M et al. 1997 *Z. Phys.* A **359** 243
[10] Brown B A, Etchegoyen A, Rae W D M, Godwin N S 1984 OXBASH (unpublished).
[11] Kuo T T S, Brown G E 1966 *Nucl. Phys.* **85** 40
[12] Kuo T T S, Herling G H 1975 US Naval Research Laboratory Report No. 2258 (unpublished)
[13] Warburton E K, Brown B A 1991 *Phys. Rev.* C **43** 602
[14] Rydström L, Blomqvist J, Liotta R J, Pomar C 1990 *Nucl. Phys.* A **512** 217
[15] Rejmund M 1998 *PhD thesis* (Warsaw :Institute of Experimental Physics, Warsaw University)
[16] Warburton E K 1991 *Phys. Rev.* C **44** 233
[17] Engeland T, Hjorth-Jensen M and Jansen G R, CENS, a Computational Environment for Nuclear Structure, in preparation.
[18] Hjorth-Jensen M, Kuo T T S, Osnes E 1995 *Physics Reports* **261** 125
[19] Bogner S, Kuo T T S, Coraggio L 2001 *Nucl. Phys.* A **684** 432c
[20] Bogner S, Kuo T T S, Coraggio L, Covello A and Itaco N 2002 *Phys. Rev.* C **65** 051301(R)
[21] Coraggio L, Covello A, Gargano A, Itaco N and Kuo T T S 2009 *Prog. Part. Nucl. Phys.* **62** 135 and references therein.

Structure of neutron-rich nuclei near N=40

2011 J. Phys.: Conf. Ser. 267 012036

(http://iopscience.iop.org/1742-6596/267/1/012036)

Structure of neutron-rich nuclei near N=40

Silvia M Lenzi

Dipartimento di Fisica dell'Università and INFN, Sezione di Padova, Padova, Italy

Email: lenzi@pd.infn.it

Abstract. Recent experimental results obtained for neutron-rich nuclei of mass A~60-70 at LNL are presented together with the shell model interpretation. The development of a new region of deformation approaching the neutron number N = 40 in Cr, Fe and Co isotopes is discussed. A new effective interaction (LNPS) that considers the full *fp* shell model space for protons and the $p_{3/2}$, $p_{1/2}, f_{5/2}$, $g_{9/2}$, $d_{5/2}$ orbitals for the neutrons allows to reproduce, for the first time, the different structure properties of these nuclei. The configurations of the calculated states suggest the development of a new *island of inversion* at N ~ 40.

1. Introduction

For many decades, theoretical models in nuclear structure were constructed relying on the experimental information obtained for nuclear systems near the stability line. The continuous experimental developments allow nowadays the study of exotic nuclei far from stability. In particular, neutron-rich nuclei are of current interest because of clear indications that the traditional magic numbers of the shell structure near stability are not always preserved in this region of the table of isotopes, in particular in light and medium-light nuclei [1-4]. The observed changes help to shed light on specific terms of the effective nucleon-nucleon interaction and to improve our knowledge of the nuclear structure evolution towards the drip lines. In particular, it has been shown that the monopole part of the tensor force of the proton-neutron interaction gives the main contribution to the shell evolution [2].

Among other interesting phenomena recently observed far from stability, the development of deformation in neutron-rich nuclei at N=20 has been observed. This mass region, is known as the *Island of Inversion* because the ground-state wave functions are dominated by intruder configurations, while "natural" states ($0\hbar\omega$ excitations) lie higher in energy (Fig. 1 in ref. [5]). Recent experimental and theoretical studies have suggested that neutron-rich Chromium and Iron isotopes around N=40 would lie in a new *Island of Inversion*. In this case, the deformation is produced by the erosion of the harmonic oscillator gap at N=40 and the quadrupole correlations induced by the excitation of neutrons from the *pf* shell to both the $g_{9/2}$ and $d_{5/2}$ orbitals of the upper main shell [3]. These latter two orbitals form a quasi-SU3 space able to generate quadrupole collectivity, as discussed in Ref. [6].

In this work we present recent data on neutron-rich even-even Cr and Fe isotopes and odd-mass Co isotopes, obtained in the Legnaro National Laboratories using the CLARA-PRISMA detector complex. Data are compared to recent shell model calculations obtained by diagonalizing the new effective interaction LNPS [7]. The adopted shell model space is the *pf* shell for protons and the $f_{5/2}$ $p_{3/2}$ $p_{1/2}$ $g_{9/2}$ $d_{5/2}$ for neutrons. This interaction is able to describe both collective and single-particle behavior near N=40 and to follow the rapid evolution of the nuclear structure along the isotopic and isotonic chains.

In Sect. 2 we present the experimental method and the data. Sect. 3 is devoted to the discussion of the results. Conclusions are given in Sect. 4.

2. Experimental method and results

The knowledge of excited states in neutron-rich isotopes is rather limited due to the difficulties in the production and identification of isotopes by conventional means. In recent years, the use of binary reactions between stable neutron-rich nuclei, such as multi-nucleon transfer and deep-inelastic collisions, combined with modern gamma-ray arrays, has increased substantially the amount of information on the structure of previously inaccessible nuclei far from stability.

To identify online a gamma-ray emitter produced through direct or deep-inelastic reactions, it is necessary to use sophisticated instruments, especially if dealing with reaction channels that have relatively low cross section and a broad velocity distribution. In the last years different high-efficiency gamma-ray detector systems have been used in combination with ancillary detectors of increasing complexity. A research program on nuclear structure of slightly neutron-rich nuclei has been carried out at LNL with the gamma-ray array CLARA [8], coupled to the magnetic spectrometer PRISMA [9] to study the gamma decay of excited nuclei populated by multi-nucleon transfer and deep-inelastic collisions [10-13].

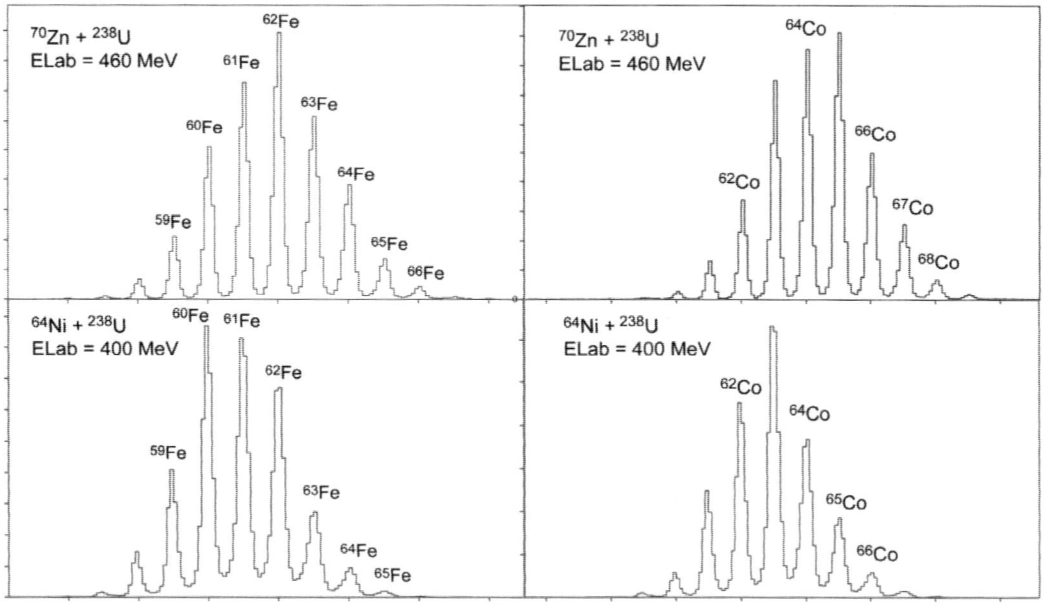

Figure 1. Mass spectrum for Fe and Co isotopes obtained with the reaction $^{70}Zn+^{238}U$ (upper panels) and $^{64}Ni+^{238}U$ (lower panels).

Neutron-rich nuclei in the mass region A~60-70 were produced in two experiments at the Tandem-ALPI accelerator complex of LNL. In the first experiment the beam was ^{64}Ni at 400 MeV bombarding energy and in the second one a ^{70}Zn beam was used. In both cases the target was a self supporting Uranium target, 400 μg/cm^2 thick. For each projectile-like ion detected in PRISMA, we determined the atomic number Z, the mass A, the initial direction of the ion flying away from the target and the absolute value of its velocity. The γ-rays following the de-excitation of the reaction products were detected with the CLARA array, placed in the hemisphere opposite to the PRISMA spectrometer, 29.5 cm from the target, uniformly covering the azimuthally angles from 98° to 180°. The Doppler correction for the photons in coincidence with the ions detected in PRISMA was performed on an event-by-event basis, using the recoil velocity vector obtained after trajectory reconstruction in the spectrometer. The γ-ray energy resolution obtained was 0.8% FWHM over the whole broad velocity distribution of the projectile-like products, ranging from 4.5% to 10% of the speed of light. In several cases it was possible to observe gamma lines in nuclei where no excited states were known. In the most exotic nuclei produced, the statistics was not enough to allow a γ-γ coincidence analysis. In this type of reactions, the population of yrast states is favored. Based on this assumption, the level schemes of the neutron-rich isotopes were proposed taking into account the relative yields and the systematic of neighboring nuclei. A complementary method that produces higher statistics is to use the same type of reaction but with a much thicker target (~50mg/cm^2) and a very high efficiency γ array. The knowledge of some γ lines in a nucleus allows a γ-γ-γ coincidence analysis and therefore the construction of the level scheme on a more solid basis [14].

In the first experiment reported in this paper, neutron-rich isotopes of V [13], Cr [10], Mn [12] and Fe [11] were produced with good statistics, allowing to follow the evolution of the nuclear structure along the isotopic chains. The second experiment was due to the study of the heavier Fe, Co and Ni isotopes [15]. The relative yields observed in both cases for Fe and Co isotopes can be seen in figure 1.

A rapid change of structure is observed in Cr and Fe isotopes. The nucleus ^{56}Cr behaves as a semi-magic nucleus with the new shell closure at N=32, ^{58}Cr shows the characteristic excitation-energy sequence of a critical point of the E(5) shape-phase transition from spherical to γ-unstable, while ^{60}Cr [10,14] shows a level scheme consistent with a deformed nucleus. A rapid shape evolution is also observed in Fe isotopes, as reported in Ref. [11]. For the case of ^{66}Fe, only one excited level was observed in the first experiment, but in the second one three γ lines were observed, which were interpreted as the cascade from the 6^+ to the ground state. In addition the transition from the first 2^+ state in ^{68}Fe (N = 42) was observed. Both spectra are reported in figure 2.

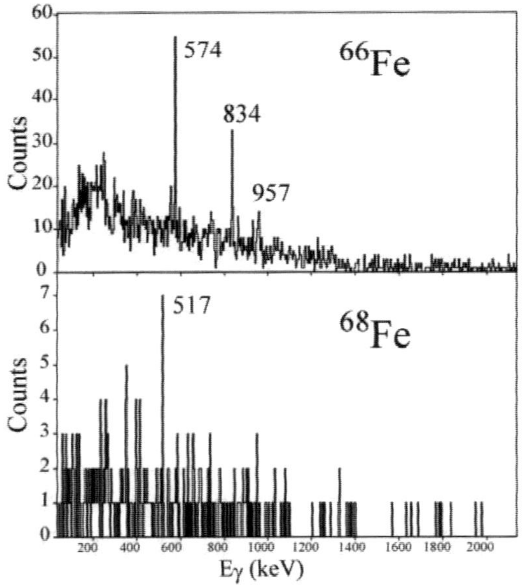

Figure 2. Spectra of ^{66,68}Fe isotopes obtained with the multinucleon transfer reaction ^{70}Zn+^{238}U at 460 MeV bombarding energy with the CLARA + PRISMA experimental setup at LNL.

Cobalt isotopes were also populated in the reaction. Yrast transitions were observed for the first time in the N=40 ^{67}Co. Spectra corresponding to the decay of $^{65\text{-}67}$Co are reported in figure 3. The transition of 2273 keV in ^{67}Co is interpreted as the decay to the ground state from the second excited state. Such a high energy transition is only observed in this isotope. This state is interpreted as resulting from the coupling of the proton hole to the 2^+ state in ^{68}Ni [16]. A γ line of 190 keV, feeding an intruder $J^\pi = 1/2^-$ state at 491 keV was also observed. This state is very interesting as it corresponds to the excitation of a proton across the Z=28 gap.

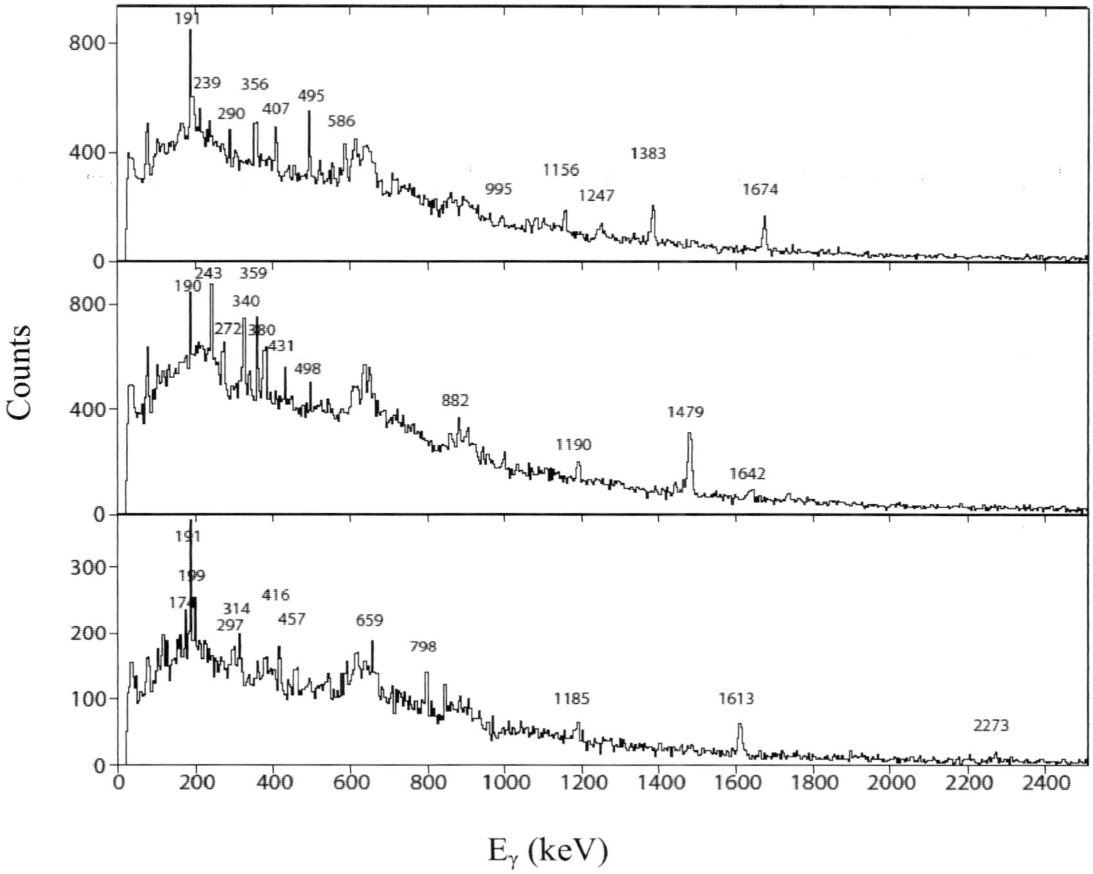

Figure 3. Spectra of Co isotopes: ^{63}Co (upper panel), ^{65}Co (central panel) and ^{67}Co in the lower panel [16].

3. Shell model description and discussion

The variety of the structure and its rapid evolution along the isotopic and isotonic chains when approaching N=40 is a challenge for any nuclear model, and in particular for the shell model. In fact, such a description implies to consider a wide shell model space and the development of an effective interaction that involves more than one main shell. While Cr isotopes up to N=34 can be very well described within the fp shell, neutron-rich Fe isotopes up to N=36 need the inclusion of the $g_{9/2}$ neutron orbital in the model space (^{48}Ca core) [12]. This latter model space is also good enough to account for the spectroscopy of moderately neutron-rich Co and Mn [14] isotopes. However, when approaching N=40 these *fpg* calculations fail to describe the structure of neutron-rich nuclei, in particular, $^{60\text{-}64}$Cr [10] and $^{66\text{-}68}$Fe [11,19].

It has been pointed out that to reproduce the quadrupole collectivity in this mass region, the inclusion of the neutron $d_{5/2}$ orbital is needed [3]. This can be explained in terms of the quasi-SU3 approximate symmetry: The deformation can be generated by the interplay between the quadrupole force and the central field in the subspace consisting on the lowest $\Delta j = 2$ orbitals of a major shell [6]. Recently, a new interaction for this model space (LNPS) has been developed that can reproduce the different phenomena suggested by the available data in this mass region [5]. The LNPS interaction has been obtained by adapting a realistic CD-Bonn potential to this model space after many-body perturbation techniques [20] and monopole modifications. The quality of the results of shell model calculations with the new LNPS interaction for the neutron rich isotopes of Cr and Fe can be appreciated in figure 4. In particular, large transition probabilities are predicted for Cr and Fe isotopes, consistent with the development of a new region of deformation at N~40. Moreover, the configuration of the wave functions of the ground states indicates the development of a new *island of inversion* in this mass region, similar to that observed already in Na and Mg nuclei at N=20.

The B(E2) values reported in the lower panel of figure 4, show a sudden increase in Cr isotopes at N=38 that remains almost stable up to N=42. For the case of Fe isotopes, the increase of collectivity continues increasing and reaches its maximum at N=42. For Cr isotopes, no experimental information on transition probabilities are yet available. For ^{62,64}Fe, on the other hand, recent data has been obtained in multi-nucleon transfer reactions at GANIL [21]. At MSU, lifetimes have been measured in $^{62\text{-}66}$Fe, produced after knock-out reactions [22]. The present theoretical predictions are in good agreement with the experimental findings.

The new effective interaction can describe these *collective* properties as well as *single-particle* behavior, including shape coexistence near Z=28. This is observed for the case of Co isotopes. The yrast structure is not collective while, as stated above, for the case of ^{67}Co, a $1/2^-$ state, that results from the excitation of a proton from the $f_{7/2}$ to the $p_{3/2}$, comes down in energy to 491 keV [17]. This is possible due to the quadrupole correlations generated in the excitation of neutrons to the $g_{9/2}$ and $d_{5/2}$ across N=40. The shell model can reproduce this shape coexistence in ^{67}Co [7,16].

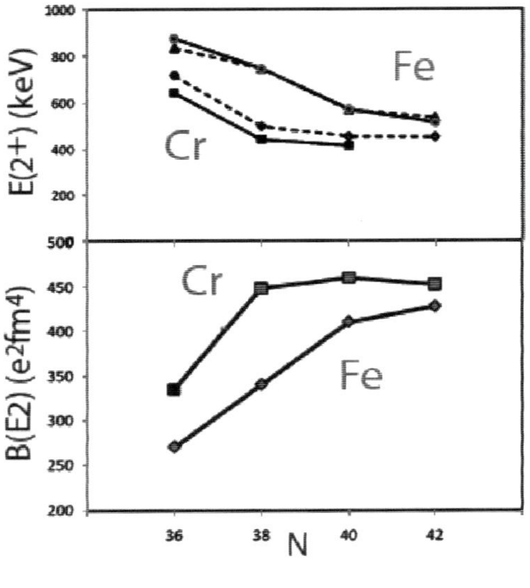

Figure 4. Excitation energy of the first 2^+ state (upper panel). Full lines connect the experimental values, open symbols connected with dashed lines correspond to the theoretical calculations with the LNPS effective interaction. In the lower panel, the calculated transition probabilities for the decay to the ground state are reported.

4. Conclusions

The development of deformation in the neutron-rich nuclei around N=40 for Z<28, is a subject of current interest from both the experimental and the theoretical points of view. Data are being obtained by means of different experimental techniques. In particular, data on neutron-rich Fe and Co isotopes recently obtained with the CLARA-PRISMA setup at LNL have been presented and discussed here. Such a degree of collectivity is very well reproduced by recent shell model calculations in a model space that includes the *pf* shell for protons and the $f_{5/2}\, p_{3/2}\, p_{1/2}\, g_{9/2}\, d_{5/2}$ for neutrons. The new effective interaction for this space, LNPS [7], is able to reproduce the rapid changes of structure in this mass region, together with the phenomena of shape coexistence.

With the AGATA Demonstrator coupled to the PRISMA magnetic spectrometer at LNL, experiments due to measure the lifetimes in these nuclei are in progress. This will allow to have a better understanding on the nuclear structure properties of these nuclei and will test the new effective interaction LNPS developed for this mass region.

Acknowledgements
This work is the result of an experimental and a theoretical study done in collaboration with several colleagues. In particular I want to thank for the experimental part F. Recchia, S. Lunardi, N. Marginean, D.R. Napoli, A. Gadea, E. Farnea and J.J. Valiente-Dobón. For the shell model part I want to thank F. Nowacki, A. Poves and K. Sieja.

References

[1] Otsuka T, Fujimoto R, Utsuno Y, Brown B A, Honma M and Mizusaki T 2001 *Phys. Rev. Lett.* **87** 082502
[2] Otsuka T, Suzuki T, Fujimoto R Grawe H and Akaishi Y 2005 *Phys. Rev. Lett.* **95** 232502
[3] Caurier E, Nowacki F and Poves A 2002 *Eur. Phys. J.* A **15** 145
[4] Sorlin O and Porquet M G 2008 *Prog. Part. Nucl. Phys.* **61** 602
[5] Brown B A 2001 *Prog. Part. Nucl. Phys.* **47** 517
[6] Zuker A P, Retamosa J, Poves A and Caurier E 1995 *Phys. Rev* C **52** R1751
[7] Lenzi S M, Nowacki F, Poves A, Sieja K, to be submitted.
[8] Gadea A *et al.* 2004 *Eur. Phys. J.* A **20** 193
[9] Stefanini A M *et al.* 2002 *Nucl. Phys.* A **701** 217c
[10] Marginean N *et al.* 2006 *Phys. Lett.* B **633** 696
[11] Lunardi S *et al.* 2007 *Phys. Rev.* C **76** 034303
[12] Valiente-Dobón J J *et al.* 2008 *Phys. Rev.* C **78** 024302
[13] Napoli D R 2006 *J. of Phys. Conf. Series* **49** 91
[14] Zhu S *et al.* 2006 *Phys. Rev.* C **74** 064315
[15] Lenzi S M et al. 2008 INFN-LNL-222 ISSN 1828-8545 15
[16] Recchia F *et al.* to be published.
[17] Pauwels D *et al.* 2009 *Phys. Rev.* C **79** 044309
[18] Aoi N *et al.* 2009 *Phys. Rev. Lett.* **102** 012502
[19] Adrich P *et al.* 2008 *Phys. Rev.* C **77** 054306
[20] Hjorth-Jensen M, Kuo T T S and Osnes E 1995 *Phys. Rep.* **261** 125
[21] Ljungvall J *et al.* 2010 *Phys. Rev.* C **81** 061301
[22] Rother W *et al.* 2010 *http://arxiv.org/abs/1006.5297*

Relativistic U(3) symmetry and pseudo-U(3) symmetry of the Dirac Hamiltonian

2011 J. Phys.: Conf. Ser. 267 012037

(http://iopscience.iop.org/1742-6596/267/1/012037)

Relativistic U(3) symmetry and pseudo-U(3) symmetry of the Dirac Hamiltonian

J N Ginocchio

Theoretical Division, Los Alamos National Laboratory, Los Alamos, NM, 87545, USA

E-mail: gino@lanl.gov

Abstract. The Dirac Hamiltonian with relativistic scalar and vector harmonic oscillator potentials has been solved analytically in two limits. One is the spin limit for which spin is an invariant symmetry of the Dirac Hamiltonian and the other is the pseudo-spin limit for which pseudo-spin is an invariant symmetry of the the Dirac Hamiltonian. The spin limit occurs when the scalar potential is equal to the vector potential plus a constant, and the pseudospin limit occurs when the scalar potential is equal in magnitude but opposite in sign to the vector potential plus a constant. Like the non-relativistic harmonic oscillator, each of these limits has a higher symmetry. For example, for the spherically symmetric oscillator, these limits have a U(3) and pseudo-U(3) symmetry respectively. We shall discuss the eigenfunctions and eigenvalues of these two limits and derive the relativistic generators for the U(3) and pseudo-U(3) symmetry. We also argue, that, if an anti-nucleon can be bound in a nucleus, the spectrum will have approximate spin and U(3) symmetry.

1. Introduction

Pseudopsin symmetry is approximately conserved in nuclei [1]. Pseudospin doublets have single nucleon quantum numbers $[n, \ell, \ell + \frac{1}{2}]$ and $[n-1, \ell+2, \ell+\frac{3}{2}]$ where n, ℓ are the radial and orbital quantum numbers, respectively, and the last quantum number in the brackets is the total angular momentum, j. For example, in the usual notation, $(1s_{\frac{1}{2}}, 0d_{\frac{3}{2}}), (1p_{\frac{3}{2}}, 0f_{\frac{5}{2}})$, etc, are pseudospin doublets. These doublets are approximately degenerate in energy. Some beautiful examples have been shown in this conference; for example, in the talk by O. Sorlin [2]. Most of the properties of nuclei are explained by the non-relativistic shell model [3]. However, to explain pseudospin symmetry we need the relativistic Dirac Hamiltonian which has two types of mean fields, a Lorentz scalar $V_S(\vec{r})$ and Lorentz vector $V_V(\vec{r})$. Pseudospin symmetry occurs when the sum of the vector and scalar potentials in the Dirac Hamiltonian is a constant , $V_S(\vec{r})$ + $V_V(\vec{r}) = C_{ps}$ [4]. Hence pseudospin symmetry is a relativistic symmetry. The approximate equality in magnitude of the vector and scalar fields in nuclei and their opposite sign have been confirmed in relativisitc mean field theories [1] and in QCD sum rules [5].

On the other hand hadrons [1] and anti-nucleons in a nuclear environment have spin symmetry [6]. These are relativistic systems and normally, in such systems, we would expect large spin-orbit splittings. However spin symmetry occurs when the difference of the vector and scalar potentials in the Dirac Hamiltonian is a constant [7], $V_S(\vec{r})$ - $V_V(\vec{r}) = C_s$, even though these systems are very relativistic.

The non relativistic harmonic oscillator has been useful in the shell model of nuclei. The relativistic mean field has been important for pseudospin in nuclei and for mesons with one heavy quark [8] and for an anti-nucleon in a nuclear environment [9]. For these reasons the relativistic harmonic oscillator has been solved in the tri-axial, axially deformed and spherical limit [10]. Like the non-relativistic harmonic oscillator, each of these limits has a higher symmetry. For example, for the spherically symmetric oscillator, these limits have a U(3) and pseudo-U(3) symmetry respectively. We shall discuss the eigenfunctions and eigenvalues of these two limits and derive the relativistic generators for the U(3) and pseudo-U(3) symmetry.

2. The Spherically Symmetric Dirac Hamiltonian with Spin Symmetry

The Dirac Hamiltonian for a spherical harmonic oscillator with spin symmetry is

$$H = \vec{\alpha} \cdot \vec{p} + \beta M + (1 + \beta) V(r), \tag{1}$$

where $\vec{\alpha}$, β are the Dirac matrices, $\vec{p}$ is the momentum, M is the mass, $\vec{r}$ is the radial coordinate, r its magnitude, and the velocity of light is set equal to unity, c =1. The generators for the spin SU(2) algebra and the orbital angular momentum SU(2) algebra, $\vec{S}, \vec{L}$, which commute with the Dirac Hamiltonian with any potential $V(r)$, $[H\,, \vec{S}] = [H\,, \vec{L}] = 0$, are given by [11]

$$\vec{S} = \begin{pmatrix} \vec{s} & 0 \\ 0 & U_p\, \vec{s}\, U_p \end{pmatrix}, \; \vec{L} = \begin{pmatrix} \vec{\ell} & 0 \\ 0 & U_p\, \vec{\ell}\, U_p \end{pmatrix}, \tag{2}$$

where $\vec{s} = \vec{\sigma}/2$ are the usual spin generators, $\vec{\sigma}$ the Pauli matrices, $\vec{\ell} = \frac{(\vec{r} \times \vec{p})}{\hbar}$, and $U_p = \frac{\boldsymbol{\vec{\sigma}} \cdot \boldsymbol{\vec{p}}}{p}$ is the helicity unitary operator introduced in [12].

2.1. The Dirac Hamiltonian with Spin Symmetry and Harmonic Oscillator Potential

With the harmonic oscillator potential $V(r) = \frac{M\omega^2}{2}\, r^2$ the eigenvalue equation is [10]

$$\sqrt{E_N + M}\; (E_N - M) = \sqrt{2\, \hbar^2 \omega^2 M}\; (N + \frac{3}{2}) \tag{3}$$

where $N = 2n + \ell$, is the total harmonic oscillator quantum number, n is the radial quantum number and ℓ is the orbital angular momentum. Hence the eigenenergies have the same degeneracies as the non-relativistic harmonic oscillator. This eigenvalue equation is solved with Mathematica,

$$E_N = \frac{M}{3} \left[3B(A_N) + 1 + \frac{4}{3\; B(A_{n_1,n_2,n_3})} \right], \tag{4}$$

where $B(A_N) = \left[\frac{A_N + \sqrt{A_N^2 - \frac{32}{27}}}{2} \right]^{\frac{2}{3}}$, and $A_N = \frac{\sqrt{2}\hbar\omega}{M}(N + \frac{3}{2})$. The spectrum is non-linear in contrast to the non-relativistic harmonic oscillator; i.e., the relativistic harmonic oscillator is not harmonic. However for small A_N (large M), the binding energy,

$$E_N - M \approx M\; (\frac{A_N}{\sqrt{2}} + \cdots) \approx \hbar\omega\; (N + \frac{3}{2}), \tag{5}$$

in agreement with the non-relativistic harmonic oscillator. For large A_N (small M) the spectrum goes as

$$E_N \approx M\; (A_N^{\frac{2}{3}} + \frac{1}{3} + \cdots), \tag{6}$$

which, in lowest order, agrees with the spectrum for $M \rightarrow 0$ [13].

2.2. U(3) Generators

The relativistic energy spectrum has the same degeneracies as the non-relativistic spectrum, even though the dependence on N is different. This suggests that the relativistic harmonic oscillator has a higher U(3) symmetry. The non-relativistic U(3) generators are the orbital angular momentum $\vec{\ell}$, the quadrupole operator $q_m = \frac{1}{\hbar M\omega}\sqrt{\frac{3}{2}}(2M^2\omega^2[rr]^{(2)}_m + [pp]^{(2)}_m)$, where $[rr]^{(2)}_m$ means coupled to angular momentum rank 2 and projection m, and the total oscillator quantum number operator, $\mathcal{N}_{NR} = \frac{1}{2\sqrt{2}\hbar M\omega}(2M^2\omega^2 r^2 + p^2) - \frac{3}{2}$. They form the closed U(3) algebra

$$[\mathcal{N}_{NR}, \vec{\ell}] = [\mathcal{N}_{NR}, q_m] = 0, \tag{7}$$

$$[\vec{\ell}, \vec{\ell}]^{(t)} = -\sqrt{2}\,\vec{\ell}\,\delta_{t,1},\ \ [\vec{\ell}, q]^{(t)} = -\sqrt{6}\,q_m\,\delta_{t,2},\ \ [q, q]^{(t)} = 3\sqrt{10}\,\vec{\ell}\,\delta_{t,1}, \tag{8}$$

with $\mathcal{N}_{NR}$ generating a U(1) algebra whose eigenvalues are the total number of quanta N and $\vec{\ell}, q_m$ generating an SU(3) algebra. In the above we use the coupled commutation relation between two tenors, $T_1^{(t_1)}, T_2^{(t_2)}$ of rank t_1, t_2 which is $[T_1^{(t_1)}, T_2^{(t_2)}]^{(t)} = [T_1^{(t_1)} T_2^{(t_2)}]^{(t)} - (-1)^{t_1+t_2-t}[T_2^{(t_2)} T_1^{(t_1)}]^{(t)}$ [14].

The relativistic orbital angular momentum generators $\vec{L}$ are given in Eq. (2). We shall now determine the the quadrupole operator Q_m and monopole operator $\mathcal{N}$ that commute with the Hamiltonian in Eq.(1). In order for the quadrupole generator

$$Q_m = \begin{pmatrix} (Q_m)_{11} & (Q_m)_{12}\vec{\sigma}\cdot\vec{p} \\ \vec{\sigma}\cdot\vec{p}\,(Q_m)_{21} & \vec{\sigma}\cdot\vec{p}\,(Q_m)_{22}\,\vec{\sigma}\cdot\vec{p} \end{pmatrix}, \tag{9}$$

to commute with the Hamiltonian, $[Q_m, H] = 0$, the matrix elements must satisfy the conditions,

$$(Q_m)_{12} = (Q_m)_{21}, \tag{10}$$

$$2[(Q_m)_{11}, V] + [(Q_m)_{12}, p^2] = 0, \tag{11}$$

$$2[(Q_m)_{12}, V] + [(Q_m)_{22}, p^2] = 0, \tag{12}$$

$$(Q_m)_{11} = (Q_m)_{12}\ 2(V + M) + (Q_m)_{22}\ p^2. \tag{13}$$

One solution is

$$Q_m = \lambda_2 \begin{pmatrix} M\omega^2\ (M\omega^2\ r^2 + 2M)[rr]^{(2)}_m\ + [pp]^{(2)}_m & M\omega^2\ [rr]^{(2)}_m\ \vec{\sigma}\cdot\vec{p} \\ \vec{\sigma}\cdot\vec{p}\ M\omega^2\ [rr]^{(2)}_m & [pp]^{(2)}_m \end{pmatrix}, \tag{14}$$

where λ_2 is an overall constant undetermined by the commutation of Q_m with the Dirac Hamiltonian.

For this quadrupole operator to form a closed algebra, the commutation with itself must be the orbital angular momentum operator as in Eq. (8). This commutation relation gives

$$[Q, Q]^{(t)} = \sqrt{10}\ \lambda_2^2\ M\omega^2\hbar^2 \begin{pmatrix} (M\omega^2\ r^2 + 2M)\ \vec{\ell} & \vec{\ell}\,\vec{\sigma}\cdot\vec{p} \\ \vec{\sigma}\cdot\vec{p}\,\vec{\ell} & 0 \end{pmatrix} = \sqrt{10}\ \lambda_2^2\ M\omega^2\hbar^2(H + M)\ \vec{L}\ \delta_{t,1}, \tag{15}$$

and we get the desired result if $\lambda_2 = \sqrt{\frac{3}{M\omega^2\hbar^2(H+M)}}$. The quadrupole operator then becomes

$$Q_m = \sqrt{\frac{3}{M\omega^2\hbar^2(H + M)}} \begin{pmatrix} M\omega^2\ (M\omega^2\ r^2 + 2M)[rr]^{(2)}_m\ + [pp]^{(2)}_m & M\omega^2\ [rr]^{(2)}_m\ \vec{\sigma}\cdot\vec{p} \\ \vec{\sigma}\cdot\vec{p}\ M\omega^2\ [rr]^{(2)}_m & [pp]^{(2)}_m \end{pmatrix} \tag{16}$$

which can also be written as

$$Q_m = \sqrt{\frac{3}{M\omega^2\hbar^2(H+M)}}\,\left(M\omega^2\,(H+M)\,[rr]^{(2)}_m + [pp]^{(2)}_m\right), \tag{17}$$

which reduces to the non-relativistic quadrupole generator for $H \to M$. In the original paper that derives the quadrupole generators there are two typos. In Eq (6) of that paper [15], $\frac{M\omega^2}{2}\,r^2$ should be replaced by $M\omega^2\,r^2$ and in the non-relativistic quadrupole operator $M^2\omega^2[rr]^{(2)}_m$ should be replaced by $2M^2\omega^2[rr]^{(2)}_m$. Also, the expression for $B(A_N)$ in that paper has a misplaced factor of 2 in the denominator.

For the monopole generator, we can solve the same equations.This has been done [15]. But there is a simpler way. From Eq (3) we get,

$$\mathcal{N} = \frac{\sqrt{H+M}(H-M)}{\hbar\sqrt{2M\omega^2}} - \frac{3}{2}. \tag{18}$$

In the non-relativisitc limit, $H+M \to 2M$ and the non-relativistic Hamiltonian $(H-M) \to \hbar\omega(N+\frac{3}{2})$ which gives the correct result.

The commutation relations are then those of the U(3) algebra,

$$[\mathcal{N},\vec{L}] = [\mathcal{N},Q_m] = 0. \tag{19}$$

$$[\vec{L},\vec{L}]^{(t)} = -\sqrt{2}\,\vec{L}\,\delta_{t,1},\; [\vec{L},Q]^{(t)} = -\sqrt{6}\,Q\,\delta_{t,2},\; [Q,Q]^{(t)} = 3\sqrt{10}\,\vec{L}\,\delta_{t,1}. \tag{20}$$

The spin generators in Eq.(2), $\vec{S}$, commute with the U(3) generators as well as the Dirac Hamiltonian, and so the invariance group is U(3)×SU(2), where the SU(2) is generated by the spin generators, $[\vec{S},\vec{S}]^{(t)} = -\sqrt{2}\,\vec{S}\,\delta_{t,1}$.

3. The Spherically Symmetric Dirac Hamiltonian with Pseudospin Symmetry

The Dirac Hamiltonian with pseudospin symmetry is [4]

$$\tilde{H} = \vec{\alpha}\cdot\vec{p} + \beta M + (1-\beta)V(r), \tag{21}$$

which explains the pseudospin doublets observed in nuclei [1]. This pseudospin Hamiltonian can be obtained from the spin Hamiltonian with a transformation

$$\gamma_5 = \begin{pmatrix} 0 & 1 \\ 1 & 0 \end{pmatrix} \textit{and } M \to -M, \tag{22}$$

which gives the pseudospin and pseudo-orbital angular momentum generators [11]

$$\vec{\tilde{S}} = \begin{pmatrix} U_p\,\vec{s}\,U_p & 0 \\ 0 & \vec{s} \end{pmatrix},\; \vec{\tilde{L}} = \begin{pmatrix} U_p\,\vec{\ell}\,U_p & 0 \\ 0 & \vec{\ell} \end{pmatrix}. \tag{23}$$

3.1. The Dirac Hamiltonian with Pseudospin Symmetry and Harmonic Oscillator Potential

With the harmonic oscillator potential $V(r) = \frac{M\omega^2}{2}\,r^2$ the eigenvalue equation in the pseudospin limit is [10]

$$\sqrt{E_{\tilde{N}} - M}\,(E_{\tilde{N}} + M) = \sqrt{2\,\hbar^2\omega^2 M}\,(\tilde{N} + \frac{3}{2}) \tag{24}$$

where $\tilde{N} = 2\tilde{n} + \tilde{\ell}$, is the pseudo total harmonic oscillator quantum number, $\tilde{n}$ is the pseudo radial quantum number and $\tilde{\ell}$ is the pseudo-orbital angular momentum. While n is the number of radial nodes and ℓ the rank of the spherical harmonic of the upper Dirac radial amplitude, $\tilde{n}$ is the number of radial nodes and $\tilde{\ell}$ the rank of the spherical harmonic of the lower Dirac radial amplitude. Again the eigenenergies have the same degeneracy pattern as the non-relativistic harmonic oscillator in the spin symmetry limit. This eigenvalue equation is solved on Mathematica,

$$E_{\tilde{N}} = \frac{M}{3}\left[3B(A_{\tilde{N}}) - 1 + \frac{4}{3\,B(A_{\tilde{N}})}\right], \tag{25}$$

where $B(A_N) = \left[\frac{A_{\tilde{N}} + \sqrt{A_{\tilde{N}}^2 + \frac{32}{27}}}{2}\right]^{\frac{2}{3}}$, and $A_{\tilde{N}} = \frac{\sqrt{2}\hbar\omega}{M}(\tilde{N} + \frac{3}{2})$. The spectrum is non-linear in contrast to the non-relativistic harmonic oscillator; i.e., the relativistic harmonic oscillator is not harmonic in either limit. Even for small $A_{\tilde{N}}$ (large M), the binding energy,

$$E_{\tilde{N}} - M \approx M(\frac{A_{\tilde{N}}^2}{4} + \cdots), \tag{26}$$

and hence goes quadratically with the the total pseudo-number of quanta and is non-linear even for large M. For large $A_{\tilde{N}}$ (small M) the spectrum goes as

$$E_{\tilde{N}} \approx M\ (A_{\tilde{N}}^{\frac{2}{3}} - \frac{1}{3} + \cdots), \tag{27}$$

which, in lowest order, agrees with the spectrum for spin symmetry.

3.2. Pseudo-U(3) Generators

The pseudo-U(3) generators which commute with the Dirac Hamiltonian, $[\tilde{H}, \vec{\tilde{S}}] = [\tilde{H}, \vec{\tilde{L}}] = [\tilde{H}, \tilde{Q}_m] = [\tilde{H}, \tilde{\mathcal{N}}] = 0$, are then obtained by the transformation in Eq (22) and are given by

$$\tilde{Q}_m = \sqrt{\frac{3}{M\omega^2\hbar^2(\tilde{H} - M)}}\left(M\omega^2\ (\tilde{H} - M)\ [rr]^{(2)}_m + [pp]^{(2)}_m\right) \tag{28}$$

$$\tilde{\mathcal{N}} = \frac{\sqrt{\tilde{H} - M}(\tilde{H} + M)}{\hbar\sqrt{2M\omega^2}} - \frac{3}{2} \tag{29}$$

The commutation relations are then those of the U(3) algebra,

$$[\tilde{\mathcal{N}}, \vec{\tilde{L}}] = [\tilde{\mathcal{N}}, \tilde{Q}_m] = 0. \tag{30}$$

$$[\vec{\tilde{L}}, \vec{\tilde{L}}]^{(t)} = -\sqrt{2}\ \vec{\tilde{L}}\ \delta_{t,1},\ [\vec{\tilde{L}}, \tilde{Q}]^{(t)} = -\sqrt{6}\ \tilde{Q}\ \delta_{t,2},\ [\tilde{Q}, \tilde{Q}]^{(t)} = 3\sqrt{10}\ \vec{\tilde{L}}\ \delta_{t,1}. \tag{31}$$

The pseudospin generators in Eq.(23), $\vec{\tilde{S}}$, commute with the U(3) generators as well as the Dirac Hamiltonian, and so the invariance group is pseudo-U(3)×pseudo-SU(2), where the pseudo-SU(2) is generated by the pseudospin generators, $[\vec{\tilde{S}}, \vec{\tilde{S}}]^{(t)} = -\sqrt{2}\ \vec{\tilde{S}}\ \delta_{t,1}$.

4. The Axially Deformed Dirac Hamiltonian with a Harmonic Oscillator Potential

The relativistic non-spherical harmonic oscillator has also been solved analytically in these two limits [10]. The non-relativistic axially symmetric deformed harmonic oscillator will have a U(2)×U(1) symmetry. Likewise the relativistic axially symmetric deformed harmonic oscillator will have a U(2)×U(1) symmetry in the spin symmetry limit and a pseudo-U(2)× pseudo-U(1) symmetry in the pseudospin limit. This will be discussed in a forthcoming paper.

5. Pseudospin Symmetry Limit and Perturbation Theory

Even though the harmonic oscillator eigenenergies for the pseudospin symmetry limit, Eq(25), are positive, the eigenstates [10] are not those of Dirac valence states but of Dirac hole states. This follows from the fact that the eigenstates have the radial nodal structure of hole states rather than valence states. For example, the state with $\tilde{n} = \tilde{\ell} = 0$, has by definition a lower amplitude with zero radial nodes. This state corresponds to a $1p_{\frac{1}{2}}$ state. Using a theorem which relates the radial nodes of upper and lower amplitudes [16], the lower amplitude of a $1p_{\frac{1}{2}}$ state would have to have one radial node if it were a valence state. Hence this state is a hole state. This means that perturbation theory can not be used to describe nuclei even though nuclei have approximate pseudospin symmetry. In general, the pseudospin limit does not have bound valence states independent of the form of $V(r)$ [16] and hence perturbation theory is not possible. For this reason, the analytical solution of the Dirac Hamiltonian with general scalar and vector harmonic oscillator potentials would be useful.

6. Spin Symmetry Limit and Anti-nucleons

The fact that a nucleon moving in the mean field of a nucleus has a vector and scalar potential almost equal in magnitude but opposite in sign has been substantiated by experimental evidence in nuclei [1]. An anti-nucleon moving in the mean field of a nucleus will then have a vector and scalar potential almost equal in magnitude because charge conjugation leaves the Lorentz scalar field invariant while changing the sign of the Lorentz vector field. Thus pseudospin in nuclei predicts spin symmetry for an anti-nucleon in a nuclear environment [6, 1].

7. Summary

We have shown that a Dirac Hamiltonian with equal scalar and vector harmonic oscillator potentials has an U(3)$\times$ SU(2) symmetry and with scalar and vector harmonic oscillator potentials equal in magnitude but opposite in sign has a pseudo-U(3) $\times$ pseudo-SU(2) symmetry and we have derived the corresponding generators for each case. If speculation that an anti-nucleon can be bound inside a nucleus is valid [9], the anti-nucleon spectrum will have an approximate spin symmetry and, most likely, an approximate U(3) symmetry, because the vector and scalar potentials are approximately equal and very large [1].

Acknowledgments

This research was supported by the US Department of Energy, Contract No. W-7405-ENG-36.

References

[1] Ginocchio J N 2005 *Phys.Rep.* **414** 165
[2] Sorlin O and Porquet M G 2008 *Prog. Part. Nucl. Phys.* **61** 602
[3] Talmi I 1993 *Simple Models of Complex Nuclei*, (Switzerland: Hardwood)
[4] Ginocchio J N 1997 *Phys.Rev. Lett.* **78** 436
[5] Cohen T D, *et al* 1995 *Prog. Part. Nucl. Phys.* **35** 221
[6] Ginocchio J N 1999 *Phys.Rep.* **315** 231
[7] Bell J S and Ruegg H 1997 *Nucl.Phys.* B **98** 151
[8] Page P R, Goldman T and Ginocchio J N 2001 *Phys.Rev. Lett.* **86** 204
[9] Burvenich T, *et al* 2002 *Phys.Lett.* B **542** 261
[10] Ginocchio J N 2004 *Phys.Rev.* C **69** 034318
[11] Ginocchio J N and Leviatan A 1998 *Phys. Lett.* B **425** 1
[12] Blokhin A L, Bahri C and Draayer J P 1995 *Phys.Rev. Lett.* **74** 4149
[13] Bhaduri R K 1988 *Models of the Nucleon: From quarks to Soliton*, (Addison-Wesley)
[14] French J B 1966 *Many-Body Description of Nuclear Structure and Reactions*, ed C. Bloch (New York: Academic Press).
[15] Ginocchio J N 2005 *Phys.Rev.Lett.* **95** 252501
[16] Leviatan A and Ginocchio J N 2001 *Phys.Lett.* B **518** 214

High resolution particle spectroscopy in ^{208}Pb with the Q3D magnetic spectrograph of the Maier-Leibnitz-Laboratorium at München

2011 J. Phys.: Conf. Ser. 267 012038

(http://iopscience.iop.org/1742-6596/267/1/012038)

High resolution particle spectroscopy in ^{208}Pb with the Q3D magnetic spectrograph of the Maier-Leibnitz-Laboratorium at München

A. Heusler

Max-Planck-Institut für Kernphysik, Heidelberg, Germany

E-mail: `A.Heusler@mpi-hd.mpg.de`

Abstract. By using the Q3D magnetic spectrograph of the Maier-Leibnitz-Laboratorium at München, particle-hole states in ^{208}Pb are investigated. The reaction ^{208}Pb(p,p') via seven known isobaric analog resonances in ^{209}Bi and the reaction ^{207}Pb(d,p) were studied. An instrumental resolution of about 3 keV for protons without energy loss in the target is achieved; depending on the effective target thickness the peak shape becomes asymmetric with a tail of 3-8 keV width. A dozen doublets of states at less than 2 keV or even vanishing distance are resolved by the selective excitation in different analog resonances. For excitation energies $3.9 < E_x \lesssim 6.1$ MeV in ^{208}Pb, 80 states with negative parity and 30 states with positive parity are predicted by the shell model. Most states are identified and work is going on to determine their structure.

In the doubly magic nucleus ^{208}Pb the lowest states are expected to be well described by one-particle one-hole configurations with neutron holes for $82 \leq N \leq 126$, neutron particles for $126 \leq N \leq 184$, and proton holes for $50 \leq Z \leq 82$, proton particles for $82 \leq Z \leq 126$. In each group of particles or holes one intruder exists, hence about one third of the states are expected to have positive parity.

The level density is rather high. For the region $4.8 < E_x < 6.1$ MeV, the shell model predicts a mean distance of 15 keV between two states with any spin or parity. Gamma-spectroscopy with a resolution of 2 keV may resolve most states. In particle spectroscopy, the resolution has been improved only slowly, from 15 keV in 1969 to 4 keV ten years ago (see Table and [5]).

Experiments of ^{208}Pb(p,p') and ^{207}Pb(d,p) were performed with the Q3D magnetic spectrograph of the Maier-Leibnitz-Laboratorium at München [3, 4, 6, 7, 8] (Fig. 1). In the reaction ^{208}Pb(p,p'), proton energies $E_p = 14.8 - 18.2$ MeV were chosen to cover the known isobaric analog resonances (IAR) [1]. Targets of $100 - 300\,\mu$g/cm^2 thickness were used. The protons were detected by the Q3D magnetic spectrograph at scattering angles $\Theta = 20^\circ - 140^\circ$. The multiwire detector gathers protons in a solid angle of $3^\circ \times 3^\circ$. A spectrum of 1 MeV length (highly linear in the momentum space) is gathered in typically half an hour with sufficient statistics. The reaction ^{207}Pb(d,p) was performed with $E_d = 22.0$ MeV, too [3, 4].

The line shape is asymmetric (Fig. 2). With thin targets, an instrumental resolution for protons of 3 keV (corresponding to a relative energy resolution of $3 \cdot 10^{-4}$) is achieved. (The target must be correctly placed to avoid the protons crossing the carbon backing.) Because of the energy loss by straggling in the target, an exponential tail appears, linear on the logarithmic scale. The width of the tail depends on the target thickness, the scattering angle, the target

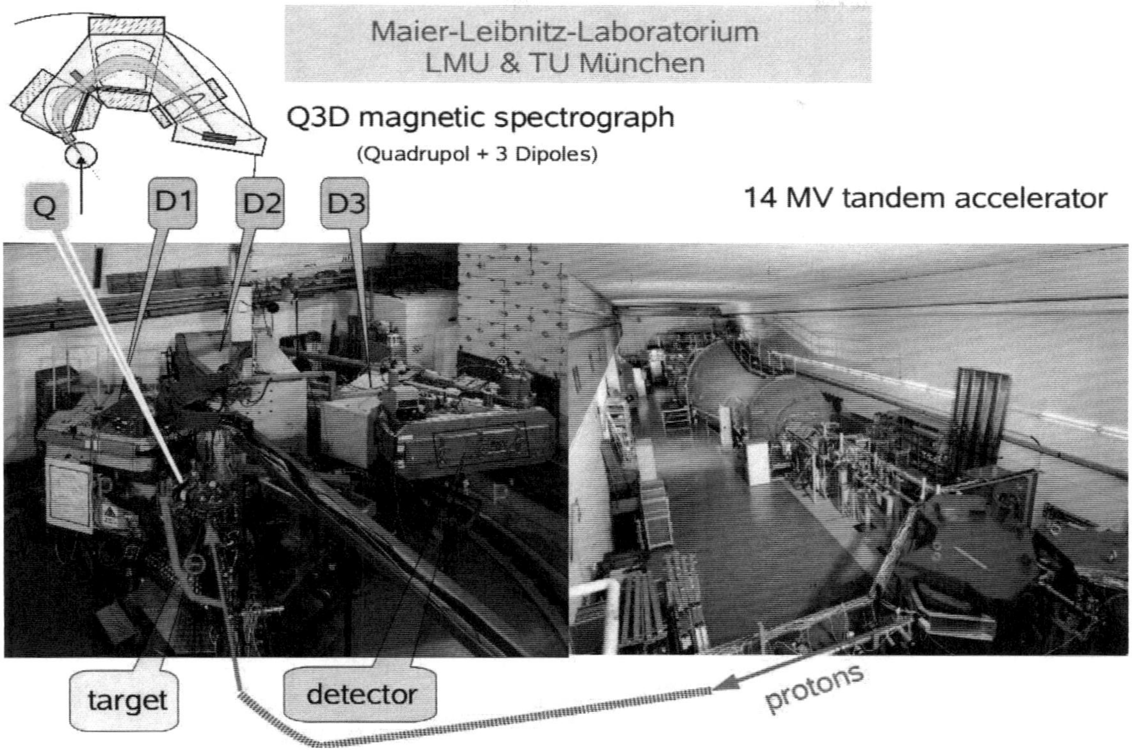

Figure 1. The MP accelerator and the Q3D magnetic spectrograph of the Maier-Leibnitz-Laboratorium at München. With a resolution of down to 3 keV, experiments for the reactions ^{208}Pb(p, p') with proton energies $E_p = 15 - 18$ MeV [3, 6, 7, 8] and ^{207}Pb(d, p) with deuteron energy $E_d = 22$ MeV [3, 4] were performed.

angle, and whether forward or backward scattering is used. In addition, the knockout of $L_{I,II,III}$-electrons with $13 - 15$ keV produces satellites to each peak with intensities in the order of a few percent.

particle-spectroscopy				**resolution**		**gamma-spectroscopy**	
			MeV	keV			
1967	^{208}Pb(p, p')	$E_p =$	$14 - 18$	9	foto-emulsion	**resolution $\approx$ 2 keV**	
1969	^{208}Pb(p, p')	$E_p =$	$14 - 18$	15	semiconductor		
1975	^{208}Pb(p, p')	$E_p =$	35	$5 - 8$	foto-emulsion	1996	^{207}Pb$(d, p\gamma)$
1982	^{209}Bi$(d, {}^3$He$)$	$E_d =$	45	15		1996	^{208}Pb$(p, p'\gamma)$ via IAR
2001	^{208}Pb(p, p')	$E_p =$	22	4		1997	^{209}Bi$(t, \alpha\gamma)$
2001	^{207}Pb(d, p)	$E_d =$	22	4		1997	^{208}Pb$(n, n'\gamma)$

In the following, the inelastic proton scattering on the doubly magic nucleus ^{208}Pb via isobaric analog resonances in ^{209}Bi [1, 3, 7] is shortly explained (Fig. 3). In the doubly magic nucleus ^{208}Pb there are 44 excess neutrons. By adding one neutron a particle state with orbital angular momentum L and spin J in ^{209}Pb is created. The isobaric analog of such a state consists of 45

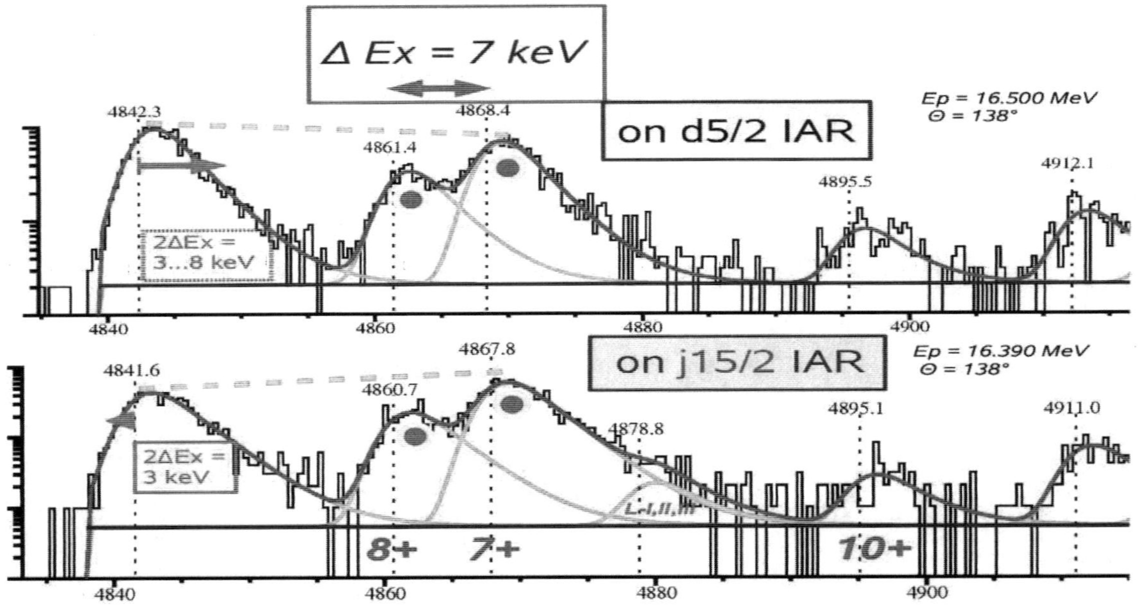

Figure 2. Extract of a ^{208}Pb(p, p') spectrum of 1 MeV length taken on the $j_{15/2}$ and $d_{5/2}$ IAR. The 4861 8^+, 4868 7^+ states with dominant structure $j_{15/2}p_{1/2}$ are resolved. On the logarithmic scale the line shape is asymmetric. *(lower panel, magenta left arrow.)* The instrumental resolution is 3 keV as determined for protons not suffering an energy loss by straggling in the target. Satellites produced by the knockout of $L_{I,II,III}$-electrons are seen near $E_x = 4.88$ MeV. *(upper panel, magenta right arrow.)* The width of the exponential tail depends on the effective target thickness.

components (left part of Fig. 3),

$$|\Psi^{IAR}_{LJ}(^{209}\mathrm{Bi})\rangle = \frac{1}{\sqrt{2T_0+1}} T_- \left(|LJ,\nu\rangle \otimes |^{208}\mathrm{Pb}(0^+\, g.s.)\rangle\right). \tag{1}$$

In each component one excess neutron is converted into a proton,

$$|\Psi^{IAR}_{LJ}(^{209}\mathrm{Bi})\rangle = \frac{1}{\sqrt{2T_0+1}}|LJ,\pi\rangle \otimes |^{208}\mathrm{Pb}(0^+\, g.s.)\rangle +$$

$$\sum_{lj} \sqrt{\frac{2j+1}{2T_0+1}}(|lj^{+1},\pi\rangle \otimes |lj^{-1},\nu\rangle)_{0^+} \otimes |LJ,\nu\rangle \otimes |^{208}\mathrm{Pb}(0^+\, g.s.)\rangle. \tag{2}$$

In the proton decay of such an analog resonance (right part of Fig. 3) either the proton escapes with the unchanged energy (this corresponds to the elastic proton scattering), or ^{208}Pb is left in an excited state and a superposition of up to 30 neutron particle-hole configurations is created. The proton energy can be precisely measured; relative distances between two states can thus be determined with an uncertainty of 100 eV up to excitation energies of 8 MeV.

Inelastic proton scattering via IAR has many advantages.

(i) By adjusting the proton energy to a certain analog resonance, the neutron particle of a particle-hole configuration can be chosen. Doublets may be resolved by selective excitation in different IAR. Examples are the doublets $E_x = 4.71$ and 5.40 [3], 4.93, 5.19 and 5.99 [7], 5.65 [5, 6], 5.69 [6], 6.42 MeV [8].

(ii) The particle of a particle-hole configuration is determined from excitation functions (Fig. 4). Only the $j_{15/2}$ intruder IAR excites positive parity states. Among more than 150 identified states, the parity of 30 states is determined to be positive [7].

(iii) Amplitudes of particle-hole configurations are deduced from angular distributions. The relevant single particle widths can be determined in a self-consistent manner, since on each of the seven IAR, they must be the same for the holes $p_{1/2}$, $p_{3/2}$, $\cdots$.

(iv) If mainly a single configuration contributes to the cross section of some state, the shape of the angular distribution is given by pure geometry and the spin can be determined [3, 6].

Almost all states below $E_x \lesssim 6.1$ MeV and a few states up to $E_x = 7.0$ MeV are identified. For many states spin and parity is determined [3, 6, 7, 8]. Fig. 6 compares the excitation energies of positive parity states below $E_x = 6.0$ MeV to predictions of the shell model without residual interaction [3, 7]; their structure is determined [7]. Similarly, work is going on to determine the structure of negative parity states.

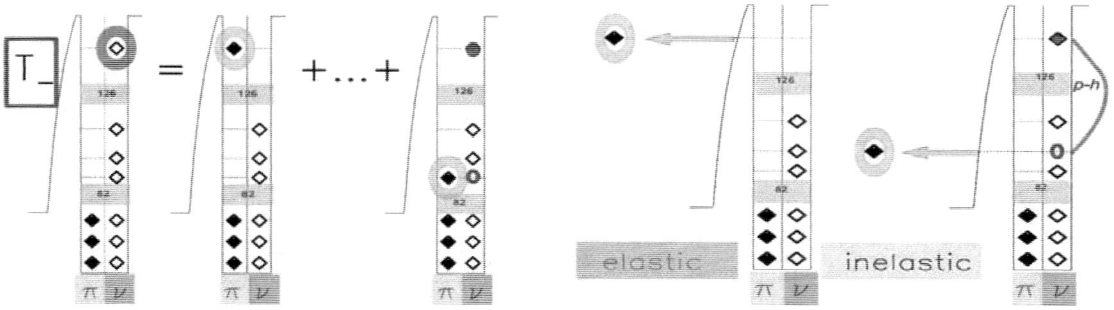

Figure 3. Schematic description of the inelastic proton scattering via isobaric analog resonances in ^{209}Bi. *(left panel)* The creation of an IAR and *(right panel)* the proton decay of an IAR exciting a coherent superposition of up to 30 neutron particle-hole configurations in states of ^{208}Pb.

The goal of the work is to determine the residual interaction from experiment [2, 4]. Within the shell model without residual interaction [3, 7], we consider a set of configurations $|n, I^\pi\rangle$ and a set of physical states $\underline{|n, I^\pi\rangle}$ with spin I and parity π and consecutive number $n = 1, 2, \cdots$. In the shell model the states consist of particle-hole configurations $|k, I^\pi\rangle$ and inversely particle-hole configurations consist of states,

$$\underline{|n, I^\pi\rangle} = \sum_k t_{nk}(I^\pi)|k, I^\pi\rangle, \qquad |k, I^\pi\rangle = \sum_n t^\dagger_{nk}(I^\pi)\underline{|n, I^\pi\rangle}. \tag{3}$$

Here $t_{nk}(I^\pi)$ denotes the unitary transformation matrix of the Hilbert space of physical states $\underline{|n, I^\pi\rangle}$ into the configuration space $|k, I^\pi\rangle$. The transformation matrix is real because of the time reversal invariance of the Hamiltonian. The amplitudes of the transformation matrix t_{nk} can be determined from angular distributions for ^{208}Pb(p, p') via IAR, ^{208}Pb(d, p), and ^{209}Bi$(d, {}^3$He$)$.

The Hamiltonian acting on the the states $\underline{|n, I^\pi\rangle}$ and the Hamiltonian acting on the configurations $|k, I^\pi\rangle$ have the eigenvalues

$$H_{nm}(I^\pi) = \delta_{nm}E_n(I^\pi), \qquad h_{kl}(I^\pi) = \delta_{kl}e_k(I^\pi). \tag{4}$$

The energies E of the physical states are deduced from experiment. By assuming the lowest states in the nuclei ^{209}Bi, ^{209}Pb to be single particle states and the lowest states in the nuclei ^{207}Tl, ^{207}Pb to be single hole states and from the mass differences of the four neighbor nuclei to ^{208}Pb with $A = 208 \pm 1$, the configuration energies e are derived [3, 7].

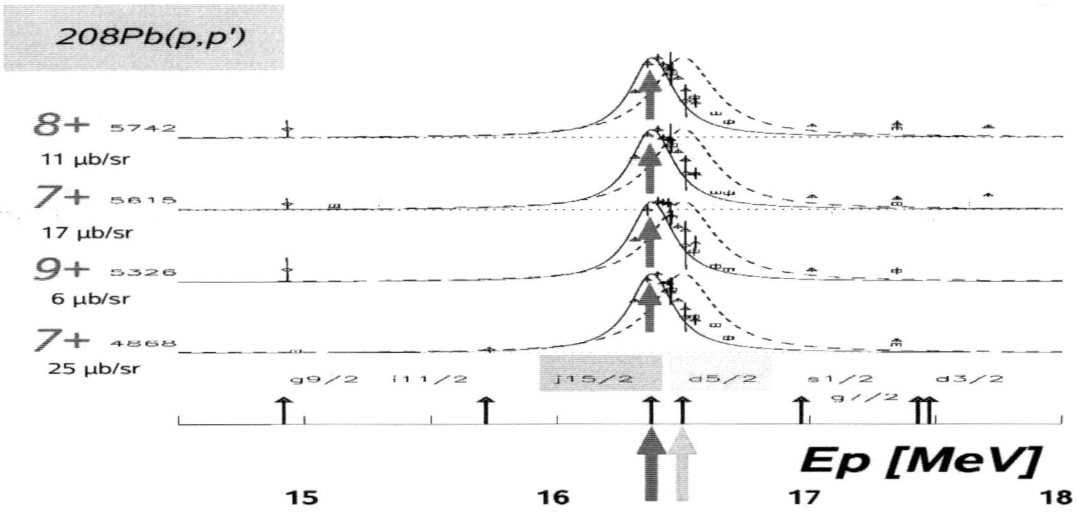

Figure 4. Excitation functions of ^{208}Pb(p, p') via IAR in ^{209}Bi (from [7]). All seven known IAR between $E_p = 14.9 - 17.5$ MeV are covered *(black arrows)*. The width of the $j_{15/2}$ IAR *(red arrow)* is 210 keV, the distance to the $d_{5/2}$ IAR *(green arrow)* is 100 keV. For the four states in ^{208}Pb with the given excitation energies in keV, the particle of the particle-hole configuration is determined to be $j_{15/2}$, hence the parity is positive.

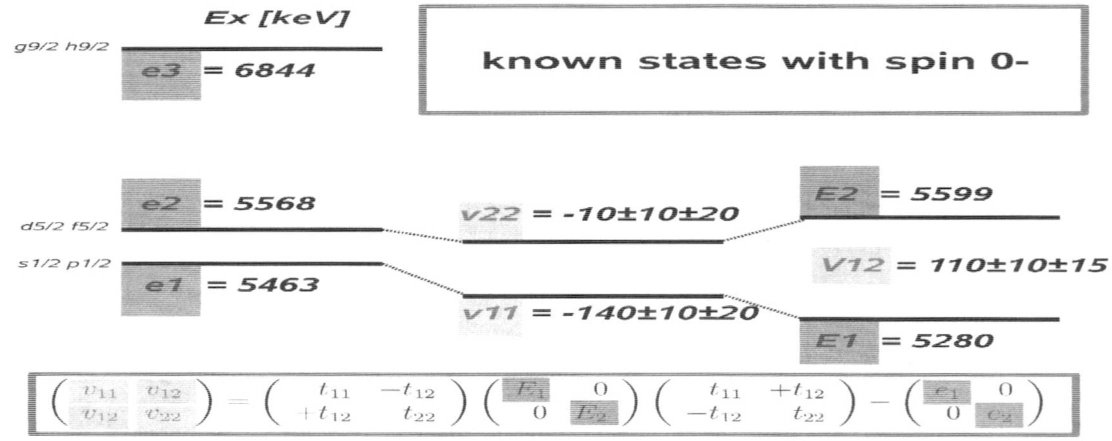

Figure 5. Determination of matrix elements of the residual interaction from experiment [Eq. (5)]. The two 0^- states in ^{208}Pb are the only known bound 0^- states in the whole nuclear chart. They offer an excellent example of a two-level scheme since the next configurations are predicted at a distance of ten times the off-diagonal matrix element determined as $v_{12} = v_{21} = 110 \pm 10$ *(expt.)* ± 15 *(syst.)* keV [4]. The systematic uncertainty of 15 keV derives from the estimate of the deviation of the matrix t_{nk} [Eq. (3)] from unitarity by comparing admixtures of $s_{1/2}p_{1/2}$ in the two 0^- states to nine 1^- states.

The two Hamiltionians H, h differ by the residual interaction v,

$$v_{kl}(I^\pi) = \sum_{nm} t_{kn}(I^\pi) H_{nm}(I^\pi) t^\dagger_{ml}(I^\pi) - h_{kl}(I^\pi). \tag{5}$$

Similarly as done for the two 0^- states [4] (Fig. 5), for all states below $E_x \lesssim 6.1$ MeV with spins $1^- - 8^-$ and $5^+ - 10^+$, the amplitudes of the transformation matrix [Eq. (3)] are being derived from experimental data. Most states (see Fig. 6 for positive parity states) are identified and their dominant configurations are already determined. Some doublets have still to be discussed and some spin assignments are not yet settled.

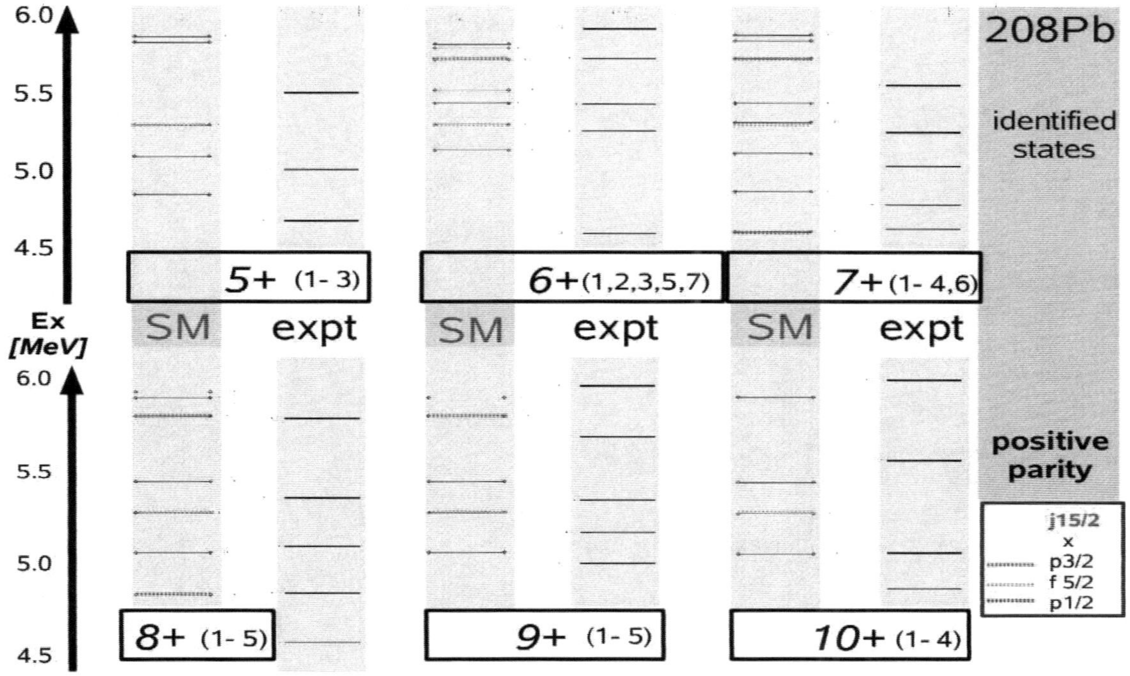

Figure 6. For spins 5^+, 6^+, 7^+, 8^+, 9^+, 10^+ and $E_x < 6.0$ MeV, the excitation energies E_x of states in ^{208}Pb predicted by the shell model without residual interaction [3, 7] are shown (marked blue). Almost all predicted states are identified [7] (marked magenta). The configuration numbers correspond to the shell model predictions [Eq. (3)]. The configurations $j_{15/2}p_{1/2}$, $j_{15/2}f_{5/2}$, $j_{15/2}p_{3/2}$ are marked by colors.

Acknowledgments

The experiments were performed with the assistance of G. Graw, R. Hertenberger, F. Riess, H.-F. Wirth, T. Faestermann, R. Krücken, T. Behrens, V. Bildstein, K. Eppinger, C. Herlitzius, O. Lepyoshkina, M. Mahgoub, A. Parikh, S. Schwertel, K. Wimmer, J. Jolie, D. Mücher, N. Pietralla, C. Scholl, P. von Brentano.

References

[1] Moore C F, Kulleck J G, von Brentano P and Rickey F 1967 *Phys. Rev.* **164** 1559 – 66
[2] Heusler A and von Brentano P 1973 *Ann. Phys. (NY)* **75** 381 – 404
[3] Heusler A et al 2006 *Phys. Rev. C* **74** 034303-1 – 03-15
[4] Heusler A et al 2007 *Phys. Rev. C* **75** 024312-1 – 12-8
[5] Martin M J 2007 *Nucl. Data Sheets* **108** 1583 – 1806
[6] Heusler A et al 2010 *Eur. Phys. J. A* **44** 233 – 8
[7] Heusler A et al 2010 *Phys. Rev. C* **82** 014316-1 – 16-18
[8] Heusler A et al 2010 *Eur. Phys. J. A* http://dx.doi.org/10.1140/epja/i2010-11019-8

Tensor effects in shell evolution using non-relativistic and relativistic mean field

Tensor effects in shell evolution using non-relativistic and relativistic mean field

M. Moreno-Torres[1], M. Grasso[2], H. Liang[2,3], V. De Donno [4], M. Anguiano[1] and N. Van Giai[2]

[1] Departamento de Física Atómica, Molecular y Nuclear, Universidad de Granada, E-18071 Granada, Spain

[2] Institut de Physique Nucléaire, IN2P3-CNRS, Université Paris-Sud, F-91406 Orsay Cedex, France

[3] State Key Laboratory of Nuclear Physics and Technology, School of Physics, Peking University, Beijing 100871, People's Republic of China

[4] Dipartimento di Fisica, Università del Salento and Istituto Nazionale di Fisica Nucleare sez. di Lecce, I-73100 Lecce, Italy

E-mail: `grasso@ipno.in2p3.fr`

Abstract. Tensor effects in shell evolution are studied within the mean-field approach. Particular attention is paid to the analysis of the proton magic gap $Z = 8$. Hartree-Fock calculations with Skyrme and Gogny interactions are performed where the tensor term has a zero and finite range, respectively. Results obtained with and without the tensor component are compared between them and with the experimental data. To complete this analysis, the tensor effect is also investigated within the relativistic Hartree-Fock model.

It turns out that the tensor effect can be easily identified in the evolution of the proton magic gap $Z = 8$. We suggest to explicitly include the data associated to this gap evolution as new constraints in the fitting procedures when the tensor contribution is taken into account at the mean-field level.

1. Introduction

The tensor component is a non-central contribution of the nucleon-nucleon interaction which has been extensively discussed in the last years. It is reasonable to expect that some nuclear observables are strongly sensitive to the tensor force owing to the rich spin-isospin structure of finite nuclei. For instance, the shell evolution and the magicity modification far from stability are often interpreted in terms of tensor effects [1]. In the framework of the shell model, the tensor contribution in shell evolution has been explored by Otsuka and collaborators [2-4]. Within the mean-field models, the study of the nuclear structure at the ground-state level is usually done either by using effective nuclear interactions (non-relativistic framework), like the zero-range Skyrme [5-7] or the finite range Gogny [8,9] forces, or by employing covariant effective Lagrangians (relativistic framework). In general, in all these models, the tensor component has been systematically neglected and considered only in very recent investigations [10-20].

In this work, we discuss how the tensor component contributes to the shell evolution at the mean-field level, using non relativistic Skyrme- and Gogny-Hartree-Fock (HF) models as well as relativistic HF (RHF) models. In particular, we are interested in the behavior of the proton gap at the magic number $Z = 8$. We are mainly interested in the effects due to the neutron-proton interaction related to the tensor contribution: i.e., the effects on proton levels due to the filling of neutron orbits. The theoretical gaps obtained with and without the tensor contribution are compared with the experimental ones to evidentiate the cases where the tensor effects are unambigously important in determining the shell evolution.

The article is organized as follows. In Sec. 2 some formal aspects related to the tensor component are briefly recalled for Gogny (2.1), Skyrme (2.2) and RHF (2.3). Results are illustrated in Sec. 3. Conclusions are drawn in Sec. 4.

2. Models

We work in spherical symmetry and consider only even nuclei in all the adopted models.

2.1. Tensor force in the Gogny case

We use the GT2 parametrization [10] that has been introduced by adding to the standard Gogny interaction a term depending on the tensor force only in the isospin-isospin channel. Thus, the new channel that is included in the effective interaction GT2 is:

$$v_{\mathrm{T}} = F_{\mathrm{T}}\, \tau_1 \cdot \tau_2\, 3 \left(\frac{(\sigma_1 \cdot r_{12})(\sigma_2 \cdot r_{12})}{(r_{12})^2} - \sigma_1 \cdot \sigma_2 \right) f_{\mathrm{G}}(r), \tag{1}$$

where $F_{\mathrm{T}} = 50.79506$ MeV and $f_{\mathrm{G}}(r)$ is a Gaussian function, with a range of 1.2 fm. The parameter F_{T} has been fixed assuming that the volume integral reproduces that of the AV8'. For simplicity, the range of $f_{\mathrm{G}}(r)$ has been chosen equal to the longest range of the central part.

2.2. Tensor force in the Skyrme case

The tensor interaction was already written as a zero-range force in the original articles of Skyrme [5,6]. It modifies the spin-orbit potential by

$$\Delta W_{\mathrm{so}}^{(q)} = \alpha J_q + \beta J_{q'}, \tag{2}$$

where q (q') indicates protons (neutrons) or neutrons (protons) and J_q is the spin-orbit density:

$$J_q(r) = \frac{1}{4\pi r^3} \sum_i (2j_i + 1) \left[j_i(j_i + 1) - l_i(l_i + 1) - \frac{3}{4} \right] R_i^2(r). \tag{3}$$

In the above expression, R_i are the radial wave functions. The parameters α and β are expressed as $\alpha = \alpha_{\mathrm{C}} + \alpha_{\mathrm{T}}$ and $\beta = \beta_{\mathrm{C}} + \beta_{\mathrm{T}}$, where α_{C} and β_{C} are related to the central exchange part of the interaction. They can be written in terms of the usual parameters of the Skyrme interaction; α_{T} and β_{T} are related to the tensor part. If $\beta > 0$, when neutrons (protons) fill a $j_> = l + 1/2$ orbital, the proton (neutron) spin-orbit splitting is reduced because the neutron (proton) spin-orbit density constructed with the $j_>$ orbital is positive. This mechanism reproduces the picture of Otsuka and collaborators [3] that implies an attractive (repulsive) neutron-proton interaction between $j_>$ and $j'_<$ ($j'_>$). The mechanism would be the opposite if $\beta < 0$. In this work, we use as an illustration the values of α_{T} and β_{T} fitted in Ref. [11] on top of the SLy5 parametrization which already contains the central exchange part, i.e., $(\alpha_{\mathrm{T}}, \beta_{\mathrm{T}}) = (-170, 100)$ MeV fm^5. With the parametrization SLy5, α_{C} and β_{C} are equal to 80.2 and -48.87 MeV fm^5, respectively (this means that $\beta = \beta_{\mathrm{C}} + \beta_{\mathrm{T}} > 0$ in this case).

2.3. Tensor in RHF

In the relativistic framework, the tensor correlations are deduced from the pion-nucleon and ρ-nucleon tensor interactions. The part of the Lagrangian containing the pion and ρ-meson fields is written as [20,21]

$$\begin{aligned}\mathcal{L}_{\pi+\rho} &= +\frac{1}{2}\partial_\mu\vec{\pi}\cdot\partial^\mu\vec{\pi} - \frac{1}{2}m_\pi^2\vec{\pi}\cdot\vec{\pi} - \frac{f_\pi}{m_\pi}\bar{\psi}\gamma_5\gamma^\mu\partial_\mu\vec{\pi}\cdot\vec{\tau}\psi \\ &\quad -\frac{1}{4}\vec{R}^{\mu\nu}\cdot\vec{R}_{\mu\nu} + \frac{1}{2}m_\rho^2\vec{\rho}^\mu\cdot\vec{\rho}_\mu - g_\rho\bar{\psi}\gamma^\mu\vec{\rho}_\mu\cdot\vec{\tau}\psi \\ &\quad +\frac{f_\rho}{2M}\bar{\psi}\sigma_{\mu\nu}\partial^\nu\vec{\rho}^\mu\cdot\vec{\tau}\psi, \end{aligned} \tag{4}$$

where $\vec{R}^{\mu\nu} \equiv \partial^\mu\vec{\rho}^\nu - \partial^\nu\vec{\rho}^\mu$. In the parametrization PKA1 [20], the coupling strengths f_π, g_ρ, and f_ρ are exponentially density-dependent.

In principle, we should perform similar calculations as in the Skyrme and Gogny cases where we switch off the tensor terms without changing the rest of the force. However, using the relativistic Lagrangian, it is difficult to isolate the tensor contributions from the central ones. Setting $f_\pi = f_\rho = 0$ would lead to huge changes also in the central part of the mean field and in most cases the mean-field calculation would not even converge. To compare with results without tensor, we thus perform RMF calculations with DDME2 [22] where the pion and ρ-tensor interactions are not included. The results obtained with these two parametrizations do not coincide for the spin saturated nuclei because the two relativistic models PKA1 and DDME2 differ not only by the tensor couplings.

3. Results

The theoretical gaps are calculated as differences of HF single-particle energies. For the experimental data, we have used an approximation (the same procedure is also adopted in Ref. [1]). For the proton gap at Z_{magic}, we have calculated the single-particle energies of the last occupied (below) and the first unoccupied (above) orbit, ϵ_b and ϵ_a, as:

$$\begin{aligned}\epsilon_b(Z_{\text{magic}},N) &= -(E(Z_{\text{magic}},N) - E(Z_{\text{magic}}-1,N)) \\ &= -S_p(Z_{\text{magic}},N) \\ \epsilon_a(Z_{\text{magic}},N) &= -(E(Z_{\text{magic}}+1,N) - E(Z_{\text{magic}},N)) \\ &= -S_p(Z_{\text{magic}}+1,N), \end{aligned} \tag{5}$$

where E are the binding energies and S_p the proton separation energies. The energy of the gap is then evaluated as $\epsilon_a - \epsilon_b$. All the experimental values are taken from Ref. [23]. The separation energies are supposed to be very similar to the single-particle energies if one assumes that the core Z_{magic} remains almost unchanged when one nucleon is added to or removed from the magic core. This assumption is of course more adapted for treating spherical nuclei rather than deformed ones where the presence of strong correlations may deteriorate the quality of this schematic picture. We stress that we consider here the experimental values just as indications to provide qualitative (and not precise) experimental trends to compare with the theoretical results.

We calculate the difference of the HF single-particle energies between the first empty level (proton $1d_{5/2}$) and the last occupied one (proton $1p_{1/2}$). It is worth noting that $d_{5/2}$ is the $j_> = l+1/2$ orbit of the d spin-orbit pair while $p_{1/2}$ is the $j_< = l-1/2$ orbit of the p spin-orbit partners. The two orbitals will be thus affected in the opposite way by the tensor force. Going from ^{16}O to ^{22}O, the neutron $d_{5/2}$ orbital ($j'_>$) is filled. Owing to this, one expects that the energy of the proton $p_{1/2}$ state is lowered whereas that of the proton $d_{5/2}$ one is pushed up by the tensor force. The values of the gaps obtained for ^{16}O and ^{22}O are shown in the three panels of Fig. 1 for the Gogny (a), Skyrme (b) and relativistic (c) cases, respectively. Results obtained with the tensor contribution (full lines) are compared with the corresponding results obtained without tensor (dotted lines) and with the experimental values (dashed lines). In this figure (and in the following ones) the lines are plotted only to guide the eye and see more clearly the trend from one nucleus to the other.

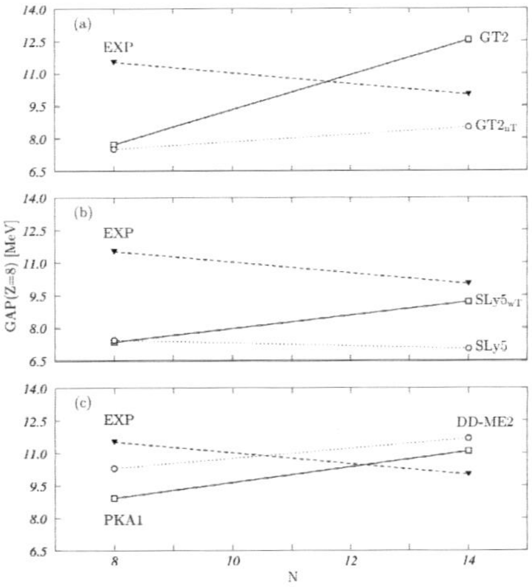

Figure 1. Proton gap $Z = 8$ with Gogny (a), Skyrme (b) and relativistic calculations (c). In $GT2_{nT}$ the tensor part of GT2 has been suppressed. In $SLy5_{wT}$ the tensor part has been added on top of the SLy5 parametrization.

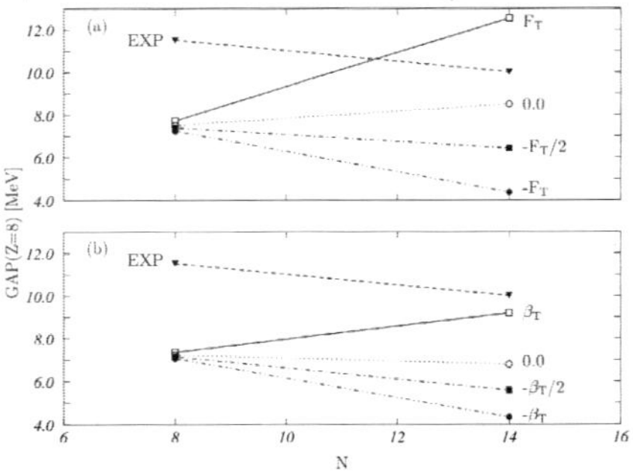

Figure 2. (a) Proton gap $Z = 8$ with GT2 for different choices of the parameter F_T. The current choice is indicated as F_T in the figure. The other choices have been drawn relative to it. (b) Proton gap $Z = 8$ with Skyrme SLy5 for different choices of the parameter β_T. The current choice is indicated as β_T in the figure. The other choices have been expressed in terms of it. The parameter α_T is kept equal to -170 MeV fm^5 in all cases.

Going from ^{16}O to ^{22}O, the tensor produces, as expected, an increase of the gap energy of 4 MeV approximately for the GT2 interaction and 2 MeV for the Skyrme and the PKA1-RHF cases (one observes that the gap increases more strongly with PKA1 than with DDME2 going from ^{16}O to ^{22}O). In the relativistic case, it is important to notice that the two calculations with and without tensor are not based on the same Lagrangian. This explains why the results obtained with DDME2 (no tensor) and PKA1 (with tensor) do not coincide for the spin saturated nucleus ^{16}O. The effect of the tensor force is very clear in all the calculations: the slope is always increased by tensor going from ^{16}O to ^{22}O and the gap $Z = 8$ is more strongly enhanced with respect to the results without tensor. This is coherent with what expected according to the mechanism described in Ref. [3]. However, the experimental gap decreases when going from ^{16}O to ^{22}O: the magicity at $Z = 8$ is reduced when one goes towards the neutron drip line. This means that the tensor contribution acts in the opposite way with respect to what would be necessary to reproduce the experimental trend.

To analyze this more in detail, we consider as an illustration the case of Skyrme: it has been underlined that the predicted trend depends on the parameters, namely on the sign of the parameter β which is positive in the present calculations ($\beta = \beta_T + \beta_C = 51.13$ MeV fm^5). Since oxygen isotopes are spin saturated for protons, the tensor effect is practically due only to the filling of neutron orbitals and is only sensitive to the parameter β. A value of $\beta < 0$ would trivially lead to a decreasing of the gap from ^{16}O to ^{22}O: we plot in Fig. 2 (b) the gap $Z = 8$ for the nuclei ^{16}O and ^{22}O for different values of β while α is kept fixed at the value $\alpha_T + \alpha_C$ = -89.8 MeV fm^5. For β_T = 0, -50 and -100 MeV fm^5, β is negative and the slope for the gap can be modified. For Gogny, the tensor parameter (and its sign) is fixed on a realistic case. However, if F_T in Eq. (1) is treated as a free parameter one can obtain similar results as in the Skyrme case by changing its sign (see Fig. 2(a)).

In the relativistic case the scenario is more complicated and the tensor contribution cannot be easily isolated.

^{16}O is a $Z = N$ nucleus. To extract the experimental single-particle energies, we have used an approximate procedure where masses of nuclei are employed. This is of course an approximation which is valid in nuclei where correlations are quite weak, as it has been already emphasized. Magic nuclei are thus expected to be typically good systems where this approximation can be used. However, the case of $N = Z$ nuclei deserves a special attention. In these nuclei, the last orbitals for protons and neutrons are quite similar and this feature strongly favors correlations associated to neutron-proton pairing. The problem of extracting single-particle energies from the experimental masses in this case is thus very crucial. The so-called Wigner energy is used in many cases to actually correct these values, even if a consensus on this specific point and on the procedures to use is still missing. Several procedures have been adopted (see, e.g., Ref.[24]). Since the problem is still open, we have decided not to include this correction in the previous figures but we evaluate here the importance of this eventual correction on the gap for ^{16}O. By using as an illustration one of the prescriptions introduced in Ref. [25], we have estimated a correction of -3.7 MeV for ^{16}O. Fig. 3 shows the same quantities as Fig. 1 but the experimental data corrected by the Wigner energy are also plotted. This figure gives an idea of how much delicate is actually this point. The experimental slope is strongly changed if the correction is applied. Before making any fit on experimental values in these region, one should thus finally clarify this point on the way of extracting the experimental energies.

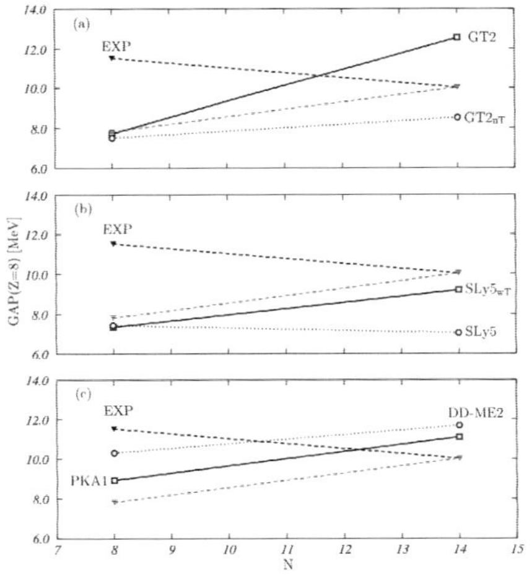

Figure 3. The same as Fig. 1, but including the Wigner correction for the experimental data (red dashed line).

4. Conclusions

We have analyzed the effects of the tensor force in the shell evolution at the magic number $Z = 8$ within non-relativistic and relativistic mean-field approaches. A more systematic analysis extended also at the cases Z = 20 and 28 and N = 8, 20 and 28 is done in Ref. [26].

The experimental values have been estimated by using the same approximated procedure as that of Ref. [1]. The issue of the Wigner correction has been discussed. We have neglected pairing correlations in our analysis and performed only Hartree-Fock calculations for all nuclei, both closed and open-shell. The main reason for this is that the GT2 parametrization which we adopt in the Gogny case has been fitted at the Hartree-Fock level. It is also important mentioning that our mean field predictions for the gaps are expected to be modified by beyond-mean-field effects which we are not taking into account, e.g., the particle-vibration coupling.

The considered nuclei are spin saturated for protons and the only tensor contribution comes from the filling of neutrons (neutron-proton interaction). The tensor effect is easily identified in the evolution of the $Z = 8$ magic gap. In the two non relativistic cases, we have observed that it is possible to modify the theoretical trends by changing the signs of the parameter β in the Skyrme case and F_{T} in the Gogny case. This means that, in principle, a fitting procedure including the experimental gap evolution at the magic number 8 could generate some sets of parameters able to reproduce these experimental trends owing to the tensor contribution.

To conclude, since the evolution of the gaps is an important feature characterizing exotic nuclei we strongly suggest that the observables related to this feature should be included in the fit procedures when the tensor terms have to be constrained. It is also important to properly choose the regions where to perform these fits and we recommend $Z = 8$ as a suitable region where the role played by the tensor force can be identified and is not mixed with other mean-field effects.

Acknowledgments

We thank G. Co' and A.M. Lallena for useful discussions. M.M.T aknowledges support by the Spanish Ministerio de Ciencia e Innovación (contract FPA2009-14091-C02-02) and by the Junta de Andalucía (FQM0220).

References

[1] Sorlin O and Porquet M.G. 2008 *Prog. Part. Nucl. Phys.* **61** 602
[2] Otsuka T et al. 2001 *Phys. Rev. Lett.* **87** 082502
[3] Otsuka T et al. 2005 *Phys. Rev. Lett.* **95** 232502
[4] Otsuka T et al. 2010 *Phys. Rev. Lett.* **104** 012501
[5] Skyrme T H R 1956 *Phil. Mag.* **1** 1043
[6] Skyrme T H R 1959 *Nucl. Phys.* **9** 615
[7] Vautherin D and Brink D M 1972 *Phys. Rev.* C **5** 626
[8] Dechargé J and Gogny D 1980 *Phys. Rev.* C **21** 1568
[9] Berger J F, Girod M and Gogny D 1991 *Comput. Phys. Commun.* **63** 365
[10] Otsuka T et al. 2006 *Phys. Rev. Lett.* **97** 162501
[11] Colò G et al. 2007 *Phys. Lett.* B **646** 227
[12] Brink D M and Stancu F 2007 *Phys. Rev.* C **75** 064311
[13] Grasso M et al 2007 *Phys. Rev.* C **76** 044319
[14] Zou W et al 2008 *Phys. Rev.* C **77** 014314
[15] Brown B A et al. 2006 *Phys. Rev.* C **74** 061303(R)
[16] Lesinski T et al 2007 *Phys. Rev.* C **76** 014312
[17] Bender M et al., 2009 *Phys. Rev.* C **80** 064302
[18] Long W H et al., 2006 *Phys. Lett.* B **640** 150
[19] Long W H et al. 2008 *Europhys. Lett.* **82** 12001
[20] Long W H et al. 2007 *Phys. Rev.* C **76** 034314
[21] Bouyssy A et al. 1987 *Phis. Rev.* C **36** 380
[22] Lalazissis D V G A et al. 2005 *Phys. Rev.* C **71** 024312
[23] Audi G et al. 2003 *Nucl. Phys.* A **729** 337
[24] Chasman R T 2007 *Phys. Rev. Lett.* **99** 082501
[25] Goriely S et al. 2002 *Phys. Rev.* C **66** 024326
[26] Moreno-Torres M et al. 2010 *Phys. Rev.* C **81** 064327

New phenomenon in exotic neutron-rich Sn isotopes: role of 3-body force

S. Sarkar[1] **and M. Saha Sarkar**[2]

[1]Department of Physics, Bengal Engineering and Science University, Shibpur, Howrah - 711103, INDIA
[2] Nuclear Physics Division, Saha Institute of Nuclear Physics, Kolkata 700064, INDIA

E-mail: `ss@physics.becs.ac.in; maitrayee.sahasarkar@saha.ac.in`

Abstract. The excitation energies of the first 2^+ states ($E(2_1^+)$) of even-even $^{134-140}$Sn have been calculated using shell model with empirical SMPN and realistic CWG interactions in the $\pi(gdsh) \times \nu(hfpi)$ valence space above ^{132}Sn core. The CWG predicts nearly constant $E(2_1^+)$ for $^{134-138}$Sn isotopes, normally expected for singly-magic nuclei, and a weak shell closure at ^{140}Sn. The prediction of SMPN differs dramatically, decreasing $E(2_1^+)$ with increasing valence neutron number for $^{134-138}$Sn and a strong shell closure at ^{140}Sn. This variation predicted by SMPN has striking similarity with the variation of experimental $E(2_1^+)$ of even - even $^{18-22}O$ and $^{42-48}Ca$, clearly showing that the N =84-88 spectra exhibit the effect of gradually filling up the $\nu(2f_{7/2})$ orbital which finally culminates in a new shell closure at ^{140}Sn (N= 90). Spin-tensor decomposition of SMPN and CWG and the variation of their components with valence neutron number reveals that the origin of the decreasing $E(2_1^+)$ and shell closure at ^{140}Sn might lie in the three-body effects. Calculations with CWG3M interaction, which is obtained by including a simple 3-body monopole term in the CWG predict decreasing $E(2_1^+)$ for $^{134-138}$Sn and a shell closure at ^{140}Sn.

1. Introduction

Several experimental and theoretical investigations in the close-to-neutron-drip-line region above the doubly magic ^{132}Sn nucleus have recently revealed many intriguing issues concerning some newer aspects of nuclear structure and N-N interaction in such exotic environments. As the nuclei with $50 \leq Z \leq 56$ and $82 \leq N \leq 88$ lie on or close to the path of the astrophysical r-process flow, extensive knowledge about their structure, particularly the binding energies, low-lying excited states, and beta - decay rates of these nuclei at finite temperatures, are important ingredients for astrophysical calculations. Sn isotopes are of particular importance in this respect. Even Sn isotopes, *viz.* ^{136}Sn, is known to be the classical "waiting point" nucleus in A=130 solar system abundance peak under typical r-process condition [1]. However, many of these important nuclei are yet to be studied experimentally. Some spectroscopic information exists only for ^{134}Sn. Recently, the beta -decay half-lives via delayed neutron emission have been measured for ^{135}Sn to ^{138}Sn nuclei. Their small production rates and lifetimes are the main causes of the severe limitations in acquiring spectroscopic information on them. Therefore, reliable theoretical calculations and their comparisons with the hitherto available experimental data are not only important for extending our knowledge of N-N interaction and nuclear structure in the exotic neutron-rich environment, but they also provide very useful ingredients for reviewing the problems related to the nuclear astrophysics in general.

Until recently the nuclear magic numbers at N, Z = 2, 8, 20, 28, 50, 82 and 126 were considered to be valid throughout the entire nuclear chart. However, in the last two decades advent of radioactive ion beams and advancement of detection techniques have made it possible to study the evolution of the shells far from stability. New results for light nuclei showed some unexpected changes in the nuclear shell structure as a function of proton and neutron numbers. At least four doubly magic oxygen isotopes have been observed [2, 3, 4]. They are ^{14}O, ^{16}O, ^{22}O and ^{24}O. Brown and Richter [4] have framed a generalised new rule for magic numbers, valid specially for lighter nuclei. Several theoretical investigations have been pursued to interpret these observations. New predictions [5] were made that some magic numbers will disappear and new shell gaps will appear in certain regions of the nuclear chart.

In this background, we present in this theoretical work, some new features and shell evolution in the neutron-rich semi - magic Sn isotopes above the doubly magic ^{132}Sn core. Our analysis of the Hamiltonians used for studying nuclear structure in the shell model formalism also points to the fact that the 3-body force might have a role in the shell evolution in this region too.

2. Formalism: model space and modified Hamiltonian

We have generated a new empirical interaction SMPN [6] by modifying CW5082 [7, 8] nuclear Hamiltonian derived from Kuo-Herling interaction of the ^{208}Pb region in the light of the recent data for this region. We assume ^{132}Sn as the inert core. The valence space consists of $\pi(1g_{7/2}, 2d_{5/2}, 2d_{3/2}, 3s_{1/2}, 1h_{11/2})$ and $\nu(1h_{9/2}, 2f_{7/2}, 2f_{5/2}, 3p_{3/2}, 3p_{1/2}, 1i_{13/2})$ orbitals. Few-valence-particle nuclei above the strong shell closure at ^{132}Sn furnish very useful information on the N-N interaction in the n-rich environment. We have used available recent experimental information on the binding energy of ^{132}Sn and binding energies and level spectra of only A=134 isobars of Sn, Sb, Te. These A=134 isobars of Sn, Sb and Te provide information on the nn, np and pp two - body matrix elements, respectively. Motivation for construction of this new empirical Hamiltonian, SMPN, and the details of changes made are given in detail in Ref. [6] and the changed tbmes have been tabulated in Ref. [9].

In the present work, CWG, the realistic interaction obtained starting with a G matrix derived from the CD-Bonn nucleon-nucleon potential using the Q-box method [10] and the empirical SMPN [6] interactions have been used. The Hamiltonians have the same set of single-particle energies [11] of the valence orbitals but different sets of two-body interaction matrix elements (tbmes). Both these interactions predict excitation spectra of neutron - rich nuclei like ^{138}Te remarkably well [11] along with those of other A=138 isobars like ^{138}Xe and ^{138}Ba, for which experimental data are available. The shell model codes OXBASH and NUSHELL@MSU have been used [12] in the present work.

3. Results and Discussions

3.1. $^{136,138}Sn$ and a probable new feature above the ^{132}Sn core

The binding energies predicted by these two calculations for the Sn isotopes till N=88 are compared in Ref. [6, 11]. It is remarkable that for $^{136,138}Sn$, where the experimental level schemes are not known, the excitation spectra predicted by two interactions differ dramatically [11]. Calculated energies for the first 2^+ levels [E(2_1^+)] in $^{136,138}Sn$, using SMPN, fit nicely to the respective systematics of isotones for N >82 (Fig.1). The E(2_1^+) values calculated for Sn isotopes above N = 82 decreases with increasing neutron number. But predicted E(2_1^+) values from the calculations using the CWG Hamiltonian deviate sharply from the systematics of isotones for N >82. For CWG, the E(2_1^+) excitation energies for all the even-even Sn isotopes with N > 84 are almost constant, normally expected for singly - magic nuclei, similar to those below N = 82, for Sn isotopes with A=102-130. The only difference is that the bunching of E(2_1^+) values now occurs at around 750 keV for N≥84 instead of 1200 keV for N <82. The 2_1^+, 4_1^+, and 6_1^+ states of

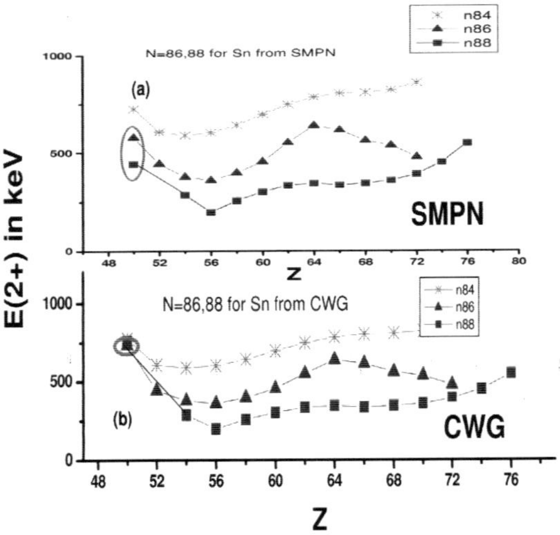

Figure 1. The experimental E(2_1^+) energies for N=84, 86 and 88 isotones are compared with theoretical results for ^{136,138}Sn with CWG and SMPN.

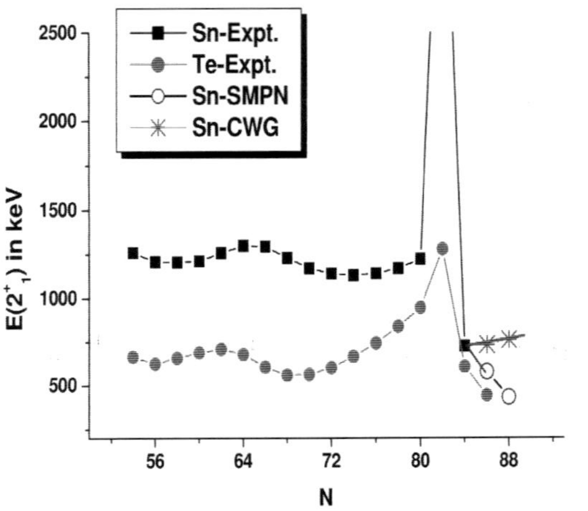

Figure 2. The Casten - Sherrill Systematics showing the differences of E(2_1^+) values for isotones of Sn and Te nuclei. See text for details.

semi-magic nuclei on and near the stability line also show almost constant energies. The isotopes of Sn (A = 134-138) are very neutron rich having N/Z ratios between 1.68 and 1.76. There are examples of semi-magic nuclei where similar feature of reduced 2_1^+ energy has been observed, viz., $^{32}_{12}Mg_{20}$ and $^{42}_{14}Si_{28}$. ^{32}Mg (N/Z = 1.67) and ^{42}Si (N/Z = 2.0) are also very neutron-rich species having similar N/Z ratio as those of $^{136,138}Sn$. Such large N/Z ratios in many cases have shown new exotic phenomena in nuclear structure. So deviation from expectations derived from the empirical knowledge of the stability and near stability domain may be a reality. The deviation of the E(2_1^+) energy for ^{134}Sn from that shown by other lighter isotopes (N <82) of Sn was also thought to be a remarkable result [13]. Now the CD-Bonn based realistic interactions can reproduce the binding energy and spectra of ^{134}Sn quite well [14]. Casten and Sherill [13] have pointed out that, although [E(2_1^+)Sn -E(2_1^+)Te] = 400 keV for a given neutron number over most of the N = 50-82 shell, the difference [E(2_1^+)Sn -E(2_1^+)Te] is only 119 keV for N = 84 [as, E(2_1^+)= 606 keV for ^{136}Te]. It is indeed remarkable that the difference between the calculated values of [E(2_1^+)(^{136}Sn) - E(2_1^+)(^{138}Te)] for N = 86 is 108 keV with SMPN (Fig.2). It is consistent with the trend discussed by Casten and Sherill . For CWG, this difference is 733 - 356 = 377 keV for N = 86, which deviates from the trend. Thus the systematic behaviour of E(2_1^+) and the E(2_1^+) differences between isotones of Sn and Te are reproduced well by SMPN. Therefore the decreasing trend of E(2_1^+) for $^{134-138}Sn$ obtained with SMPN might be a strong possibility and constitutes a new feature for semi-magic nuclei in this mass region. This is similar to that observed in even -even (e-e) Ca isotopes in the pf shell and that observed very recently in the e-e O isotopes in the sd shell (Fig.3).

3.2. The new shell closure at N=90 and comparison with other neutron-rich domains

The SMPN results for $^{136,138}Sn$ show that for increasing neutron number (>82), the Z = 50 shell closure weakens and Sn isotopes behave very similar to the other Z = 52-60 nuclei (Fig.1). The E(2_1^+) energies of isotopes of Sn for A=132-140 have been shown in Fig. 3 as a function of valence neutrons above ^{132}Sn core. The same figure contains variation of the experimental E(2_1^+) energies of even $_{20}Ca$ isotopes from A=40 to 48 as a function of valence neutrons above ^{40}Ca core. Similarly, the variation of the experimental E(2_1^+) energies of even $_8O$ isotopes from A=16 to 24 as a function of valence neutron numbers above ^{16}O core are also shown. The variations of experimental E(2_1^+) with the valence neutron number for two different mass regions and shells, show striking similarity with the theoretical predictions with SMPN in the ^{132}Sn region. The gradual filling up the $\nu(2f_{7/2})$ orbital by neutrons is very distinctively shown by the variation of

$E(2_1^+)$ from $^{134-140}Sn$. The $E(2_1^+)$ for ^{140}Sn is 1949 keV showing a sudden increase for N=90, indicating a closed shell structure for ^{140}Sn.

The trend is very similar to that observed for neutron - rich isotopes of Ca while filling up the $\nu(1f_{7/2})$ orbital and that shown by neutron - rich oxygen isotopes while filling up the $\nu(1d_{5/2})$ single particle orbital (Fig.3). The shell closure at N=14 for ^{22}O with 6 neutrons in $\nu(1d_{5/2})$ is indicated by the fact that 2_1^+ state energy (=3.199 MeV) in it is almost twice of that in the adjacent N=10 and 12 isotopes of oxygen. With filling up the $\nu(2s_{1/2})$ orbital, the increase in 2_1^+ energy ($E(2_1^+)$ = 4.72 MeV) in ^{24}O is much more dramatic indicating another new shell gap at N=16 for Z=8 which is not present for Z=6,10 and 12. It has been discussed [3] that the mechanism behind the appearance of these new shell gaps for the oxygen isotopes is the tensor force. With protons present in the $\pi(1d_{5/2})$ orbital in Ne and Mg isotopes, the $\nu(1d_{3/2})$ orbital is drawn down in energy via the isospin component of the tensor force, thereby reducing the N = 16 shell gap. Similar situation also prevails around ^{140}Sn. It has been observed experimentally

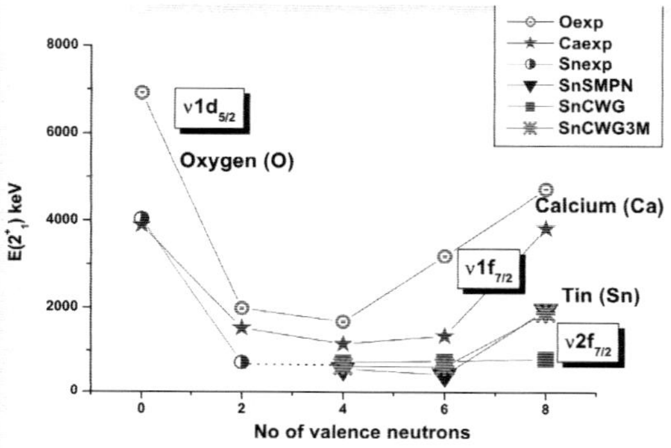

Figure 3. Comparison of the variation of calculated $E(2_1^+)$ values of Sn isotopes with the experimental $E(2_1^+)$ values of neutron rich isotopes of O and Ca as functions of valence neutron numbers with respect to ^{132}Sn, ^{16}O and ^{40}Ca cores, respectively.

that N=90 is suitable for onset of deformation for nuclei above Sn (like Xe, Ba etc.) [2]. But for N=90 isotope of Sn nucleus, a distinct shell closure is predicted by the SMPN results, while CWG denotes it to be a weak one. The presence of valence protons above the inert core is essential [15, 16] for onset of collectivity. The SMPN interaction which indicate the features of a shell closure for Z=50 at N=90, also indicate the possibility of onset of deformation at N=90 with increasing Z. This has been discussed in Ref. [17].

3.3. The effective single particle energies and the spin-tensor decomposition

In order to understand the shell closure at ^{140}Sn more precisely, the variation of the effective single-particle energies (ESPE) [18] of the neutron orbitals as functions of valence neutron numbers for the two Hamiltonians have been compared. The ESPE is defined as bare single particle energy (spe) added with the monopole part of the diagonal two body matrix elements (tbme). The bare spe is originated from the interaction of a valence nucleon with the doubly closed core. The monopole interaction contribution is the $(2J + 1)$ weighted average of the diagonal tbme, which arises from the interaction of a valence nucleon with the other valence nucleons. The ESPE for the configurations $\nu(2f_{7/2})^n$ in $^{132-140}Sn$ with neutron number varying from 82-90 (valence neutron number n varying from 0 to 8) are shown in Fig. 4. For both SMPN and CWG, the energy gap between $\nu(2f_{7/2})$ and $\nu(3p_{3/2})$ single particle orbitals is 854 keV for ^{132}Sn core. But the gap between the corresponding ESPEs increases to 2.246 MeV at N=90 with SMPN. This gap is sufficient to make ^{140}Sn a doubly- closed shell nucleus. For CWG this gap does not show any increase but instead decreases slightly to 826 keV. To analyse the origin of this new shell closure, the important physical aspects of both the residual interactions are extracted by a spin-tensor decomposition [19] of the two body matrix elements (tbmes) in to central, antisymmetric spin-orbit (als), spin-orbit (ls) and tensor components. Fig.5 shows this decomposition. From this figure, for SMPN results, one can find that the variation in als part

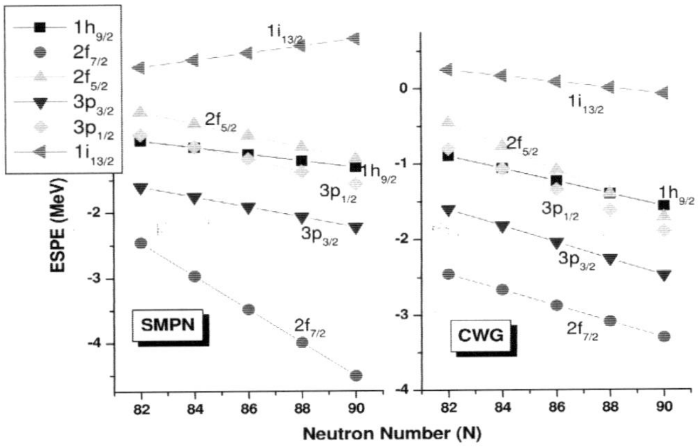

Figure 4. The ESPE's for neutrons orbitals shown as functions of neutron numbers.

is primarily responsible for this observed shell gap at N=90. The *als* component in the tbmes for an empirical interaction usually arises from inadequate constraint by the data indicating important contributions from higher order renormalisation or many body forces to the effective interactions. In empirical SMPN such many - body effects might have been included in some way through the modification of important tbmes. Fig. 6 shows the spin tensor decomposition of the USD interaction which reprodcues the experimentally observed new shell closures for oxygen isotopes. It is quite revealing that the decomposition of the empirical USD interaction also indicates that the variation in *als* part is also responsible for the observed shell gap at N=14. Thus it seems natural to conjecture that the realistic CWG differs from SMPN owing to the absence of any many (three) - body effects in the CWG tbmes.

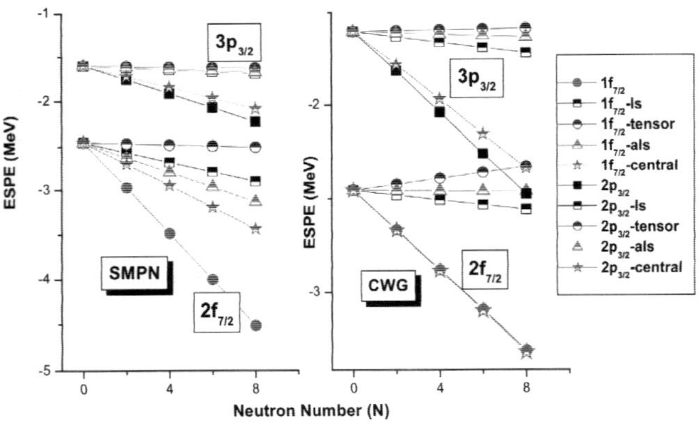

Figure 5. Variation of the contributions to ESPEs by the various components of the Hamiltonian obtained by the spin-tensor decomposition of the CWG and SMPN interactions as functions of valence neutron numbers.

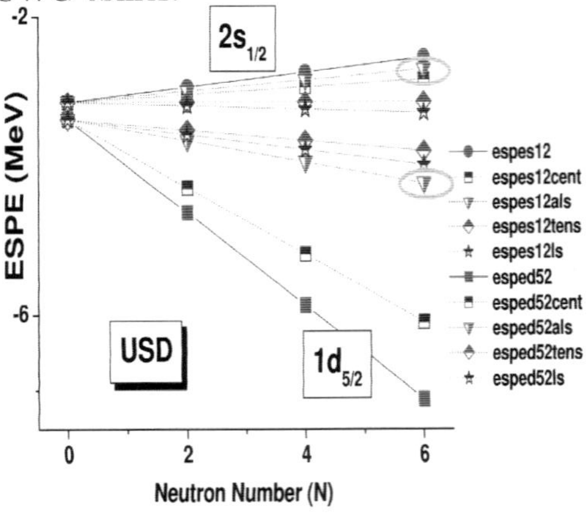

Figure 6. The spin-tensor decomposition of the USD interaction. See also caption of Figure 5.

3.4. The three-body effects

At this point it is important to note that shell model calculations using two-body realistic interactions derived from the free nucleon-nucleon force fail to reproduce some shell closures [20]. It is now rather well established that increase of the $1d_{5/2}$ - $2s_{1/2}$ gap for Z=8 and $1f_{7/2}$ - $2p_{3/2}$ gap for Z=20 (as a function of neutron number), required to explain empirical data are not obtained in the calculations with these interactions. It has been shown that the three-body forces have to be taken into account to reproduce these shell gaps [20, 21, 22]. As a next step therefore, we have incorporated a simple three-body monopole term in CWG to construct CWG3M as prescribed in Ref.[21] for KB3 interaction. However as the mass regions considered in KB3 (A>40) and the present case (A> 132) are quite far off, we have included the effect of mass scaling by a factor of $(40/132)^{(1/3)}$. It should be noted that the correction factor will be effective for nuclei for which the valence neutron number n $\geq$ 3. A shell gap for N=90 now appears with CWG3M which is very close to that with SMPN. The $E(2_1^+)$ energies of $^{136,138}Sn$ are 0.639 and 0.633 MeV, respectively. The $E(2_1^+)$ energy of ^{140}Sn predicted by CWG3M (1.889 MeV) is also close to that predicted by SMPN (1.949 MeV) (Fig.3).

4. Conclusions

The SMPN results for $^{136,138}Sn$ show that for increasing neutron number (>82), the Z = 50 shell closure weakens and Sn isotopes behave very similar to the other Z = 52-60 nuclei. However the new result on ^{140}Sn, its high 2_1^+ energy and its comparison with examples from other neutron - rich domains clearly show that N=84-88 spectra with SMPN show the effect of gradual filling up the $\nu(2f_{7/2})$ orbital which finally culminates in a new shell closure at N=90. A large contribution of als term in the ESPE of $2f_{7/2}$ in empirical interaction SMPN has been found to be responsible for the shell gap similar to that in sd shell. A simple three-body monopole term has been included in CWG to get CWG3M, which predicts a shell gap at N=90 for Sn isotopes as well as decreasing 2_1^+ energies for $^{136,138}Sn$, similar to that from SMPN. This indicates that three body effect plays an important role for shell evolution in neutron rich Sn isotopes above ^{132}Sn, as also observed in *sd* and *pf* shells. The anomalously depressed 2_1^+ states in Sn isotopes having N=84-88, and the new shell closure for N=90, might have interesting consequences for the r - process nucleosynthesis.

References

[1] Wöhr A *et al.* 2003 *Fission and properties of neutron-rich nuclei, Proc. of the 3rd Int. Conf.* 2002 Sanibel Island, Florida, ed J. H. Hamilton, A. V. Ramayya, and H. K. Carter (World Scientific) p 337.

[2] http://www.nndc.bnl.gov.

[3] Górska M *et al.* 2009 *Phys. Lett. B* **672** 313; Thirolf P G *et al.* 2000 *Phys. Lett. B* **485** 16; Kanungo R *et al.* 2009 *Phys. Rev. Lett.* **102** 152501 ; Hoffman C R *et al.* 2009 *Phys. Lett. B* **672** 17.

[4] Brown B Alex and Richter W A 2005 *Phys. Rev. C* **72** 057301.

[5] Dobaczewski J, Hamamoto I, Nazarewicz W and Sheikh J A 1994 *Phys. Rev. Lett.* **72** 981; Dobaczewski J, Michel N, Nazarewicz W, Ploszajczak M and Rotureau J 2007 *Prog. Part. Nucl. Phys.* **59** 432.

[6] Sarkar Sukhendusekhar and Saha Sarkar M 2004 *Eur. Phys. Jour. A* **21** 61.

[7] Chou W T and Warburton E K 1992 *Phys. Rev. C* **45** 1720.

[8] Sarkar Sukhendusekhar and Saha Sarkar M 2001 *Phys. Rev. C* **64** 014312.

[9] Sarkar S and Saha Sarkar M 2008 Structure of A=138 isobars above the ^{132}Sn core and the N-N interaction in the neutron-rich environment *arXiv*:0802.3740v1.

[10] Brown B A, Stone N J, Stone J R, Towner I S and Hjorth-Jensen M 2005 *Phys. Rev. C* **71** 044317.

[11] Sarkar S and Saha Sarkar M 2008 *Phys. Rev. C* **78** 024308.

[12] Brown B A *et al.* 2004 Oxbash for Windows PC, *MSU-NSCL Report No.* **1289**; Brown B A and Rae W D M 2007 Nushell@MSU *MSU-NSCL report.*

[13] Casten R F and Sherrill B M 2000 *Prog. Part. Nucl. Phys.* **45** 171.

[14] Coraggio L, Covello A, Gargano A and Itaco N 2002 *Phys. Rev. C* **65** 051306(R).

[15] Casten R F, Warner D D, Brenner D S and Gill R L 1981 *Phys. Rev. Lett.* **47** 1433.

[16] Kshetri Ritesh, Saha Sarkar M and Sarkar S 2006 *Phys. Rev. C* **74** 034314.

[17] Sarkar S and Saha Sarkar M 2010 *Phys. Rev. C* **81** 064328.

[18] Otsuka Takaharu, Suzuki Toshio, Fujimoto Rintaro, Grawe Hubert and Akaishi Yoshinori 2005 *Phys. Rev. Lett.* **95** 232502.

[19] Kirson M W 1973 *Phys. Lett.* **47B**, 110; Yoro Kenji 1980 *Nucl. Phys. A* **333** 67.

[20] Sorlin O and Porquet M G 2008 *Prog. Part. and Nucl. Phys.* **61** 602.

[21] Poves A and Zuker A 2003 *Phys. Rep.* **70** 235; Zuker A P 2003 *Phys. Rev. Lett.* **90** 042502.

[22] Otsuka Takaharu, Suzuki Toshio, Holt Jason D, Schwenk Achim and Yoshinori Akaishi 2010 *Phys. Rev. Lett.* **105** 032501.

Deformed pseudospin doublets as a fingerprint of a relativistic supersymmetry in nuclei

Deformed pseudospin doublets as a fingerprint of a relativistic supersymmetry in nuclei

A. Leviatan

Racah Institute of Physics, The Hebrew University, Jerusalem 91904, Israel

E-mail: ami@phys.huji.ac.il

Abstract. The single-particle spectrum of deformed shell-model states in nuclei, is shown to exhibit a supersymmetric pattern. The latter involves deformed pseudospin doublets and intruder levels. The underlying supersymmetry is associated with the relativistic pseudospin symmetry of the nuclear mean-field Dirac Hamiltonian with scalar and vector potentials.

1. Introduction

The concept of pseudospin doublets [1, 2] is based on the empirical observation in nuclei of quasi-degenerate pairs of certain normal-parity shell-model orbitals with non-relativistic single-nucleon quantum numbers

$$(n, \ell, j = \ell + 1/2) \quad \text{and} \quad (n-1, \ell+2, j = \ell + 3/2)\ . \tag{1}$$

The doublet structure is expressed in terms of a "pseudo" orbital angular momentum, $\tilde{\ell} = \ell + 1$, and "pseudo" spin, $\tilde{s} = 1/2$, which are coupled to $j = \tilde{\ell} \pm \tilde{s}$. For example, $(1d_{5/2}, 0g_{7/2})$ and $(2s_{1/2}, 1d_{3/2})$ will have $\tilde{\ell} = 3$ and $\tilde{\ell} = 1$, respectively. The states $(n = 0, \ell, j = \ell + 1/2)$, with aligned spin and no nodes, are not part of a doublet. For large j they are the "intruder" abnormal-parity states, *i.e.*, $0g_{9/2}$, $0h_{11/2}$, $0i_{13/2}$, which are unique in the major shell. In the presence of axially-symmetric deformation, the doublets persist [3] with asymptotic (Nilsson) quantum numbers

$$[N, n_3, \Lambda]\Omega = \Lambda + 1/2 \quad \text{and} \quad [N, n_3, \Lambda + 2]\Omega = \Lambda + 3/2\ . \tag{2}$$

Here N, n_3, are harmonic oscillator quantum numbers, Λ and Ω are, respectively, the components of the orbital and total angular momentum along the symmetry (3rd) axis. In this case, the doublets can be expressed in terms of pseudo-orbital, $\tilde{\Lambda} = \Lambda + 1$, and pseudospin, $\tilde{\mu} = \pm 1/2$, projections, to obtain $\Omega = \tilde{\Lambda} \pm 1/2$. For example, ([420]1/2, [422]3/2) and ([411]3/2, [413]5/2) will have $\tilde{\Lambda} = 1$ and $\tilde{\Lambda} = 2$, respectively. The intruder nodeless states, $[N, n_3, \Lambda = N - n_3]\Omega = \Lambda + 1/2$, are not part of a doublet. Pseudospin doublets play a central role in explaining features of nuclei [3], including superdeformation [4] and identical bands [5, 6]. They have been shown to originate from a relativistic symmetry of the Dirac Hamiltonian in which the sum of the scalar and vector mean-field potentials cancel [7]. The generators of this relativistic SU(2) symmetry have been identified [8] and consequences for radial nodes have been considered [9]. Recently, it has been shown that a pseudospin-invariant Dirac Hamiltonian with spherical potentials, gives

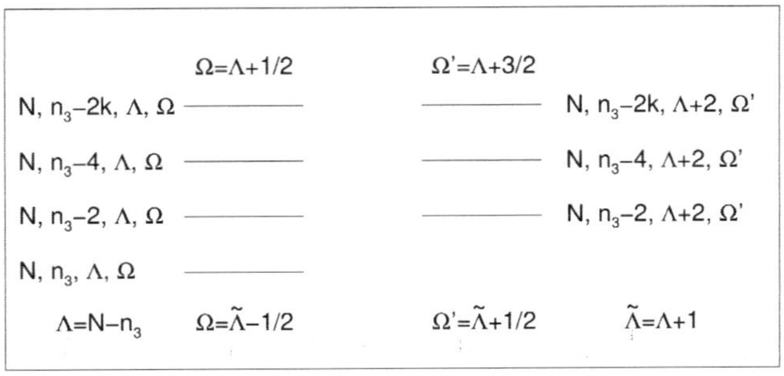

Figure 1. Nuclear single-particle spectrum composed of deformed pseudospin doublets and an intruder level. All states share a common $\tilde{\Lambda}$.

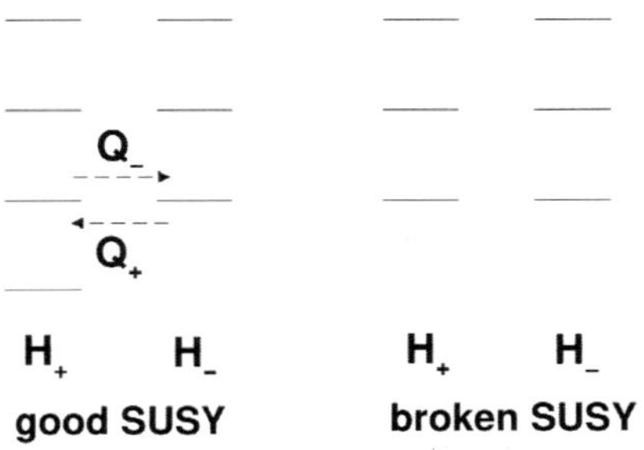

Figure 2. Typical spectra of good and broken SUSY. The operators Q_- and Q_+ connect degenerate states in the H_+ and H_- sectors.

rise to a supersymmetric pattern [10]. In the present contribution we extend this study and show that a supersymmetric pattern occurs also when the relevant potentials are axially deformed [11].

Fig. 1 portrays the level scheme of a subset of deformed pseudospin doublets, Eq. (2), with fixed N, Λ, Ω, Ω' and $n_\perp = (N - \Lambda - n_3)/2 = 1,2,3,\ldots$ together with the intruder level $[N, n_3, \Lambda = N - n_3]\Omega = \Lambda + 1/2$, (or equivalently, $n_\perp = 0$). The single-particle spectrum exhibits towers of pair-wise degenerate states, sharing a common $\tilde{\Lambda}$, and an additional non-degenerate nodeless intruder state at the bottom of the spin-projection aligned tower. A comparison with Fig. 2 reveals a striking similarity with a supersymmetric pattern of good supersymmetry (SUSY). In what follows we identify the underlying supersymmetric structure associated with a Dirac Hamiltonian possessing a relativistic pseudospin symmetry.

2. Dirac Hamiltonian with axially-deformed potentials

A relativistic mean-field description of nuclei employs a Dirac Hamiltonian, $H = \hat{\boldsymbol{\alpha}} \cdot \hat{\boldsymbol{p}} + \hat{\beta}(M + V_S) + V_V$ for a nucleon of mass M moving in external scalar, V_S, and vector, V_V, potentials. When the potentials are axially-symmetric, *i.e.*, independent of the azimuthal angle ϕ, $V_{S,V} = V_{S,V}(\rho, z)$, $\rho = \sqrt{x^2 + y^2}$, then the z-component of the angular momentum operator, $\hat{J}_z$, commutes with H and its half-integer eigenvalues Ω are used to label the Dirac wave functions

$$\Psi_\Omega(\rho, \phi, z) = \begin{pmatrix} g^+(\rho, z)\, e^{i(\Omega - 1/2)\phi} \\ g^-(\rho, z)\, e^{i(\Omega + 1/2)\phi} \\ if^+(\rho, z)\, e^{i(\Omega - 1/2)\phi} \\ if^-(\rho, z)\, e^{i(\Omega + 1/2)\phi} \end{pmatrix} . \tag{3}$$

Here $g^\pm \equiv g^\pm(\rho, z)$ and $f^\pm \equiv f^\pm(\rho, z)$ are the radial wave functions of the upper and lower components, respectively. The Dirac equation, $H\Psi = E\Psi$, leads to a set of four coupled partial differential equations for these components. For each solution with $\Omega > 0$, there is a degenerate time-reversed solution with $-\Omega < 0$, hence, we confine the discussion to solutions with $\Omega > 0$. Of particular interest are bound Dirac valence states with $0 < E < M$ and normalizable wave functions. The potentials enter the Dirac equation through the combinations

$$\begin{aligned} A(\rho, z) &= E + M + V_S(\rho, z) - V_V(\rho, z) \ , \\ B(\rho, z) &= E - M - V_S(\rho, z) - V_V(\rho, z) \ . \end{aligned} \tag{4}$$

For relativistic mean-fields relevant to nuclei, V_S is attractive and V_V is repulsive with typical values $V_S(0) \sim -400$, $V_V(0) \sim 350$, MeV. The potentials satisfy $\rho V_S(\rho, z), \rho V_V(\rho, z) \to 0$

for $\rho \to 0$ and $V_S(\rho,z), V_V(\rho,z) \to 0$ for $\rho \to \infty$ or $z \to \pm\infty$. The boundary conditions imply that the radial wave functions fall off exponentially for large distances and behave as a power law for $\rho \to 0$. Furthermore, for $z = 0$ and $\rho \to \infty$, $f^-/g^+ \propto (M - E) > 0$ and $g^-/f^+ \propto (M + E) > 0$, while for $z = 0$ and $\rho \to 0$, $f^-/g^+ \propto B(0)\rho$ and $g^-/f^+ \propto -A(0)\rho$. These properties have important implications for the structure of radial nodes. In particular, it follows that for potentials with the indicated asymptotic behaviour and $A(0)$, $B(0) > 0$, as encountered in nuclei, a necessary condition for a nodeless bound valence eigenstate of a Dirac Hamiltonian is [11]

$$g^- = 0 \text{ or } f^+ = 0 \; . \tag{5}$$

3. Relativistic pseudospin symmetry in nuclei

A relativistic pseudospin symmetry occurs when the sum of the scalar and vector potentials is a constant [7]

$$V_S(\rho,z) + V_V(\rho,z) = \Delta_0 \; . \tag{6}$$

The symmetry generators, $\hat{\tilde{S}}_i$, commute with the Dirac Hamiltonian and span an SU(2) algebra [8]

$$\hat{\tilde{S}}_i = \begin{pmatrix} U_p\, \hat{s}_i\, U_p & 0 \\ 0 & \hat{s}_i \end{pmatrix} \qquad i = x,y,z \qquad U_p = \frac{\boldsymbol{\sigma}\cdot\boldsymbol{p}}{p} \; . \tag{7}$$

Here $\hat{s}_i = \sigma_i/2$ are the usual spin operators, defined in terms of Pauli matrices. The Dirac eigenfunctions in the pseudospin limit satisfy

$$\begin{aligned} \hat{\tilde{S}}_z \Psi^{(\tilde{\mu})}_{\Omega} &= \tilde{\mu}\Psi^{(\tilde{\mu})}_{\Omega} \qquad \tilde{\mu} = \pm 1/2 \\ (\hat{\tilde{S}}_x + i\hat{\tilde{S}}_y)\Psi^{(-1/2)}_{\Omega} &= \Psi^{(1/2)}_{\Omega} \end{aligned} \tag{8}$$

and form degenerate $SU(2)$ doublets. Their wave functions have been shown to be of the form [12]

$$\Psi^{(-1/2)}_{\Omega_1=\tilde{\Lambda}-1/2} = \begin{pmatrix} g^+\, e^{i(\tilde{\Lambda}-1)\phi} \\ -g\, e^{i\tilde{\Lambda}\phi} \\ 0 \\ if\, e^{i\tilde{\Lambda}\phi} \end{pmatrix} \; , \quad \Psi^{(1/2)}_{\Omega_2=\tilde{\Lambda}+1/2} = \begin{pmatrix} g\, e^{i\tilde{\Lambda}\phi} \\ g^-\, e^{i(\tilde{\Lambda}+1)\phi} \\ if\, e^{i\tilde{\Lambda}\phi} \\ 0 \end{pmatrix} \tag{9}$$

where $\tilde{\Lambda} = \Omega - \tilde{\mu} \geq 0$ is the eigenvalue of $\hat{J}_z - \hat{\tilde{S}}_z$. Near the pseudospin limit, eigenstates with $\tilde{\Lambda} \neq 0$ occur in degenerate doublets. An exception to this rule are eigenstates whose radial components have no nodes and, therefore, are not part of a doublet. The latter property follows from the fact that a nodeless bound Dirac state satisfies the criteria of Eq. (5), hence has a wave function $\Psi^{(-1/2)}_{\Omega_1=\tilde{\Lambda}-1/2}$, as in Eq. (9), with g^+, g, $f \neq 0$ and $f/g^+ > 0$. Its pseudospin partner state has a wave function $\Psi^{(1/2)}_{\Omega_2=\tilde{\Lambda}+1/2}$. The radial components satisfy $Bg^- = [B - 2(\tilde{\Lambda}/\rho)f/g^+]g^+$, where B is defined in Eq. (4). This relation is satisfied, to a good approximation, for mean-field potentials relevant to nuclei, and the r.h.s. is non-zero and, consequently, $g^- \neq 0$. If so, then the partner state, $\Psi^{(1/2)}_{\Omega_2=\tilde{\Lambda}+1/2}$, is also nodeless, but it cannot be a bound valence eigenstate since its radial components do not fulfill the condition of Eq. (5).

The relativistic pseudospin symmetry has been tested in numerous realistic mean-field calculations of nuclei and was found to be obeyed to a good approximation, especially for

[411]1/2	[400]1/2	[402]3/2						
[431]1/2	[420]1/2	[422]3/2	[411]3/2	[413]5/2	[402]5/2	[404]7/2		
	[440]1/2		[431]3/2		[422]5/2		[413]7/2	[404]9/2
$\tilde{\Lambda}=0$	$\tilde{\Lambda}=1$		$\tilde{\Lambda}=2$		$\tilde{\Lambda}=3$		$\tilde{\Lambda}=4$	$\tilde{\Lambda}=5$

Figure 3. Grouping of deformed shell-model states $[N = 4, n_3, \Lambda]\Omega$, exhibiting a pattern of good SUSY, relevant to the pseudospin symmetry limit [11].

doublets near the Fermi surface [12, 13, 14, 15, 16, 17, 18]. The dominant upper components of the states in Eq. (9), involving g^+ and g^-, correspond to the non-relativistic pseudospin doublets of Eq. (2). The nodeless Dirac eigenstates correspond to the non-relativistic intruder states, $[N, n_3, \Lambda = N - n_3]\Omega = \Lambda + 1/2$, mentioned above. In what follows we show that the ensemble of deformed pseudospin doublets and intruder levels, forms a supersymmetric pattern whose origin can be traced to the relativistic pseudospin symmetry of the Dirac Hamiltonian.

4. Relativistic pseudospin symmetry and supersymmetric quantum mechanics

The essential ingredients of supersymmetric quantum mechanics [19] are the supersymmetric Hamiltonian, $\mathcal{H}$, and charges Q_+, $Q_- = Q_+^\dagger$, which generate the supersymmetry (SUSY) algebra

$$[\,\mathcal{H}\,,\,Q_\pm\,] = 0 \quad , \quad \{\,Q_\pm\,,\,Q_\pm\,\} = 0 \quad , \quad \{\,Q_-\,,\,Q_+\,\} = \mathcal{H}\ . \tag{10}$$

Accompanying this set is a Hermitian Z_2-grading operator satisfying

$$\mathcal{P}^\dagger = \mathcal{P} \quad , \quad [\,\mathcal{H}\,,\mathcal{P}\,] = 0 \quad , \quad \{\,Q_\pm\,,\,\mathcal{P}\,\} = 0 \quad , \quad \mathcal{P}^2 = \mathbb{1}\ . \tag{11}$$

The $+1$ and -1 eigenspaces of $\mathcal{P}$ define the "positive-parity", H_+, and "negative-parity", H_-, sectors of the spectrum, with eigenvectors $\Psi^{(+)}$ and $\Psi^{(-)}$, respectively. The SUSY algebra imply that if $\Psi^{(+)}$ is an eigenstate of $\mathcal{H}$, then also $\Psi^{(-)} = Q_-\Psi^{(+)}$ is an eigenstate of $\mathcal{H}$ with the same energy, unless $Q_-\Psi^{(+)}$ vanishes or produces an unphysical state, (*e.g.*, non-normalizable). The resulting spectrum consists of pairwise degenerate levels with a non-degenerate single state (the ground state) in one sector when the supersymmetry is exact. If all states are pairwise degenerate, the supersymmetry is said to be broken. Typical spectra for good and broken SUSY are shown in Fig. 2.

Degenerate doublets, signaling a supersymmetric structure, can emerge in a quantum system with a Hamiltonian H, from the existence of two Hermitian, conserved and anticommuting operators, $\hat{R}$ and $\hat{B}$

$$[H, \hat{R}] = [H, \hat{B}] = \{\hat{R}, \hat{B}\} = 0\ . \tag{12}$$

The operator $\hat{R}$ has non-zero eigenvalues, r, which come in pairs of opposite signs. $\hat{B}^2 = \hat{B}^\dagger\hat{B} = f(H)$, is a function of the Hamiltonian. A Z_2-grading operator, $\mathcal{P}_r$, and Hermitian supersymmetric charges, Q_1, Q_2 and Hamiltonian $\mathcal{H}$ can now be constructed as

$$\mathcal{P}_r = \hat{R}/|r| \quad , \quad Q_1 = \hat{B} \quad , \quad Q_2 = iQ_1\mathcal{P}_r \quad , \quad \mathcal{H} = Q_1^2 = f(H)\ . \tag{13}$$

The triad of operators $Q_\pm = (Q_1 \pm iQ_2)/2$ and $\mathcal{H}$ form the standard SUSY algebra of Eq. (10).

Focusing the discussion to the relativistic pseudospin limit, the elements of the SUSY algebra are

$$\begin{aligned} &Q_1 = 2\,(\,M + \Delta_0 - H\,)\,\hat{\tilde{S}}_x \ , \quad Q_2 = 2\,(\,M + \Delta_0 - H\,)\,\hat{\tilde{S}}_y \ , \quad \mathcal{P}_r = 2\hat{\tilde{S}}_z \ , \\ &\mathcal{H} = Q_1^2 = (\,M + \Delta_0 - H\,)^2 \ , \end{aligned} \tag{14}$$

Table 1. Conserved, anticommuting operators, $\hat{R}$ and $\hat{B}$, for Dirac Hamiltonians (H), with axially-deformed scalar and vector potentials, exhibiting a supersymmetric structure [11]. Here $\hat{\tilde{S}}_i = \begin{pmatrix} U_p\,\hat{s}_i\,U_p & 0 \\ 0 & \hat{s}_i \end{pmatrix}$ and $\hat{S}_i = \begin{pmatrix} \hat{s}_i & 0 \\ 0 & U_p\,\hat{s}_i\,U_p \end{pmatrix}$ are the pseudospin and spin generators, respectively. $\hat{R}_z = [\,M + V_S(z)\,]\,\hat{\beta}\,\hat{\Sigma}_3 + \gamma_5\hat{p}_z$ and $\hat{R}_\rho = [\,M + V_S(\rho)\,]\hat{\Sigma}_3 - i\hat{\beta}\,\gamma_5(\,\hat{\boldsymbol{\Sigma}} \times \hat{\boldsymbol{p}}\,)_3$, where $\hat{\Sigma}_i = \begin{pmatrix} \sigma_i & 0 \\ 0 & \sigma_i \end{pmatrix}$.

Dirac Hamiltonian	$\hat{R}$	$\hat{B}$	$\hat{B}^2 = f(H)$	SUSY
$V_S(\rho,z) + V_V(\rho,z) = \Delta_0$	$\hat{\tilde{S}}_z$	$2(M + \Delta_0 - H)\hat{\tilde{S}}_x$	$(M + \Delta_0 - H)^2$	good
$V_S(\rho,z) - V_V(\rho,z) = \Xi_0$	$\hat{S}_z$	$2(\,M + \Xi_0 + H)\,\hat{S}_x$	$(M + \Xi_0 + H)^2$	broken
$V_S = V_S(z),\ V_V = \frac{\alpha_V}{\rho}$	$\hat{R}_z$	$\hat{\beta}\,\hat{\Sigma}_3\{i\hat{J}_z\gamma_5[\,H - \hat{\Sigma}_3\hat{R}_z] - \frac{\alpha_V}{\rho}(\hat{\boldsymbol{\Sigma}}\cdot\boldsymbol{\rho})\,\hat{R}_z\}$	$\hat{J}_z^2(H^2 - \hat{R}_z^2) + \alpha_V^2\hat{R}_z^2$	good
$V_S = \frac{\alpha_S}{\rho},\ V_V = V_V(z)$	$\hat{R}_\rho$	$\hat{\Sigma}_3\{i\hat{J}_z\gamma_5[M - \hat{\Sigma}_3\hat{R}_\rho] - \frac{\alpha_S}{\rho}(\hat{\boldsymbol{\Sigma}}\cdot\boldsymbol{\rho})\hat{\beta}\hat{R}_\rho\}$	$\hat{J}_z^2(\hat{R}_\rho^2 - M^2) + \alpha_S^2\hat{R}_\rho^2$	good

Table 2. As in Table 1 but for spherical scalar and vector potentials [10, 20] in the Dirac Hamiltonian (H). Here $\hat{K} = -\hat{\beta}\left(\hat{\boldsymbol{\Sigma}}\cdot\hat{\boldsymbol{\ell}} + 1\right)$, with $\hat{\boldsymbol{\ell}} = \boldsymbol{r}\times\hat{\boldsymbol{p}}$, $\hat{\tilde{T}}_z = (\hat{\tilde{\boldsymbol{S}}}\times\hat{\tilde{\boldsymbol{L}}})_z$, $\hat{T}_z = (\hat{\boldsymbol{S}}\times\hat{\boldsymbol{L}})_z$. $\hat{\tilde{Y}} = \hat{\tilde{\boldsymbol{L}}}\cdot\hat{\tilde{\boldsymbol{L}}} + 1/4 - \hat{J}_z^2$ and $\hat{Y} = \hat{\boldsymbol{L}}\cdot\hat{\boldsymbol{L}} + 1/4 - \hat{J}_z^2$. The operators $\hat{\Sigma}_i$, $\hat{\tilde{S}}_i$ and $\hat{S}_i$ are defined in the caption of Table 1 and $\hat{\tilde{L}} = \hat{J}_i - \hat{\tilde{S}}_i$, $\hat{L} = \hat{J}_i - \hat{S}_i$.

Dirac Hamiltonian	$\hat{R}$	$\hat{B}$	$\hat{B}^2 = f(H)$	SUSY
$V_S(r) + V_V(r) = \Delta_0$	$\hat{K} - 1/2$	$2(\,M + \Delta_0 - H)\,\hat{\tilde{T}}_z$	$(M + \Delta_0 - H)^2\,\hat{\tilde{Y}}$	good
$V_S(r) - V_V(r) = \Xi_0$	$\hat{K} + 1/2$	$2(\,M + \Xi_0 + H)\,\hat{T}_z$	$(M + \Xi_0 + H)^2\,\hat{Y}$	broken
$V_S = \frac{\alpha_S}{r},\ V_V = \frac{\alpha_V}{r}$	$\hat{K}$	$-i\hat{K}\gamma_5\,(H - \hat{\beta}M) + \frac{1}{r}(\hat{\boldsymbol{\Sigma}}\cdot\boldsymbol{r})(\alpha_V M + \alpha_S H)$	$(\alpha_V M + \alpha_S H)^2 + \hat{K}^2(H^2 - M^2)$	good

where H is the pseudospin-invariant Dirac Hamiltonian with potentials satisfying the condition of Eq. (6). The operator $\hat{B} = Q_1$ connects the doublet states of Eq. (9). The spectrum, for each $\tilde{\Lambda} \neq 0$, consists of twin towers of pairwise degenerate pseudospin doublet states, with $\Omega_1 = \tilde{\Lambda} - 1/2$ and $\Omega_2 = \tilde{\Lambda} + 1/2$, and an additional non-degenerate nodeless state at the bottom of the $\Omega_1 = \tilde{\Lambda} - 1/2$ tower. Altogether, the ensemble of Dirac states with $\Omega_2 - \Omega_1 = 1$ exhibits a supersymmetric pattern of good SUSY, as illustrated in Fig. 3.

The supersymmetry discussed above, is one of several possible classes of supersymmetries of Dirac Hamiltonians in 3+1 dimensions, with cylindrically–deformed scalar and vector potentials [11]. The supersymmetries associated with the pseudospin and spin limits arise when the potentials obey a constraint on their sum or difference, respectively. Additional supersymmetries arise when one of the potentials has a $1/\rho$ dependence and the second potential depends on z. The relevant $\hat{R}$ and $\hat{B}$ operators (12) of the various supersymmetries, are listed in Table 1. A similar analysis can be done for spherical scalar and vector potentials and, as shown in Table 2, the corresponding supersymmetries are associated with the pseudospin ($V_S + V_V = \Delta_0$), spin ($V_S - V_V = \Xi_0$) and Coulomb ($V_S, V_V \propto 1/r$) limits of the Dirac Hamiltonian [10, 20]. The spin (Coulomb) limits are relevant to quark (electron) dynamics in heavy mesons (atoms) [18].

5. Summary

As previously noted, the single-particle spectrum of spherical shell-model states in nuclei, exhibits characteristic features of a supersymmetric pattern. The latter consists of twin towers of pair-wise degenerate spherical pseudospin doublets (with $n = 1, 2 \ldots$) sharing a common $\tilde{\ell}$, and an additional non-degenerate nodeless ($n = 0$) intruder state at the bottom of the spin-aligned tower

$$\begin{array}{ccc} (n,\, \ell,\, j = \ell + 1/2\,) & (n-1,\, \ell + 2,\, j = \ell + 3/2\,) & n = 1, 2, \ldots \\ (n = 0,\, \ell,\, j = \ell + 1/2\,) & & n = 0 \\ & & \\ j = \tilde{\ell} - 1/2 & j = \tilde{\ell} + 1/2 & \tilde{\ell} = \ell + 1 \end{array} \tag{15}$$

As discussed in the present contribution, a similar situation is observed in the single-particle spectrum of deformed shell model states. The corresponding supersymmetric pattern consists of twin towers of pair-wise degenerate deformed pseudospin doublets (with $n_\perp = 1, 2 \ldots$) sharing a common $\tilde{\Lambda}$, and an additional non-degenerate nodeless ($n_\perp = 0$) intruder state at the bottom of the spin-projection aligned tower

$$\begin{array}{ccc} [N, n_3 - 2n_\perp, \Lambda]\, \Omega = \Lambda + 1/2 & [N, n_3 - 2n_\perp, \Lambda + 2]\, \Omega = \Lambda + 1/2 & n_\perp = 1, 2, \ldots \\ [N, n_3, \Lambda]\, \Omega = \Lambda + 1/2 & & n_\perp = 0 \\ & & \\ \Omega = \tilde{\Lambda} - 1/2 & \Omega = \tilde{\Lambda} + 1/2 & \tilde{\Lambda} = \Lambda + 1 \end{array} \tag{16}$$

with $\Lambda = N - n_3$. In both cases, the underlying supersymmetric algebra is associated with the relativistic pseudospin symmetry of the nuclear mean-field Dirac Hamiltonian, involving scalar and vector potentials. Additional supersymmetries of Dirac Hamiltonians with such mixed Lorentz structure can be identified.

Acknowledgments

This work is supported by the Israel Science Foundation.

References

[1] Hecht K T and A. Adler A 1969 *Nucl. Phys.* A **137** 129.
[2] Arima A, Harvey M and Shimizu K 1969 *Phys. Lett.* B **30** 517.
[3] Bohr A, Hamamoto I and Mottelson B R 1982 *Phys. Scr.* **26** 267.
[4] Dudek J, Nazarewicz W, Szymanski Z and Leander G A 1987 *Phys. Rev. Lett.* **59** 1405.
[5] Nazarewicz W, Twin P J, Fallon P and Garrett J D 1990 *Phys. Rev. Lett.* **64** 1654.
[6] Stephens F S *et al.* 1990 *Phys. Rev. Lett.* **65** 301.
[7] Ginocchio J N 1997 *Phys. Rev. Lett.* **78** 436.
[8] Ginocchio J N and Leviatan A 1998 *Phys. Lett.* B **425** 1.
[9] Leviatan A and Ginocchio J N 2001 *Phys. Lett.* B **518** 214.
[10] Leviatan A 2004 *Phys. Rev. Lett.* **92** 202501.
[11] Leviatan A 2009 *Phys. Rev. Lett.* **103** 042502.
[12] Ginocchio J N 2002 *Phys. Rev. C* **66** 064312.
[13] Ginocchio J N and Madland D G 1998 *Phys. Rev. C* **57** 1167.
[14] Lalazissis G A, Gambhir Y K, Maharana J P, Warke C S and Ring P 1998 *Phys. Rev. C* **58** R45.
[15] Meng J, Sugawara-Tanabe K, Yamaji S and Arima A 1999 *Phys. Rev. C* **59** 154.
[16] Ginocchio J N and Leviatan A 2001 *Phys. Rev. Lett.* **87** 072502.
[17] Ginocchio J N, Leviatan A, Meng J and Zhou S G 2004 *Phys. Rev.* C **69** 034303.
[18] Ginocchio J N 2005 *Phys. Rep.* **414** 165.
[19] Junker G 1996 *Supersymmetric Methods in Quantum and Statistical Physics*, (Berlin: Springer).
[20] Leviatan A and Ginocchio J N 2010 in preparation.

Random phase approximation in a self-consistent covariant approach: recent applications

2011 J. Phys.: Conf. Ser. 267 012042

(http://iopscience.iop.org/1742-6596/267/1/012042)

Random phase approximation in a self-consistent covariant approach: recent applications

Haozhao Liang[1,2], Yifei Niu[1,3], Jie Meng[4,1,5] and Nguyen Van Giai[2]

[1]State Key Laboratory of Nuclear Physics and Technology, School of Physics, Peking University, Beijing 100871, China

[2]Institut de Physique Nucléaire, IN2P3-CNRS and Univ. Paris-Sud, 91406 Orsay, France

[3]Physics Department, Faculty of Science, University of Zagreb, Croatia

[4]School of Physics and Nuclear Energy Engineering, Beihang University, Beijing 100191, China

[5]Department of Physics, University of Stellenbosch, Stellenbosch, South Africa

E-mail: `nguyen@ipno.in2p3.fr`

Abstract. The relativistic Hartree-Fock (RHF) and Random Phase Approximation (RPA) methods are self-consistently applied to two issues of current interest. The first application is related to the isospin mixing corrections in the problem of super-allowed $0^+ \rightarrow 0^+$ β-transitions and the unitarity of the CKM matrix. The second application concerns the prediction of inclusive neutrino-nucleus cross-sections, where the results of the present model are compared with other approaches.

1. Introduction

The self-consistent covariant mean field method is well known for its numerous applications to the description of nuclear ground and excited states. The most widely known versions of the method are within the context of the relativistic Hartree approximation: the Relativistic Mean Field (RMF) method [1, 2] and the Random Phase Approximation (RPA) derived from it [3]. In the recent years there have been significant advances in the covariant approaches dealing with the effects of exchange interactions and leading to the Relativistic Hartree-Fock (RHF) and Relativistic Hartree-Fock-Bogoliubov (RHFB) theories [4, 5] as well as the RHF-RPA description of nuclear excitations [6]. One interesting finding is that, the effects of the tensor ρ-N coupling which contributes mostly in the exchange (Fock) terms can improve somewhat the spurious shell gaps at $N, Z = 56$ and 92 found in RMF parametrizations [7]. Another effect of exchange terms in the particle-hole (p-h) channel is that the isoscalar σ- and ω-couplings play an important role in the description of isovector excitations like Gamow-Teller states whereas they are absent by definition in the RMF-RPA approach.

In this presentation we shall focus on two recent studies [8, 9] where the RHF-RPA method is used with a degree of self-consistency which cannot be met by the RMF-RPA approach. The first subject concerns the theoretical evaluation of isospin impurity effects in the super-allowed $0^+ \rightarrow 0^+$ β-transitions. It turns out that the exchange Coulomb effects cannot be disregarded in that problem. The second subject is the prediction of neutrino-nucleus cross sections using RPA wave functions. The important spin-isospin modes (Gamow-Teller, spin dipole, etc...) which determine the cross-sections can be calculated here in a fully self-consistent manner [6] whereas the RMF-RPA must introduce an additional adjusted parameter in the π-N couplings.

2. Isospin corrections for superallowed β decays

The key issue of the validity of the Standard Model of electro-weak interactions is linked to the unitarity of the Cabbibo-Kobayashi-Maskawa (CKM) matrix. One can have an accurate experimental determination of the leading matrix element V_{ud} by studying the $0^+ \to 0^+$ superallowed Fermi β decays which lead to the value of the vector coupling constant G_V of semi-leptonic weak interactions. One has:

$$V_{ud} = \frac{G_V}{G_F}, \tag{1}$$

where G_F is the Fermi coupling constant for purely leptonic decays.

We concentrate on the β-transitions between a parent state (N, Z) and daughter states $(N \pm 1, Z \mp 1)$, both parent and daughter states having $J^\pi = 0^+$ and $T = 1$. One has the general relation for the transition strength:

$$|\langle \text{daughter}|T_+|\text{parent}\rangle|^2 \equiv 2(1 - \delta_c) , \tag{2}$$

where δ_c represents the effect of isospin mixing caused by Coulomb and other isospin-breaking forces. Here, we restrict the discussion to the Coulomb force. For each nucleus, one can deduce from the *measured* ft value a *nucleus-independent* $\mathcal{F}t$ value:

$$\mathcal{F}t = ft(1 + \delta'_R)(1 + \delta_{NS} - \delta_c) , \tag{3}$$

where δ'_R and δ_{NS} are radiative corrections which can be evaluated [10]. Then, the value of G_V is obtained as:

$$G_V^2 = \frac{K}{2(1 + \Delta_R)\mathcal{F}t} , \tag{4}$$

K being a known numerical constant and Δ_R comes from radiative corrections.

We have calculated the values of δ_c for a number of $T = 1$ nuclei by charge-exchange RHF-RPA. In all cases the spherical symmetry is assumed and the filling approximation is applied to the last partially occupied orbitals. It must be noted that all meson-induced interactions (isoscalar as well as isovector) participate to the RPA p-h interaction whereas the Coulomb interaction contributes only to the single-particle energies but not to the p-h interaction which is of neutron-proton nature. The δ_c values calculated in RHF-RPA and RMF-RPA are compared in Table 1. The parameter set DD-ME2 [11] was determined for RMF calculations while PKO1 [4] was constructed for RHF calculations, and they are consequently used here for this comparison. It is seen that a systematic difference is predicted by the two models, the deviations being more pronounced in lighter nuclei. It is easy to understand where the deviations come from by repeating the RHF-RPA calculations without the Coulomb exchange contributions in the mean field (PKO1* column of Table 1). The corresponding results for δ_c coincide closely with those predicted by DD-ME2. We have concentrated the discussion here to a comparison of PKO1 for the RHF side and DD-ME2 for the RMF side, but similar conclusions are reached [8] if one compares with other RMF models. Thus, one clearly sees that Coulomb exchange contributions are important for this kind of study while the values of δ_c are rather insensitive to the particular effective Lagrangians as long as they predict nuclear ground states reasonably well.

In Fig.1 the nucleus-independent $\mathcal{F}t$ values are plotted as a function of the charge Z of the daughter nucleus. For comparison the uncorrected experimental ft values [12] and the partially corrected values (including only radiative corrections) are also shown. This figure shows the respective importance of radiative and isospin mixing corrections. The latter corrections become more important as the charge Z increases. We find an average value $\langle \mathcal{F}t \rangle$=3081.4(7)s leading to $|V_{ud}|$=0.97273(27). Combining with the values of $|V_{us}|$ and $|V_{ub}|$ [13] the sum of squared matrix elements of the CKM matrix top row is displayed in Fig.2 for the different cases. Our prediction deviates significantly from the unitarity condition while it agrees with those obtained by other experimental methods, especially with the most recent mirror transition results [14].

Table 1. Isospin symmetry-breaking corrections δ_c for the $0^+ \to 0^+$ superallowed transitions obtained by RHF-RPA with PKO1 [4] and RMF-RPA with DD-ME2 [11]. The column PKO1* presents the results obtained with PKO1 without the Coulomb exchange (Fock) term. All values are expressed in %

	PKO1	PKO1*	DD-ME2
$^{10}C \to {}^{10}B$	0.082	0.148	0.150
$^{14}O \to {}^{14}N$	0.114	0.178	0.197
$^{18}Ne \to {}^{18}F$	0.270	0.357	0.430
$^{26}Si \to {}^{26}Al$	0.176	0.246	0.252
$^{30}S \to {}^{30}P$	0.497	0.625	0.633
$^{34}Ar \to {}^{34}Cl$	0.268	0.359	0.376
$^{38}Ca \to {}^{38}K$	0.313	0.406	0.441
$^{42}Ti \to {}^{42}Sc$	0.384	0.460	0.523
$^{26}Al \to {}^{26}Mg$	0.139	0.193	0.198
$^{34}Cl \to {}^{34}S$	0.234	0.298	0.307
$^{38}K \to {}^{38}Ar$	0.278	0.344	0.371
$^{42}Sc \to {}^{42}Ca$	0.333	0.395	0.448
$^{54}Co \to {}^{54}Fe$	0.319	0.392	0.393
$^{66}As \to {}^{66}Ge$	0.475	0.571	0.572
$^{70}Br \to {}^{70}Se$	1.140	1.234	1.268
$^{74}Rb \to {}^{74}Kr$	1.088	1.230	1.258

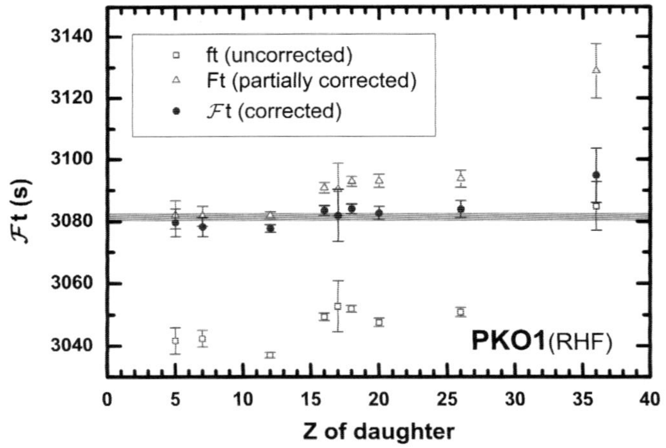

Figure 1. Corrected $\mathcal{F}t$ values by RHF-RPA with PKO1 (full circles) as a function of the charge Z for the daughter nucleus. The shaded horizontal band gives one standard deviation around the average $\mathcal{F}t$ value. The uncorrected experimental ft values [12] (open squares) and partially corrected ($\delta_c = 0$) Ft values (open triangles) are shown for comparison.

3. Inclusive neutrino-nucleus cross sections

As a second example of RHF-RPA application we consider the charged-current neutrino-nucleus reactions

$$\nu_l +_Z X_N \to_{Z+1} X^*_{N-1} + l^-, \tag{5}$$

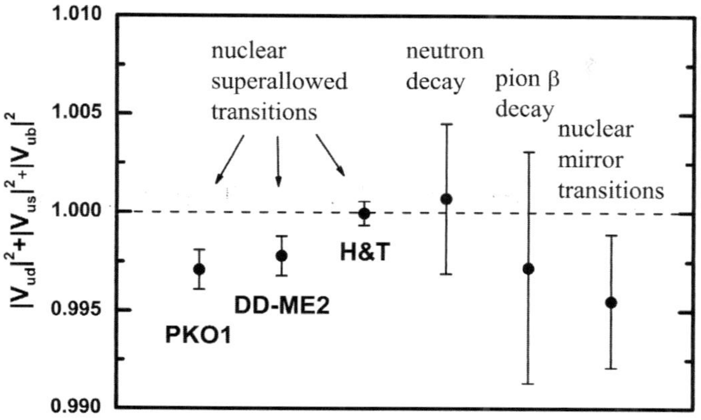

Figure 2. The sum of squared top row elements of the CKM matrix from super-allowed transitions with corrections calculated by RHF-RPA (PKO1), RMF-RPA (DD-ME2), and shell model (H&T) [12]. Values deduced from measurements of neutron decay [13], pion β decay [15] and nuclear mirror transitions [14] are also shown. Taken from Ref. [16].

where l denotes the charged lepton, e.g., electron or muon. The charged-current neutrino-nucleus cross section reads [17, 18]

$$\frac{d\sigma_\nu}{d\Omega} = \frac{V^2}{(2\pi)^2} p_l E_l \sum_{\text{lepton spins}} \frac{1}{2J_i+1} \sum_{M_i M_f} \left| \langle f | \hat{H}_W | i \rangle \right|^2, \tag{6}$$

where p_l and E_l are the momentum and energy of the outgoing lepton, respectively. For charged current reactions this cross section has to be corrected for the distortion of the outgoing lepton wave function by the Coulomb field of the daughter nucleus. In the low energy region this can be done by multiplying the cross section by a Fermi function whereas in the high energy region this effect is handled by the effective momentum approximation (EMA) [19].

The Hamiltonian $\hat{H}_W$ of the weak interaction is expressed in the standard current-current form, i.e., in terms of the nucleon $\mathcal{J}_\lambda(\boldsymbol{x})$ and lepton $j_\lambda(\boldsymbol{x})$ currents [17, 18]

$$\hat{H}_W = -\frac{G}{\sqrt{2}} \int d\boldsymbol{x} \mathcal{J}^\lambda(\boldsymbol{x}) j_\lambda(\boldsymbol{x}), \tag{7}$$

Denoting the leptonic matrix current as $l_\lambda e^{-i\boldsymbol{q}\cdot\boldsymbol{x}}$, the transition matrix elements read

$$\langle f | \hat{H}_W | i \rangle = -\frac{G}{\sqrt{2}} l_\lambda \int d\boldsymbol{x} e^{-i\boldsymbol{q}\cdot\boldsymbol{x}} \langle f | \mathcal{J}^\lambda(\boldsymbol{x}) | i \rangle, \tag{8}$$

with the four-momentum transfer $(q_0, \boldsymbol{q}) = (E_l, \boldsymbol{p}_l) - (E_\nu, \boldsymbol{p}_\nu)$, which must be space-like, i.e., $q^2 = q_0^2 - \boldsymbol{q}^2 \leqslant 0$.

Using the set of excited states of the final nucleus calculated with our charge-exchange RHF-RPA model we can obtain the differential neutrino-nucleus cross sections in the extreme relativistic limit (ERL) where the energy of the outgoing lepton is much larger than its rest mass. This leads to the total cross sections if one integrates the differential cross sections over all directions.

There are two issues that we wish to discuss in this short presentation. The first one has to do with which states one should include in the summation over final states. In the daughter nucleus spectrum the excitation energy of the final state is:

$$E^* = E_{\mathrm{RPA}} - [(m_f - m_i) + (m_n - m_p)] \ , \tag{9}$$

where E_{RPA} is the excitation energy with respect to the ground state of the parent nucleus, m_i (m_f) is the calculated mass of the parent (daughter) nucleus and $(m_n - m_p)$ is the neutron-proton mass difference. Some authors choose to include in their summation the RPA states whose RPA energies are larger than the threshold value

$$Q^{\mathrm{exp}}_{\mathrm{thr}} = [(m_f^{\mathrm{exp}} - m_i^{\mathrm{exp}}) + (m_n - m_p)] \ . \tag{10}$$

However, in microscopic models the predicted values $m_f - m_i$ may differ somehow from $m_f^{\mathrm{exp}} - m_i^{\mathrm{exp}}$ and we prefer not to use this empirical threshold value in the RPA summations to keep the consistency of the model. Furthermore, it would be impossible to apply this recipe in exotic regions of the nuclear chart where no experimental data on masses are available.

Table 2. Inclusive cross sections of the reactions $^{16}\mathrm{O}(\nu_e,e^-)^{16}\mathrm{F}$ averaged over Michel flux from the DAR of μ^+ [20]. The RHF-RPA results are compared with those from Refs. [21, 22, 23, 24, 25, 26]. All units are in 10^{-42} cm^2.

	$g_A = 1.262$		$g_A^{\mathrm{eff}} = 1.000$	
	all E_{RPA}	$E_{\mathrm{RPA}} \geqslant Q^{\mathrm{exp}}_{\mathrm{th}}$	all E_{RPA}	$E_{\mathrm{RPA}} \geqslant Q^{\mathrm{exp}}_{\mathrm{th}}$
PKO1	12.19	8.67	8.31	6.00
PKO2	12.12	8.77	8.24	6.05
DD-ME2 [21, 22]				6.00
shell model [23]	16.9		10.8	
SIII [24]	13.1			
HFSk [25]		9.43		
WSSk [25]		6.67		
RPA [26]		14.55		

The second issue concerns the value of g_A adopted in different calculations. While the empirical value is $g_A = 1.262$ some authors use the effective value $g_A^{\mathrm{eff}} = 1$ introduced by Bohr and Mottelson. There are mainly two reasons for adopting an effective value but we feel that they are not appropriate in the present circumstances. An effective coupling constant might be necessary in a shell model description due to the limited model space but in our RPA calculations the space is large enough to exhaust the total transition strength. The other reason has to do with the long-lasting quenching factor (of about 2/3) observed in the Gamow-Teller strengths. However, the multipole decomposition analysis of Yako et al. [27] has shown that about 88% of Gamow-Teller strength is detected in (p,n) and (n,p) reactions on ^{90}Zr. Furthermore, it is well known that the RPA can reproduce satisfactorily the total μ^- capture rate in ^{16}O without any quenching of g_A [28, 29]. Thus, we favor the use of g_A over g_A^{eff} in applying our RPA model to predicting neutrino-nucleus cross-sections. We nevertheless mention in Table 2 the results we obtain when applying the other prescriptions in order to make meaningful comparisons with other predictions.

In Table 2 we present our results of inclusive cross sections of the reaction $^{16}\mathrm{O}(\nu_e, e^-)^{16}\mathrm{F}$. They correspond to neutrinos from the decay at rest of μ^+ and averaged over the Michel flux [20]. We compare our predictions with those of other models. The strongest disagreements occur when comparing with those of Refs. [23, 26]. The RHF-RPA results are not very sensitive to the particular parametrization - PKO1 or PKO2 - used.

4. Conclusion

The RHF-RPA is a new tool for exploring in a consistent way many different aspects of nuclear excitations. Its main merit is that, once the density-dependent couplings are fixed by the overall ground state properties there is no additional degree of freedom in the effective Lagrangians and the description of the excited states remains consistent with that of the ground states within the random phase approximation.

We have presented here two examples of studies performed with RHF-RPA, choosing to concentrate on the charge-exchange channels and applying the method to problems of current interest. On the question of isospin mixing in super-allowed transitions there should be little room left for model dependence in RPA predictions if the Coulomb force provides the main effects. Our results still do not contain contributions from charge-symmetry and charge-independence breaking forces which are present and are known to affect the isospin mixing [30]. This should be further studied in the RHF-RPA context. The second example shows that it is possible to apply the present approach to evaluate the neutrino-nucleus cross-sections, and this should be quite useful for future experimental studies. In both examples the RHF-RPA method allows one to describe the isospin and spin-isospin excited states more consistently than in the RMF-RPA approach.

Acknowledgements

The authors are grateful to Nils Paar for helpful discussions. This work is partly supported by State 973 Program 2007CB815000, the NSF of China under Grant Nos. 10775004, 10947013 and 10975008.

References

[1] Ring P 1996 *Prog. Part. Nucl. Phys.* **37** 193
[2] Meng J, Toki H, Zhou S G, Zhang S Q, Long W H and Geng L S 2006 *Prog. Part. Nucl. Phys.* **57** 470
[3] Paar N, Vretenar D, Khan E and Colò G 2007 *Rep. Prog. Phys.* **70** 691
[4] Long W H, Van Giai N and Meng J 2006 *Phys. Lett. B* **640** 150
[5] Long W H, Ring P, Van Giai N and Meng J 2010 *Phys. Rev. C* **81** 024308
[6] Liang H, Van Giai N and Meng J 2008 *Phys. Rev. Lett.* **101** 122502
[7] Long W H, Sagawa H, Van Giai N and Meng J 2007 *Phys. Rev. C* **76** 034314
[8] Liang H, Van Giai N and Meng J 2009 *Phys. Rev. C* **79** 064316
[9] Liang H 2010 Ph.D. thesis Université Paris-Sud (unpublished)
[10] Towner I S and Hardy J C 2008 *Phys. Rev. C* **77** 025501
[11] Lalazissis G A, Nikšić T, Vretenar D and Ring P 2005 *Phys. Rev. C* **71** 024312
[12] Hardy J C and Towner I S 2009 *Phys. Rev. C* **79** 055502
[13] Amsler C *et al.* 2008 *Phys. Lett. B* **667** 1
[14] Naviliat-Cuncic O and Severijns N 2009 *Phys. Rev. Lett.* **102** 142302
[15] Počanić D *et al.* 2004 *Phys. Rev. Lett.* **93** 181803
[16] Liang H, Van Giai N and Meng J 2010 *J. Phys. Conf. Ser.* **205** 012028
[17] O'Connell J S, Donnelly T W and Walecka J D 1972 *Phys. Rev. C* **6** 719
[18] Walecka J D 1975 *Muon Physics* (Academic Press, New York)
[19] Engel J 1998 *Phys. Rev. C* **57** 2004
[20] Krakauer D A *et al.* 1992 *Phys. Rev. C* **45** 2450
[21] Paar N, Vretenar D, Marketin T and Ring P 2008 *Phys. Rev. C* **77** 024608
[22] Paar N *private communication*
[23] Auerbach N and Brown B A 2002 *Phys. Rev. C* **65** 024322
[24] Lazauskas R and Volpe C 2007 *Nucl. Phys. A* **792** 219
[25] Jachowicz N, Heyde K, Ryckebusch J and Rombouts S 2002 *Phys. Rev. C* **65** 025501
[26] Sajjad Athar M, Ahmad S and Singh S 2006 *Nucl. Phys. A* **764** 551
[27] Yako K *et al.* 2005 *Phys. Lett. B* **615** 193
[28] Van Giai N, Auerbach N and Mekjian A Z 1981 *Phys. Rev. Lett.* **46** 1444
[29] Marketin T, Paar N, Nikšić T and Vretenar D 2009 *Phys. Rev. C* **79** 054323
[30] Sagawa H, Van Giai N and Suzuki T 1996 *Phys. Rev. C* **53** 2163

Covariant density functional theory with spectroscopic properties and a microscopic theory of quantum phase transitions in nuclei

2011 J. Phys.: Conf. Ser. 267 012043

(http://iopscience.iop.org/1742-6596/267/1/012043)

Covariant density functional theory with spectroscopic properties and a microscopic theory of quantum phase transitions in nuclei.

Ring P[1], Lalazissis G A [2] Li Z P[3], Meng J[3], Nikšić T[4], Próchniak L[5], Vretenar D[4], and Yao J M[6]

[1] Physik-Department der Technischen Universität München, D-85748 Garching, Germany
[2] Department of Theoretical Physics, Aristotle University Thessaloniki, GR-54124, Greece
[3] School of Physics, Peking University, Beijing 100871, China,
[4] Physics Department, Faculty of Science, University of Zagreb, 10000 Zagreb, Croatia,
[5] Institute of Physics, Maria Curie-Sklodowska University, Lublin, Poland.
[5] School of Physical Science and Technology, Southwest University, Chongqing 400715, China.

E-mail: `ring@ph.tum.de`

Abstract. Covariant Density functional theory is used as a basis for a microscopic description of spectroscopic properties of quantum phase transitions in nuclei. Since it is well known that the mean field approximation breaks down in transitional nuclei, where configuration mixing and fluctuations connected with broken symmetries play an important role, a theory is developed which uses the Relativistic Generator Coordinate Method to perform configuration mixing calculations of angular momentum and particle number projected wave functions. As applications with show 3D-calculations of the spectrum of low-lying collective states in the nucleus ^{24}Mg. This method can also be used to study the behavior of characteristic physical quantities as a function of the physical control parameter, the number of nucleons, in the region of quantum phase transitions.

1. Introduction

Theoretical approaches based on Covariant Density Functional Theory (CDFT) remain undoubtedly among the most successful microscopic descriptions of nuclear structure. The CDFT approaches are derived from a Lorentz invariant density functional which connects in a consistent way the spin and spatial degrees of freedom in the nucleus. Therefore, it needs only a relatively small number of parameters which are adjusted to reproduce a set of bulk properties of spherical closed-shell nuclei [1, 2] and it is valid over the entire periodic table. Over the years, Relativistic Mean-Field (RMF) models based on the CDFT have been successfully applied to describe ground state properties of finite spherical and deformed nuclei over the entire nuclear chart [3] from light nuclei [4] to super-heavy elements [5, 6], from the neutron drip line where halo phenomena are observed [7], to the proton drip line [8] with nuclei unstable against the emission of protons [9]. The relativistic cranking approximation has been developed to calculate rotational bands [10, 11]. For a description of nuclear excited states, the Relativistic Random Phase Approximation (RRPA) [12] and the quasiparticle RRPA (RQRPA) [13] have been formulated as the small amplitude limit of the time-dependent RMF models. These models have provided a very good description for the positions of giant resonances and a theoretical interpretation of the low-lying dipole [13] and quadrupole [14, 15] excitations. Proton-neutron versions of the RRPA and the RQRPA have been developed and successfully applied to the description of spin/isospin excitations as the Isobaric Analog Resonance (IAR) or the Gamow-Teller Resonance (GTR) [16].

This framework has, however, its limitations. It is based on the mean-field concept which is characterized by symmetry breaking. and can only provide an approximate description of specific nuclear excitations, such as collective rotations in the framework of the Cranking model or vibrations in the framework of time-dependent mean field theory, which corresponds in the small amplitude limit to RPA or QRPA. To calculate excitation spectra with specific quantum numbers and electromagnetic transition rates in individual nuclei, it is necessary to extend this framework to include collective correlations related to the restoration of broken symmetries and to fluctuations of collective coordinates. Collective correlations are sensitive to shell effects, display pronounced variations with particle number, and cannot be incorporated in a single Slater determinant.

On this level density functional theory takes into account collective correlations through the restoration of broken symmetries and configuration mixing of symmetry-breaking product states. In recent years several accurate and efficient models and algorithms have been developed that perform the restoration of symmetries broken by the static nuclear mean field, and take into account fluctuations around the mean-field minimum. The most effective approach to configuration mixing calculations is the generator coordinate method (GCM) [17, 18]. With the simplifying assumption of axial symmetry, GCM configuration mixing of angular-momentum, and even particle-number projected quadrupole-deformed mean-field states, has become a standard tool in nuclear structure studies with non-relativistic [19, 20], and relativistic density functionals [21, 22].

Much more involved and technically difficult is the description of intrinsic quadrupole modes including triaxial deformations. Only very recently fully microscopic three-dimensional calculations have been presented for Skyrme [23] and Gogny [24] mean field models generated by triaxial quadrupole constraints that are projected and mixed by the generator coordinate method. This method is actually equivalent to a seven-dimensional GCM calculation, mixing all five degrees of freedom of the quadrupole operator and the gauge angles for protons and neutrons. Here we report on configuration mixing of angular-momentum projected triaxial relativistic mean-field wave functions based on covariant density functionals [25].

2. The relativistic generator coordinate method.

The generator coordinate method is based on the assumption that, starting from a set of mean-field states $|\Phi(q)\rangle$ which depend on collective coordinates q, one can build approximate eigenstates of the nuclear Hamiltonian by linear combination of such states:

$$|\Psi_\alpha\rangle = \int dq f_\alpha(q) \, |\Phi(q)\rangle \, . \tag{1}$$

Detailed discussions of the GCM method can be found in Ref. [17]. To be able to compare theoretical predictions with data, it is necessary to construct states with good angular momentum

$$|\Psi_\alpha^{JM}\rangle \;=\; \int dq \sum_{K\geq 0} f_\alpha^{JK}(q) \frac{1}{(1+\delta_{K0})} |JMK+, q\rangle \tag{2}$$

α labels collective eigenstates for a given angular momentum J. The details of the 3D angular-momentum projection in the relativistic case are given in Ref. [26]. The basis states $|JMK+, q\rangle$

are projected from the intrinsic wave functions $|\Phi(q)\rangle$:

$$|JMK+,q\rangle = [\hat{P}^J_{MK} + (-1)^J \hat{P}^J_{M-K}]|\Phi(q)\rangle, \tag{3}$$

where $\hat{P}^J_{MK}$ is the angular-momentum projection operator.

The set of intrinsic wave functions $|\Phi(q)\rangle$, with the generic notation for quadrupole deformation parameters $q \equiv (\beta, \gamma)$, is generated by imposing constraints on the axial q_{20} and triaxial q_{22} mass quadrupole moments in self-consistent RMF+BCS calculations. The following calculations are based on the relativistic point coupling model PC-F1 [27]. Pairing correlations are taken into account in the simple BCS approximation, using a δ-pairing force.

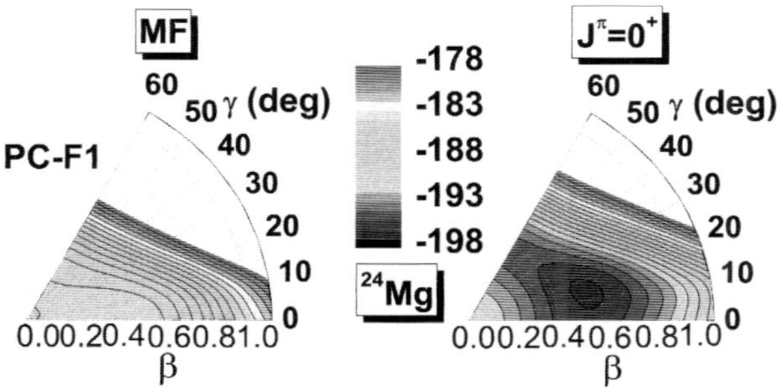

Figure 1. (Color online) Self-consistent RMF+BCS energy surface (left panel) of ^{24}Mg in the β-γ plane ($0 \leq \gamma \leq 60^0$), and angular momentum projected energy surface with $J^\pi = 0^+$ (right panel). The contours join points on the surface with the same energy. The difference between neighboring contours is 1.0 MeV.

The weight functions $f^{JK}_\alpha(q)$ in the collective wave function Eq. (2) are determined from the variation of the energy by the Hill-Wheeler-Griffin (HWG) integral equation:

$$\int dq' \sum_{K' \geq 0} \left[\mathscr{H}^J_{KK'}(q,q') - E^J_\alpha \mathscr{N}^J_{KK'}(q,q')\right] f^{JK'}_\alpha(q') = 0, \tag{4}$$

where $\mathscr{H}$ and $\mathscr{N}$ are the angular-momentum projected GCM kernel matrices of the Hamiltonian and the norm, respectively.

As an example for an illustrative configuration mixing calculation we investigate the spectrum of the nucleus ^{24}Mg. The intrinsic wave functions have been obtained as solutions of the self-consistent relativistic mean-field equations, subject to constraint on the triaxial mass quadrupole moments.

Fig. 1 shows the self-consistent RMF+BCS triaxial energy surface in the $\beta - \gamma$ plane, The panel on the right displays the projected energy surface with J = 0. The energy surfaces nicely illustrate the effects of including triaxial shapes and of the restoration of rotational symmetry. The mean-field energy surfaces are found to be quite soft with a minimum at an axial prolate deformation $\beta = 0.5$. Projection shifts the minimum to a triaxial shape with $\beta = 0.5$ and $\gamma = 0.20°$. The gain in energy from the restoration of rotational symmetry is 4.266 MeV. The fact that angular momentum projection leads to triaxial minima in the PES was already been

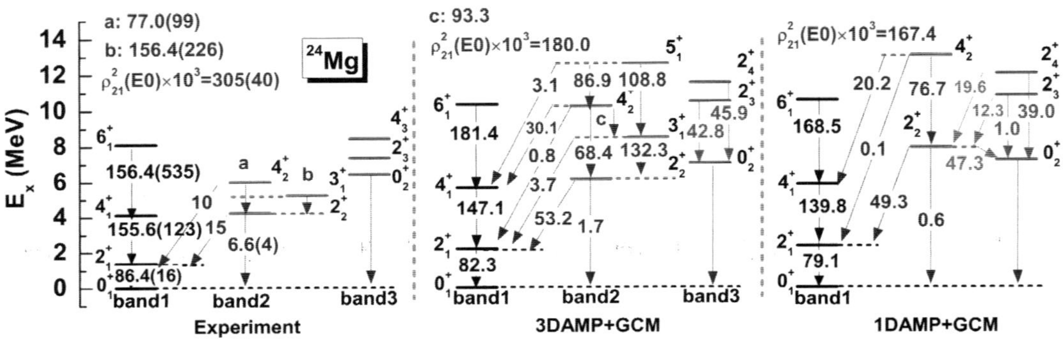

Figure 2. (Color online) The low-spin level scheme of ^{24}Mg calculated using the 3DAMP+GCM model and 1DAMP+GCM model with the PC-F1 relativistic density functional, in comparison with data [28, 29, 30]. The $B(E2)$ values are given in units of e^{2}fm^4.

noted in 3DAMP calculations in the eighties [31] and very similar results have been obtained recently using the Skyrme functional SLy4 [23].

Fig. 2 displays the spectrum of the nucleus ^{24}Mg, calculated with the 3DAMP+GCM and 1DAMP+GCM codes. The 3DAMP+GCM and 1DAMP+GCM calculations produce virtually identical results. The level scheme is in rather good agreement with data, but in both cases the calculated spectra are systematically stretched as compared to experimental bands. This is because angular-momentum projection is performed only after variation and, therefore, time-odd components and alignment effects are neglected. Cranking calculations, for instance, correspond to an approximate angular-momentum projection before variation [32], and lead to an enhancement of the moments of inertia in better agreement with data [11, 33]. The agreement of the calculated quadrupole transition probabilities with data in Fig. 2 is remarkable, especially considering that the calculation of B(E2) values is parameter-free, i.e. the transitions are calculated employing bare proton charges.

The nucleus ^{24}Mg presents an illustrative test case for the 3DAMP+GCM approach to low-energy nuclear structure. However, collective phenomena are, of course, much more pronounced in heavy nuclei and, therefore, the goal is to eventually apply the present approach to rare earth nuclei and the actinide region. At present a straightforward application of the existing 3DAMP+GCM codes to heavy nuclei is not possible. On the other hand, the introduction of additional approximations can considerably reduce the computing requirements. For instance, the overlap functions are strongly peaked at $q = q'$, and the use of the Gaussian overlap approximation has produced excellent results in many cases. These approximations form the basis for a microscopic derivation of a collective Bohr Hamiltonian for quadrupole degrees of freedom [34, 35] based on covariant density functionals.

3. A microscopic theory of quantum phase transitions

Quantum phase transitions from spherical to deformed shapes have recently gained considerable interest. However, most of these investigations have carried out within simple boson models [37]. The Generator Coordinate Method has recently been applied for a microscopic theory of quantum phase transitions in finite nuclei. Starting from constrained self-consistent mean-field calculations of potential energy curves, the one-dimensional generator coordinate

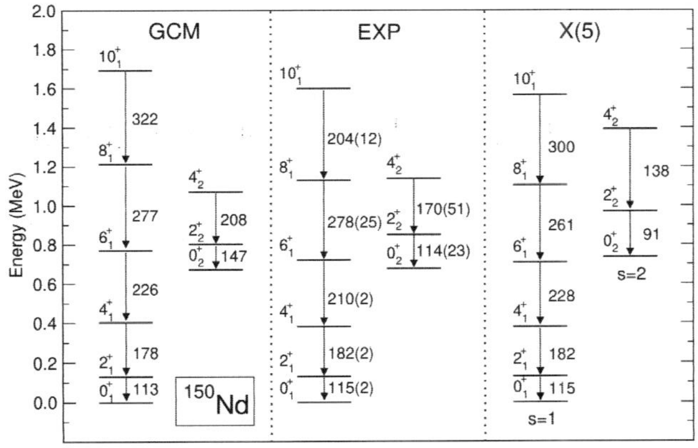

Figure 3. The particle-number projected GCM spectrum of ^{150}Ne (left), compared with the data [36] (middle), and the X(5)-symmetry predictions (right) for the excitation energies, and B(E2) values (in Weisskopf units) of the ground-state and the β-band. The theoretical spectra are normalized to the experimental energy of the state 2_1^+, and the X(5) transition strengths are normalized to the experimental B(E2;$2_1^+ \rightarrow 0_1^+$).

method (GCM) is used to perform configuration mixing of angular-momentum and particle-number projected relativistic wave functions [38]. It is shown that the microscopic framework based on universal density functionals, adjusted to nuclear ground-state properties and extended to take into account correlations related to symmetry restoration and fluctuations of collective variables, describes not only general features of shape transitions but also the unique behavior of excitation spectra and transition rates at the critical point of quantum shape phase transition. In particular the shape phase transition occurs precisely at the isotope ^{150}Nd, in agreement with empirical evidence for the realization of X(5) in the N = 90 rare-earth isotones. As shown in Fig. 3 the calculated spectrum and the intra-band and inter-band B(E2) transitions for ^{150}Nd are in excellent agreement with data [36] and close to the X(5)-model predictions for the nucleus ^{150}Nd [37].

Because of the considerable numerical effort 3D angular momentum projected GCM calculations are at present not yet possible for such heavy nuclei as the Ne-chain. Therefore this phase transition has been investigated recently [39] on the basis of a 5-dimensional Bohr hamiltonian, whose parameters were derived microscopically form the covariant density functional PC-F1. The resulting spectrum and the electromagnetic transition probabilities are more or less the same as in the 1D-AMP GCM calculation shown in Fig. 3.

4. Summary and outlook.

Among the microscopic approaches to the nuclear many-body problem, the framework of nuclear density functional theory provides the most complete description of ground-state properties and collective excitations over the whole nuclide chart. The self-consistent mean-field models based on the density functional framework describe not only general features of shape transitions but also particular properties of spectra and transition rates at the critical point of the QPT.

However, to calculate excitation spectra and transition probabilities, the self-consistent mean-field approach has to be extended to include correlations related to restoration of broken symmetries and fluctuations of collective variables. This can be done either by performing GCM configuration mixing calculations of projected wave functions or by constructing collective Bohr-type Hamiltonians with deformation-dependent parameters determined from self-consistent mean-field calculations. The possibility to perform self-consistent microscopic studies of shape transitions opens a new perspective on the origin of nuclear quantum phase transitions in various mass regions. It is therefore important to systematically analyze these methods, also employing different energy-density functionals, various types of shape phase transitions that have been predicted in several regions of medium-heavy and heavy nuclei.

Acknowledgments

This work was partly supported by the DFG cluster of excellence "Origin and Structure of the Universe" (www.universe-cluster.de).

References

[1] Ring P 1996 *Prog. Part. Nucl. Phys.* **37** 193
[2] Vretenar D, Afanasjev A V, Lalazissis G A and Ring P 2005 *Phys. Rep.* **409** 101
[3] Gambhir Y K, Ring P and Thimet A 1990 *Ann. Phys. (N.Y.)* **198** 132
[4] Lalazissis G A, Vretenar D and Ring P 2004 *Euro. Phys. J.* **A22** 37
[5] Lalazissis G A, Sharma M M, Ring P and Gambhir Y K 1996 *Nucl. Phys.* **A608** 202
[6] Bürvenich T, Bender M, Maruhn J A and Reinhard P G 2004 *Phys. Rev.* **C69** 014307
[7] Meng J and Ring P 1996 *Phys. Rev. Lett.* **77** 3963
[8] Lalazissis G A, Vretenar D and Ring P 2004 *Phys. Rev.* **C69** 017301
[9] Lalazissis G A, Vretenar D and Ring P 1999 *Nucl. Phys.* **A650** 133
[10] Koepf W and Ring P 1989 *Nucl. Phys.* **A493** 61
[11] Afanasjev A V, Ring P and König J 2000 *Nucl. Phys.* **A676** 196
[12] Ring P, Ma Z Y, Van Giai N, Vretenar D, Wandelt A and Cao L G 2001 *Nucl. Phys.* **A694** 249
[13] Paar N, Ring P, Nikšić T and Vretenar D 2003 *Phys. Rev.* **C67** 034312
[14] Ansari A 2005 *Phys. Lett.* **B623** 37
[15] Ansari A and Ring P 2006 *Phys. Rev.* **C74** 054313
[16] Paar N, Nikšić T, Vretenar D and Ring P 2004 *Phys. Rev.* **C69** 054303
[17] Ring P and Schuck P 1980 *The Nuclear Many-Body Problem* (Berlin: Springer-Verlag)
[18] Blaizot J P and Ripka G 1986 *Quantum Theory of Finite Systems* (Cambridge Mass.: MIT Press)
[19] Valor A, Heenen P H and Bonche P 2000 *Nucl. Phys.* **A671** 145
[20] Rodríguez-Guzmán R, Egido J L and Robledo L M 2002 *Nucl. Phys.* **A709** 201
[21] Nikšić T, Vretenar D and Ring P 2006 *Phys. Rev.* **C73** 034308
[22] Nikšić T, Vretenar D and Ring P 2006 *Phys. Rev.* **C74** 064309
[23] Bender M and Heenen P H 2008 *Phys. Rev.* **C78** 024309
[24] Rodriguez T R and Egido J L 2010 *Phys. Rev.* **C81** 064323
[25] Yao J M, Meng J, Ring P and Vretenar D 2010 *Phys. Rev.* **C81** 044311
[26] Yao J M, Meng J, Ring P and Peña Arteaga D 2009 *Phys. Rev.* **C79** 044312
[27] Bürvenich T, Madland D G, Maruhn J A and Reinhard P G 2002 *Phys. Rev.* **C65** 044308
[28] Endt P M 1993 *At. Data Nucl. Data Tables* **55** 171
[29] Branford D, McGough A C and Wright I F 1975 *Nucl. Phys.* **A241** 349
[30] Keinonen *et al* J 1989 *Nucl. Phys.* **A493** 124
[31] Hara K, Hayashi A and Ring P 1982 *Nucl. Phys.* **A385** 14
[32] Beck R, Mang H J and Ring P 1970 *Z. Phys.* **231** 26
[33] Afanasjev A V, König J, Ring P, Robledo L M and Egido J L 2000 *Phys. Rev.* **C62** 054306
[34] Próchniak L and Ring P 2004 *Int. J. Mod. Phys.* **E13** 217
[35] Nikšić T, Li Z P, Vretenar D, Próchniak L, Meng J and Ring P 2009 *Phys. Rev.* **C79** 034303
[36] Kruecken *et al* R 2002 *Phys. Rev. Lett.* **88** 232501
[37] Iachello F 2001 *Phys. Rev. Lett.* **87** 052502
[38] Nikšić T, Vretenar D, Lalazissis G A and Ring P 2007 *Phys. Rev. Lett.* **99** 092502
[39] Li Z P, Nikšić T, Vretenar D, Meng J, Lalazissis G A and Ring P 2009 *Phys. Rev.* **C79** 054301

Progression towards an effective shell-model theory of rotational nuclei

2011 J. Phys.: Conf. Ser. 267 012044

(http://iopscience.iop.org/1742-6596/267/1/012044)

Progression towards an effective shell-model theory of rotational nuclei

David J Rowe

Department of Physics, University of Toronto, Toronto, ON M5S 1A7, Canada

E-mail: rowe@physics.utoronto.ca

Abstract. A goal of nuclear physics is to determine the validity of the nuclear shell model and to express the various models of nuclear phenomena as sub-models of this model. It is also profitable to start from a phenomenological model interpretation of nuclear data and use the results to design an effective shell-model description of the data. This paper suggests a method for deriving effective shell-model spaces for rotational nuclei and indicates that new methods are needed for deriving effective interactions for rotational nuclei.

1. Introduction

As a practical matter, it is necessary to restrict nuclear shell-model calculations to finite-dimensional subspaces of the full many-nucleon Hilbert space and to treat the nucleons as elementary particles. Thus, the interactions between the nucleons of the shell model are effective interactions and the properties of calculated shell-model states must be interpreted in terms of appropriate effective operators.

An A-nucleon Hilbert space is an anti-symmetrised, i.e., exterior, product,

$$\mathbb{H}^{(A)} = \mathbb{H}^{(1)} \wedge \mathbb{H}^{(1)} \wedge \cdots \wedge \mathbb{H}^{(1)}, \quad A \text{ copies}, \tag{1}$$

of single-nucleon Hilbert spaces. To facilitate the separation of the centre-of-mass component of $\mathbb{H}^{(A)}$, and exploit the symmetry properties of the many sub-models of the shell model, it is convenient to define basis wave functions for $\mathbb{H}^{(A)}$ in terms of anti-symmetric products of single-nucleon harmonic-oscillator wave functions.

Derivations of effective shell model operators on a finite-dimensional subspace of $\mathbb{H}^{(A)}$ proceed in essentially two steps. The first step restricts the single-nucleon space, $\mathbb{H}^{(1)}$, to a (very) large but finite-dimensional subspace, $\tilde{\mathbb{H}}^{(1)} \subset \mathbb{H}^{(1)}$. Because two- and three-nucleon problems can be solved to a high degree of accuracy with bare-nucleon interactions, it is possible to derive an effective Hamiltonian, $\tilde{H}^{(A)}_{\rm eff}$, appropriate for the truncated many-nucleon space, $\tilde{\mathbb{H}}^{(A)} = \tilde{\mathbb{H}}^{(1)} \wedge \tilde{\mathbb{H}}^{(1)} \wedge \ldots$, with the understanding that the effective three- and more-nucleon interactions generated from bare two-nucleon interactions are negligible provided the truncated subspace $\tilde{\mathbb{H}}^{(1)}$ is sufficiently large (cf. [1] for a review of methods). The second step is to derive the effective interactions and operators appropriate for the chosen effective A-nucleon shell-model space, $\mathbb{H}^{(A)}_{\rm eff}$, for which calculations are to be carried out.

This talk addresses the question of how the effective space $\mathbb{H}^{(A)}_{\rm eff}$ should be chosen. Calculations to date have generally adopted the strategy of ordering the many-particle configurations by increasing energy with respect to the conventional spherical harmonic-oscillator independent-particle shell model Hamiltonian with a spin-orbit interaction. The effective shall-model space is then chosen to be that spanned by as many of the lowest-energy configurations as can be handled. This make sense for many states of light nuclei and medium-light singly-closed-shell nuclei which appear to be spherical in their lowest-energy states. However, there is much evidence to show that even the so-called spherical nuclei exhibit highly-deformed rotational states at low excitation energies [2, 3]. These states are commonly described as *intruder states* and are associated with higher-energy spherical harmonic-oscillator configurations that are not included in the standard effective shell-model space. In fact, it appears that such higher-energy harmonic oscillator configurations become the lowest-energy states in doubly-open shell nuclei. Thus, ordering shell-model subspaces by increasing energies with respect to a spherical harmonic-oscillator Hamiltonian with spin-orbit interactions appears to be inappropriate for the majority of doubly-open shell nuclei, which are deformed and exhibit rotational structure.

Much of what I have to say in addressing these questions has been known for 30 years. However, only recently has its relevance to effective interaction theory [4] and the widely observed phenomenon of shape coexistence [2, 3] become apparent.

2. The Bohr model and its microscopic extension

Collective vibrational and rotational states have been successfully interpreted in terms of collective models based on the Bohr model [5]. The latter is an incompressible liquid drop with a sharply-defined surface defined by a set of deformation parameters, $\alpha_{\lambda\nu}$, in an expansion

$$R(\theta, \varphi) = R_0[1 + \sum_{\lambda\nu} \alpha^*_{\lambda\nu} Y_{\lambda\nu}(\theta, \varphi) + \dots], \tag{2}$$

of its radius as a function of orientation angles, where $Y_{\lambda\nu}(\theta, \varphi)$ are standard spherical harmonics. Thus, it has only shape-vibrational and rotational degrees of freedom. For present purposes we restrict to quadrupole shape degrees of freedom for which $\lambda = 2$ and for which we define $\alpha_\nu \equiv \alpha_{2\nu}$. The dynamics of the Bohr model is described by a Hamiltonian

$$\hat{H} = \frac{1}{2B} \sum_\nu \hat{\pi}^\nu \hat{\pi}_\nu + V(\alpha), \tag{3}$$

where $\pi_\nu = B\dot{\alpha}_\nu$ is the momentum associated with the coordinate α_ν and $\hat{\pi}_\nu = -i\hbar\frac{\partial}{\partial \alpha^*_\nu}$ is its quantisation; $\hat{\pi}^\nu = -i\hbar\frac{\partial}{\partial \alpha_\nu}$. The Hilbert space for the Bohr model is then that of a five-dimensional harmonic oscillator. The spectra and properties of such a Hamiltonian can be calculated quickly and easily using the algebraic methods of ref. [6].

Our strategy is now to extend the Bohr model to a microscopic collective model that has representations carried by subspaces of shell-model states. These subspaces are then identified as the appropriate shell-model spaces needed for a microscopic description of collective-model states. A first step towards construction of such a model is achieved by replacing the quadrupole shape coordinates, $\{\alpha_\nu\}$, by the quadrupole and monopole moments of a system of N point nucleons with position coordinates $\{x_{ni}\}$:

$$Q_{ij} = \sum_{n=1}^{N} x_{ni} x_{nj}, \quad i, j = 1, 2, 3. \tag{4}$$

These moments have time derivatives $\dot{Q}_{ij} = \sum_{n=1}^{N} (x_{ni}\dot{x}_{nj} + x_{ni}\dot{x}_{nj})$. Hence, the canonical momenta are

$$P_{ij} = \sum_{n=1}^{N} (x_{ni} p_{nj} + p_{ni} x_{nj}), \quad i, j = 1, 2, 3, \tag{5}$$

where $\{p_{ni}\}$ are the single-nucleon momenta. The quantisation of a collective model with these coordinates is now given by $Q_{ij} \to \hat{Q}_{ij}$ and $P_{ij} \to \hat{P}_{ij}$, where $\hat{Q}_{ij}$ and $\hat{P}_{ij}$ act on many-nucleon wave functions according to standard quantum mechanics according to the equations

$$\hat{Q}_{ij}\Psi(x) = \sum_n x_{ni}x_{nj}\Psi(x), \qquad \hat{P}_{ij}\Psi(x) = \sum_n (x_{ni}\hat{p}_{nj} + \hat{p}_{ni}x_{nj})\Psi(x), \tag{6}$$

with $\hat{p}_{nj} = -\mathrm{i}\hbar\partial/\partial x_{nj}$. Thus, whereas the Bohr model observables satisfy the Heisenberg commutation relations

$$[\hat{\alpha}_\mu, \hat{\pi}^\nu] = \mathrm{i}\hbar\delta^\nu_\mu, \tag{7}$$

those of the microscopic collective model, satisfy the relations

$$[\hat{Q}_{ij}, \hat{P}_{kl}] = \mathrm{i}\hbar\big[\delta_{il}\hat{Q}_{ik} + \delta_{ik}\hat{Q}_{il} + \delta_{jl}\hat{Q}_{jk} + \delta_{jk}\hat{Q}_{jl}\big]. \tag{8}$$

The collective model group, generated by the operators $\{\hat{Q}_{ij}, \hat{P}_{ij}\}$, is known by the acronym GCM(3), where G (for general) signifies the inclusion of monopole degrees of freedom.

Whereas the Bohr model has only one unitary representation, the GCM(3) model has many [7], each of which is characterised by an intrinsic vorticity quantum number which can take any non-negative integer value. Thus, the microscopic collective model corrects one of the deficiencies of the Bohr model which is its restriction to irrotational quantum-fluid flows.

3. Embedding the microscopic collective model in the shell model

We now consider the microscopic collective model as a submodel of the shell model and relate the basis states for its irreps to those of a shell-model coupling scheme. This is achieved by extending the GCM(3) model to a larger symplectic model with an $\mathrm{Sp}(3,\mathbb{R}) \supset \mathrm{GCM}(3) \supset \mathrm{SO}(3)$ dynamical group [8]. This extension is useful because the Lie algebra of $\mathrm{Sp}(3,\mathbb{R})$ contains the moments of momenta,

$$\hat{K}_{ij} = \sum_n \hat{p}_{ni}\hat{p}_{nj}, \tag{9}$$

as well as the elements of the GCM(3) Lie algebra. In particular, it contains the many nucleon kinetic energy operator $\hat{T} = \frac{1}{2m}\sum_{ni}\hat{p}^2_{ni}$. Moreover, $\mathrm{Sp}(3,\mathbb{R})$ is a dynamical group for the spherical harmonic oscillator Hamiltonian and its representations are easy to construct within the harmonic-oscillator shell-model space. The construction is greatly facilitated by the classification of shell-model states by means of the subgroup chain

$$\mathrm{Sp}(3,\mathbb{R}) \times \mathrm{U}(4) \supset \mathrm{U}(3) \times \mathrm{U}(2)_S \times \mathrm{U}(2)_T \supset \mathrm{SO}(3) \times \mathrm{SU}(2)_S \times \mathrm{SU}(2)_T, \tag{10}$$

where $\mathrm{U}(4) \supset \mathrm{SU}(2)_S \times \mathrm{SU}(2)_T$ is Wigner's supermultiplet group; it contains the spin and isospin groups as subgroups. The group $\mathrm{U}(3) \supset \mathrm{U}(1) \times \mathrm{SU}(3)$ is the symmetry group of the spherical harmonic oscillator Hamiltonian and contains SU(3), the dynamical group of Elliott's model [9], as a subgroup. An important observation is that *the whole (multi-shell) Hilbert space of the shell model is spanned by irreducible representations of this chain of subgroups.* In fact, this subgroup chain defines an LST coupling scheme.

$\mathrm{Sp}(3,\mathbb{R})$ irreps can be labelled by $\mathrm{U}(1) \times \mathrm{SU}(3)$ quantum numbers $N(\lambda\mu)$ where $N\hbar\omega$ is the harmonic oscillator energy of the lowest-weight state; e.g., $N=36$ for the closed-shell state of $^{16}\mathrm{O}$ before removal of the centre-of-mass energy. They are simply constructed from the observation that the nuclear (mass) quadrupole tensor $\hat{Q}$ can be separated into three components:

$$\hat{Q}_{ij} = \hat{\mathcal{Q}}^{(0)}_{ij} + \hat{\mathcal{Q}}^{(+2)}_{ij} + \hat{\mathcal{Q}}^{(-2)}. \tag{11}$$

When acting on a shell-model state, $\hat{\mathcal{Q}}_{ij}^{(0)}$ conserves the quantum number N and $\hat{\mathcal{Q}}_{ij}^{(\pm2)}$ change N to $N\pm2$, respectively. Observe also that $\hat{Q}$ is a sum of irreducible $L=0$ and $L=2$ spherical tensors, $\hat{Q}_0$, and $\{\hat{Q}_{2\nu}\}$. Thus, $\hat{\mathcal{Q}}^{(0)}$ is a $\Delta N=0$, U(3) tensor of the Elliott type, whereas the components of $\hat{\mathcal{Q}}_{0,2}^{(\pm2)}$ are, respectively, $\Delta N=\pm2$ raising and lowering operators for giant monopole and quadrupole excitations. We refer to the U(3) irreps that are annihilated by the $\mathcal{Q}^{(-2)}$ lowering operators as *lowest-grade* representations. Sp(3, $\mathbb{R}$) has three kinds of irreps:

(i) *Spherical representations*

These representations are based on a closed-shell state, which spans a trivial one-dimensional representation of U(3). The $\hat{\mathcal{Q}}^{(-2)}$ operators annihilate this state and the $\hat{\mathcal{Q}}^{(+2)}$ operators act on it to create giant monopole and giant quadrupole vibrational excitations. Acting twice, they create two phonon giant resonance excitations, etc., as illustrated in figure 1.

(ii) *Axially symmetric representations*

These representations are based on a non-trivial U(3) representation $N(\lambda,\mu{=}0)$ or $N(\lambda{=}0,\mu)$.

(iii) *Triaxial representations*

These representations are based on a generic U(3) representation $N(\lambda\mu)$. For a Hamiltonian of the Elliott type

$$\hat{H}=\hat{H}_{\mathrm{HO}}-\chi\hat{\mathcal{Q}}_2^{(0)}\cdot\hat{\mathcal{Q}}_2^{(0)}, \tag{12}$$

with no mixing between harmonic-oscillator shells, one obtains a low-lying triaxial SU(3)-rotational bands with higher-lying giant monopole and giant quadrupole vibrational bands built upon it, as illustrated in figure 2.

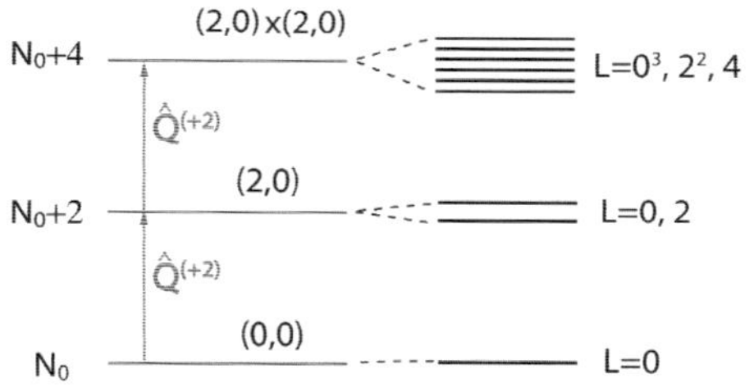

Figure 1. Basis states for an Sp(3, $\mathbb{R}$) irrep $N(0,0)$. The lowest-weight state is a closed-shell $L=0$ state. The one-phonon giant monopole and giant quadrupole states span a (2,0) SU(3) irrep. The two-phonon excitations span a reducible $(2,0)\times(2,0)$ SU(3) representation.

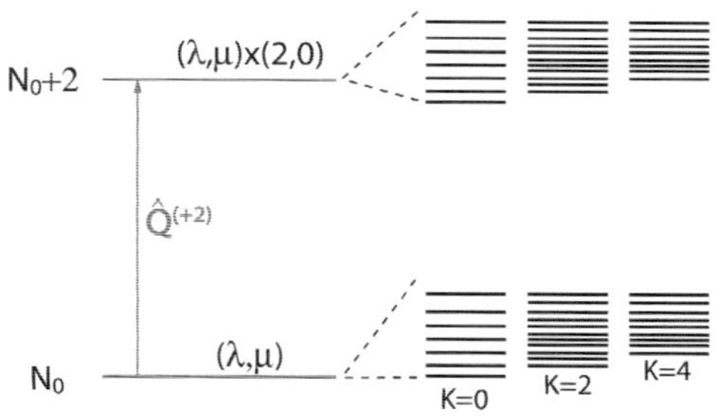

Figure 2. Basis states for an Sp(3, $\mathbb{R}$) irrep $N(\lambda,\mu)$. The lowest-grade states spans a (λ,μ) SU(3) irrep. The one-phonon giant monopole and giant quadrupole states now span a reducible $(\lambda,\mu)\times(2,0)$ SU(3) representation.

4. Selecting an effective shell model space in an SU(3) basis

In adapting the strategy outlined in the introduction to an A-nucleon nucleus in which monopole and quadrupole shape fluctuations dominate over other correlations, one should start with basis states for the Hilbert space $\tilde{\mathbb{H}}^{(A)}$ classified by means of the subgroup chain (10). If the Sp(3, $\mathbb{R}$) irreps were ordered such that the low-energy states of the A-nucleon nucleus were expected to have decreasing overlaps with the correspondingly ordered representation spaces, it would then be natural to choose the space spanned by the direct sum of lowest-grade U(3) sub-states for a finite number of the leading Sp(3, $\mathbb{R}$) irreps in this ordered sequence as the Hilbert space for an effective shell-model of the A-nucleon nucleus. It would then remain to derive the effective interactions and effective operators appropriate for such a choice of shell-model space.

As shown by Jarrio et al. [10], the dominant $\mathrm{Sp}(3,\mathbb{R})$ irreps needed to describe observed rotational states are determined by experimental observations of quadrupole moments and E2 transition rates. For example, they show that the irrep needed to describe the ground-state band of ^{168}Er has $N \sim N_0+16$, $\lambda \sim 88$, and $\mu \sim 11$, where N_0 is the minimum possible value of N for an $\mathrm{Sp}(3,\mathbb{R})$ irrep of ^{168}Er states. The results of ref. [10] prove to be consistent with the following self-consistent method based on a generalisation [11] of the Nilsson model.

Recall that an $\mathrm{Sp}(3,\mathbb{R})$ irrep is labelled by the quantum numbers $N(\lambda,\mu)$ of its lowest-grade U(3) irrep, defined as follows. A spherical harmonic-oscillator Hamiltonian is a sum of three simple harmonic-oscilllator Hamiltonians, $\sum_i \hat{h}_i$, with common frequencies, $\omega_1 = \omega_2 = \omega_3$. A U(3) highest-weight state is an eigenstate of each $\hat{h}_i$ with eigenvalue $N_i\hbar\omega_i$ and the standard labels of an $\mathrm{Sp}(3,\mathbb{R})$ irrep are defined by $N = N_1+N_2+N_3$, $\lambda = N_1 - N_2$, and $\mu = N_2 - N_3$. Now, because the group $\mathrm{Sp}(3,\mathbb{R})$ contains quadrupole shape transformations, it can transform the highest-weight state of its lowest-grade U(3) irrep into an eigenstate of a triaxial harmonic oscillator Hamiltonian with arbitrary values of ω_1, ω_2, and ω_3, without changing the values of N_1, N_2, and N_3. Any such state is then a deformed lowest-weight state for the $\mathrm{Sp}(3,\mathbb{R})$ irrep $N(\lambda,\mu)$. Thus, Hartree-Fock methods could be used to determine the minimum energy, $E_{N(\lambda\mu)}$, for the Hamiltonian $\tilde{\mathbb{H}}^{(A)}$, among the set of such deformed lowest-weight states. More simply, self-consistency arguments can be used to determine the triaxial harmonic oscillator energy $\sum_i N_i\hbar\omega_i$ for which the deformed lowest-weight state has the same quadrupole shape as the triaxial harmonic oscillator potential of which it is an eigenstate [11]. These energies can be used to order the $\mathrm{Sp}(3,\mathbb{R})$ irreps.

5. Concluding comments

The above analysis gives an outline of what is required for a realistic effective shell-model description of deformed nuclei. In particular, it gives criteria for identifying a meaningful effective shell-modell space in an *LST* coupling scheme. The proposed scheme is based on an understanding of the character of nuclear rotational correlations gained from experimentally-based model calculations. However, proceeding on to actually deriving effective shell-model interactions and other operators for doubly-open shell nuclei poses numerous challenges and will undoubtedly require major approximations and further guidance from model investigations.

For example, for light nuclei one might initially set aside the shape coexisting states for which higher shells in the model space are required and define the active shell-model space to be the lowest open spherical harmonic-oscillator shell. Standard methods could then be used to renormalise the interactions restricted to this space by taking account of the coupling of each SU(3) subspace in this shell to the many-phonon giant monopole/quadrupole states of the $\mathrm{Sp}(3,\mathbb{R})$ irrep built upon it. This approach is being pursued by Draayer and his group [4].

A simple example illustrates how a single effective SU(3) model can reproduce the result of a complicated $\mathrm{Sp}(3,\mathbb{R})$ calculation. Figure 3 shows the U(3) amplitudes for the lowest-energy states of ^{168}Er of angular momentum $L = 0$ to 10 for a Hamiltonian $\hat{H} = \frac{1}{2m}\sum_n p_n^2 + V(Q)$ on an $\mathrm{Sp}(3,\mathbb{R})$ irrep $826\frac{1}{2}(78,0)$ (with centre-of-mass degrees of freedom removed) obtained by Bahri [12]. The potential $V(Q)$ is a rotationally-invariant collective model potential, expressed in terms of the many-nucleon quadrupole moments, chosen to have a minimum at the experimentally observed mean deformation of the ^{168}Er ground-state rotational band. Note that the lowest-grade U(3) irrep contributes an almost negligible amount to the states calculated, and suggests that a perturbative renormalisation of the lowest-grade U(3) states by coupling to the giant resonance excitations would be unlikely to converge. Nevertheless, as figure 4 shows, an SU(3) model calculation with this irrep and just a two-body $Q \cdot Q$ interaction and an effective charge of $e_{\text{eff}} \sim 2$ fits the energy levels and E2 transition rates extraordinarily well.

A few observations give hope that shell-model calculations of rotational states in heavy nuclei with realistic effective interactions might be possible in representative examples. If $\mathrm{Sp}(3,\mathbb{R})$

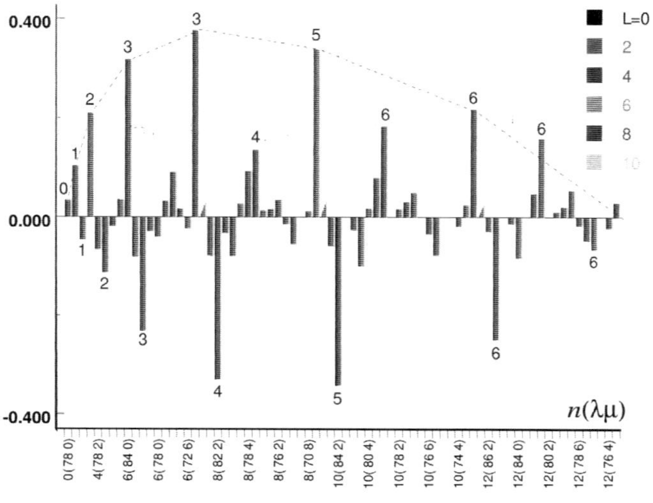

Figure 3. U(3) amplitudes for the $L = 0, \ldots, 10$ states of the ground-state rotational band of ^{168}Er calculated with a phenomenogical interaction. The number of monopole/quadrupole giant resonance phonons of the basis states are shown above each amplitude.

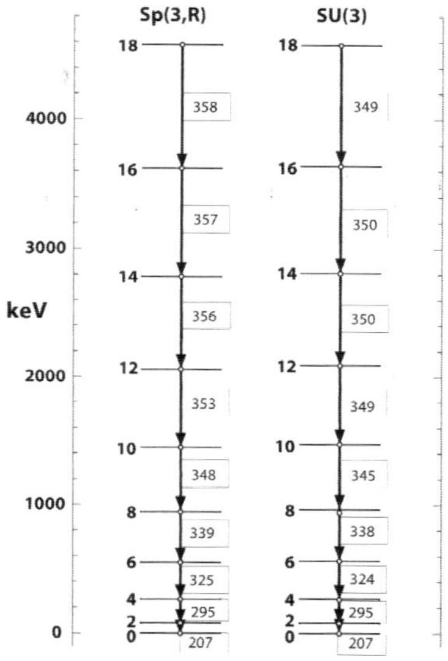

Figure 4. Energy levels and E2 transitions obtained in the $\mathrm{Sp}(3,\mathbb{R})$ calculation of ^{168}Er and fits to these results with a two parameter SU(3) model.

symmetry were conserved for heavy deformed nuclei, it is feasible (cf. LSU workshop [13]) that such calculations could be done in the not too distant future. From numerous model studies of systems with competing dynamical symmetries, it invariably transpires (except in phase transition regions) that one symmetry is dominant [14]. Thus, for example, one could start with phenomenological calculations with single $\mathrm{Sp}(3,\mathbb{R})$ irreps to describe the rotational bands of ^{168}Er. If successful, corresponding shell-model calculations could then be attempted with a realistic interaction and finally perturbative corrections might be added perturbatively.

References

[1] Rowe D J and Wood J L 2010 *Fundamentals of Nuclear Models: Foundational Models* (Singapore: World Scientific)

[2] Wood J L, Heyde K, Nazarewicz W, Huyse M and van Duppen P 1992 *Phys. Repts.* **215** 101

[3] Heyde K and Wood J L 2010 *Rev. Mod. Phys.* (in press)

[4] Dytrych T, Sviratcheva K, Bahri C, Draayer J and Vary J 2007 *Phys. Rev. C* **76** 014315

[5] Bohr A 1952 *Mat. Fys. Medd. Dan. Vid. Selsk.* **26**

[6] D J Rowe T A W and Caprio M A 2009 *Phys. Rev. C* **79** 054304

[7] Rosensteel G and Rowe D J 1977 *Phys. Rev. Lett.* **38** 10

[8] Rosensteel G and Rowe D J 1977 *Phys. Rev. Lett.* **38** 10

[9] Elliott J P 1958 *Proc. Roy. Soc. London* **A245** 128

[10] Jarrio M, Wood J L and Rowe D J 1991 *Nucl. Phys.* **A528** 409

[11] Carvalho J and Rowe D J 1992 *Nucl. Phys.* **A548** 1

[12] Bahri C and Rowe D J 2000 *Nucl. Phys.* **A662** 125

[13] Draayer J *et al.* 2009 *Mardi Gras Workshop at LSU on Special Symmetries and ab initio Methods for Light Nuclei* URL `www.phys.lsu.edu/lsunpw/index.php?pageID=schedule`

[14] Rowe D J 2004 *Int. Conf. on Computational and Group-Theoretical Methods in Nuclear Physics* ed Escher J *et al.* (Singapore: World Scientific)

Valence shell excitations in even-even spherical nuclei within a microscopic model

2011 J. Phys.: Conf. Ser. 267 012045

(http://iopscience.iop.org/1742-6596/267/1/012045)

Valence shell excitations in even-even spherical nuclei within a microscopic model

Lo Iudice N[1], **Stoyanov Ch**[2]

[1]Dipartimento di Scienze Fisiche Università di Napoli Federico II and INFN, Sezione di Napoli, Napoli, Italy

[2]Institute for Nuclear Research and Nuclear Energy, Bulgarian Academy of Sciences, 1784 Sofia, Bulgaria

E-mail: loiudice@na.infn.it

Abstract. The structure of some low-lying states in the N=80 and N=84 isotones is investigated within the quasiparticle-phonon model. The calculations reproduce the experimental levels, the electromagnetic transition strengths, and the selection rules. They account also for the M1 splittings observed in ^{138}Ce,^{144}Nd and ^{142}Ce and relate them to correlations specific of each nucleus.

1. Introduction

Low-energy spectra in nuclei are of crucial importance for understanding the correlations among valence nucleons. They include elementary modes, like the collective lowest 2^+, but also excitations of complex nature.

These spectra fit well in the proton-neutron interacting boson model (IBM-2), which classifies the states according to the number of s ($L = 0$) an d ($L = 2$) bosons and the symmetry with respect to the exchange between proton and neutron bosons [1]. Strong $E2$ transitions connect states of the same symmetry (the same F spin) differing by one boson, while strong $M1$ transitions connect symmetric $F = F_{max}$ to mixed symmetry ($F = F_{max} - 1$) states with the same number of bosons.

More detailed investigations have been carried out within the shell model (SM) [2, 3] and, more thoroughly, the quasiparticle-phonon model (QPM) [4, 5, 6, 7]. The focus of our QPM studies here will be the nuclei around $N = 82$, object of recent detailed experimental investigations [8, 9, 10, 11].

2. A sketch of the QPM

The QPM [12] adopts a Hamiltonian of the form

$$H = H_{sp} + V_{pair} + V_M^{ph} + V_{SM}^{ph} + V_M^{pp} \ . \quad (1)$$

H_{sp} is a Woods-Saxon one-body Hamiltonian, V_{pair} the monopole pairing, V_M^{ph} and V_{SM}^{ph} are respectively sums of separable multipole and spin-multipole particle-hole potentials, and V_M^{pp} is the sum of multipole pairing potentials.

In the QPM, one first expresses the Hamiltonian in terms of the quasiparticle creation and annihilation operators $\alpha^{\dagger}_{jm}$ (α_{jm}) and, then, generates the QRPA energies $\omega_{i\lambda}$ and the corresponding phonons

$$Q^{\dagger}_{i\lambda\mu} = \frac{1}{2}\sum_{jj'}\left\{\psi^{i\lambda}_{jj'}[\alpha^{\dagger}_{j}\alpha^{\dagger}_{j'}]_{\lambda\mu} - (-1)^{\lambda-\mu}\varphi^{i\lambda}_{jj'}[\alpha_{j'}\alpha_{j}]_{\lambda-\mu}\right\}. \tag{2}$$

These allow to reduce the quasiparticle separable Hamiltonian to the phonon form

$$H_{QPM} = \sum_{i\mu}\omega_{i\lambda}Q^{\dagger}_{i\lambda\mu}Q_{i\lambda\mu} + H_{vq}, \tag{3}$$

where the first term is the unperturbed phonon Hamiltonian and H_{vq} is a phonon-coupling piece whose exact expression can be found in Ref. [12].

The phonon Hamiltonian is accordingly diagonalized in a space spanned by states composed of one, two, and three QRPA phonons. The eigenfunctions have the structure

$$\begin{aligned}\Psi_{\nu}(JM) &= \sum_{i} R_i(\nu J)Q^{\dagger}_{iJM}|0> + \sum_{\substack{i_1\lambda_1\\ i_2\lambda_2}} P^{i_1\lambda_1}_{i_2\lambda_2}(\nu J)\left[Q^{\dagger}_{i_1\lambda_1}\otimes Q^{\dagger}_{i_2\lambda_2}\right]_{JM}|0> \\ &+ \sum_{\substack{i_1\lambda_1 i_2\lambda_2\\ i_3\lambda_3 I}} T^{i_1\lambda_1 i_2\lambda_2 I}_{i_3\lambda_3}(\nu J)\times\left[\left[Q^{\dagger}_{i_1\lambda_1}\otimes Q^{\dagger}_{i_2\lambda_2}\right]_I\otimes Q^{\dagger}_{i_3\lambda_3}\right]_{JM}|0>,\end{aligned} \tag{4}$$

where ν labels the specific QPM excited state of total spin JM. The above wave functions are properly antisymmetrized according to the procedure outlined in [5].

3. Numerical details

All the strength parameters of the different pieces of the Hamiltonian have been fixed by an independent fit procedure [7]. In this respect, the calculation is to be considered parameter free.

The occurrence of at least one low-energy isovector collective QRPA 2^+ , in addition to the lowest isoscalar 2^+, is a preliminary condition for getting QPM states that can be considered the counterpart of the IBM symmetric and mixed symmetry (MS) states. To test such a symmetry we have computed the ratio

$$\beta(2^+) = \frac{|\langle 2^+\|\sum^{\mathrm{p}}_k r^2_k Y_{2\mu}(\Omega k) - \sum^{\mathrm{n}}_k r^2_k Y_{2\mu}(\Omega k)\|g.s.\rangle|^2}{|\langle 2^+\|\sum^{\mathrm{p}}_k r^2_k Y_{2\mu}(\Omega k) + \sum^{\mathrm{n}}_k r^2_k Y_{2\mu}(\Omega k)\|g.s.\rangle|^2}. \tag{5}$$

Clearly, a state is isoscalar $([2^+_{is}]_{RPA})$ or isovector $([2^+_{iv}]_{RPA})$according that $\beta(2^+) < 1$ or $\beta(2^+) > 1$. The value of $\beta(2^+)$ depends critically on the ratio $G^{(2)}/\kappa_0^{(2)}$ between the coupling constants of the quadrupole pairing and particle-hole interactions. For realistic $G^{(2)}$ and $\kappa_0^{(2)}$, at least one isovector $[2^+_{iv}]_{RPA}$ occurs at sufficiently low energy.

3.1. N = 80 isotones

The second lowest 2^+_2 in ^{134}Xe and ^{136}Ba is fairly collective and has a MS character, with the main proton and neutron amplitudes in opposition of phase. In ^{138}Ce, instead, there are two 2^+ states which get an appreciable $E2$ strength and have a dominant MS character. Both 2^+_2 and 2^+_3 states play a role and are to be treated on equal footing.

The different behavior of ^{138}Ce with respect to the other two isotones is understood if we analyze the energies of the three lowest two-quasiparticle proton states. These are all closely spaced in ^{138}Ce. In ^{134}Xe and ^{136}Ba, instead, one level is well above the other two. This is a

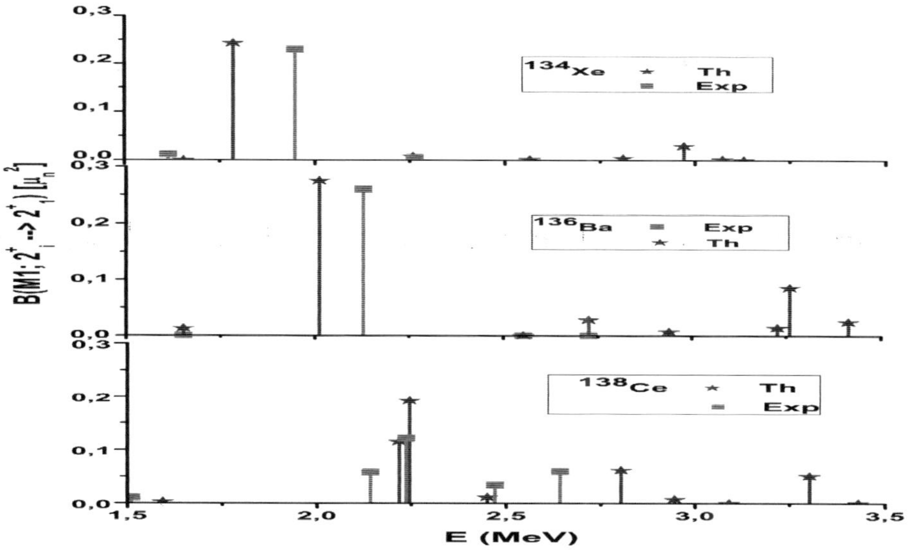

Figure 1. QPM versus Experimental $M1$ strength distribution in ^{134}Xe, ^{136}Ba, and ^{138}Ce.

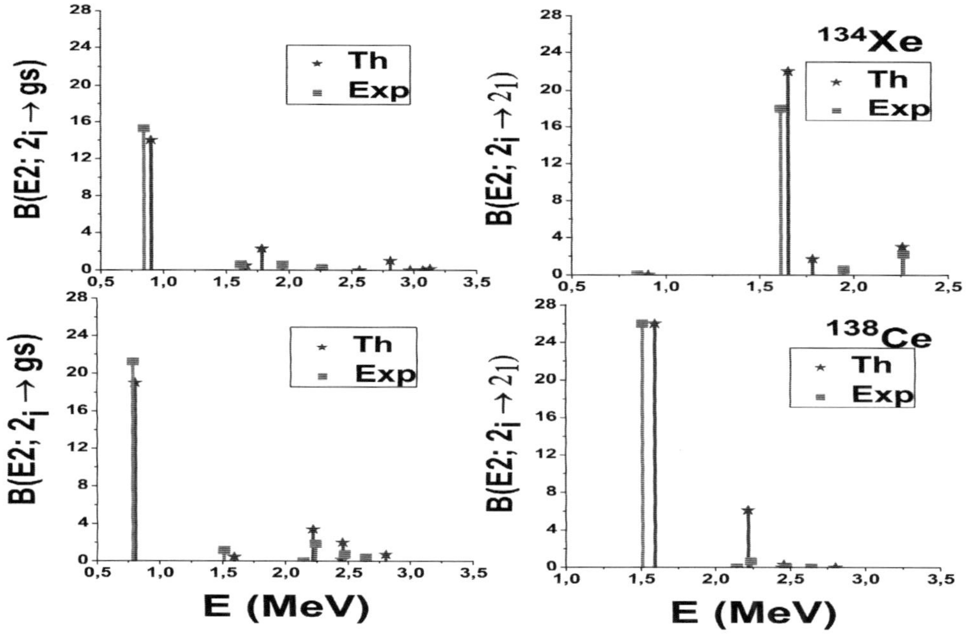

Figure 2. QPM $E2$ strength distribution in ^{134}Xe and ^{138}Ce. ^{136}Ba has a similar behavior.

clear shell effect. In ^{138}Ce, indeed, the $1g7/2$ proton subshell is filled. Because of the diffuse Fermi surface, due to pairing, several proton configurations can enter the 2_3^+. This does not occur in ^{134}Xe and ^{136}Ba, where the chemical potential is within the $1g7/2$ proton subshell, which is only partially filled.

In all three nuclei, the first isoscalar 2_1^+ has a prominent QRPA one-phonon component, with an appreciable two-phonon piece, especially in ^{136}Ba. The second QPM state has a dominant two-phonon component in all three nuclei.

A marked difference between the ^{138}Ce and the other two isotones, ^{134}Xe and ^{136}Ba, is registered for the other states. In ^{134}Xe and ^{136}Ba, the third 2_3^+ is dominated by the MS $[2_{iv}^+]_{RPA}$. In ^{138}Ce, both 2_3^+ and the 2_4^+ contain the MS QRPA $[2_2^+]_{RPA}$ with an appreciable amplitude. We have pointed out already that in ^{138}Ce the residual quadrupole collectivity was shared by the second and third QRPA 2^+ states.

Apparently, the interaction between these two fairly collective states lead to the generation of two QPM states with MS character. This is reflected in the $M1$ transitions. While in ^{134}Xe and ^{136}Ba we have one strong $M1$ peak (first and second panels of Fig. 1), in ^{138}Ce the $M1$ strength splits into two peaks (third panel of Fig. 1). This small splitting in ^{138}Ce is related to the filling of the proton $1g_{7/2}$ subshell closure in correspondence of $Z = 58$. Because of the gap with the other subshells, the low-lying proton excitations are made possible only because of the diffuse Fermi surface induced by pairing.

3.2. $N = 84$ isotones

A more pronounced $M1$ splitting is observed in ^{142}Ce, having the same proton number as ^{138}Ce, and also in ^{144}Nd, with two more protons. Since ^{142}Ce and ^{144}Nd have the same neutron number, it is tempting to relate the alike $M1$ responses in these two nuclei to the two valence neutrons in excess with respect to the $N = 82$ shell closure.

A QPM study [14], which refines and expands a previous investigation [15], has been performed on these nuclei. In both ^{142}Ce and ^{144}Nd, the QRPA yields two low-lying collective 2^+ states. The lowest one is np symmetric, while the second has mixed symmetry. The proton and neutron components are in phase in the first 2^+ state and out of phase in the second.

Few QPM observables are compared with the experimental data in Figs. 3 and 4. As shown pictorially in Fig. 3, the strength distribution of the $E2$ transitions to the ground and 2_1^+ states is quite similar in ^{142}Ce and ^{144}Nd. Concerning the $E2$ transitions to the ground state, most of the strength is concentrated into the lowest 2_1^+ state of each nucleus at 0.641 MeV and 0.696 MeV, respectively. The remaining $E2$ strength, about 5.4 W.u. and 4.2 W.u., respectively, is distributed mainly among two or three 2^+ states in the energy range $2 - 2.6$ MeV.

The QPM underestimates the second large peak in ^{142}Ce and overestimates the lowest large peak in ^{144}Nd. The overall strength is comparable with experiments in both nuclei. It is remarkable, in any case, that the calculation accounts for the main properties of the spectra.

The immediate reason of the $M1$ splitting is to be found in the mentioned composite phonon structure of the QPM states. The first prominent peak is due to the collective MS $[2_2^+]_{RPA}$ component present with large amplitude in 2_3^+ state. The second arises from the combined presence in the fourth 2_4^+ of the same one-phonon MS $[2_2^+]_{RPA}$ plus the two-phonon MS component $[2_1^+ \otimes 2_2^+]_{RPA}$. The two components are respectively coupled by the M1 operator to the symmetric one-phonon $[2_1^+]_{RPA}$ and $[2_1^+ \otimes 2_1^+]_{RPA}$, the two main components of the isoscalar QPM 2_1^+ state.

The one- to two-phonon matrix elements of the coupling Hamiltonian are comparable in the N=80 isotone ^{138}Ce and the N=84 isotones ^{142}Ce and ^{144}Nd. In ^{142}Ce the energy of the second RPA $[2_2^+]_{RPA}$ is almost twice the energy of the lowest RPA quadrupole phonon. This small energy difference between the MS one-phonon and the symmetric two-phonon components, in both nuclei ^{142}Ce and ^{144}Nd, enhances considerably the coupling between these configurations and, therefore, their admixture in the QPM states leading finally to the fragmentation of the $M1$ strength. In ^{138}Ce, instead, the energy difference is slightly larger. Moreover, the collectivity is

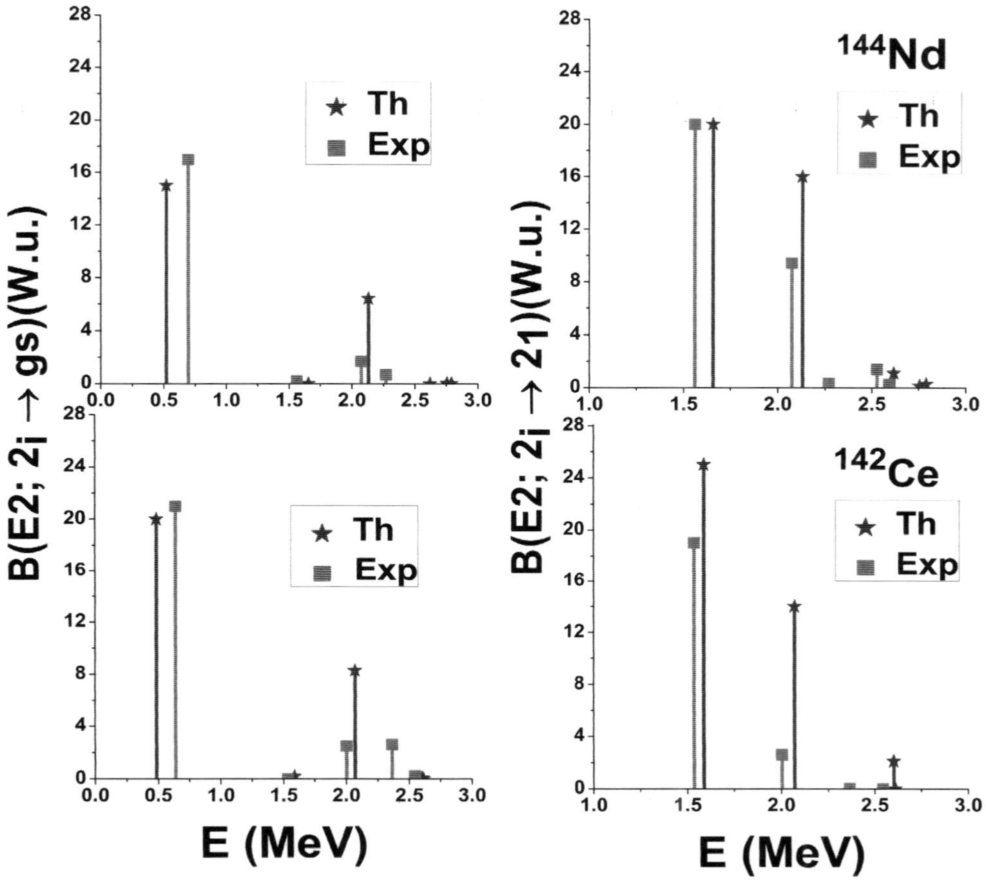

Figure 3. Experimental [9, 10] and QPM strength distributions of the $2_i^+ \to 0_{gs}^+$ and $2_i^+ \to 2_1^+$ $E2$ transitions in ^{142}Ce and ^{144}Nd.

shared among the second and third RPA phonons $[2_2^+]_{RPA}$ and $[2_3^+]_{RPA}$, very close in energy and, therefore, strongly interacting. This mitigates their coupling with the two-phonon configuration $[2_1^+ \times 2_1^+]_{RPA}$.

In summary, the splitting mechanism in the N=84 isotones ^{142}Ce and ^{144}Nd is different from the one in ^{138}Ce. In the latter case is associated to the filling of the proton $g_{7/2}$ subshell and the diffuse Fermi surface induced by pairing, while in ^{142}Ce and ^{144}Nd is a result of the phonon coupling.

The different strengths fragmentation in $N = 80$ and $N = 84$ isotones, respectively, are ultimately to be ascribed to the asymmetry between particle and hole spectra with respect to the closed shell $N = 82$.

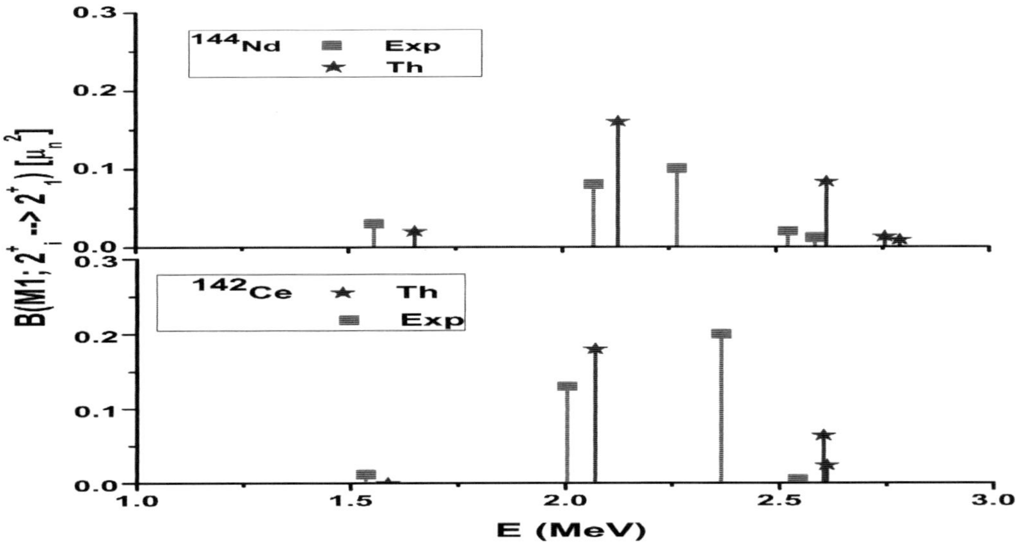

Figure 4. QPM versus Experimental [9, 10] strength distributions of the $2_i^+ \rightarrow 2_1^+$ $M1$ transitions in ^{142}Ce and ^{144}Nd.

3.3. Acknowledgments

One of the author (Ch.S.) acknowledges the financial support within the agreement between INFN and the Bulgarian Science Foundation.

References

[1] Iachello F and Arima A 1987 *The Interacting Boson Model* (Cambridge: Cambridge University Press)
[2] Werner V *et al.* 2002 *Phys. Lett.* **B550** 140
[3] Lisetskiy A F, N. Pietralla, Fransen C, Jolos R V and von Brentano P 2000 *Nucl. Phys.* **A677** 1000
[4] Lo Iudice N and Stoyanov Ch 2000 *Phys. Rev. C* **62** 047302
[5] Lo Iudice N and Stoyanov Ch 2002 *Phys. Rev. C* **65** 064304
[6] Lo Iudice N and Stoyanov Ch 2004 *Phys. Rev. C* **69** 044312
[7] Lo Iudice N and Stoyanov Ch 2006 *Phys. Rev. C* **73** 037305
[8] Rainovski G *et al.* 2006 *Phys. Rev. Lett.* **96** 122501
[9] Vanhoy J R *et al.* 1995 *Phys. Rev. C* **52** 2387
[10] Hicks S F, Davoren C M, Faulkner W M and Vanhoy J R 1998 *Phys. Rev. C* **57** 2264
[11] Pietralla N, von Brentano P, Lisetskiy A F 2008 *Progr. Part. Nucl. Phys.* **60** (2008) 225
[12] Soloviev V G 1992 *Theory of Atomic Nuclei: Quasiparticles and Phonons* (Bristol: Institute of Physics Publishing)
[13] Lo Iudice N, Stoyanov Ch and Tarpanov D 2008 *Phys. Rev. C* **77** 044310
[14] Lo Iudice N, Stoyanov Ch and Pietralla N 2009 *Phys. Rev. C* **80** 024311
[15] Thai Khac Dinh Grinberg M and Stoyanov Ch, 1992 *J. Phys. G: Nucl. Part. Phys.* **18** 329

The structure of ^{12}C and stellar helium burning

2011 J. Phys.: Conf. Ser. 267 012046

(http://iopscience.iop.org/1742-6596/267/1/012046)

The structure of ^{12}C and stellar helium burning *

Moshe Gai
for the UConn-Yale-Duke-Weizmann-PTB-UCL Collaboration
Laboratory for Nuclear Science at Avery Point
University of Connecticut, 1084 Shennecossett Rd, Groton, CT 06340-6097, USA
E-mail: moshe.gai@yale.edu

Abstract. The rate of stellar formation of carbon at high temperatures (T > 3 GK) may increase beyond that which is expected from the Hoyle state at 7.654 MeV due to contributions from higher lying states in ^{12}C. The long sought for second 2^+ state predicted at 9 - 10 MeV excitation energy in ^{12}C was predicted to significantly increase the production of ^{12}C. An Optical Readout Time Projection Chamber (O-TPC) operating with the gas mixture of $CO_2(80\%)$ + $N_2(20\%)$ at 100 torr with gamma beams from the HIγS facility of TUNL at Duke was used to study the formation of carbon (and oxygen) during helium burning. Preliminary measurements were carried out at beam energies: E = 9.51, 9.61, 9.72, 10.00, 10.54, 10.84 and 11.14 MeV. Extra attention was paid for separating the carbon dissociation events, $^{12}C(\gamma,3\alpha)$, from the oxygen dissociation events, $^{16}O(\gamma,\alpha)^{12}C$. Complete angular distributions were measured giving credence to a newly identified 2^+ state just below 10.0 MeV.

1. Introduction

Carbon is formed during stellar helium burning in the triple-alpha process, the $^8Be(\alpha,\gamma)^{12}C$ reaction, that is mostly governed by the contribution of the 0^+ Hoyle state at 7.654 MeV. At high temperatures (T > 3 GK) higher lying states in ^{12}C may contribute. Indeed a broad (Γ = 560 keV, $\Gamma_\gamma = 0.2$ eV) 2^+ state at 9.11 MeV in ^{12}C was included in the NACRE compilation [1] following theoretical prediction [2] for the 2^+ member of the rotational band built on top of the 0^+ Hoyle state at 7.654 MeV. It increases the production of carbon at temperatures in excess of 1 GK by up to a factor of 15. A larger production of ^{12}C at high temperatures increases the neutron density as required for an r-process, due to the competition between the $^8Be(\alpha,\gamma)$ reaction and the $^8Be(n,\gamma)$ reaction [3, 4]. A 2^+ member of the rotational band build on top of the Hoyle state is not predicted in the conjectured alpha condensate [5] which predicts a spherical Hoyle state. An evidence for the broad second 2^+ state at 9.6 MeV was found in a $^{12}C(\alpha,\alpha')$ [6] and a $^{12}C(p,p')$ measurement [7] but such a state was not observed in the beta-decay of ^{12}B and ^{12}N [8].

2. Measurement of the $^{12}C(\gamma,3\alpha)$ Reaction With O-TPC.

We used our Optical-Readout Time Projection Chamber (O-TPC) [9] operating with CO_2 gas with gamma beams extracted from the HIγS facility of TUNL at Duke University [10] to search for such states via the identification of triple alpha events from the $^{12}C(\gamma,3\alpha)$ reaction as shown in Fig. 1. We have studied this reaction at E = 9.51, 9.61, 9.72, 10.00, 10.54, 10.84 and 11.14 MeV [11].

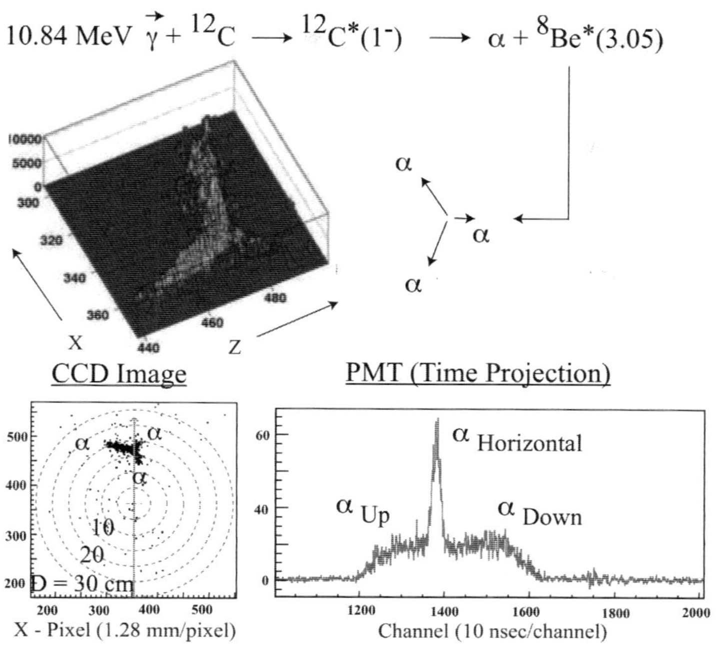

Figure 1. A typical three alpha dissociation event detected by the O-TPC, with the γ-beam (indicated by a green arrow) along the positive z-axis. The large opening angle between the two emitted alpha-particle indicates dissociation to $\alpha +^8 Be^*(3.0)$; into the first excited state of ^8Be.

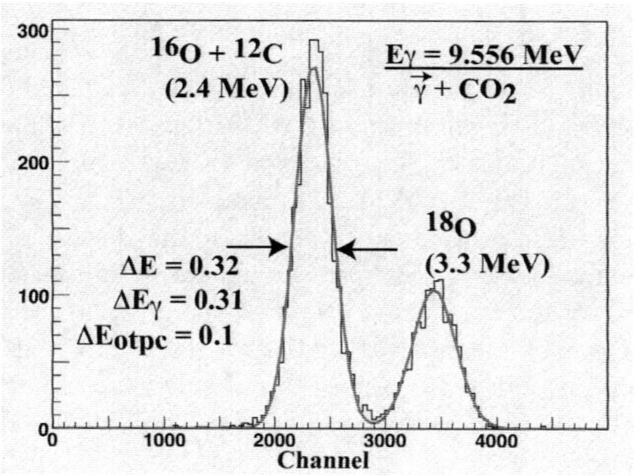

Figure 2. A typical total energy spectrum (grid-charge signal) recorded by the O-TPC.

2.1. Event Identification

The outgoing particle resulting from the photo-disscoiation of target nuclei are fully determined by the tracks recorded in the O-TPC in three dimensions. Thus all relevant kinematical variables are measured by the O-TPC [9]. The total energy deposited (grid charge-signal) in the O-TPC detector is determined by the Q-value for the dissociation event; $E_{total} = E_\gamma - Q$; Q = 6.227, 7.162 and 7.275 for the dissociation of ^{18}O, ^{16}O and ^{12}C, respectively. The Q-values for ^{16}O and ^{18}O dissociations are sufficiently different (935 keV) and differ considerably more than the beam width of FWHM $\approx$ 300 keV, hence these events are well separated in the total energy spectrum shown in Fig 2. However one major draw back for using CO_2 gas is that the difference of Q-values (112 keV) for the dissociation of ^{16}O and ^{12}C is considerably smaller than the beam width and comparable to the detector resolution of approximately 80 keV [9]. In addition the larger quenching factor for the low energy ^{12}C projectiles leads to a smaller grid charge-signal from the dissociation of ^{16}O with very similar energy as for the dissociation of ^{12}C, as can be seen in Fig. 3. Hence the total energy deposit cannot be used to separate (and thus identify) ^{12}C and ^{16}O dissociation events.

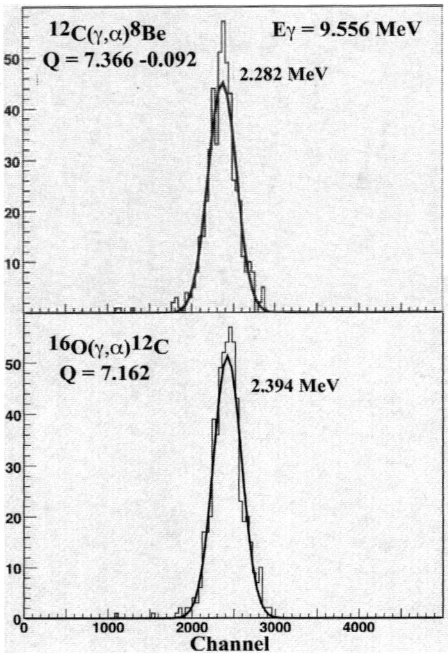

Figure 3. A typical total energy spectrum (grid-charge signal) recorded by the O-TPC for well identified ^{12}C and ^{16}O dissociation events.

To identify and distinguish ^{12}C dissociation events we relied on the line shape of the PMT signal. Unfortunately the noise level in the CCD camera was too high and it did not permit line shape analysis of the pixel-content. Hence only out of plane events with out of plane angle β larger than 20° (approximately 40% of the data) could be analyzed in the current setup. A new cleaner camera is being installed that will permit including all data. In addition the resolution of the optical system did not permit resolving the two outgoing alphas emitted from the decay of the ground state of ^{8}Be. Due to the poor resolution such decays most of the time appear co-linear in the image recorded by the CCD camera but are clearly distinguished from ^{16}O events in the PMT signal. In contrast the two outgoing alphas emitted in the decay of the first excited state of $^{8}Be^{*}(3.0)$ are well resolved as shown in Fig. 1. The very low energy of the $^{8}Be^{*}(3.0)$ yield a decay pattern which are almost as for a decay in rest.

The line shape of the (PMT) signal is very well determined by the calculated dE/dX along the track. In Fig. 4 we compare the observed PMT signal to the calculated line shape. In this calculation we used the drift velocity that we measured with a ^{148}Gd source of 1.05 cm/μs at 100 torr [9]. A similar value of the drift velocity is calculated by MAGBOLTZ [12]. Hence the calculated line shape shown in Fig. 4 has essentially only one free parameter, the out of plane angle beta. A good "effective reduced χ^2" is found for ^{16}O dissociation events as shown in Fig. 4. In this calculation we did not include the single photo-electron calibration for the PMT signal and the error is considered to be the square root of the signal. Hence the deduced χ^2 is determined up to an over all (calibration) constant and must be considered "effective χ^2". As can be seen from Fig. 4 the χ^2 for ^{16}O events is of the order of unity. The line shape of

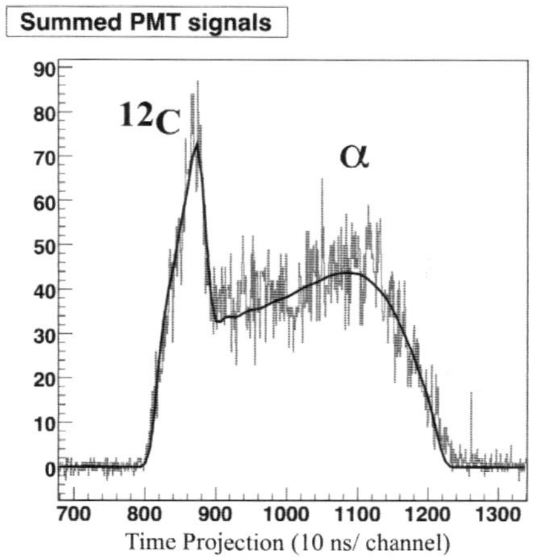

Figure 4. The measured PMT signal compared to the predicted line shape for an α + ^{12}C dissociation event.

^{12}C dissociation events arise from a considerably more complicated dE/dX of the three body (non-colinear) decay pattern. Hence the "effective χ^{2}" for ^{12}C dissociation events fitted with the line shape calculated for an ^{16}O dissociation is seldom unity.

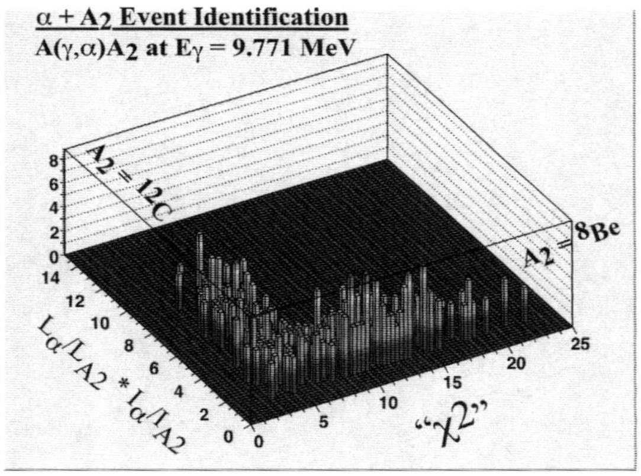

Figure 5. Event identification surface plot (1 count suppressed) for all events as discussed in the text.

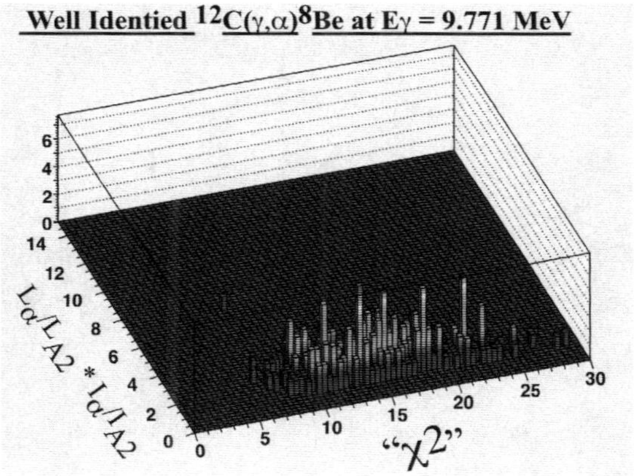

Figure 6. Event identification surface plot (1 count suppressed) for well identified ^{12}C dissociation events as discussed in the text.

The energy shared by the two body α + A_2 decay differs for ^{16}O and ^{12}C events. We define the ratio: $R = \frac{E(\alpha)}{E(A_2)}$. This ratio is 3 for $16O$ events and 2 for ^{12}C events. Due to the attenuation factor of the outgoing ^{12}C the measured ratio R for ^{16}O events is in fact closer to 4. Thus the ratio R can be used in addition to the χ^2 to distinguish and identify ^{12}C events. The ratio R can be measured by measuring the ratio of the emitted light (integrated yield of the signal shown in Fig. 4) as well as the ratio of the track length in time. In addition a similar measurement can be performed using the pixel content of the CCD camera as long as the CCD camera provides clean signals. Thus the ratio R can be measured four times. A multiplication of the measured R values yield a separation which is approximately a factor of 2 better since the fluctuation increase by approximately the square root of 4.

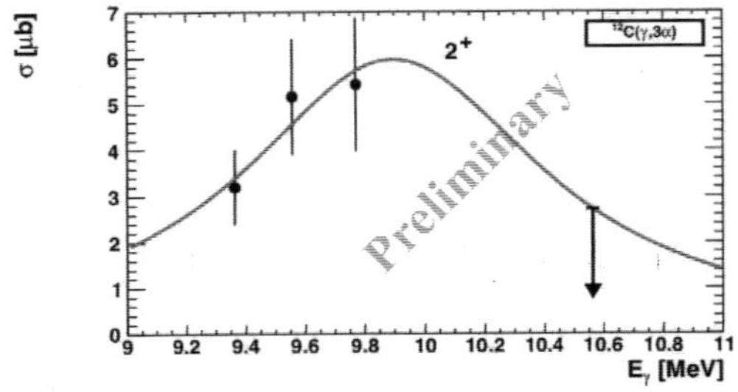

Figure 7. Preliminary measured excitation curve.

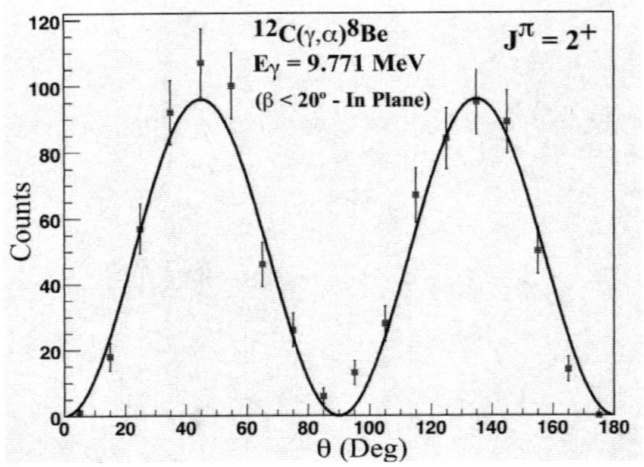

Figure 8. Measured angular distribution for in plane ($\beta < 20°$) $^{12}C(\gamma, 3\alpha)$ events compared to the prediction for a pure $0^+ \rightarrow 2^+$ E2 transition.

In Fig. 5 we show the so obtained surface plot of event identification using the "effective χ^2" and the multiplication of the ratio R as obtained from the PMT signal; the ratio of the integrated light (I) times the track length (L). In Fig. 6 we show the same surface plot for events that were clearly identified as ^{12}C events by observing kinks in the PMT signal or in the CCD image. In both Figs. 5 and 6 we show data bins including at least two events. The separation of ^{12}C events using the above discussed method is estimated to yield an uncertainty which is smaller than 10%. The so obtained ^{12}C events allow us to measure the excitation curve shown in Fig. 7.

3. Angular Distribution

The in plane angle (α) measured by the track registered in the CCD image and the out-of-plane angle (β) measured by the Time projection signal of the PMT allow us to deduce for each event the scattering angle (θ) and the azimuthal angle (ϕ) of the polar coordinate system used in scattering theory: $\cos\theta = \cos\beta x \cos\alpha$ and $\tan\phi = \tan\beta/\sin\alpha$. The so obtained angular distribution is shown in Fig. 8 together with that predicted for a pure $0^+ \rightarrow 2^+$ E2 transition. For these data we used only in plane ($\beta < 20°$) data for which the scattering angle (θ) is determined with high accuracy.

4. Conclusions

Dissociation events from the $^{12}C(\gamma, 3\alpha)$ reaction were identified in our measurement using an O-TPC detector and clear evidence is observed for a pure E2 angular distribution most likely arising from a 2^+ state just below 10.0 MeV. These data are being remeasured with an improved setup including a CCD camera with lower background. This study is in progress.

5. References

[1] Angulo C *et al.* 1999 *Nucl. Phys.* **A656** 3.

[2] Descouvemont P and Baye D 1987 *Phys. Rev. C* **36** 54.

[3] Delano M D and Cameron A G W 1971 *Astr Space Sci.* **10** 203.

[4] Pruet J *et al.* 2005 *Astr. Jour.* **623** 325.

[5] Funaki Y *et al.* 2009 *Phys. Rev. C* **80** 64326 and references therein.

[6] Itoh M *et al.* 2004 *Nucl. Phys.* **A738** 268.

[7] Freer M *et al.* 2009 *Phys. Rev. C* **80** 041303(R).

[8] Hyldegaard S *et al.* 2009 *Phys. Lett.* **B678** 459.

[9] Gai M, Ahmed M W, Stave S C, Zimmerman W R, Breskin A, Bromberger B, Chechik R, Dangendorf V, Delbar Th, France R H III, Henshaw S S, Kading T J, Martel P P, McDonald J E R, Seo P-N, Tittelmeier K, Weller H R, and Young A H to be submitted to *Jour Instr.*

[10] Weller H R, Ahmed M W, Gao H, Tornow W, Wu Y, Gai, and Miskimen M R 2009 *Prog. Part. Nucl. Phys.* **62** 257.

[11] Zimmerman W R, Seo P-N, Gai M, Ahmed M W, Hinshaw S S, Howell C R, Stave S C, Weller H R, Martel P P 2009 *Bull. Amer. Phys. Soc.* **54#4** 171.

[12] Biagi S F 1999 *Nucl. Instr. Meth.* **A421** 234; http://ref.web.cern.ch/ref/CERN/CNL/2000/001/magboltz.

Collective excitations in Random Phase Approximation and beyond

2011 J. Phys.: Conf. Ser. 267 012047

(http://iopscience.iop.org/1742-6596/267/1/012047)

Collective excitations in Random Phase Approximation and beyond

F Catara[1,2], D Gambacurta,[1,2] M Grasso [3] and M Sambataro[2]

[1] Dipartimento di Fisica e Astronomia dell'Universitá di Catania, Via S.Sofia 64, I-95123 Catania, Italy

[2] Istituto Nazionale di Fisica Nucleare, Sezione di Catania, Via S.Sofia 64, I-95123 Catania, Italy

[3]Institut de Physique Nucléaire, Université Paris-Sud, IN2P3-CNRS, F-91406 Orsay Cedex, France

E-mail: francesco.catara@ct.infn.it

Abstract. Collective excitations are one of the most common and interesting features of many-body systems. Of particular interest are those collective modes which can be interpreted in terms of vibrations. Nuclei show a large variety of such modes, both low-lying and high-lying. In particular the giant dipole resonance is due to the coherent motion of protons against neutrons. Its analogue in metal clusters is the dipole plasmon excitation, due to the oscillation of electrons against the positive ions. The Random Phase Approximation (RPA) is extensively used as a microscopic theory to study the basic properties of these collective excitations. In the derivation of RPA use is made of the Quasi Boson Approximation (QBA). It consists in replacing the expectation values in the correlated ground state with those in the uncorrelated (Hartree-Fock) one. Strictly related to QBA is the feature of RPA of predicting a harmonic spectrum. On the other hand, the existence of anharmonicities in the multiphonon spectra of nuclei and their relevance in various physical processes are well established. Overcoming QBA has represented the starting point of many attempts aiming at improving RPA. A recently developed approach in this line will be presented and discussed. A very nice property of RPA is to satisfy Energy Weighted Sum Rules (EWSR). However, some violations are present in several extensions of RPA. We will show how to cure this shortcoming.

1. Introduction

A common feature of finite-size many-body systems is the existence of collective excitations. The Random Phase Approximation (RPA) is the microscopic approach most currently used to study such modes, described as superpositions of particle-hole and hole-particle elementary configurations [1]. To derive RPA in its standard form, use is made of the Quasi Boson Approximation (QBA) which amounts to consider the Hartree-Fock ground state, rather than the correlated one, as reference state. This is justified only if ground state correlations are not too strong. Otherwise a better and, if possible self-consistent, treatment of these correlations becomes necessary. For example, it has been shown in the context of metal clusters that correlations in the ground state can be quite strong and induce important deviations from 1 and 0 in the occupation numbers [2]. One possible way to explicitly introduce correlations in the ground state consists in constructing a RPA-like formalism where the violations of the Pauli principle related to QBA are cured. This has been done along two main lines, by using either boson expansion methods [3] or remaining in the fermionic space [4]. The so-called renormalized RPA (RRPA) method, starting from the early works of Hara and Rowe in the 1960's [5] belongs to the latter. In RRPA the one body density matrix is assumed diagonal but, due to the g.s. correlations, its eigenvalues are different from 0 and 1 and are calculated self consistently. In the derivation of the equations of motion no use is made of QBA. A very important feature of standard RPA is that it preserves the energy weighted sum rules (EWSR) [1]. This property is lost within all the above approaches since only particle-hole excitations are considered. An attempt to overcome this problem has been discussed in Ref. [6], where an extended RRPA approach has been introduced, including particle-particle and hole-hole configurations in addition to the standard p-h ones. The method has been applied to a three-level Lipkin model, with one single particle level below the Fermi energy and two above, and it has been shown that the EWSR are exactly satisfied. However, a new problem arises related to the existence of a non physical mode that does not correspond to any level in the exact excitation spectrum of the model. This unpleasant problem makes unfeasible the application of this method to a realistic case because it would be impossible to reveal the existence of such mode and to identify it. A way to overcome this difficulty has been recently proposed and tested within the Lipkin model [7]. It is based on a generalization of the approach introduced in Ref. [8] in the context of metal clusters, which has been shown to be self-consistent within the p-h space. The generalization with respect to [8] consists in enlarging the space following the line suggested in Ref. [6]. Similarly to what is done in Ref. [8], no renormalization is introduced and the one-body density matrix (OBDM) is not assumed to be diagonal.

2. Formalism and Results

Let us now sketch the derivation of the main equations and show the most relevant results. After the single particle basis is fixed by solving the HF equations, we introduce phonon operators Q having the same form as the RPA ones,

$$Q_\nu^\dagger = \sum_{ph}(X_{ph}^\nu a_p^\dagger a_h - Y_{ph}^\nu a_h^\dagger a_p) \tag{1}$$

avoiding thus the use of renormalized operators. The X and Y amplitudes are solutions of

$$\begin{pmatrix} A & B \\ B^* & A^* \end{pmatrix} \begin{pmatrix} X^{(\nu)} \\ Y^{(\nu)} \end{pmatrix} = \omega_\nu \begin{pmatrix} G & 0 \\ 0 & -G^* \end{pmatrix} \begin{pmatrix} X^{(\nu)} \\ Y^{(\nu)} \end{pmatrix}, \tag{2}$$

where

$$A_{ph,p'h'} = \langle 0 \mid [a_h^\dagger a_p, H, a_{p'}^\dagger a_{h'}] \mid 0 \rangle, \qquad B_{ph,p'h'} = -\langle 0 \mid [a_h^\dagger a_p, H, a_{h'}^\dagger a_{p'}] \mid 0 \rangle \tag{3}$$

$$G_{ph,p'h'} = \langle 0 \mid [a_h^\dagger a_p, a_{p'}^\dagger a_{h'}] \mid 0 \rangle \tag{4}$$

and satisfy the orthonormality conditions

$$\sum_{ph,p'h'} (X_{ph}^\nu X_{p'h'}^{\nu'} - Y_{ph}^\nu Y_{p'h'}^{\nu'}) G_{ph,p'h'} = \delta_{\nu\nu'}. \tag{5}$$

In Eq. (3) the symmetrized double commutators are defined as

$$[A, B, C] = \frac{1}{2}\{[A, [B, C]] + [[A, B], C]\}. \tag{6}$$

The standard RPA equations can be obtained by replacing the state $\mid 0\rangle$ with the uncorrelated HF one in the matrices A, B and G. This substitution introduces a visible inconsistency since, on one hand, the definition of the ground state as the vacuum of the Q operators [?] is used to derive the Eqs. (2) , while, on the other hand, the HF state is used in calculating the expectation values appearing in those equations. Furthermore, it introduces some violations of the Pauli principle.

In general one and two-body density matrices appear in the double commutators. By using the method of the linearization of the equations of motion,one gets the matrices A, B and G expressed in terms of the OBDM only (see ref. [8]). Its matrix elements can be evaluated by using the number operator method [9] in terms of the X and Y amplitudes. Therefore the equations of motion (2) are nonlinear. In order to solve them we use an iterative procedure. At the n-th iterative step, we compute A, B and G by using the X and Y amplitudes of the (n-1)-th step. As starting point we take the solutions of the standard RPA equations. This iterative procedure is carried on until convergence is reached. The above described method is quite general and can be applied to any finite Fermi system. In order to analyze how well it works, an application has been done to study metallic clusters which, besides being realistic systems having some similarities with nuclei, have the great advantage that the basic interaction is well known, the Coulomb one. In Fig. 1 we show the $S = 0$ dipole strength distribution in several Na clusters. As it is apparent, the method works quite well in improving over standard RPA.

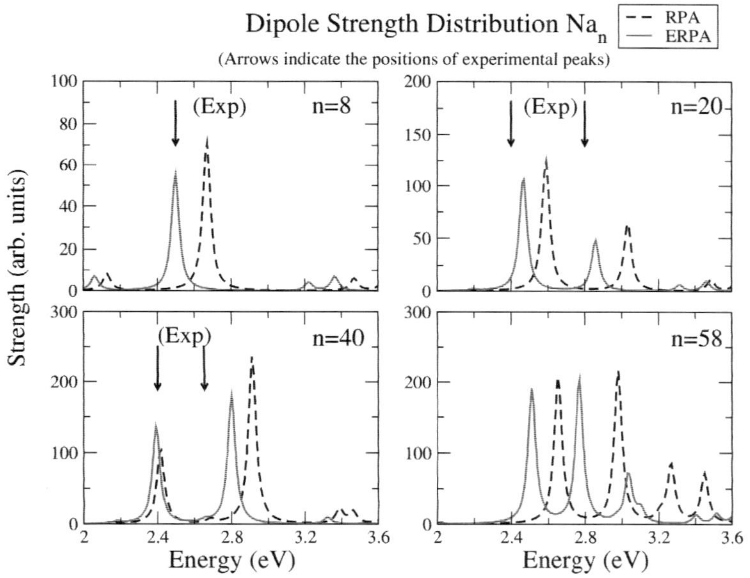

Figure 1. (Color online) Spin=0 dipole strength for Na_N metal cluster with N =8, 20, 40 and 58. Solid line refers to calculations performed within the present approach, while the dashed line refers to standard RPA calculations

As it is well known, if $\mid 0\rangle$ and$\mid \nu\rangle$ are a complete set of exact eigenstates of the Hamiltonian with eigenvalues E_0 and E_ν , the following identity holds:

$$\sum_\nu \omega_\nu |\langle \nu \mid F \mid 0\rangle|^2 = \frac{1}{2}\langle 0 \mid [F,[H,F]] \mid 0\rangle, \tag{7}$$

where $\omega_\nu = E_\nu - E_0$

The above equality is in general violated to some extent when $| 0\rangle, | \nu\rangle$, and ω_ν are calculated with some approximation. To which extent it is satisfied is a measure of the adequacy of the approximation. Let us examine the transition amplitudes induced by a one-body operator,

$$F = \sum_{\alpha,\beta} \langle \alpha \mid F \mid \beta \rangle a_\alpha^\dagger a_\beta \tag{8}$$

between the ground and excited states . By using the definition (1) and the vacuum property of $| 0\rangle$, one gets

$$\langle \nu \mid F \mid 0 \rangle = \langle 0 \mid [Q_\nu, F] \mid 0 \rangle. \tag{9}$$

The above expression is general and it is valid independently of the explicit form of the Q operators. When the latter have the form (1) , only the p -h components of the transition operator F are selected. A very important feature of RPA, known as Thouless theorem [10] can be described as follows. If in both sides of Eq. (7) , the HF state is used, from Eq. (9) one gets

$$\langle \nu \mid F \mid 0 \rangle = \sum_{ph} \Big\{ X_{ph}^{\nu *} \langle p \mid F \mid h \rangle + Y_{ph}^{\nu *} \langle h \mid F \mid p \rangle \Big\}. \tag{10}$$

i.e., the standard RPA expression.The right-hand side can be completely expressed in terms of the A and B matrices of RPA. It is then easy to show that equality (7) is exactly satisfied. This result is very important also because it guarantees that spurious excitations corresponding to broken symmetries as, for example, the translational invariance, separate out and are orthogonal to the physical states. We remark that, when the right-hand side is evaluated in the HF state, only the p -h components of the transition operator F appear in it. The same happens in the l.h.s. but it is essentially related to the p-h nature of the Q operators. In fact, when the correlated $| 0\rangle$ is maintained, it is still true that only the p-h components of the transition operator F appear in the l.h.s. and one has

$$\langle \nu \mid F \mid 0 \rangle = \sum_{php'h'} \Big\{ X_{ph}^{\nu *} \langle p' \mid F \mid h' \rangle + Y_{ph}^{\nu *} \langle h' \mid F \mid p' \rangle \Big\} G_{ph,p'h'}, \tag{11}$$

while this is no more the case in the r.h.s., where the whole structure of F appears. This is the reason why all extension of RPA, with only p-h excitations, violate Eq. (7).

We thus introduce more general Q operators

$$Q_\nu^\dagger = \sum_{\alpha > \beta} (X_{\alpha\beta}^\nu a_\alpha^\dagger a_\beta - Y_{\alpha\beta}^\nu a_\beta^\dagger a_\alpha), \tag{12}$$

where α and β are single-particle states and $\alpha > \beta$ means that $n_\alpha < n_\beta$ (n being the occupation number of the state).

The new A, B and G matrices have the same expressions of the so far discussed approach but the $p's$ and $h's$ indices are replaced now by the $\alpha's$ and $\beta's$ ones, respectively (see Ref. [7]). The strong advantage of this method with respect to that of Ref. [6] is that, by diagonalizing the norm matrix and by looking at the overlaps of the excited states with the ground state the non-physical states can be isolated and eliminated from the spectrum. In order to analyze the advantages of this procedure we have applied it to a 3-level Lipkin model and the results are shown in Fig. 2 where the excitation spectrum (panel (a)) and the eigenvalues of the norm matrix (panel(b)) are displayed as a function of the strength parameter. The present calculation is denoted in the figure with the acronym EERPA (extended and enlarged RPA). EERPA results (solid lines) are compared with the exact corresponding values (dashed lines) and with the standard RPA energies (dotted lines). The first two RPA excited states are indicated with (1) and (3). A much better agreement with the exact results is found for EERPA with respect to RPA. This is mostly evident for the first RPA excited state already before the RPA collapse point, $\tau = 0.026$. With EERPA the collapse point is located much further and the good agreement with the exact values is kept in the whole region where the EERPA results exist. The state (2), whose energy is equal to $\epsilon_2 - \epsilon_1$ when τ approaches 0, does not correspond to any exact solution. Thus it is non-physical. This straightforward identification would not be possible when the EERPA is applied to the study of realistic systems since exact results are not available in that case. In general, however, one can proceed as follows. In the panel (b) of the same figure we plot the eigenvalues of the norm matrix. One can observe that even beyond the RPA collapse point one eigenvalue is much smaller than the others, its value being equal to 0.03 at the RPA collapse. At $\tau = 0.035$, that is far beyond the collapse, the three eigenvalues are still well separated: 0.09, 0.69 and 0.84, respectively. This tells us that one of the three solutions is non-physical. In order to identify it we look at the overlap of the ground state with the three excited ones, which are shown in the panel (c) of Fig. 2. As it is very clearly visible, such overlap for the state labelled as (2) is strongly different from zero, while for the others it is much smaller. Thus we can conclude that indeed such state is the non-physical one. Of course, in the case of a schematic, exactly solvable model such identification can be done directly by looking at the exact energy spectrum while this is not possible in realistic cases. On the other hand, the appearance of a (almost) zero eigenvalue of the norm matrix together with a large overlap with g.s. is an unambiguous criterion to single out the states to be eliminated and it is viable also in realistic systems. The fact that the overlaps of the physical states with the ground state are not exactly zero, especially for large values of the strength, is related to the approximations present in the approach. For example, the OBDM is evaluated by using the number operator method truncated at a certain order, which probably is not enough for values of the strength well beyond the RPA collapse point. However, in physical cases we expect to be in situations corresponding to smaller values of the strength or, in worst cases, around the RPA collapse point. We remark that the above discussed criterion, to disentangle physical from non-physical states, could not be applied in the previous approach [6] since there the OBDM is assumed diagonal.

3. Conclusions and Perspectives

In conclusion, we have presented an extension of RPA in which ground state correlations are treated in a self-consistent way and no use of the quasiboson approximation is made. An application to metallic clusters shows that an important improvement with respect to standard RPA is achieved. However, a drawback of that approach, common to all extensions of RPA, is that EWSR are violated. In order to overcome this difficulties we have further generalized the approach by enlarging the configuration space beyond the p-h excitations, including also the particle-particle and hole-hole ones. The method has been applied to a three-level Lipkin model and we have shown that possible non-physical states can be isolated and eliminated from the spectrum by diagonalizing the norm matrix and by looking at the overlaps with the ground state. The obtained results strongly encourage the application of the present approach to more realistic systems. Work in this direction is in progress.

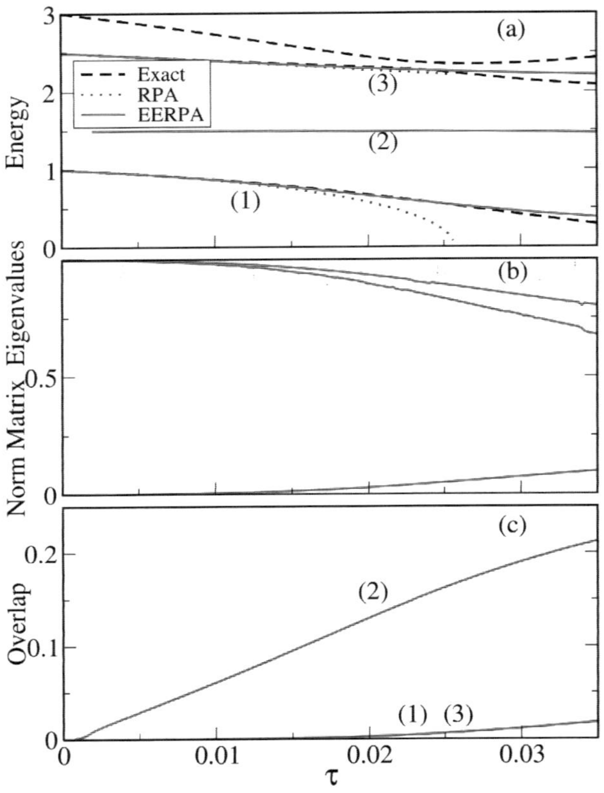

Figure 2. (Color online) Panel (a): Excitation energies of the states (1), (2) and (3) and the eigenvalues of the norm matrix (panel (b)) as a function of the strength parameter $\tau = \chi/\epsilon$. The energies in the Y axis are expressed in units of ϵ. In the panel (c) the corresponding overlaps of the three states with the ground state are shown. See Ref. [7] for more details.

References

[1] Ring P and Schuck P, The Nuclear Many-Body Problem (Springer-Verlag, 1980).

[2] Catara F, Piccitto G, Sambataro M, and Van Giai N, *Phys. Rev.* B 54, 17536 (1996).

[3] Catara F, Chomaz Ph and Van Giai N, *Phys. Lett.* B233, 6(1989); Beaumel D and Chomaz Ph, *Ann.Phys (NY)* 213, 405 (1992); Sambataro M and Catara F, *Phys. Rev.* C 51, 3066 (1995); Sambataro M and Suhonen J, ibid. 56, 782 (1997); Volpe C et al., *Nucl. Phys.* A647, 246 (1999); Sambataro M and Dinh Dang N, *Phys. Rev.* C 59, 1422 (1999); Grasso M, Catara F and Sambataro M, ibid. 66, 064303 (2002); Gambacurta D, Sambataro M, and Catara F, ibid. 73, 014310 (2006).

[4] Dukelsky J and Schuck P, *Nucl. Phys.* A512, 466 (1990); Klein A,Walet N R and Do Dang G, *Nucl. Phys.* A535, 1 (1991);Karadjov D, Voronov V V and Catara F, *Phys.Lett.* B306, 197 (1993); Catara F, Dinh Dang N and Sambataro M, *Nucl. Phys.* A579, 1 (1994); Dukelsky J and Schuck P, *Phys. Lett.* B387, 233 (1996); Catara F, Grasso M, Piccitto G and Sambataro M, *Phys. Rev.* B 58, 16070 (1998); Krmpotic F et al., ibid. A637, 295 (1998); Simkovic F et al., *Phys. Rev.* C 61, 044319 (2000); Takahara S, Tohyama M and Schuck P, ibid. 70, 057307 (2004); Gambacurta D, Grasso M, Catara F and Sambataro M, ibid. 73,024319 (2006); Tohyama M, ibid. 75, 044310 (2007).

[5] Hara K, *Prog. Theor. Phys.* 32, 88 (1964); Rowe D J, *Rev. Mod. Phys.* 40, 153 (1968).

[6] Grasso M and Catara F, *Phys. Rev.* C 63, 014317 (2000).

[7] Gambacurta D, Catara F and Grasso M, *Phys. Rev.* C 80, 014303 (2009).

[8] Gambacurta D and Catara F, *Phys. Rev.* B 77, 205434 (2008).

[9] Rowe D J, *Phys. Rev.* 175, 1283 (1968).

[10] Thouless D J, *Nucl. Phys.* 21, 225 (1960).

Lifetime measurements in mirror nuclei ^{31}S and ^{31}P: A test for isospin mixing

2011 J. Phys.: Conf. Ser. 267 012048

(http://iopscience.iop.org/1742-6596/267/1/012048)

Lifetime measurements in mirror nuclei ^{31}S and ^{31}P: A test for isospin mixing

D Tonev[1,2], G de Angelis[2], P Petkov[1], S Iliev[1], R Orlandi[2,4], C Ur[3], N Goutev[1], M S Yavahchova[1], R Menegazzo[3], D Bazzacco[3], P G Bizzeti[5], A M Bizzeti-Sona[5], S Brant[6], D Bucurescu[7], I Deloncle[8], F Della Vedova[2], E Farnea[3], A Gadea[2,9], C He[10], A Iordachescu[7], M Ionescu-Bujor[7], H Laftchiev[1], S M Lenzi[3], G Lo Bianco[11], S Lunardi[3], N Marginean[2,7], R Marginean[3,7], D Mengoni[3], D R Napoli[2], H Penttilä[12], F Recchia[3], E Sahin[2], R P Singh[13], A Saltarelli[11], J J Valiente-Dobón[2], H -F Wirth[14] and Q Zhong[10]

[1] INRNE, Bulgarian Academy of Science, Sofia, Bulgaria
[2] INFN, Laboratori Nazionali di Legnaro, Legnaro, Italy
[3] Dipartimento di Fisica and INFN Padova, Padova, Italy
[4] The University of the West of Scotland, UK,
[5] Dipartimento di Fisica and INFN Firenze, Florence, Italy
[6] University of Zagreb, Croatia
[7] NIPNE, Bucharest, Romania
[8] CSNSM, CNRS/Université, Paris-Sud XI, 91405 Orsay Campus, France
[9] IFIC, Valencia, Spain
[10] Department of Nuclear Physics, CIAE, Beijing, P.R. China
[11] Dipartimento di Fisica, Università di Camerino and INFN Perugia, Italy
[12] University of Jyväskylä, Jyväskylä, Finland
[13] Inter-University Accelerator Center, New Delhi, India
[14] Ludwig-Maximilians Universität München, Germany

E-mail: `mitko@lnl.infn.it`

Abstract. Using the ^{20}Ne + ^{12}C fusion-evaporation reaction at $E_{^{20}Ne} = 33$ MeV and the multidetector array GASP in conjuction with the EUCLIDES charged particle detector, angular correlations of coincident pairs of γ transitions and lifetimes in mirror nuclei ^{31}S and ^{31}P have been measured at the Piave-Alpi accelerator of the Laboratori Nazionali di Legnaro. A comparison of the determined B(E1) strengths of the analog mirror $7/2^- \rightarrow 5/2^+$ transitions indicates the presence of a violation of isospin symmetry.

1. Introduction

The investigation of mirror nuclei along the N=Z line is of considerable interest since it directly addresses the validity of the charge symmetry of the nuclear forces and the role of the Coulomb effects on nuclear structure. If the charge symmetry of the nuclear force is exact, then in the limit of long wavelengths, the E1 transition operator is purely isovector and therefore E1 transitions are forbidden in N=Z nuclei between states of equal isospin and they have equal strength in mirror nuclei. An approach to investigate the isospin symmetry breaking is to observe experimental deviations from the two rules above. The aim of the present experiment is to verify whether the transition strengths of the E1 transitions depopulating the $7/2^-$ analog states in the mirror couple ^{31}S and ^{31}P are equal. Using the same reaction in Ref. [1] the level schemes of ^{31}S and ^{31}P were obtained and a difference of the branching ratios of the analog transitions depopulating the $7/2^-$ state was observed (see Fig. 1). Can the different pattern of the decay of the $7/2^-$ to the $5/2^+$ states in both nuclei lead correspondingly to different B(E1) values? In order to answer this question we need to know the lifetimes of the two states and the M2/E1 mixing ratios in the transitions. Lifetimes of the analog $7/2^-$ states in ^{31}S and ^{31}P [2] were recently investigated with the Doppler-shift attenuation method (DSAM), the data being analyzed using gates from below and making some hypothesis for the unknown feeding, an approach which is generally limited by intrinsic uncertainties. The purpose of the present experiment is to extract precise lifetime values of the excited states in the mirror A=31 pair using more advanced methods, together with performing angular correlation analysis in order to determine the multipole mixing ratios of the transitions of interest.

2. Experiment and data analysis

Excited states in ^{31}S and ^{31}P were populated using the $1n$ and $1p$ exit channels, respectively, of the reaction ^{20}Ne + ^{12}C. The beam of ^{20}Ne, with an energy of 33 MeV, was delivered for the first time by the Piave-Alpi accelerator of the LNL. In order to obtain a thick Carbon target needed for a DSAM measurement we used a two steps procedure. The first step was to evaporate a 10 mg/cm^2 gold layer on a 0.6 mg/cm^2 ^{12}C foil. In the second step 0.15 mg/cm^2 ^{12}C were evaporated onto the carbon foil in order to reach 0.75 mg/cm^2 final thickness. We note that this was the maximum Carbon thickness we could reach keeping at the same time a high quality of the target. The deexciting γ rays were registered with the GASP array [3] in its configuration II.

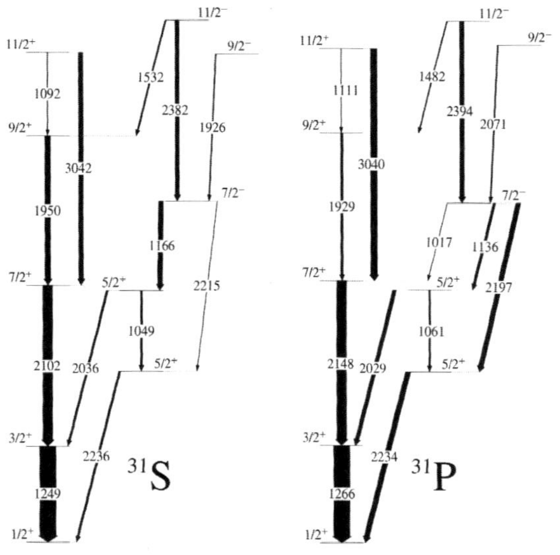

Figure 1. Partial level-schemes of ^{31}S and ^{31}P from Ref. [1]. The different pattern of decay of the $7/2^-$ state in both nuclei is illustrated by different arrow thickness.

Charged particles were detected with the EUCLIDES silicon ball [4]. Gain matching and efficiency calibration of the Ge detectors were performed using ^{152}Eu and ^{56}Co radioactive sources. The data were sorted into coincidence γ - γ matrices whereby the detection of protons was required to construct the matrices for ^{31}P.

The angular correlation function W(θ_1, θ_2, ϕ) for a cascade of two successive transitions from oriented states is presented by

$$W(\theta_1, \theta_2, \phi) = \sum_{\lambda_1, \lambda, \lambda_2} B_{\lambda_1}(I_1) A_{\lambda}^{\lambda_1 \lambda_2}(\gamma_1) A_{\lambda_2}(\gamma_2) H_{\lambda_1, \lambda, \lambda_2}(\theta_1, \theta_2, \phi) \tag{1}$$

This function mainly depends on the spins of the initial, intermediate, and final states, on the multipolarities of the transitions as well as on the two multiple mixing ratios of the two coincident transitions. The first γ transition is detected at an angle θ_1 with respect to the beam axis, the second γ transition at an angle θ_2, and ϕ is the difference of the azimuthal angles of the corresponding detectors. The term $B_{\lambda_1}(I_1)$ describes the orientation of the upper nuclear state, the term $A_{\lambda}^{\lambda_1 \lambda_2}(\gamma_1)$ the orientation of the intermediate state due to emission of γ_1, and $A_{\lambda_2}(\gamma_2)$ the emission of γ_2. With the $H_{\lambda_1, \lambda, \lambda_2}(\theta_1, \theta_2, \phi)$ is denoted the angular function. It is reduced to an ordinary Legendre polynomial if any of the λ's vanish. Detailed information about the angular correlation function $W(\theta_1, \theta_2, \phi)$ is presented in Ref. [5]. The symmetries of the coincident radiation event of the two γ rays lead to symmetries of the function W. They can be used to establish the independent angular correlation groups for a given setup. The coincidence efficiency can be represented to a good approximation as a product of the efficiencies of the two detectors. For the angular correlation analysis, we used the code CORLEONE [6]. The relative efficiencies of the detector groups were adjusted by requiring a reasonable reproduction of the properties of known $4_1^+ \rightarrow 2_1^+ \rightarrow 0_1^+$ cascades of the even-even nuclei: ^{24}Mg, ^{28}Si and ^{30}Si. The quality of the experimental data as well as the excellent agreement with the theoretical predictions of CORLEONE is shown for the cases of ^{24}Mg (Fig. 2) and ^{30}Si (Fig. 3).

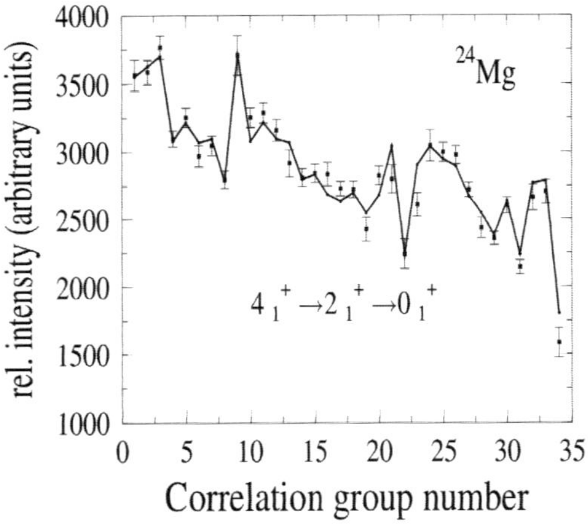

Figure 2. Angular correlation pattern for the cascade involving $4_1^+ \rightarrow 2_1^+ \rightarrow 0_1^+$ transitions of ^{24}Mg. For this spin hypothesis, the best fit shows clearly the pure E2 character of the investigated transitions.

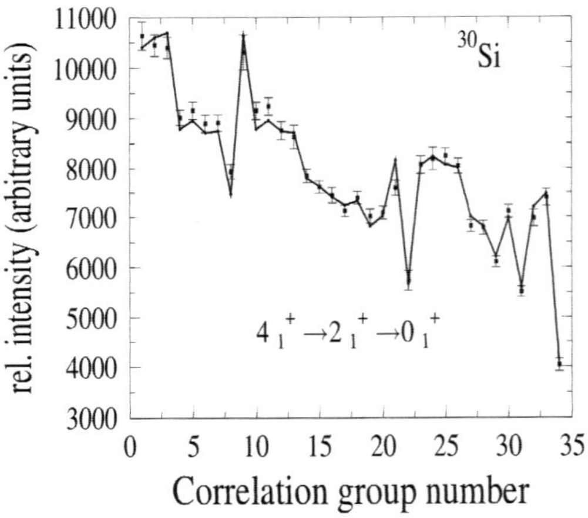

Figure 3. Angular correlation pattern for the cascade involving $4_1^+ \rightarrow 2_1^+ \rightarrow 0_1^+$ transitions of ^{30}Si. For this spin hypothesis, the best fit shows clearly the pure E2 character of the investigated transitions.

For a given spin hypothesis, the data analysis consists of fitting the intensity of the cascade by adjusting the parameter σ characterizing the distribution of the magnetic sub-states m of the spin of the first oriented level and the multiple mixing ratios δ_1 and δ_2 of the two successive transitions. Usually, the analysis is simpler if the spins of the cascade are known and we concentrate our work on the determination of the δ_1 and δ_2. This procedure was successfully used to derive the multiple mixing ratios for the transitions depopulation the $7/2^-$ analog states in the ^{31}P and ^{31}S. Due to the good statistics of the experiment we constructed 34 detector correlation groups, which ensures very precise determination of the multiple mixing coefficients.

The next step in our analysis was to derive precisely the lifetimes of the states of interest. For this reason a DSAM measurements was utilized. For the purposes of a such measurement, the information on the detection angles of the γ rays is of primordial importance because the Doppler shifts of their energies depend on the angle between the direction of the recoil velocity and the direction of observation. Therefore the Ge detectors of GASP were grouped into rings corresponding to approximately the same polar angle with respect to the beam axis. In both cases of ^{31}P and ^{31}S the statistics allowed to use γ-γ matrices where the angular information was conserved on both axes of the matrix, i.e., the events represented the registration of two coincident γ rays by the detectors of two particular rings.

In order to determine a lifetime in a DSAM measurement we need to know exactly the velocity histories of recoiling nuclei when they are slowing down in the target and stopper, until the moment they stop. The best way to obtain this information is to perform a Monte-Carlo (MC) simulation. For the MC simulation of the slowing-down histories of the recoils we used a modified [7, 8] version of the program DESASTOP [9] written by G.Winter. In this version, the time-evolution of the recoil velocities in the target and stopper is followed in three dimensions. The electronic stopping powers used were obtained from the Northcliffe and Schilling tables [10] with corrections for the atomic structure of the medium along the lines discussed in Ref. [11]. As suggested in Ref. [12], an empirical reduction of $f_n = 0.7$ was applied to down-scale the nuclear stopping power predicted by the theory of Lindhard, Scharff and Schiøtt [13]. According to the calculations performed, the mean velocity of the recoils when leaving the target was about 3.7 % of the velocity of light, and they needed in average 1.1 ps to come to rest. The evaporation of charged particles, which is of importance for the velocity distribution of light residual nuclei, was taken into account in the MC simulation and led to better fits of the spectra. The database of about 10000 velocity histories was additionally randomized with respect to the experimental setup by taking into account the exact position of the GASP detectors and their finite size. Complementary details on our approach for Monte Carlo simulation can be found in Refs. [7, 8, 14, 15].

The strength of the ^{31}P and ^{31}S reaction channels made it possible to apply the newly developed procedure for analysis of coincidence DSAM data where the gate is set on the shifted portion of the line shape of a transition directly feeding the level of interest [8]. Within this approach, the timing quality of the gated line shape is improved compared to the case where the gate includes also fully stopped events because they do not bring lifetime information. Moreover, gating from above allows the elimination of the uncertainties related to the unobserved feeding of the level of interest which perturb singles measurements and coincidence measurements where the gate is set on a transition deexciting a level fed by the level of interest.

Fits of the line shapes obtained using the approach [8] and applied to determine the lifetime of the $3/2^+_1$ states in ^{31}P and ^{31}S are presented correspondingly in Figs. 4 and 5. (See also the captions to the Figs. 4 and 5.)

We estimate the uncertainty due to the imprecise knowledge of the stopping powers to 10% and include it in the final errors of the lifetimes. It should be noted that the derivation of lifetimes in ^{31}S and ^{31}P in the same experiment makes the determination of the ratios of the corresponding transition strengths very precise since uncertainties related to the stopping powers nearly cancel.

3. Results and discussion

Following the procedure described in the previous section and using the code CORLEONE by I. Viedenhöver [6] we succeeded to derive the multipole mixing ratios for the transitions depopulating the $7/2^-$ states in both mirror nuclei. We checked how the approach works by an

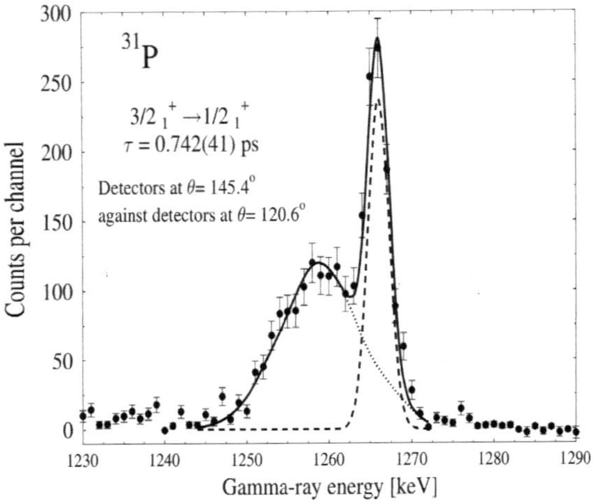

Figure 4. Example of the line-shape analysis of the 1264 keV γ-ray transition and determination of the lifetime of the $I^\pi = 3/2_1^+$ in ^{31}P. The spectrum measured with the detectors of ring 7, the fit (full line) the DSA portion of the fit(dotted line) and the unshifted portion (dashed line) are presented.

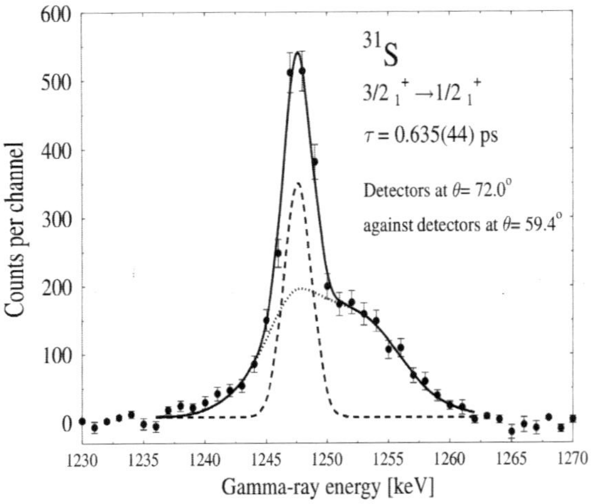

Figure 5. Example of the line-shape analysis of the 1249 keV γ-ray transition and determination of the lifetime of the $I^\pi = 3/2_1^+$ in ^{31}S. The spectrum measured with the detectors of ring 3, the fit (full line) the DSA portion of the fit(dotted line) and the unshifted portion (dashed line) are presented.

investigation of already known and reported δ values in Ref. [16]. A very good agreement was found.

The results for the transitions $7/2^- \rightarrow 5/2_2^+$ in both mirror nuclei show dominantly an E1 character. In the present contribution we report this result for the first time. The exact values for the multiple mixing ratios will be reported in forthcoming publication [17].

As a test we determined some lifetimes in ^{31}P which are already known and our values are in a very good agreement with the results from earlier measurements [16, 18]. For example we could compare the value of the lifetime of the $3/2_1^+$ state reported in the literature [18] $\tau =$ 745(35) fs with the value derived by us of τ = 736(24) fs. The agreement is perfect and in our case due to the excellent statistics the error is smaller. The quality of the data is illustrated in Fig. 4.

The lifetimes previously measured in ^{31}S were obtained with an experimental setup which contained only one detector and have been analyzed with a gate from below [19]. Often, due to the fact that the side feeding is not exactly determined, the lifetime values are overestimated. The lifetime of the $3/2_1^+$ state in ^{31}S is reported to be 720(180) fs [19]. Then within the frames of the errors the lifetimes of the analog states in both nuclei are the same.

With significantly better statistics and using much more detectors and a procedure with a gate from above we derived a shorter the lifetime value of 624(24) fs for the $3/2_1^+$ level in ^{31}S.

As it is for the cases of A = 47 [14] and A = 51 [20] mirror pairs, most of the lifetimes in the nucleus with one proton more have shorter lifetimes. This is the conclusion also for the $3/2_1^+$ levels in the A = 31 mirror couple.

The quality of the data from the present experiment is good enough in order to determine the lifetime values of the $7/2^-$ state in both mirror nuclei, using a gate on the shifted component of a feeding transition. The lifetime values determined from the present experiment are with smaller errors than previously determined and allow us to distinguish a difference in the B(E1) values of the analog transitions in ^{31}P and ^{31}S.

For the $7/2^-$ excited state of ^{31}P we obtained a value which is not different from that reported in the literature [16, 18], but its error is four times smaller. The value derived by us for the $7/2^-$ state of ^{31}S is different from that reported in the Ref. [2]. It is about twice shorter than 1.03(21) ps.

Using precisely determined branching ratios, multipole mixing ratios and lifetimes we could compare corresponding B(E1) values for the analog transitions depopulating the $7/2^-$ state of the A=31 mirror couple. The B(E1) value derived by us for the $7/2^-$ state of ^{31}S is about two and a half times larger than the already known B(E1) value characterizing the analog state in ^{31}P. The observed significant difference between the B(E1) values is an indication for a symmetry violation component. Such behavior was observed in the mirror couples A = 35 [21] and A = 67 [22] and it was explained by the presence of a large isoscalar component. This component provides evidence for a coherent contribution to isospin mixing, probably involving the isovector giant monopole resonance [22]. The presence of a large (induced) isoscalar component could be the reason for the different B(E1) values in the case of ^{31}P and ^{31}S [17]. An extensive discussion on the "isoscalar" term will be presented in the forthcoming paper [23].

Acknowledgments

D.T. express his gratitude to Ivanka Necheva for her outstanding support. D T, I D and P P are indebted to the French Egide Organism and the French Ministère des Affaires Etrangères et Européennes under ECO-NET contract number 23331RH. M.Y. and D.T. acknowledge support from the European Operational Programme HRD, contract BGO 051 P O 001/07/3.3 - 02/53 with the Bulgarian Ministry of Education.

References

[1] Jenkins D G *et al* 2005 *Phys. Rev.* C **72** 031303(R)
[2] Pattabiraman N S *et al* 2008 *Phys. Rev.* C **78** 024301
[3] Bazzacco D Chalk River Report **(AECL 10613)**, p. 386.
[4] Farnea E *et al* 1987 *Nucl. Instr. and Meth.* A **400** 87
[5] Krane K S, Steffen R M and Wheeler R M 1973 *Nucl. Data Tables* **11** 351
[6] Wiedenhöver I *et al* 1998 *Phys. Rev.* C **58** 721
[7] Petkov P *et al* 1998 *Nucl. Phys.* A **640** 293
[8] Petkov P, Tonev D, Gableske J, Dewald A, von Brentano P, 1999 *Nucl. Instr. and Meth.* A **437** 274
[9] Winter G 1983 *Nucl. Instr. Meth.* **214** 537
[10] Northcliffe L C and Schilling R F 1970 *Nucl. Data* **Sect. 7** 233
[11] Ziegler J F and Biersack J P, in *Treatise on Heavy-Ion Science*, edited by D. A. Bromley (Plenum Press, New York, 1985), Vol. 6, p. 95.
[12] Keinonen J, in *Capture Gamma-Ray Spectroscopy and Related Topics-1984*, Proc. of the Fifth International Symposium, Knoxville, TN, 1984, edited by S. Raman, AIP Conf. Proc. No. 125 (AIP, New York, 1985), p. 557.
[13] Lindhard J, Scharff M, and Schiøtt H E, 1963 *K. Dan. Vidensk. Selsk. Mat. Fys. Medd.* **33** 14
[14] Tonev D *et al* 2002 *Phys. Rev.* C **65** 034314
[15] Tonev D *et al* 2007 *Phys. Rev.* C **76** 044313
[16] Ionescu-Bujor M *et al* 2006 *Phys. Rev.* C **73** 024310
[17] Tonev D *et al to be published*
[18] Endt P M, 1998 *Nucl. Phys.* A **633** 1
[19] Engmann R, Ehrmann E, Brandolini F and Signorini C 1971 *Nucl. Phys.* A **162** 295
[20] du Rietz R *et al* 2004 *Phys. Rev. Lett.* **93** 222501
[21] Ekman J *et al* 2004 *Phys. Rev. Lett.* **92** 132502
[22] Orlandi R *et al* 2009 *Phys. Rev. Lett.* **103** 052501
[23] Bizzeti P G *et al to be published*

Thermal nuclear pairing within the self-consistent quasiparticle RPA

2011 J. Phys.: Conf. Ser. 267 012049

(http://iopscience.iop.org/1742-6596/267/1/012049)

Thermal nuclear pairing within the self-consistent quasiparticle RPA

N. Dinh Dang[1,2] **and N. Quang Hung**[1,3]

[1] Theoretical Nuclear Physics Laboratory, RIKEN Nishina Center for Accelerator-Based Science, 2-1 Hirosawa, Wako City, 351-0198 Saitama, Japan

[2] Institute for Nuclear Science and Technique, Hanoi, Vietnam

[3] Institute of Physics, Hanoi, Vietnam

E-mail: dang@riken.jp (N.D.D.), nqhung@riken.jp (N.Q.H.)

Abstract. The self-consistent quasiparticle RPA (SCQRPA) is constructed to study the effects of fluctuations on pairing properties in nuclei at finite temperature and z-projection M of angular momentum. Particle-number projection (PNP) is taken into account within the Lipkin-Nogami method. Several issues such as the smoothing of superfluid-normal phase transition, thermally assisted pairing in hot rotating nuclei, extraction of the nuclear pairing gap using an improved odd-even mass difference are discussed. A novel approach of embedding the PNP SCQRPA eigenvalues in the canonical and microcanonical ensembles is proposed and applied to describe the recent empirical thermodynamic quantities for iron, molybdenum, dysprosium, and ytterbium isotopes.

1. Introduction

Sharp phase transitions such as the superfluid-normal (SN) or shape ones are prominent features of infinite systems such as metal superconductors, ultra-cold gases, liquid helium, etc. They are well described by many-body theories such as the BCS, RPA or quasiparticle RPA (QRPA). The situation changes in finite small systems such as atomic nuclei, where strong quantal and thermal fluctuations strongly or completely smooth out these sharp phase transitions. It is well known that the conventional BCS, RPA or QRPA theories fail in a number of cases in the description of the ground states as well as excited states of these systems. The reason is that strong fluctuations invalidate the assumptions, based on which the main equations of these theories have been derived. Amongst these assumptions are the Cooper pairs, which violate the particle-number conservation, and the closely related quasiboson-approximation (QBA) used in the (Q)RPA, which violates the Pauli principle between the fermion pairs. These assumptions cause the BCS and QRPA to break down at a certain critical value G_c of the pairing interaction parameter G, below which the BCS theory only has a trivial solution with zero pairing gap $\Delta = 0$. The same is true in the weak coupling region, where the particle-particle RPA is valid but its solution also breaks down at $G \geq G_c$. Meanwhile, the exact solution of the pairing problem exposes no singularity at any G [1]. Similarly, at finite temperature $T \neq 0$, the omission of quasiparticle-number fluctuations (QNF) within the BCS theory leads to the collapse of the pairing gap at the critical temperature T_c, corresponding to the temperature of the SN phase transition in infinite systems. Meanwhile, the exact eigenvalues of the pairing problem embedded in the canonical ensemble (CE) shows a smooth decreasing pairing energy with increasing T due to thermal fluctuations incorporated in the CE [2]. In rotating nuclei, strong fluctuations also smear out the Mottelson-Valatin effect, according to which the pairing gap, existing at zero angular momentum $M = 0$, would collapse at a certain critical angular momentum M_c. This situation means that, in order to be reliable, the BCS, RPA, and/or QRPA theories need to be corrected to include these effects of fluctuations when applied to nuclei, in particular, the light ones. This is done within the framework of the self-consistent QRPA (SCQRPA) presented in this work.

2. Formalism

We consider the pairing Hamiltonian $H = \sum_{k>0} \epsilon_k \hat{N}_{\pm k} - G\sum_{kk'} \hat{P}_k^\dagger \hat{P}_{k'}$, where $\hat{N}_{\pm k} = a_{\pm k}^\dagger a_{\pm k}$ is the particle-number operator, and $\hat{P}_k = a_k^\dagger a_{-k}^\dagger$, $\hat{P}_j = (\hat{P}_j^\dagger)^\dagger$ are the pairing operators. The operators $a_k^\dagger$ and a_k are respectively the single-particle creation and destruction operators. This Hamiltonian has been diagonalized exactly in [1]. The exact partition function is constructed by embedding the exact eigenvalues into the CE as $Z_{\rm Exact}(\beta) = \sum_S d_S \exp(-\beta\varepsilon_S^{\rm Exact})$, with the degeneracy $d_S = 2^S$, inverse temperature $\beta = 1/T$, and $S = 0, 2, ...N$ being the total seniority of the system. Knowing the partition function Z, one calculates the free energy F, entropy S, total energy $\mathcal{E}$, heat capacity C, and pairing gap Δ as $F = -T\ln Z(T)$, $S = -\partial F/\partial T$, $\mathcal{E} = F + TS$, $C = \partial\mathcal{E}/\partial T$, and $\Delta = [-G(\mathcal{E} - 2\sum_k \epsilon_k f_k + G\sum_k f_k^2)]^{1/2}$, where f_k is the single-particle occupation number on the kth level obtained by averaging the state-dependent occupation numbers $f_k^{(S)}$ within the CE [2].

The SCQRPA theory [3, 4] includes a set of BCS-based equations, corrected by the effects of QNF, namely

$$\Delta_k = \Delta + \delta\Delta_k \ , \quad \Delta = G\sum_{k'}\langle\mathcal{D}_{k'}\rangle u_{k'}v_{k'} \ , \quad \delta\Delta_k = 2G\frac{\delta\mathcal{N}_k^2}{\langle\mathcal{D}_k\rangle}u_k v_k \ . \tag{1}$$

$$N = 2\sum_k\left[v_k^2\langle\mathcal{D}_k\rangle + \frac{1}{2}(1 - \langle\mathcal{D}_k\rangle)\right] \ . \tag{2}$$

where u_k and v_k are the Bogoliubov's coefficients,

$$u_k^2 = \frac{1}{2}\left(1 + \frac{\epsilon_k' - Gv_k^2 - \lambda}{E_k}\right) \ , \quad v_k^2 = \frac{1}{2}\left(1 - \frac{\epsilon_k' - Gv_k^2 - \lambda}{E_k}\right) \ , \quad E_k = \sqrt{(\epsilon_k' - Gv_k^2 - \lambda)^2 + \Delta_k^2} \ , \tag{3}$$

with the renormalized single-particle energies ϵ_k'

$$\epsilon_k' = \epsilon_k + \frac{G}{\langle\mathcal{D}_k\rangle}\sum_{k'}(u_{k'}^2 - v_{k'}^2)\Big(\langle\mathcal{A}_k^\dagger\mathcal{A}_{k'\neq k}^\dagger\rangle + \langle\mathcal{A}_k^\dagger\mathcal{A}_{k'}\rangle\Big) \ , \tag{4}$$

$\langle\mathcal{D}_k\rangle = 1 - 2n_k$, the quasiparticle-pair operators $\mathcal{A}_k^\dagger = \alpha_k^\dagger\alpha_{-k}^\dagger$, $\mathcal{A}_k = (\mathcal{A}_k^\dagger)^\dagger$, and $\delta\mathcal{N}_k^2 \equiv n_k(1 - n_k)$ is the QNF on kth level. To avoid level-dependent gaps Δ_k, the level-weighted gap $\bar{\Delta}_k = \sum_k \Delta_k/\Omega$ (Ω is the number of levels) is considered in the numerical results. Because of coupling to collective vibrations beyond the quasiparticle mean field, the quasiparticle occupation number n_k is not given by a Fermi-Dirac distribution of free fermions, but is found from the integral equation [4]

$$n_k = \frac{1}{\pi}\int_{-\infty}^{\infty}\frac{\gamma_k(\omega)(e^{\beta\omega} + 1)^{-1}}{[\omega - E_k - M_k(\omega)]^2 + \gamma_k^2(\omega)}d\omega \ , \tag{5}$$

where the mass operator $M_k(\omega)$ and the quasiparticle damping $\gamma_k(\omega)$ are functions of n_k, SCQRPA eigenvalues ω_μ, SCQRPA $\mathcal{X}_k^\mu$ and $\mathcal{Y}_k^\mu$ amplitudes, SCQRPA phonon occupations

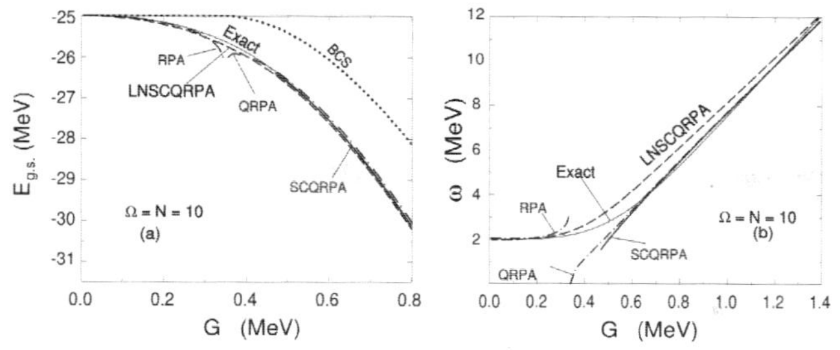

Figure 1. Energies of the ground state (a) and first excited states for the $N = \Omega = 10$ as functions of G at $T = M = 0$. $\omega_{ppRPA} = \mathcal{E}_1(N+2) - \mathcal{E}_0(N+2)$ with the ppRPA eigenvalues $\mathcal{E}_i$.

numbers ν_μ, as well as u_k and v_k. The SCQRPA submatrices A and B contain the screening factors $\langle \mathcal{A}_k^\dagger \mathcal{A}_{k'}^\dagger \rangle$ and $\langle \mathcal{A}_k^\dagger \mathcal{A}_{k'} \rangle$ so that the set of SCQRPA equations should be solved self-consistently with Eqs. (1), (2) and (5) to simultaneously determine $\bar{\Delta}$, chemical potential λ, n_k, ω_μ, $\mathcal{X}_k^\mu$ and $\mathcal{Y}_k^\mu$. To eliminate particle-number fluctuations inherent in the BCS theory, the Lipkin-Nogami (LN) particle-number projection (PNP) [5] is applied on top of Eqs. (1) and (2). The ensuing theory, called the LNSCQRPA theory, has also been extended to include the finite z-projection M of angular momentum (noncollective rotation) [6]. The set of obtained equations is formally the same except that now, depending on the single-particle spin projections $\mp\gamma m_k$ with γ being the angular velocity, one has two types of quasiparticle occupation number, $n_k^\pm$, so that $\langle \mathcal{D}_k \rangle = 1 - n_k - n_{-k}$. At $T = 0$ and $M = 0$ the SCQRPA theory reduces to its zero temperature and non-rotating limit, where $\langle \mathcal{D}_k \rangle = 1/[1 + 2\sum_\mu (\mathcal{Y}_k^\mu)^2]$ [3].

3. Numerical results and discussions

Shown in Fig. 1 are the energies of the ground state (a) and first excited state (b) obtained at $T = M = 0$ within several approximations as well as by exactly diagonalizing the pairing Hamiltonian for the schematic model, which consists of Ω doubly-folded equidistant levels with the single-particle energies chosen as $\epsilon_k = k - (\Omega + 1)/2$ MeV. The displayed results are for the half-filled case with $N = \Omega = 10$, and plotted as functions of the pairing interaction parameter G. It is seen that the LNSCQRPA describes rather well the exact energies of both the ground and first excited states without any discontinuity in the region around G_c, where all other approaches such as the RPA, QRPA, and SCQRPA collapse.

The level-weighted gap, total energy, and heat capacity obtained for the systems with $N = \Omega = 10$ and 50 are shown as functions of temperature T in Fig. 2. Beside the predictions by the SCQRPA, LNSCQRPA, as well as by their corresponding limits, FTBCS1 and FTLN1, where coupling to QRPA is omitted (i.e. n_k is described by the Fermi-Dirac distribution for free fermions), and the finite-temperature (FT) BCS results, the exact CE results are also shown. This figure clearly demonstrates how QNF smooth out the sharp SN phase transition in finite systems. The pairing gap never collapses, but decreases monotonously with increasing T, whereas the spike at T_c in the heat capacity, which serves as a signature of sharp SN phase transition within the FTBCS, becomes strongly depleted to a broad bump.

At finite angular momentum $M \neq 0$, the FTBCS theory predicts the Mottelson-Valatin effect, according to which, the zero-temperature pairing gap decreases with increasing M and collapses at $M = M_c$ because the angular momentum blocks the levels close to the Fermi surface [Fig. 3 (a) and 3 (c)]. Thermal effects relax the blocking, opening some levels around the Fermi surface for pairing. This leads to the thermally assisted pairing gap (or pairing reentrance), according to which at a certain $T = T_1$ the pairing gap becomes finite even at $M > M_c$ [7, 8]. With increasing

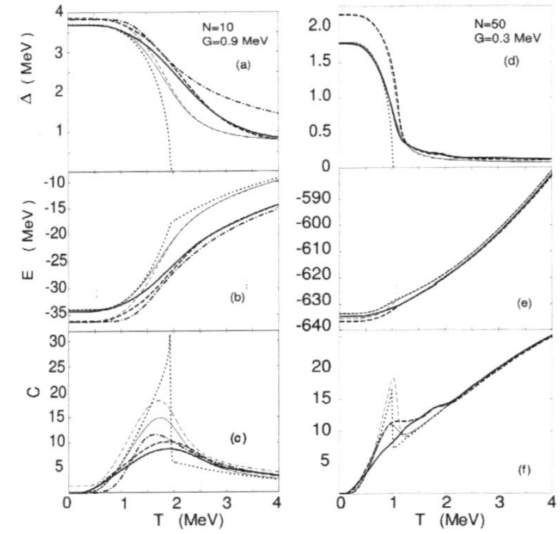

Figure 2. Level-weighted pairing gap $\bar{\Delta}$, total energy E, and heat capacity C, as functions of T for $N = \Omega = 10$ (a - c) and 50 (d - f) obtained within the FTBCS (dotted), FTBCS1 (thin solid), FTLN1 (thin dashed), SCQRPA (thick solid), LNSCQRPA (thick dashed). The dash-dotted lines for $N = 10$ are the exact CE results.

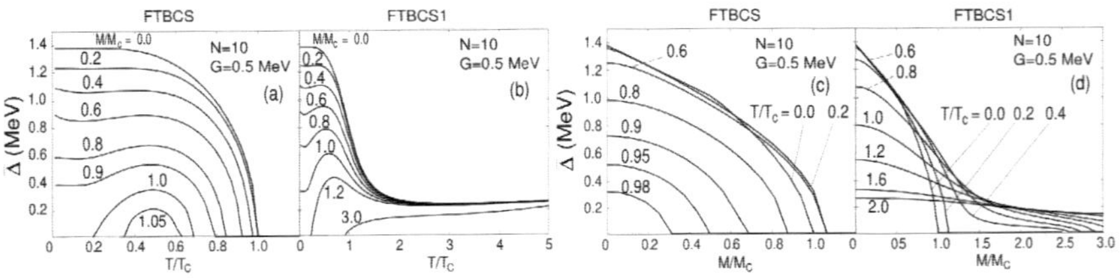

Figure 3. Level-weighted pairing gaps $\bar{\Delta}$ for $N = \Omega = 10$ as a functions of temperature T at various values of M/M_c and angular momentum M at various values of T/T_c within the FTBCS (a, c) and FTBCS1 (b, d) theories.

T thermal effects again break the pairs so that the gap disappears at $T = T_2 > T_1$ [See Fig. 3 (a) for $M/M_c \geq 1$]. In finite systems, the QNF smooth out both the Mottelson-Valatin transition and thermal assisted pairing, for instance, for $N = \Omega = 10$ with $G = 0.5$ MeV at $T/T_c \geq 1$, the gap only decreases monotonously with increasing M but never vanishes [See Fig. 3 (d) for $M/M_c \geq 1$], whereas at $M/M_c \geq 3$, the pairing gap reappears at $T > T_1$ but remains finite with further increasing T [See Fig. 3 (b) for $M/M_c \geq 3$].

The odd-even mass difference contains the admixture with the contribution from uncorrelated single-particle configurations, which increases with T. Therefore, the simple extensions of this formula to obtain the three-point and four-point gaps, in principle, do not hold at finite temperature. We propose an improved odd-even mass difference formula at $T \neq 0$, namely

$$\tilde{\Delta}^{(3)}(\beta, N) = \frac{G}{2}\left[(-1)^N + \sqrt{1 - 4\frac{S'}{G}}\right], \quad S' = \frac{1}{2}[\langle\mathcal{E}(N+1)\rangle_\alpha + \langle\mathcal{E}(N-1)\rangle_\alpha] - \langle\mathcal{E}(N)\rangle_\alpha^{(0)}, \quad (6)$$

where $\langle\mathcal{E}(N)\rangle_\alpha$ is the total energy of the system with N particles within the grand canonical ensemble (GCE) (α =GC) or CE (α =C); $\langle\mathcal{E}\rangle_\alpha^{(0)} \equiv 2\sum_k [\epsilon_k - Gf_k^{(\alpha)}/2]f_k^{(\alpha)}$ with $-G\sum_k [f_k^{(\alpha)}]^2$

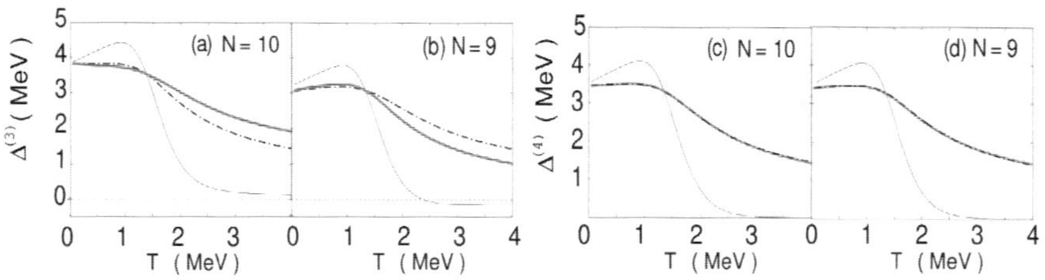

Figure 4. Pairing gaps extracted from the odd-even mass differences as functions of T for $N = 10$ (a,c) and $N = 9$ (b,d) ($\Omega = 10$, $G = 0.9$ MeV). The thin solid and thick solid lines denote the gap $\Delta^{(3)}(N)$ ($\Delta^{(4)}(N) = [\Delta^{(3)}(N) + \Delta^{(3)}(N-1)]/2$), and the improved gap $\widetilde{\Delta}^{(i)}(\beta, N)$ ($i = 3$, 4) from Eq. (6), respectively. The dash-dotted lines are the canonical gaps $\Delta_{\rm C}^{(i)}$.

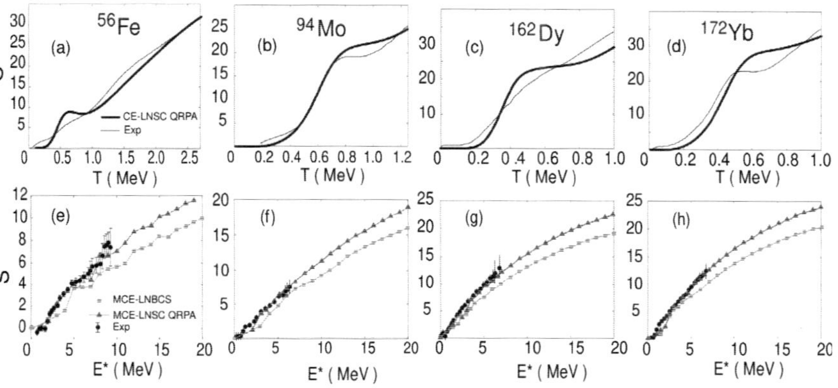

Figure 5. CE heat capacities C as functions of T and MCE entropies S as functions of excitation energy E^* for ^{56}Fe, ^{94}Mo, ^{162}Dy, and ^{172}Yb. Experimental data are taken from [11].

coming from uncorrelated single-particle configurations. Shown in Fig. 4 are the pairing gaps $\Delta^{(i)}(\beta, N)$ ($i = 3$ and 4), obtained for $N = 9$ and 10 ($\Omega = 10$) by using the simple extension of the odd-even mass formula to $T \neq 0$ as well as the modified gaps $\widetilde{\Delta}^{(i)}(\beta, N)$ from Eq. (6), and the canonical gaps $\Delta_{\rm C}^{(i)}$. It is seen in Fig. 4 that the naive extension of the odd-even mass formula to $T \neq 0$ fails to match the temperature-dependence of the canonical gap $\Delta_{\rm C}^{(i)}$. The gap $\Delta^{(3)}(\beta, N = 9)$ even turns negative at $T > 2.4$ MeV, suggesting that such simple extension of the odd-even mass difference to finite T is invalid. The modified gap $\widetilde{\Delta}^{(i)}(\beta)$ is found in much better agreement with the canonical one, whereas the modified four-point gaps $\widetilde{\Delta}^{(4)}(\beta)$ practically coincide with the canonical gaps. The comparison in Fig. 4 suggests that formula (6) is a much better candidate for the experimental gap at $T \neq 0$, rather than the simple odd-even mass difference.

In order to construct a feasible description for pairing within the CE, the eigenvalues of the LNBCS and LNSCQRPA, obtained for each total seniority S at $T = 0$, are embedded into the CE by using the partition function $Z_\gamma(\beta) = \sum_S d_S e^{-\beta \mathcal{E}_S^\gamma}$ (γ = LNBCS, LNSCQRPA). The resulting approaches are called the CE-LNBCS and CE-LNSCQRPA, respectively [10]. These solutions are also embedded into the microcanonical ensemble (MCE) by using the Boltzmann's definition for entropy, $S(\mathcal{E}) = \ln W(\mathcal{E})$, where $W(\mathcal{E})$ is the number of accessible states within the energy interval $(\mathcal{E}, \mathcal{E} + \delta\mathcal{E})$. The corresponding approaches are called the MCE-LNBCS and MCE-LNSCQRPA, respectively [10].

The CE heat capacities and MCE entropies for several nuclei are shown in Fig. 5 as functions of T and excitation energy E^*, respectively. The single-particle energies are calculated within the deformed Woods-Saxon potentials. In order to have a consistent comparison with the recent experimental data in [11], we carried out calculations by using the CE-LNBCS and CE-LNSCQRPA for ^{56}Fe, where pairing is included within the complete $pf + g_{9/2}$ shell above the ^{40}Ca core. For Mo isotopes, pairing is included in the 22 degenerated single-particle levels above the ^{48}Ca core. For Dy and Yb the same is done on top of the ^{132}Sn core. It is clear from this figure that the CE-LNSCQRPA results agree quite well with the experimental data [11], which are also deducted from the CE. The MCE entropies, obtained within the MCE-LNBCS and MCE-LNSCQRPA using $\delta\mathcal{E}$ =1 MeV, are plotted versus the experimental data. It is seen that the MCE-LNSCQRPA entropy not only offers the best fit to the experimental data but also predicts the results up to higher $E^* > 10$ MeV.

4. Conclusions

The proposed LNSCQRPA theory can describe without discontinuity the pairing properties of hot noncollectively rotating nuclei at any values of pairing interaction parameter G, temperature T, and angular momentum M. In the limit of zero temperature and zero angular momentum, it offers the best fits to the exact solutions in the weak coupling region with large particle numbers for the energy of the first excited state, whereas the SCQRPA reproduces well the exact one in the strong coupling region. In the limit of very large G all the approximations predict nearly the same value as that of the exact one. Under the effect of QNF, the paring gaps obtained at different values M of angular momentum decrease monotonously as T increases, and do not collapse even at hight T. The effect of thermally assisted pairing (pairing reentrance) shows up in such a way that the pairing gap reappears at a given $T_1 > 0$ and remains finite at $T > T_1$, in qualitative agreement with the results of Ref. [9]. We suggest a novel formula to extract the pairing gap at $T \neq 0$ from the difference of total energies of even and odd systems, where the contribution of uncorrelated single-particle motion is subtracted. Its prediction is found in much better agreement with the canonical gap than the simple extension of the odd-even mass formula to $T \neq 0$. Finally, we embedded the solutions of the LNBCS and LNSCQRPA into the CE and MCE, and found that the CE-LNSCQRPA predictions are in quite good agreements with the exact results as well as the recent experimental data. The present approach can also describe simultaneously and self-consistently the experimentally extracted total energy, heat capacity, and entropy within both CE and MCE treatments. It is simple and feasible for the application to larger finite systems, where the exact matrix diagonalization and/or solving the Richardson equation are impracticable to find all eigenvalues, whereas other methods, such as the quantum Monte-Carlo calculations, are time consuming.

References

[1] Volya A, Brown B A and Zelevinsky V 2001 *Phys. Lett.* B **509** 37.
[2] Hung N Q and Dang N D 2009 *Phys. Rev.* C **79** 054328.
[3] Hung N Q and Dang N D 2007 *Phys. Rev.* C **76** 054302; 2008 *Ibid.* **77** 029905(E).
[4] Dang N D and Hung N Q 2008 *Phys. Rev.* C **77** 064315.
[5] Lipkin H J 1960 *Ann. Phys. (NY)* **9** 272; Nogami Y 1965 *Phys. Lett.* **15** 4.
[6] Hung N Q and Dang N D 2008 *Phys. Rev.* C **78** 064315.
[7] Moretto L G 1971 *Nucl. Phys.* A **185** 145.
[8] Balian R, Flocard H and Vénéroni M 1999 *Phys. Rep.* **317** 251.
[9] Frauendorf S, Kuzmenko N K, Mikhajlov V M and Sheikh J A 2003 *Phys. Rev.* B **68** 024518.
[10] Hung N Q and Dang N D 2010 *Phys. Rev.* C **81** 057302.
[11] Melby E *et al.* 1999 *Phys. Rev. Lett.* **83** 3150; Guttormsen M *et al.* 2000 *Phys. Rev.* C **62** 024306; Schiller A *et al.* 2001 *Phys. Rev.* C **63** 021306(R); Algin E *et al.* 2008 *Phys. Rev.* C **78** 054321.

Derivation of IBM Hamiltonian for deformed nuclei

2011 J. Phys.: Conf. Ser. 267 012050

(http://iopscience.iop.org/1742-6596/267/1/012050)

Derivation of IBM Hamiltonian for deformed nuclei

Kosuke Nomura[1], **Takaharu Otsuka**[1,2,3], **Noritaka Shimizu**[1], **and Lu Guo**[4]

[1] Department of Physics, University of Tokyo, Hongo, Bunkyo-ku, Tokyo 113-0033, Japan
[2] Center for Nuclear Study, University of Tokyo, Hongo, Bunkyo-ku, Tokyo 113-0033, Japan
[3] National Superconducting Cyclotron Laboratory, Michigan State University, East Lansing, MI, USA
[4] RIKEN Nishina Center, Hirosawa, Wako-shi, Saitama 351-0198, Japan

E-mail: nomura@nt.phys.s.u-tokyo.ac.jp

Abstract.
A Hamiltonian of the Interacting Boson Model (IBM) is determined microscopically starting from the mean-field calculation. The potential energy surface (PES) by the Skyrme mean-field method is mapped onto the PES of the IBM to yield the interaction strengths of the IBM Hamiltonian, by which levels and wave functions of excited states are calculated with good angular momentum and the particle number. The response of the rotating nucleon system is mapped onto the corresponding boson system in order to determine the coefficient of the rotational kinetic-like $\hat{L} \cdot \hat{L}$ term microscopically. By the present method, not only the three dynamical symmetries of IBM but also the recently proposed critical-point symmetries of shape-phase transitions are well reproduced. In particular, the experimental rotational band of the SU(3) nucleus in Sm isotopic chain is reproduced quite nicely. We also point out the quantum fluctuation effect on the binding energies, which is included by the diagonalization of the IBM Hamiltonian determined by the comparison to the Skyrme PES. Systematic results for spectra for Os isotopes are also presented.

1. Introduction

The quadrupole collectivity in atomic nucleus exhibits distinct regularities, where the nuclear shape can be spherical, deformed and the situation in between. Like other models and theories [1, 2], the Interacting Boson Model [3] has been successful in reproducing the nuclear collective levels in terms of s and d bosons, which are essentially the collective S and D pairs of valence nucleons [4], respectively. If the interaction strengths of the IBM Hamiltonian take specific values, namely if the Hamiltonian has the so-called dynamical symmetry, then the shape of quadrupole deformation can be classified into any of spherical vibrator (U(5)), axially symmetric deformation (SU(3)), and γ-unstable deformation (O(6)). However, since most realistic nuclei are in the intermediate situations among the three limits, the spectra and the transition strengths are calculated numerically, where the IBM Hamiltonian has been determined only phenomenologically so as to reproduce the experimental data. On the other side, the IBM itself has a certain microscopic foundation starting from the nucleon degrees of freedom. It has been usually studied by means of the nuclear shell model. An well-known example can be the OAI mapping method proposed by T. Otsuka, A. Arima and F. Iachello [4], which has been, however, limited to nearly-spherical and γ-unstable nuclei. Thereby, for general situations including strong deformation, the derivation of the IBM Hamiltonian has remained open for many years.

The limitation of this type of approach is probably because of the fact that the shell-model state, which has been truncated by a generalized seniority scheme [5, 6], contains so complicated information on various many-body effects that it becomes unfeasible in practice particularly for deformed nuclei. On the other side, the potential energy surface in the β-γ plane by the mean-field methods seems to have a more direct connection to the quadrupole deformation and appears to be suitable as a starting point for constructing a model Hamiltonian for the description of low-lying states. One of the most successful and well-established mean-field models of nuclei is of the Skyrme-type, which have been useful for the description of various intrinsic-state properties of a nucleus: mass, density distributions, surface deformation, etc [2, 9, 10, 11]. While a given Skyrme interaction appears to give a basic and systematic description of matter and bulk properties of almost all nuclei on the nuclear chart, it has not been so easy in general to calculate levels and wave functions of excited states with the precise treatment of the particle number and the angular momentum.

Recently we have proposed a new approach to determine a Hamiltonian of the IBM starting from the mean-filed calculation [7, 8]. In this paper, we shall apply this method to see how the low-lying states, in particular those for deformed nuclei, and the shape-phase transitions.

2. Formalism

Now we shall illustrate the procedure and the underlying physical interpretation of the present method.

We first carry out the constrained Skyrme Hartree-Fock + BCS calculation and obtain the PES as a function of the variables denoted by $\beta_{\rm HF}$ and $\gamma_{\rm HF}$, which imply geometrical deformation parameters [1]. The constraint used here is of quadratic form. The single-particle wave functions are represented by a three-dimensional Cartesian grid and are assumed to be symmetric with respect to xyz planes [12].

Now we turn to the IBM description of the PES. The following discussion is focused on the proton-neutron IBM (or IBM-2) [4] because it is closer to the microscopic picture. We use the following boson Hamiltonian

$$\hat{H}_{\rm IBM} = E_0 + \epsilon(\hat{n}_{d\pi} + \hat{n}_{d\nu}) + \kappa\hat{Q}^{\chi_\pi}_\pi \cdot \hat{Q}^{\chi_\nu}_\nu + \alpha\hat{L}\cdot\hat{L}, \tag{1}$$

where E_0, $\hat{n}_{d\rho} = d^\dagger_\rho \cdot \tilde{d}_\rho$ and $\hat{Q}_\rho = [s^\dagger_\rho \times \tilde{d}_\rho + d^\dagger_\rho \times \tilde{s}_\rho]^{(2)} + \chi_\rho[d^\dagger_\rho \times \tilde{d}_\rho]^{(2)}$ (ρ=π or ν) stand for the constant term which contributes only to the binding energy, d-boson number operator with coefficient ϵ and the quadrupole operator with strength κ, respectively. The forth term represents the rotational kinetic term, which we denote as $\hat{L}\cdot\hat{L}$ term, hereafter, where $\hat{L}(=\hat{L}_\pi + \hat{L}_\nu)$ is the angular momentum operator with $\hat{L}_\rho$ being $\hat{L}_{\rho\mu} = \sqrt{10}[d^\dagger_\rho \times \tilde{d}_\rho]^{(1)}_\mu$.

The IBM PES is given by the expectation value $\langle \hat{H}_{\rm IBM}\rangle \equiv \langle\Phi|\hat{H}_{\rm IBM}|\Phi\rangle$. $|\Phi\rangle$ represents the boson coherent state [13, 14, 15], $|\Phi\rangle \propto \prod_\rho (s^\dagger_\rho + \sum_{\mu=\pm2,\pm1,0} a_{\rho\mu} d^\dagger_{\rho\mu})^{n_\rho}|0\rangle$. The coefficients $a_{\rho\mu}$ are given as $a_{\rho0} = \beta_\rho \cos\gamma_\rho$, $a_{\rho\pm1}$=0 and $a_{\rho\pm2} = \frac{1}{\sqrt{2}}\sin\gamma_\rho$. Thus the intrinsic shape of the nucleus is described in terms of the axially symmetric deformation β_ρ and the triaxiality γ_ρ. Here we assume that the proton and the neutron systems have the same shape, namely β_π=$\beta_\nu \equiv \beta_{\rm IBM}$ and γ_π=$\gamma_\nu \equiv \gamma_{\rm IBM}$. $\beta_{\rm IBM}$ and $\gamma_{\rm IBM}$ are interpreted as the boson images of the nucleon deformation variables and are assumed to be

$$\beta_{\rm IBM} = C_\beta \beta_{\rm HF} \text{ and } \gamma_{\rm IBM} = \gamma_{\rm HF} \tag{2}$$

for simplicity, with C_β being the coefficient.

We determine the values of the four parameters which appear in the boson Hamiltonian, together with the coefficient C_β of Eq. (2) so that the PES of IBM reproduces that of the mean-field calculation. Practically, this is done by the wavelet method [16], which is a sort of the Fourier transform but is a more efficient way to extract the characteristic features of a given signal such as the physically-relevant region of the PES. The PES is expanded in terms of a certain basis (wavelet) and can be expressed as a function of the scale value $\delta\beta$ ($\delta\gamma$) in addition to a deformation variable β (γ), which correspond to frequency and position, respectively. The wavelet transform of the PES in β direction (for fixed γ) $\tilde{E}(\delta\beta, \beta)$ is formulated as

$$\tilde{E}(\delta\beta, \beta) = \frac{1}{\sqrt{\delta\beta}} \int E(\beta', \gamma) \Psi^{\star}\Big(\frac{\beta - \beta'}{\delta\beta}\Big) d\beta', \tag{3}$$

where $\Psi^{\star}$ is a complex conjugate of a wavelet function. $E(\beta', \gamma)$ stands for the PES of either the HF or the IBM. Here the wavelet transform $\tilde{E}(\delta\beta, \beta)$ is in general a complex value. For γ direction, one has only to replace β ($\delta\beta$) with γ ($\delta\gamma$) in Eq. (3). The values of the five parameters ϵ, κ, $\chi_{\pi,\nu}$ and C_β are determined by the χ-square fit of the squared Wavelet transform of the IBM PES $|\tilde{E}|^2$ to that of the mean-field calculation. This process is nothing but the mapping of the HF PES onto the corresponding IBM PES. Here the HF PES reflects the effects of the basic features of nucleon system such as nuclear force and Pauli principle. By the mapping of the PES, such important properties of nucleon system is supposed to be simulated in terms of bosonic degrees of freedom. Using the parameters determined in the above-mentioned way, energy spectra and the transition strengths can be obtained with good angular momentum and the particle number, which enables one to make a spectroscopic predictions.

In a standard microscopic approach to collective motions, a kinetic mass term is considered independently of a potential term in order to calculate the excitation levels [2]. On the other hand, we compare the total energy of the HF+BCS with that of the IBM, where the effect of the mass term can be included to a certain extent by the diagonalization of the IBM Hamiltonian. However, this is not the case with strong deformation as illustrated by the following simple picture. If one rotates a nucleus around an axis in the intrinsic frame which is perpendicular to the symmetry axis of the nucleus, then the overlap of the intrinsic wave functions of rotated and non-rotated states can be defined. For strong deformation, such a overlap of the nucleon wave function is supposed to somewhat differ from the corresponding boson wave function. This difference of the intrinsic wave function requires one to introduce the mass term or the $\hat{L} \cdot \hat{L}$ term in the boson Hamiltonian. Here we note that the $\hat{L} \cdot \hat{L}$ term cannot be determined by the comparison of the PES because it does not contribute to the PES. Then we should formulate the response of the rotating nucleon system and map it onto the corresponding boson system in order to determine the coefficient of $\hat{L} \cdot \hat{L}$ term. We actually carry out the cranking calculation and compare the moments of inertia (MOI) in the intrinsic state between mean-field and IBM. The MOI for nucleon system is obtained by using the Inglis-Belyaev formula [17, 18], based on the Skyrme HF+BCS calculation [20, 21]. The MOI for IBM is obtained by the evaluation of the variation problem $\delta\langle\Phi|\hat{H}_{\rm IBM} - \omega\hat{L}_x|\Phi\rangle = 0$. Here ω stands for the cranking frequency and $\hat{L}_x = \hat{L}_{x\pi} + \hat{L}_{x\nu}$ represents the x component of $\hat{L}$ with $\hat{L}_{x\rho}$ being $\hat{L}_{x\rho} = (\hat{L}_{\rho,-1} - \hat{L}_{\rho,1})/\sqrt{2}$. The coherent state $|\Phi\rangle$ is generalized in such a way that $a_{\rho\pm1} \neq 0$ in accordance with the breaking of the rotational invariance, while other components, $a_{\rho\pm2} = \frac{1}{\sqrt{2}}\beta_{\rm IBM}\sin\gamma_{\rm IBM}$ and $a_{\rho 0} = \beta_{\rm IBM}\cos\gamma_{\rm IBM}$, are the same as those which give energy minimum in the non-rotational case. In addition, one can use the same values of the interaction strengths ϵ, κ, $\chi_{\pi,\nu}$ and C_β as those derived from the PES fits, which are taken from Ref. [8], because there is no need to perform the refit of the PES. Up to the lowest order in ω, the IBM MOI, $\mathcal{I}_{\rm IBM}$, is obtained [19] as a function of α.

$$\mathcal{I}_{\rm IBM} = \frac{\sum_{\rho\neq\tilde{\rho}} L_{x\rho}^2 E_{\tilde{\rho}} - 2E_{\pi\nu}L_{x\pi}L_{x\nu}}{E_\pi E_\nu - E_{\pi\nu}}, \tag{4}$$

where $L_{x\rho} \equiv \frac{\partial \langle \hat{L}_{\rho x} \rangle}{\partial a_{\rho 1}}$, $E_{\rho} \equiv \frac{\partial^2 \langle \hat{H}_{\rm IBM} \rangle}{\partial a_{\rho 1}^2}$ and $E_{\pi\nu} \equiv \frac{\partial^2 \langle \hat{H}_{\rm IBM} \rangle}{\partial a_{\pi 1} \partial a_{\nu 1}}$. We determine the strength of the $\hat{L} \cdot \hat{L}$ term for individual nucleus so that the IBM MOI of Eq. (4) reproduces that of Inglis-Belyaev formula.

3. Numerical results

3.1. Low-lying states of Sm isotopes

Figure 1 shows the evolutions of low-lying states as functions of neutron number, N. In the present calculation, which is indicated by "IBM(SkM*)" in Fig. (1), the yrast levels decrease with N, while the side-band, 0_2^+ and 2_2^+ levels are pushed up at N=88 or 90, consistently with the experiments (Expt.). One obviously sees in the behaviors of the low-lying states the phase transition from spherical to deformed shapes, together with the X(5) critical point [23].

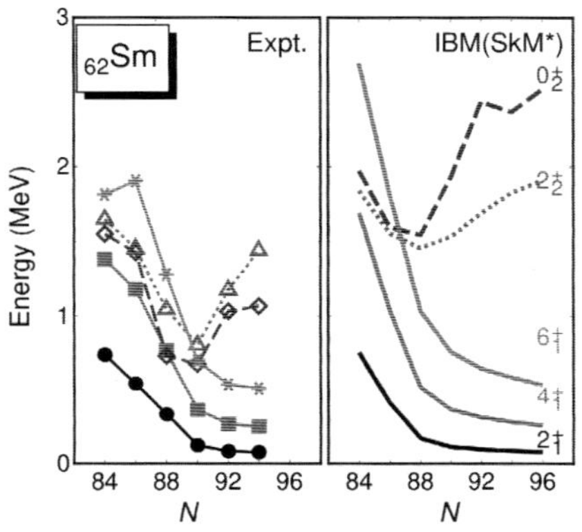

Figure 1. Low-lying spectra for Sm isotopes as functions of the neutron number N.

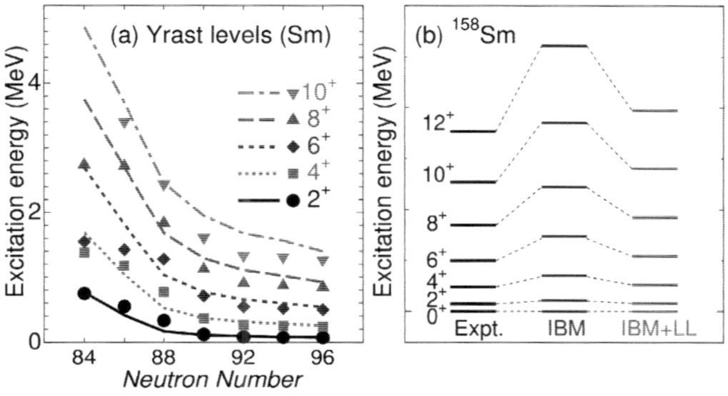

Figure 2. Same as Fig. 1, but for (a) the yrast spectra with higher angular momenta, where the curves and symbols represent the calculated and the experimental [22] values. (b) Yrast spectra for ^{158}Sm nucleus in the cases in which the $\hat{L} \cdot \hat{L}$ is (IBM+LL) and is not (IBM) included.

For higher-lying yrast levels for Sm isotopes of Fig. 2(a), one sees nice agreement between the calculation and the experimental data up to 8_1^+ state. Fig. 2(b) shows the level scheme for N=96 nucleus. This is to show how important the contribution of the $\hat{L} \cdot \hat{L}$ is on the rotational spectra by showing the cases in which the $\hat{L} \cdot \hat{L}$ term is (IBM+LL) and is not (IBM) included. The effect of the $\hat{L} \cdot \hat{L}$ term is robust for axially symmetric strong deformation, but is minor for weakly deformed and γ-soft nuclei.

3.2. Binding energy systematics

Now, we focus our discussion on the property of the binding energy. The binding energy of the IBM system is nothing but the eigenenergy of the Hamiltonian of Eq. (1), where the value of the constant term E_0 is determined for individual nucleus so that the energy minimum in the intrinsic state $\langle \hat{H}_{\rm IBM} \rangle$ is adjusted to the energy of the mean-field minimum. Figure 3 shows the deviations of the calculated binding energies for Sm isotopes from the experiments as functions of N. The results based on the Skyrme SkM* and SLy4 forces are shown. The dotted curves show the total energy or equivalently the potential minima of mean-field calculations. The solid curves represents the binding energy obtained by the diagonalization of the IBM Hamiltonian which are calibrated with the mean-field PES. Therefore, the difference between the mean-field and IBM corresponds to the quantum mechanical correlation effect which comes from the exact treatment of the angular momentum and the particle number in the IBM calculation. The correlation energy is most enhanced for nuclei corresponding to nearly-spherical to transitional regions. In addition, the binding energy of the IBM shows linear behavior as a function of N even at the shell closure. This smoothness of the binding energy due to the correlation effect enables one to describe the two-neutron separation energy better than the case otherwise, as seen in the lower panel of Fig. 3.

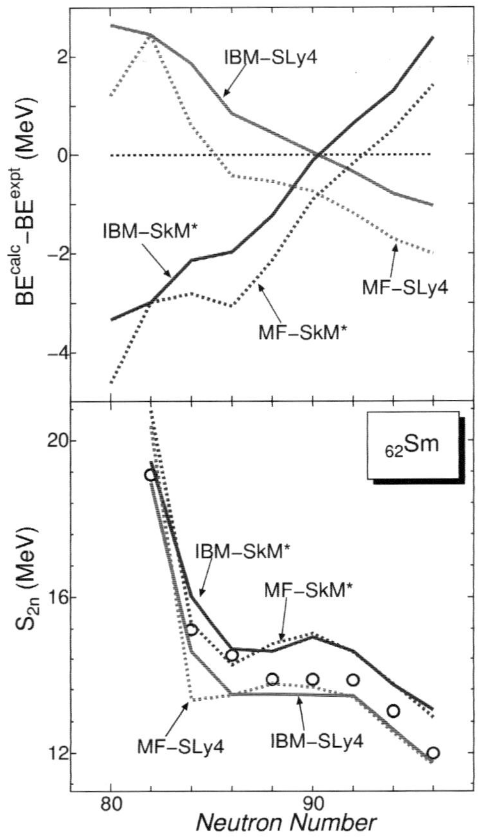

Figure 3. (Upper) Deviations of the calculated binding energies for Sm isotopes from experiments [24]. (Lower) Two-neutron separation energies (S_{2n}) for Sm isotopes. Experimental data are indicated by open circles.

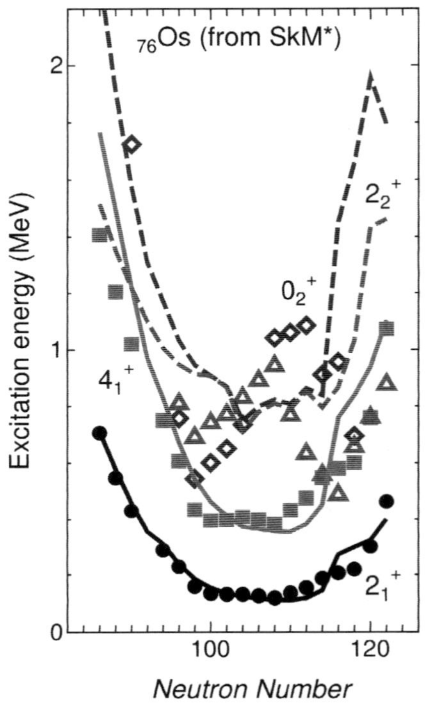

Figure 4. Same as Fig. 1, but for Os isotopes with N=86-122. Skyrme SkM* interaction is used.

3.3. Level evolution in Os isotopes

Now we show the application to other mass region. Figure 4 illustrates the level evolution of $_{76}$Os isotopes with N=86-122, where curves and symbols represent the present calculation and the experimental data [22], respectively. The Skyrme SkM* force is used. The calculated yrast levels come down as the number of valence neutrons increases, being consistent with the data. In particular, the 2_1^+ and 4_1^+ energies for N=122 nucleus in the present calculation are in rather good agreement with the recent experiment [25]. One sees similar kinks of side band levels 0_2^+

and 2_2^+ to Sm isotopes, which reflect the transition from spherical to deformed shapes. These side-band levels remain higher than 4_1^+ states around the middle of the major shell, characteristic of the deformed structure which is rather soft and weak as compared to Sm isotopes.

4. Conclusions

To summarize, we illustrated a mean-field-based formulation of the IBM, which has been developed recently, and its application to deformed nuclei. The interaction strengths of the IBM Hamiltonian relevant to the quadrupole collectivity are determined by mapping the PES of Skyrme onto the corresponding IBM PES. In addition, the rotational kinetic-like $\hat{L} \cdot \hat{L}$ term is embedded in the IBM Hamiltonian, and the coefficient of the $\hat{L} \cdot \hat{L}$ term is derived microscopically so that the response of the rotating nucleon system is reproduced by the corresponding boson states. Levels and wave functions of excited state are calculated with good angular momentum and the particle number. This allows one to describe the low-lying states of various cases of quadrupole deformation, including the relevant shape-phase transitions, as well as to evaluate the correlation effect in the binding energy.

Acknowledgments

This work has been supported by Grants-in-Aids for Scientific Research (A) 20244022 and for JSPS Fellows (No.217368).

References

[1] Bohr A and Mottelson B R 1975 *Nuclear Structure: II. Nuclear Deformations* (New York: Benjamin)
[2] Ring P and Schuck P 1980 *The Nuclear Many-Body Problem* (Berlin: Springer)
[3] Iachello F and Arima A 1987 *The interacting boson model* (Cambridge: Cambridge University Press)
[4] Otsuka T, Arima A and Iachello F 1978 *Nucl. Phys.* A **309** 1
[5] Otsuka T 1993 *Algebraic Approaches to Nuclear Structure* ed Casten R F (Harwood Academic Publishers) p 195
[6] Mizusaki T and Otsuka T 1997 *Prog. Theor. Phys. Suppl.* **125** 97
[7] Nomura K, Shimizu N and Otsuka T 2008 *Phys. Rev. Lett.* **101** 142501
[8] Nomura K, Shimizu N and Otsuka T 2010 *Phys. Rev.* C **81** 044307
[9] Skyrme T H R 1959 *Nucl. Phys.* **9** 615
[10] Vautherin D and Brink D M 1972 *Phys. Rev.* C **5** 626
[11] Bender M, Heenen P-H and Reinhard P-G 2003 *Rev. Mod. Phys.* **75** 121
[12] Bonche P, Flocard H and Heenen P-H 2005 *Comput. Phys. Commun.* **171** 49
[13] Dieperink A E L and Scholten O 1980 *Nucl. Phys.* **A346** 125
[14] Ginocchio J N and Kirson M 1980 *Nucl. Phys.* **A350** 31
[15] Bohr A and Mottelson B R 1980 *Phys. Scripta* **22** 468
[16] Kaiser G 1994 *A Friendly Guide to Wavelets* (Boston: Birkhäuser)
[17] Inglis D R 1956 *Phys. Rev.* **103** 1786
[18] Belyaev S T 1961 *Nucl. Phys.* **24** 322
[19] Schaaser H and Brink D M 1984 *Phys. Lett.* B **134** 269
[20] Guo L, Maruhn J A and Reinhard P-G 2007 *Phys. Rev.* C **76** 034317
[21] Guo L, Maruhn J A, Reinhard P-G and Hashimoto Y 2008 *Phys. Rev.* C **77** 041301(R)
[22] National Nuclear Data Center (NNDC) http://www.nndc.bnl.gov/
[23] Iachello F 2001 *Phys. Rev. Lett.* **87** 052502
[24] Table of Isotopes Atomic Mass Data http://ie.lbl.gov/
[25] Podolyak Zs *et al* 2009 *Phys. Rev.* C **79** 031305(R)

Effects of self-consistency in semiclassical pairing theory

2011 J. Phys.: Conf. Ser. 267 012051

(http://iopscience.iop.org/1742-6596/267/1/012051)

Effects of self-consistency in semiclassical pairing theory

VI Abrosimov[1], **DM Brink**[2], **A Dellafiore**[3] **and F Matera**[3,4]

[1]Institute for Nuclear Research, 03028 Kiev, Ukraine
[2]Oxford University, Oxford, U.K
[3]Istituto Nazionale di Fisica Nucleare, Sezione di Firenze, Sesto Fiorentino (Fi), Italy
[4]Dipartimento di Fisica e Astronomia, Università degli Studi di Firenze, Sesto Fiorentino (Fi), Italy

Abstract. An extended Vlasov equation, derived from the time-dependent Hartree–Fock–Bogoliubov equation of motion, is used to study the effects of pairing on the nuclear density-density linear response function in semiclassical approximation. Within the approximations adopted here, the fluctuations of the pairing field are purely imaginary and play a crucial role in eliminating a spurious mode that gives fluctuations of the number of particle and in reproducing the correct value of the energy-weighted sum rule. The semiclassical approach seems to have some difficulty in describing pairing vibrations. Further work on this problem is needed.

1. Introduction

Different approaches can be used to study the effects of pairing correlations in nuclei (see, for example, Ref.[1] and references quoted therein). Here, following [2] and [3], we use a semiclassical approximation to study the effects of pairing on the nuclear response. The semiclassical approximation is valid when

$$k_F R >> 1\,, \tag{1}$$

a condition which is well satisfied in heavy nuclei (k_F is a typical Fermi wavenumber and R is the nuclear radius).

The semiclassical approach leads to relatively simple analytical expressions for several quantities of interest and our hope is that of gaining some insight by using this approach which is certainly less accurate than a fully quantum theory.

2. Formalism

A kinetic equation valid for systems with pairing can be obtained by taking the Wigner transform of the time-dependent Hartree–Fock–Bogoliubov (TDHFB) equations of motion in the density-matrix formulation [4] and neglecting terms of order $\hbar^2$, or higher. This procedure leads to four coupled differential equations for the functions $f(\mathbf{r},\mathbf{p},t)$, $f(\mathbf{r}-\mathbf{p},t)$, $\kappa(\mathbf{r},\mathbf{p},t)$, $\Delta(\mathbf{r},\mathbf{p},t)$. The single-particle distribution $f(\mathbf{r},\mathbf{p},t)$ is real, while the Wigner-transformed pairing tensor $\kappa(\mathbf{r},\mathbf{p},t)$ and pairing field $\Delta(\mathbf{r},\mathbf{p},t)$ are complex in a time-dependent situation, thus, in this way, we have four equations of motion and six unknown quantities. However, in the TDHFB theory, the functions $\kappa(\mathbf{r},\mathbf{p},t)$ and $\Delta(\mathbf{r},\mathbf{p},t)$ are related by a self-consistency relation. This relation can be derived from the gap equation. In order to avoid introducing a cut-off parameter, we consider the gap equation written in the form [2]

$$g \int d\mathbf{p}\Big(\frac{\kappa(\mathbf{r},\mathbf{p},t)}{\Delta(\mathbf{r},t)} + \frac{1}{p^2/m}\Big) = 1\,. \tag{2}$$

Here g is a parameter determining the strength of the pairing interaction. We have assumed that the $\mathbf{p}$-dependence of the dynamic pairing field can be neglected. Moreover, in the following we will approximate the equilibrium pairing field $\Delta_0(\mathbf{r},\mathbf{p})$ with the phenomenological parameter Δ, which, in heavy nuclei takes values $\Delta \approx 1$ MeV.

Since we are interested in the linear response of nuclei, we assume that our system , which is initially in equilibrium, at time $t = 0$ is acted upon by an external weak driving field of the form

$$\delta V^{ext}(\mathbf{r}.t) = \eta\delta(t)Q(\mathbf{r}). \tag{3}$$

As a consequence of this external perturbation, the functions f, κ, Δ at $t > 0$ will oscillate about their equilibrium values and the small-amplitude oscillations $\delta f, \delta\kappa, \delta\Delta$ are the object of our study.

Thus, by considering small fluctuations of $\kappa(\mathbf{r},\mathbf{p},t)$ and $\Delta(\mathbf{r},t)$ about their equilibrium values, from Eq. (2) we get the first-order relation

$$\int d\mathbf{p}\Big(\delta\kappa(\mathbf{r},\mathbf{p},t) - \kappa_0(\mathbf{r},\mathbf{p})\frac{\delta\Delta(\mathbf{r},t)}{\Delta}\Big) = 0\,, \tag{4}$$

where $\kappa_0(\mathbf{r},\mathbf{p})$ and Δ are real equilibrium quantities, while $\delta\kappa$ and $\delta\Delta$ are their complex fluctuations. The real and imaginary parts of Eq. (4) give the two independent relations:

$$\int d\mathbf{p}\Big(\delta\kappa^r(\mathbf{r},\mathbf{p},t) - \kappa_0(\mathbf{r},\mathbf{p})\frac{\delta\Delta^r(\mathbf{r},t)}{\Delta}\Big) = 0\,, \tag{5}$$

$$\int d\mathbf{p}\Big(\delta\kappa^i(\mathbf{r},\mathbf{p},t) - \kappa_0(\mathbf{r},\mathbf{p})\frac{\delta\Delta^i(\mathbf{r},t)}{\Delta}\Big) = 0\,. \tag{6}$$

These conditions, based on Eq. (2), take into account the residual pairing interaction in a self-consistent way. Here we consider two weaker self-consistency relations that can be obtained by integrating Eqs. (5, 6) over $\mathbf{r}$. By using also the equilibrium relation

$$\kappa_0(\mathbf{r},\mathbf{p}) \approx \kappa_0(\epsilon) = -\frac{\Delta}{2E(\epsilon)}\,, \tag{7}$$

where $\epsilon = h_0(\mathbf{r},\mathbf{p}) = \frac{\mathbf{p}^2}{2m} + V_0(\mathbf{r})$ is the particle energy, $E(\epsilon) = \sqrt{\Delta^2 + (\epsilon-\mu)^2}$ the quasiparticle energy and μ the chemical potential, these weaker self-consistency relations become

$$\int d\mathbf{r}\int d\mathbf{p}\Big(\delta\kappa^r(\mathbf{r},\mathbf{p},t) + \frac{\delta\Delta^r(\mathbf{r},t)}{2E(\epsilon)}\Big) = 0\,, \tag{8}$$

$$\int d\mathbf{r}\int d\mathbf{p}\Big(\delta\kappa^i(\mathbf{r},\mathbf{p},t) + \frac{\delta\Delta^i(\mathbf{r},t)}{2E(\epsilon)}\Big) = 0\,. \tag{9}$$

The four differential equations of motion can also be linearized. In this way the problem is reduced to that of solving four coupled linear differential equations plus the two self-consistency relations (8, 9).

The mathematical problem can be simplified by using the action-angle variables $(\mathbf{I},\mathbf{\Phi})$ instead of $(\mathbf{r},\mathbf{p})$. In this way the four coupled differential equations are transformed into four coupled algebraic equations, which can be used to express the quantities $\delta\kappa^{r,i}$ in terms of $\delta\Delta^{r,i}$ and δV^{ext} in Eqs. (8, 9).

This is achieved by means of a very useful property of action-angle variables (see, for example Ref.[5], p. 467): for bound motion of a particle, any function of its coordinates $f(\mathbf{r},\mathbf{p})$ can be expanded as a multiple Fourier series of the kind

$$f(\mathbf{r},\mathbf{p}) = \sum_{\mathbf{n}} f_{\mathbf{n}}(\mathbf{I}) e^{i\mathbf{n}\cdot\mathbf{\Phi}} . \tag{10}$$

Here $\mathbf{n}$ is a three-dimensional vector with integer components: $\mathbf{n} = (n_1, n_2, n_3)$. After taking the Fourier transform in time, and changing the integration variables from $(\mathbf{r},\mathbf{p})$ to $(\mathbf{I},\mathbf{\Phi})$, Eqs. (8, 9) can be written in the form

$$\int d\mathbf{\Phi} \int d\mathbf{I} \sum_{\mathbf{n}} \Big(\delta\kappa^r_{\mathbf{n}}(\mathbf{I},\omega) + \frac{\delta\Delta^r_{\mathbf{n}}(\mathbf{I},\omega)}{2E(\epsilon)} \Big) e^{i\mathbf{n}\cdot\mathbf{\Phi}} = 0 , \tag{11}$$

$$\int d\mathbf{\Phi} \int d\mathbf{I} \sum_{\mathbf{n}} \Big(\delta\kappa^i_{\mathbf{n}}(\mathbf{I},\omega) + \frac{\delta\Delta^i_{n}(\mathbf{I},\omega)}{2E(\epsilon)} \Big) e^{i\mathbf{n}\cdot\mathbf{\Phi}} = 0 . \tag{12}$$

The orthogonality of the functions $e^{i\mathbf{n}\cdot\mathbf{\Phi}}$ gives $\int d\mathbf{\Phi} e^{i\mathbf{n}\cdot\mathbf{\Phi}} = (2\pi)^3 \delta_{\mathbf{n},0}$. Consequently, the self-consistency conditions (8, 9) give two conditions for the amplitude of the normal mode $\mathbf{n} = 0$.

$$\int d\mathbf{I} \Big(\delta\kappa^r_{\mathbf{n}=0}(\mathbf{I},\omega) + \frac{\delta\Delta^r_{\mathbf{n}=0}(\mathbf{I},\omega)}{2E(\epsilon)} \Big) = 0 , \tag{13}$$

$$\int d\mathbf{I} \Big(\delta\kappa^i_{\mathbf{n}=0}(\mathbf{I},\omega) + \frac{\delta\Delta^i_{\mathbf{n}=0}(\mathbf{I},\omega)}{2E(\epsilon)} \Big) = 0 . \tag{14}$$

In spherical nuclei, these two three-dimensional integral relations can be reduced to a system of two coupled one-dimensional integral equations. If we make the further reasonable assumption that the fluctuations of the pairing field are proportional to the external driving field,

$$\delta\Delta^r(\mathbf{r},\omega) = F^r(\omega) Q(\mathbf{r}) \tag{15}$$

$$\delta\Delta^i(\mathbf{r},\omega) = F^i(\omega) Q(\mathbf{r}) \tag{16}$$

and use the expressions of the quantities $\delta\kappa^{r,i}$ in terms of $\delta\Delta^{r,i}$ and δV^{ext}, then the two one-dimensional integral equations become two coupled algebraic equations for the unknown functions $F^{r,i}(\omega)$:

$$F^r(\omega) A_{11}(\omega) + F^i(\omega) A_{12}(\omega) = \eta B_1(\omega) , \tag{17}$$

$$F^r(\omega) A_{21}(\omega) + F^i(\omega) A_{22}(\omega) = \eta B_2(\omega) . \tag{18}$$

The functions $A_{ij}(\omega)$ are

$$A_{ij}(\omega) = \int_0^{\infty} d\epsilon g(\epsilon) a_{ij}(\epsilon,\omega) , \tag{19}$$

while

$$B_i(\omega) = \int_0^{\infty} d\epsilon g(\epsilon) b_i(\epsilon,\omega) , \tag{20}$$

$g(\epsilon)$ is the single-particle level density for the equilibrium mean field and the functions $a_{ij}(\epsilon,\omega)$, $b_i(\epsilon,\omega)$ are given by

$$\begin{aligned}
a_{11}(\epsilon,\omega) &= -i\omega\frac{\tilde{\epsilon}}{E(\epsilon)}\frac{1}{(\omega+i\varepsilon)^2-4E^2(\epsilon)}, &(21)\\
a_{12}(\epsilon,\omega) &= \frac{1}{2E(\epsilon)}\frac{\omega^2}{(\omega+i\varepsilon)^2-4E^2(\epsilon)}, &(22)\\
a_{21}(\epsilon,\omega) &= (\omega^2-4\Delta^2)\frac{a_{12}(\epsilon,\omega)}{\omega^2} &(23)\\
a_{22}(\epsilon,\omega) &= -a_{11}(\epsilon,\omega), &(24)\\
b_1(\epsilon,\omega) &= \frac{2\Delta}{i\omega}a_{12}(\epsilon,\omega), &(25)\\
b_2(\epsilon,\omega) &= \frac{2\Delta}{i\omega}a_{22}(\epsilon,\omega). &(26)
\end{aligned}$$

Here $\tilde{\epsilon}=\epsilon-\mu$, while ε is a vanishingly small positive quantity.

If the determinant

$$D(\omega)=A_{11}(\omega)A_{22}(\omega)-A_{21}(\omega)A_{12}(\omega) \qquad (27)$$

is non vanishing, the system of coupled algebraic equations (17, 18) can be easily solved, giving

$$F^r(\omega)=0\,, \qquad (28)$$

$$F^i(\omega)=\eta\frac{2\Delta}{i\omega}\,, \qquad (29)$$

however there could be also other solutions corresponding to the roots of

$$D(\omega)=0\,. \qquad (30)$$

These solutions would be particularly interesting because they correspond to collective nuclear modes related to pairing (pairing vibrations).

The determinant (27) can also be written as

$$D(\omega)=\omega^2[I_1^2(\omega)-(\omega^2-4\Delta^2)I_2^2(\omega)]\,, \qquad (31)$$

with

$$\begin{aligned}
I_1(\omega) &= \int_0^\infty d\epsilon g(\epsilon)\frac{\tilde{\epsilon}}{E(\epsilon)}\frac{1}{(\omega+i\varepsilon)^2-4E^2(\epsilon)}, &(32)\\
I_2(\omega) &= \frac{1}{2}\int_0^\infty d\epsilon g(\epsilon)\frac{1}{E(\epsilon)}\frac{1}{(\omega+i\varepsilon)^2-4E^2(\epsilon)}. &(33)
\end{aligned}$$

An obvious solution of Eq. (30) occurs for $\omega=0$. This zero-frequency solution corresponds to the Anderson-Goldstone-Nambu mode and is associated with rotations in gauge space (see, for example, Ch 4 of [1]).

The integrals (32, 33) are clearly the semiclassical versions of quantities which have been studied in analogous quantum approaches to the problem of pairing vibrations (cf., for example Eqs. (5.25) and (5.24) of [1]). In quantum theory, the integrals (32, 33) are replaced by sum over discrete nuclear levels, making the solution of Eq. (30) simpler in the quantum case. Our work on this point is still in progress, however, the solution (28, 29) is also of interest because, by using this solution, we can check that the fluctuations of the imaginary pairing field are essential to restore particle-number conservation and to give the correct value of the energy weighted sum rule (the same as in normal systems).

We conclude this communication by showing the effects of pairing on the semiclassical quadrupole and octupole density response function for a system of $A = 208$ nucleons confined to a spherical cavity of radius $R = 1.2A^{1/3}$ fm. These figures give an indication of the kind of effects that are to be expected in Pb isotopes.

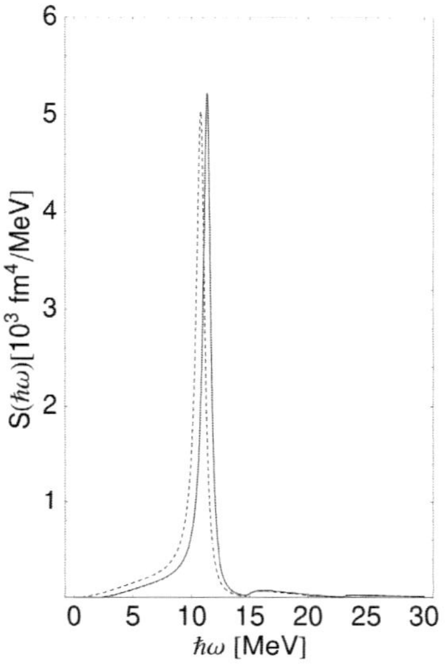

Figure 1. Quadrupole response function for $A = 208$ nucleons confined to a spherical cavity of radius R. The full curve include the effects of pairing according to Eqs. (28, 29), while the dashed curve is for a normal system. Reprinted with permission from *Nucl. Phys. A*, Ref. [3]

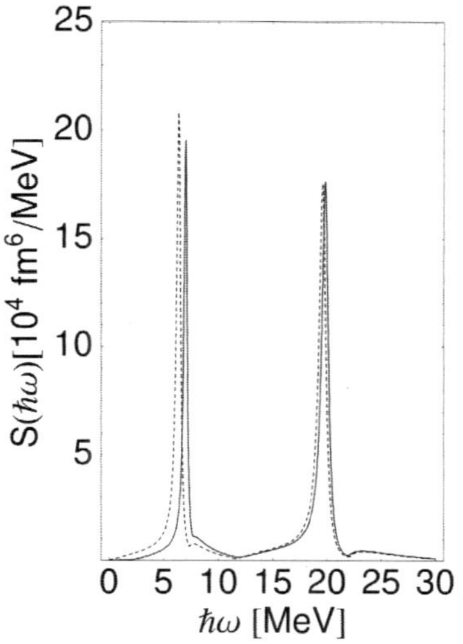

Figure 2. Same as Fig.1 for octupole response.The two peaks correspond to the low-energy and high-energy parts of the giant octupole resonance. Reprinted with permission from *Nucl. Phys. A*, Ref. [3]

Both the quadrupole and octupole response functions of Figs. 1, 2 display a gap 2Δ at low excitation energy. This fact can be appreciated more clearly in Fig. 3, where the low-energy part of Fig. 1 is shown on an enlarged scale.

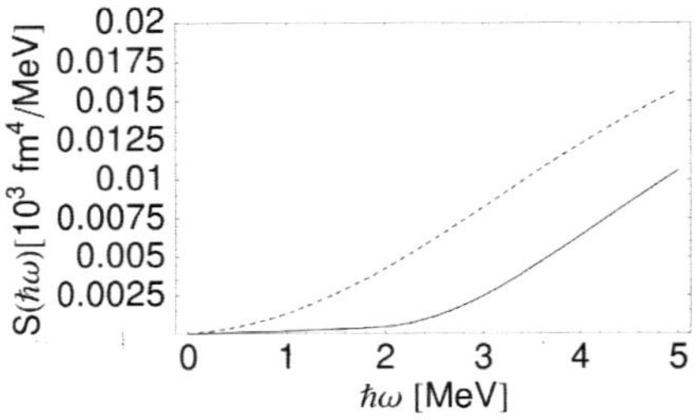

Figure 3. Same as Fig.1 in the low-energy region. The gap 2Δ is clearly visible for the full curve. Reprinted with permission from *Nucl. Phys. A*, Ref. [3]

References

[1] Brink D M and Broglia R A 2005 *Nuclear Superfluidity*, (Cambridge, UK: Cambridge University Press)
[2] Urban M and Schuck P 2006 *Phys. Rev.* A **73** 013621
[3] Abrosimov V I, Brink D M, Dellafiore A and Matera F 2008 *Nucl. Phys.* A **800** 1
[4] Ring P and Schuck P 1980 *The Nuclear Many-Body Problem* (New York: Springer)
[5] Goldstein H 1980 *Classical Mechanics* (Reading, MA: Addison-Wesley)

Second RPA with Skyrme Interaction

2011 J. Phys.: Conf. Ser. 267 012052

(http://iopscience.iop.org/1742-6596/267/1/012052)

Second RPA with Skyrme Interaction

D Gambacurta,[1,2] **M Grasso** [3] **and F Catara**[1,2]

[1] Dipartimento di Fisica e Astronomia dell'Universitá di Catania, Via S.Sofia 64, I-95123 Catania, Italy

[2] Istituto Nazionale di Fisica Nucleare, Sezione di Catania, Via S.Sofia 64, I-95123 Catania, Italy

[3]Institut de Physique Nucléaire, Université Paris-Sud, IN2P3-CNRS, F-91406 Orsay Cedex, France

E-mail: danilo.gambacurta@ct.infn.it

Abstract.
The Second Random Phase Approximation (RPA) is a natural extension of RPA obtained by introducing more general excitation operators where two particle-two hole configurations, in addition to the one particle-one hole ones, are considered. Some Second RPA results with Skyrme force in ^{16}O are presented. Different levels of approximation are compared and in particular the quality of the diagonal approximation is tested. The issue of the rearrangement terms to be used in the matrix elements beyond the standard RPA ones, when density-dependent force are used, is briefly discussed. Two approximated, and generally used, schemes are used: the rearrangement terms are neglected in the matrix elements beyond RPA or evaluated with the RPA prescription. As a general feature of Second RPA results, a several-MeV shift of the strength distribution to lower energies is systematically found with respect to RPA distributions.

1. Introduction

The Hartree-Fock (HF) plus the random phase approximation (RPA) is a widely used microscopic approach for the study of collective excitations in many-body systems. The HF method allows to obtain the self-consistent mean field in which the particles are assumed to move independently, their energies and wave functions, and thus the HF ground state. The solution of the RPA equations gives the energies and the wave functions of the collective excitations that are assumed to be superpositions of particle-hole and hole-particle configurations built on top of a correlated ground defined as the vacuum of the phonon operators describing the excited states.

In the nuclear context, the HF + RPA framework with Skyrme interaction has been extensively applied for the description of the Giant Resonances and, in general, a quite good reproduction of the energies and transition probabilities associated to these excitations is found. On the other hand, in several works (see for example [1, 2, 3]) it has been underlined that the coupling of one particle-one hole configurations with more complex states has to be taken into account in order to describe some important effects, as for example the spreading width of the excitations. In such a direction, a natural extension of RPA is the so-called Second RPA (SRPA) [4] where two particle-two hole ($2p2h$) excitations, in addition to the one particle-one hole ($1p1h$) ones, are introduced. To our knowledge, current versions of SRPA are based on non-interacting $2p2h$ configurations and only the interaction between $2p2h$ and $1p1h$ has been generally taken into account, in the so-called diagonal approximation. In this work, we present full Skyrme-SRPA results obtained for the doubly magic nucleus ^{16}O. The diagonal approximation is employed here just to analyze its quality.

It is well known that, when density-dependent forces are used, the residual interaction used in the RPA matrices contains the so-called rearrangement terms coming from second derivative of the energy density functional with respect to the density (for more detais see [6]). When dealing with the SRPA problem with density-dependent forces, a first formal aspect to consider is the determination of the residual interaction that has to be used in the new matrix elements with respect to RPA (see [5] and [6]). For the results shown here, we have explored two possibilities, often used in extensions of RPA, that we will discuss in the following.

In Section 2 the formal scheme of SRPA is briefly summarized, while in Section 3 some results for the monopole and quadrupole nuclear response obtained within SRPA are shown and compared with the RPA ones. A comparison of the full SRPA results with those obtained in the diagonal approximation is also presented. Moreover, the transition densities for the giant monopole resonances are analyzed and the radial distributions related to the main peaks are shown for RPA and SRPA. Conclusions and perspectives are finally discussed in Section 4.

2. Formal scheme

We briefly recall the main formal aspects of SRPA that may be found in several articles (see, for instance, Ref. [4]). SRPA is a natural extension of RPA where the excitation operators Q_ν^+ are a superposition of $1p1h$ and $2p2h$ configurations:

$$Q_\nu^\dagger = \sum_{ph}(X^\nu_{ph}a^\dagger_p a_h - Y^\nu_{ph}a^\dagger_h a_p) + \sum_{p<p',h<h'}(X^\nu_{php'h'}a^\dagger_p a_h a^\dagger_{p'} a_{h'} - Y^\nu_{php'h'}a^\dagger_h a_p a^\dagger_{h'} a_{p'}). \quad (1)$$

The X's and Y's are solutions of the equations,

$$\begin{pmatrix} \mathcal{A} & \mathcal{B} \\ -\mathcal{B}^* & -\mathcal{A}^* \end{pmatrix}\begin{pmatrix} \mathcal{X}^\nu \\ \mathcal{Y}^\nu \end{pmatrix} = \omega_\nu \begin{pmatrix} \mathcal{X}^\nu \\ \mathcal{Y}^\nu \end{pmatrix}, \quad (2)$$

where:

$$\mathcal{A} = \begin{pmatrix} A_{11} & A_{12} \\ A_{21} & A_{22} \end{pmatrix}, \mathcal{B} = \begin{pmatrix} B_{11} & B_{12} \\ B_{21} & B_{22} \end{pmatrix}, \mathcal{X}^\nu = \begin{pmatrix} X_1^\nu \\ X_2^\nu \end{pmatrix}, \quad \mathcal{Y}^\nu = \begin{pmatrix} Y_1^\nu \\ Y_2^\nu \end{pmatrix}.$$

The indices 1 and 2 are a short-hand notation for the $1p1h$ and $2p2h$ configurations, respectively. A_{11} and B_{11} are the usual RPA matrices, A_{12} and B_{12} are the matrices coupling $1p1h$ with $2p2h$ configurations and A_{22} and B_{22} are the matrices coupling $2p2h$ configurations among themselves. Both in RPA and SRPA, resort to the quasi boson approximation (QBA) is made, which consists in the use of the HF ground state in the evaluation of the matrix elements appearing in the equations of motion. The matrix elements of the B_{12}, B_{21} and B_{22} matrices are found to be zero within QBA [4]. The other matrix elements are:

$$A_{12} = A_{ph,p_1p_2h_1h_2} = \langle HF|[a^\dagger_h a_p, [H, a^\dagger_{p_1} a^\dagger_{p_2} a_{h_2} a_{h_1}]]|HF\rangle =$$

$$= \chi(h_1,h_2)\bar{V}_{h_1pp_1p_2}\delta_{hh_2} - \chi(p_1,p_2)\bar{V}_{h_1h_1p_1h}\delta_{pp_2} \quad (3)$$

$$A_{22} = A_{p_1h_1p_2h_2,p'_1h'_1p'_2h'_2} = \langle HF|[a^\dagger_{h_1} a^\dagger_{h_2} a_{p_1} a_{p_2}, [H, a^\dagger_{p'_2} a^\dagger_{p'_1} a_{h'_2} a_{h'_1}]]|HF\rangle =$$

$$= (\epsilon_{p_1} + \epsilon_{p_2} - \epsilon_{h_1} - \epsilon_{h_2})\chi(p_1,p_2)\chi(h_1,h_2)\delta_{h_1h'_1}\delta_{p_1p'_1}\delta_{h_2h'_2}\delta_{p_2p'_2} + \chi(h_1,h_2)\bar{V}_{p_1p_2p'_1p'_2}\delta_{h_1h'_1}\delta_{h_2h'_2}$$

$$+\chi(p_1,p_2)\bar{V}_{h_1h_2h'_1h'_2}\delta_{p_1p'_1}\delta_{p_2p'_2} + \chi(p_1,p_2)\chi(h_1,h_2)\chi(p'_1,p'_2)\chi(h'_1,h'_2)\bar{V}_{p_1h'_1h_1p'_1}\delta_{h_2h'_2}\delta_{p_2p'_2} \qquad (4)$$

where the ϵ's are the HF single particle energies, $\bar{V}$ is the residual interaction and $\chi(ij)$ is the antisymmetrizer for the indices i, j.

It can be shown [7, 8], that the SRPA problem can be reduced to an energy-dependent, RPA-like, eigenvalue problem but where the A_{11} RPA matrix depends on the excitation energies

$$A_{1,1'}(\omega) = A_{1,1'} + \sum_{2,2'} A_{1,2}(\omega + i\eta - A_{2,2'})^{-1}A_{2',1'}. \qquad (5)$$

In order to calculate this energy-dependent part one has to invert the A_{22} matrix defined in Eq. (4), whose dimensions are generally very large, requiring thus a strong numerical effort. However, if only the diagonal terms of the matrix (4) are considered, resorting to the so-called diagonal approximation, the inversion is algebraic. This approximation, often used in SRPA calculations, amounts to treat the $2p2h$ configurations as unperturbed since the residual interaction is neglected. The quality of this approximation will be analyzed in Section 3.

Expressions (3) and (4) are valid in cases where the interaction is not density dependent. Rearrangement terms should be included in the case of density-dependent forces. The results hereafter discussed have been obtained by using two approximated choices, often used in beyond RPA calculations: i) we have calculated the matrix elements (3)-(4) without any rearrangement terms, ii) we have evaluated them by adding the usual RPA rearrangement contributions also in the A_{12} and A_{22} matrices.

3. Results in ^{16}O

In this Section we present the nuclear strength distributions in ^{16}O obtained in SRPA for the monopole and quadrupole multipolarities and we compare them with the RPA ones. The SGII [9] parametrization of the effective interaction has been used in the present calculations. At SRPA level different levels of approximation will be considered. As discussed above, in the case of density-dependent interactions it is not fully clear how to define the residual interaction appearing in the matrix elements beyond RPA, in particular as far as the rearrangement terms are concerned. By following a variational approach, SRPA equations in the case of density-dependent forces can be derived [5]. It is found that rearrangement terms are also present in matrix elements beyond RPA. In the A_{22} matrix we have only the ones giving the correct HF single particle energies while in A_{12}, B_{12} matrices they appear in residual interaction but they have a different expression from the standard RPA prescription. The numerical implementation of these results is still in progress. In this work two different choices, very often used in beyond RPA-like calculations, are tested: (i) the interaction is used without rearrangement terms in matrix elements beyond RPA; (ii) rearrangement terms are included also in matrix elements beyond RPA, calculated with the usual RPA prescription. Furthermore, the full SRPA calculations are compared with the ones obtained when the diagonal approximation is used.

In order to make simpler the comparison between different results, we have folded the discrete spectra coming out from our calculations with a Lorentzian with a width of 1 MeV. The continuous strength distributions shown in this work are thus obtained by using this smoothing procedure.

In RPA calculations, $1p1h$ configurations with unperturbed energy up to 100 MeV are considered, while in the SRPA ones, we have considered all the $2p2h$ configurations with an unperturbed energy lower than an energy cutoff E_{cut}. In Figs. 1 and 2 we show the monopole and quadrupole strength distributions, respectively, for different choices of E_{cut} (indicated in MeV in parenthesis in the figures). From the figures we see that a cutoff equal to 120 MeV gives stable results. Similar stability checks have been systematically made for all the results shown in the following.

The multipole transition operators used are

$$F_\lambda^{IS} = \sum r_i^2 Y_{\lambda 0}(\hat{r}_i) \tag{6}$$

$$F_\lambda^{IV} = \sum r_i^2 Y_{\lambda 0}(\hat{r}_i)\tau_z(i) \tag{7}$$

in the isoscalar and isovector channel, respectively.

In Fig. 3 we show the RPA (dashed black lines) and SRPA (full red lines) results for the isoscalar (upper panel) and isovector (lower panel) monopole strength distributions. In both isoscalar and isovector cases the strongest effect in SRPA is a several-MeV shift of the strength distribution to lower energies with respect to RPA. This result seems to be a general feature of SRPA and has been found also in different SRPA calculations [10, 11, 12]. Looking at the figures, however, one sees that the profiles of the strength distributions are not very much changed, except for the above mentioned shift.

In the same figure we show also the SRPA results obtained with rearrangement terms in matrix elements beyond RPA (dot-dashed blue lines) are presented in order to evaluate their effect. We remark that these rearrangement terms have been calculated with the usual RPA procedure for all the matrix elements. For the isoscalar monopole case (top panel of Fig. 3) the residual interaction seems to be more repulsive when rearrangement terms are added providing a smaller energy shift to lower energies with respect to RPA. The strength distribution appears very strongly fragmented when rearrangement terms are included in all matrix elements. We know that the adopted way to evaluate the rearrangement terms in beyond RPA matrix elements is not the correct one even if it is currently used. Work is in progress to obtain the correct expressions: the fact that the SRPA$^{(*)}$ results are so different from the SRPA ones underlines that this is a very delicate point and indicates that the proper expressions are needed.

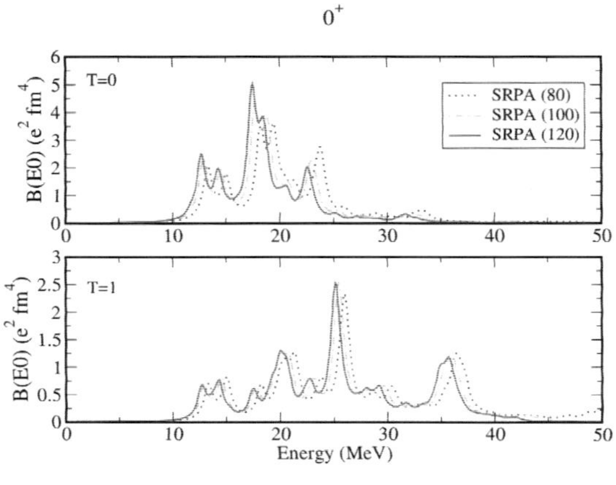

Figure 1. (Color online) Isoscalar (upper panel) and isovector (lower panel) strength distributions for monopole states obtained in SRPA for increasing values of the energy cutoff, indicated in MeV in parenthesis in the figure, on the $2p2h$ configurations.

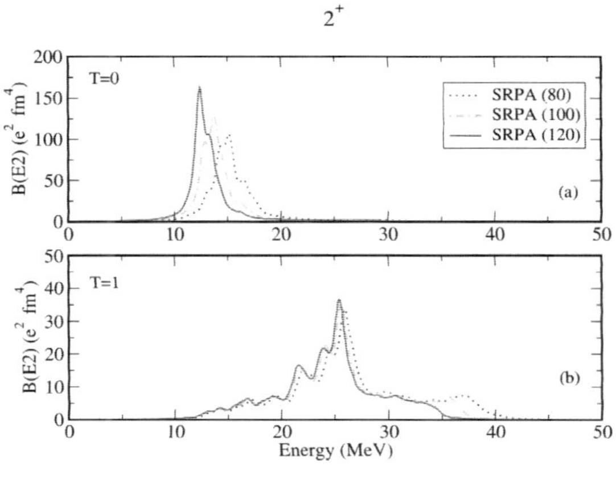

Figure 2. (Color online) Isoscalar (upper panel) and isovector (lower panel) strength distributions for quadrupole states obtained in SRPA for increasing values of the energy cutoff, indicated in MeV in parenthesis in the figure, on the $2p2h$ configurations.

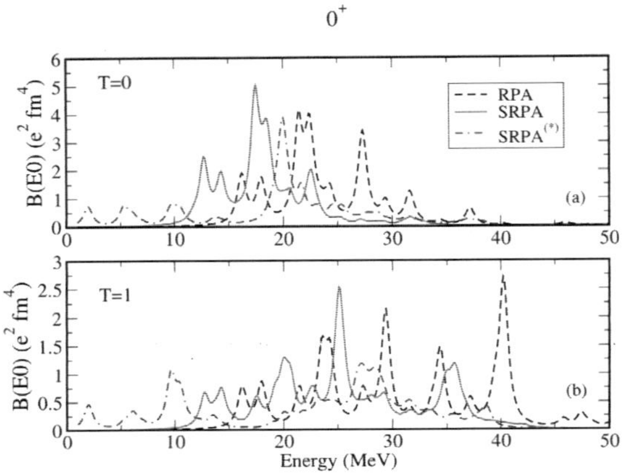

Figure 3. (Color online) Comparison between RPA, dashed (black) lines, SRPA in full (red) lines and SRPA with rearrangements terms, dot-dashed (blue) lines. See text for more details. The isoscalar (upper panel) and isovector (lower panel) monopole strength distributions are shown.

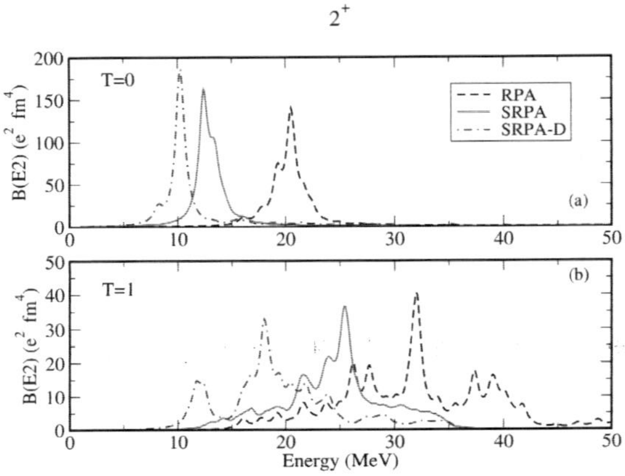

Figure 4. (Color online) Comparison between RPA, dashed (black) lines, SRPA in full (red) lines and SRPA in the diagonal approximation, dot-dashed (blue) lines. See text for more details. The isoscalar (upper panel) and isovector (lower panel) quadrupole strength distributions are shown.

In Fig. 4 the full SRPA results, full (red) lines, are compared with the ones obtained in the diagonal approximation, dot-dashed (blue) lines. The results are quite different and this suggests that the residual interaction should not be neglected in the matrix A_{22} in Skyrme-SRPA calculations. In particular, the strength distributions are pushed down towards even lower energies with respect to the complete calculations.

In Fig. 5 we show, for the monopole isoscalar case, the SRPA transition density referring to the peak at $\sim$ 17 MeV while the RPA transition density is evaluated for the peak at $\sim$ 21 MeV in top panel of Fig. 3. The profiles are not very different meaning that the nature of these RPA and SRPA excited states is not very different in terms of the spatial distributions of wave functions contributing to them.

Different is the case for the isovector responses. In Fig. 6 the RPA (SRPA) transition density, for the monopole response, is calculated for the peak at $\sim$ 29 MeV (25 MeV) shown in the bottom panel of Fig. 3. Important differences are visible between RPA and SRPA suggesting that the nature of SRPA and RPA excited states in terms of radial distribution of wave functions contributing to the excitation is quite different.

4. Conclusions and perspectives

We have performed Skyrme-SRPA calculations for describing collective states in ^{16}O. The SRPA scheme is fully treated without resorting to a general used approximated scheme, i.e. the diagonal approximations. The rearrangement terms of the residual interaction are treated in this work (i) by neglecting them in the matrix elements beyond RPA; (ii) by calculating them with the usual RPA procedure for all the matrix elements. Work is in progress to calculate the proper expressions to be used in beyond RPA matrix elements [5].

A shift towards lower energies of the strength distributions for giant resonances is generally found in SRPA with respect to RPA case. This shift being very strong, Skyrme-SRPA energies of giant resonances are typically too low with respect to the experimental response (Skyrme-RPA results are in general in good agreement with the experimental data for these excitations). This

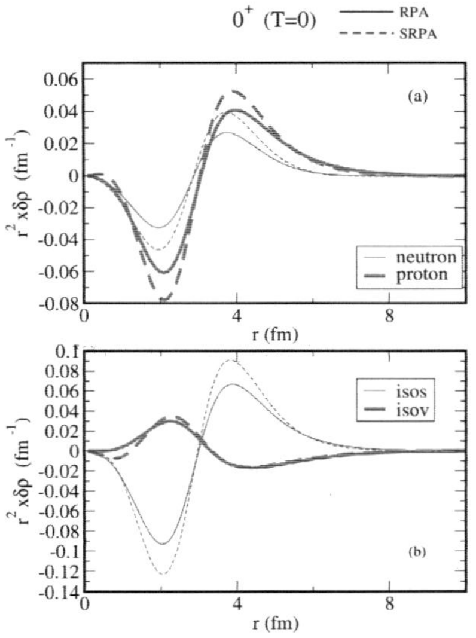

Figure 5. (Color online) (Color online) Comparison between RPA (full lines) and SRPA (dotted lines) transition densities for the monopole isoscalar states (see text): in the panel (a) neutron (thin lines) and proton (bold lines) parts, while in panel (b) the isoscalar (thin lines) and isovector (bold lines) ones. See text for more details.

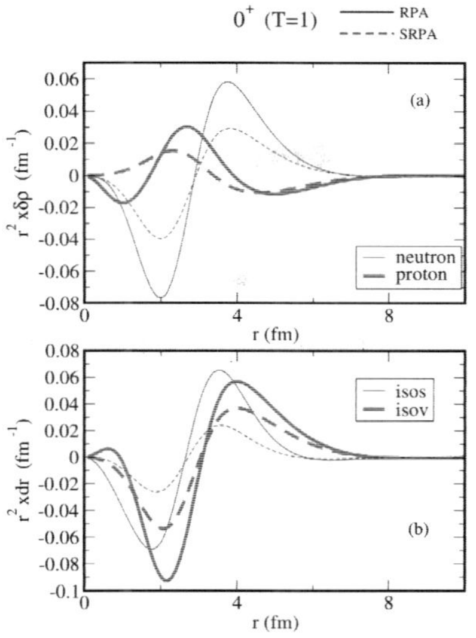

Figure 6. (Color online) Comparison between RPA (full lines) and SRPA (dotted lines) transition densities for the monopole isovector states (see text): in the panel (a) neutron (thin lines) and proton (bold lines) parts, while in panel (b) the isoscalar (thin lines) and isovector (bold lines) ones. See text for more details.

strong shift could be, at least partially, ascribed to the use of QBA. Some comments about the use of QBA in SRPA can be found in the literature [10, 13, 14]: it is said that its use in SRPA is a more drastic and severe approximation than in RPA. Extensions to cure this problem in the context of SRPA have been proposed and applied to a simple model [14] and to metal clusters [10]. Future applications to nuclei are in progress.

Finally, we plan in future works to explore with SRPA heavier or more exotic nuclei and to check the model dependence of the results by employing also other Skyrme parametrizations.

References

[1] Chomaz Ph and Frascaria N, *Phys. Rep.* **252**, 275 (1995).
[2] Bortignon PF *Nucl. Phys.* A **687**, 329c (2001)
[3] Colo G, Van Giai N , Bortignon PF and Broglia RA, *Phys. Rev.* C **50**, 1496 (1994)
[4] Yannouleas C, *Phys. Rev.* C **35**, 1159 (1986)
[5] Grasso M , Gambacura D and Catara F, arXiv:1003.2021v1 [nucl-th]
[6] Gambacurta F, Grasso M and Catara F, *Phys. Rev.* C **81**, 054312 (2010)
[7] da Providencia J, *Nucl. Phys.* **61**, 87 (1965)
[8] Drozdz S, Klent V, Speth J, and Wambach J, *Nucl. Phys.* A **451**, 11 (1986)
[9] Van Giai N and Sagawa H, *Phys. Lett.* B **106** 379 (1981); *Nucl. Phys.* A **371** 1 (1981).
[10] Gambacurta D and Catara F, *Phys. Rev.* B **79**, 085403 (2009)
[11] Papakonstantinou P and Roth R, *Phys. Lett.* B **671**, 356 (2009)
[12] Papakonstantinou P and Roth R, *Phys. Rev.* C **81**, 024317 (2010)
[13] Lauritsch G and Reinhard PG, *Nucl. Phys.* A **509**, 287 (1990); Takayanagi K, Shimizu K and Arima A, Nucl. Phys. A **477**, 205 (1988)
[14] Gambacurta D, Grasso M, Catara F, and Sambataro M, *Phys. Rev.* C **73**, 024319 (2006)
[15] Gambacurta D and Catara F, *Phys. Rev.* B **81**, 085418 (2010)

The role of isospin in fusion evaporation reactions

2011 J. Phys.: Conf. Ser. 267 012053

(http://iopscience.iop.org/1742-6596/267/1/012053)

The role of isospin in fusion evaporation reactions

A Di Nitto[1], A Brondi[1], G La Rana[1], R Moro[1], P N Nadtochy[1], E Vardaci[1], N Gelli[2], M Cinausero[3] and G Prete[3]

[1] Dipartimento di Scienze Fisiche, Università di Napoli Federico II and Istituto Nazionale di Fisica Nucleare, Complesso Universitario di Monte S. Angelo, Via Cintia, I-80126 Napoli, Italy
[2] Istituto Nazionale di Fisica Nucleare Sezione di Firenze, Via G. Sansone 1-50019 Sesto Fiorentino (FI), Italy
[3] Laboratori Nazionali di Legnaro, Viale dell' Università 2-35020 Legnaro (PD), Italy

E-mail: `dinitto@na.infn.it`

Abstract. The isospin dependence of the level density, both through the level density parameter and the symmetry energy contribution has been evaluated. The effects on the most relevant observables in fusion evaporation reactions have been estimated by using the statistical model. We present here the results for the ^{134}Sn $+^4$ He reaction, whose measurement will be possible with the second generation RIB facilities SPES and SPIRAL2. Sizeable effects are predicted on the light particle energy spectra and multiplicities and evaporation residue distribution. In particular, the measure of evaporation residue yields and light particle multiplicities are expected to unfold the two effects. In addition, the same calculations have been carried out for reactions induced by stable beams, in order to search for clues of isospin effects. In this case, the direct comparison of the experimental light charged particle energy spectra for the two reactions ^{7}Li+^{58,64}Ni at the same excitation energy and angular momentum, is proposed as a good approach to observe isospin effects.

1. Introduction

The nuclear level density is an important ingredient in the statistical model to calculate rates and other properties of compound nucleus reactions. Although much work has been devoted to study this quantity, it still remains uncertain in different regions of mass and excitation energy. Usually one starts from the Bethe formula [1] based on an energy independent level density of single particle states, where isospin effects are neglected. Interest in studying this quantity for exotic nuclei, as those that will be produced by the second generation RIB facilities, is growing up, as the level density is expected to depend on isospin. Recently, an expression of the level density parameter in terms of the distance of the nucleus from the stability line has been proposed [2, 3] and supported by the realistic calculations of Grimes and collaborators [4]. In addition, isospin effects are expected also through the symmetry energy contribution to the nuclear masses which has been usually neglected, although variations of the symmetry energy have strong implications in nuclear astrophysics. Therefore, new experiments are foreseen to look for these effects. In this framework we have undertaken a research project to study both the isospin effects in systems which will be produced by the second generation exotic beams, but also in systems now available with stable beams, using as a probe the evaporative light charged particles (LCPs). In a previous paper [5] we have shown some results on neutron rich nuclei, whose production will be possible by the SPES facility using a ^{4}He target. The ratio of the 1n to the 2n cross section appears to be sensitive to the different isospin dependences of the level density parameter. On the other hand, light particle multiplicities are sensitive to the symmetry energy contribution. Here, we will present more detailed results of the statistical model simulations for the reaction $^{134}Sn +^{4}He$ and, in addition, for the $^{7}Li +^{58,64}Ni$ reactions induced by a stable beam.

2. Isospin effects on the level density parameter a and symmetry energy

The level density parameter a appearing in the Bethe formula has been studied at low and high excitation energy. It results that this parameter changes from $A/8.5MeV^{-1}$ to $A/15MeV^{-1}$, as supported by several observations [6]. With this description of the level density, the statistical model proved to be able to reproduce most of the data of studied reactions with stable beams. In the framework of the Fermi-gas model the isospin effects on the level density can be taken into account including a factor related to the isospin component $T_3 = (N - Z)/2$ [?].

Recently, the analysis of an extended set of level density data [7] for nuclei with $20 < A < 110$ has been carried out [2, 3]. Different isospin dependences of the level density parameter a have been tested (Tab.1), where Zo is the atomic number of the $\beta-$stable isobar. The values of α, β

Table 1. Forms for the dependence of a on various parameters.

standard	$a = \alpha_1 \cdot A$	$\alpha_1 = 0.1016$	
$N - Z$	$a = \alpha_2 \cdot A/\exp[\beta(N - Z)^2]$	$\alpha_2 = 0.1062,$	$\beta = 0.00051$
$Z - Zo$	$a = \alpha_3 \cdot A/\exp[\gamma(Z - Zo)^2]$	$\alpha_3 = 0.1068,$	$\gamma = 0.0389$

and γ have been obtained from best fit to the data. The third parametrization provides a better description of the data, predicting a significant reduction of level density for nuclei far from the stability line. This prediction is supported by realistic calculations [4] of the level density of nuclei far from the stability line. The underlining assumption is that the levels of the compound system for nuclei far from the stability line have widths which are large enough to remove them from the category of compound nucleus levels. To have a greater sensitivity to this effect, low excitation energies are more suited, as they limit the contribution of particle emission from nuclei closer to the valley of stability, and can be better analyzed by the statistical model. Bombarding a ^{4}He target with medium-mass neutron-rich projectiles, we meet the condition of providing a high fusion cross section at low excitation energy of the composite system. Furthermore, the relatively low angular momenta and excitation energies involved in these reactions are expected to enhance the effects of the isospin on the level density. In order to evaluate these effects we have modified our statistical code (Lilita_N97 [8]) including the three prescriptions given in Tab.1.

As regards the symmetry energy, we used an isospin dependence given by the following formula:

$$E_{sym}(T) = b_{sym}(T)\frac{(N - Z)^2}{A} \quad (1)$$

where b_{sym} (T) is a functional derived by Donati et al. [9]. In particular, in the frame of the dynamical shell model the symmetry energy increases with increasing of temperature, for $T \leq 3$ MeV, due to the decrease of the nucleon effective mass m^*. At the same time, this implies a decrease of the level density parameter $a \propto m^*$ [9, 10, 11].

Experimentally, the effects related to the symmetry energy would appear as a change in the particle multiplicity as well as in the relative yields of the compound nucleus decay channels, as it affects the binding energies of the evaporative light particles. As this effect is of the order of a few MeV in each step of the evaporative chain, relatively low excitation energies are required in order to observe the effect and perform a test of the current models.

We have included in the statistical model the symmetry energy prescription of Ref.[9] as well as the temperature dependence of the level density parameter a, obtained in the same theoretical framework [10].

All the statistical model calculations presented in this work have been carried out adopting the optical model transmission coefficients and the yrast line calculated considering the nucleus as a rigid sphere with $r_0 = 1.2$fm.

3. ^{134}Sn $+^4$ He reaction

We report here the results obtained for the reaction ^{134}Sn $+^4$ He, which produces the ^{138}Te compound nucleus, adopting different isospin dependences of the level density parameter or including the symmetry energy contribution or both.

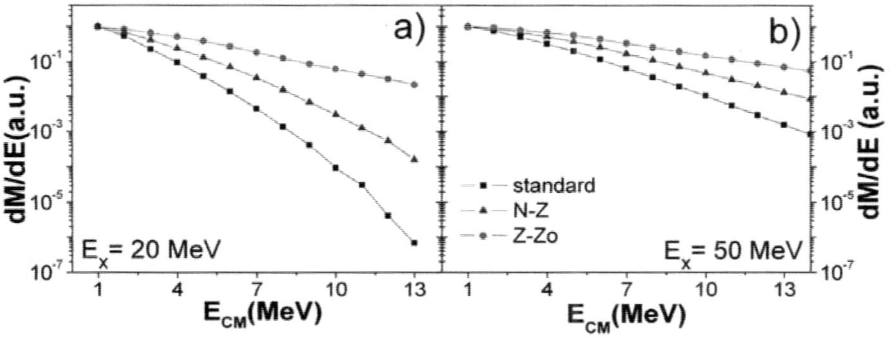

Figure 1. (Color online) Neutron energy spectra calculated with the three prescriptions of Tab. 1 for the reaction ^{134}Sn $+^4$ He at *a*) $E_x = 20$MeV and *b*) $E_x = 50$MeV. Spectra have been normalized to the maximum to compare the shapes.

In Fig.1 are reported the simulated neutron energy spectra, obtained adopting the three prescriptions given in Tab.1 at two different excitation energies of the compound nucleus (E_x = 20 MeV with $J_{MAX} = 7\hbar$ and $E_x = 50$ MeV with $J_{MAX} = 19\hbar$). Significant effects are observed, indicating the importance of measuring the evaporative neutrons. With the increase of the excitation energy, channels with a larger number of neutrons will open, and a reduction of isospin effects will result.

A significant sensitivity to the symmetry energy contribution is observed for the neutron energy spectra, Fig.2. We can note that the effects on the neutron spectra of the Z-Zo prescription and symmetry energy are comparable.

A similar behaviour is also observed for the LCP spectra, but it will be hard to measure them with the intensity of the second generation RIB facilities.

Predicted multiplicities at $E_x = 50$ MeV for neutrons, protons and alpha particles are shown in Tab.2. The effect of Z-Zo prescription is to reduce the neutron multiplicity increasing the proton one and leaving unchanged the alpha particle multiplicity. As regards the symmetry energy contribution, the multiplicities show a reduction similar to that produced by Z-Zo for what it concerns the neutrons, while a very strong enhancement is observed for protons and alpha particles. In particular, the $E_{sym}(T)$ + a(T) option promotes the opening of channels involving LCPs [5], which otherwise would be closed assuming the standard prescription.

Finally, including both effects in the statistical code the difference with respect to the standard calculations increases for neutrons and protons, but not for alpha particles.

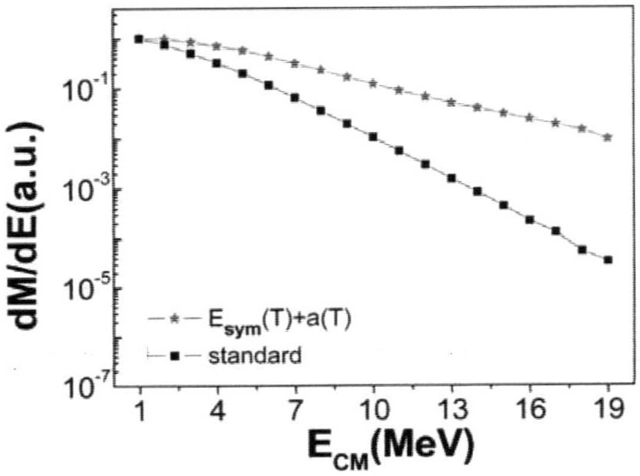

Figure 2. (Color online) Neutron energy spectra calculated with the standard and $E_{sym}(T) + a(T)$ prescriptions, for the reaction $^{134}Sn + ^{4}He$ at $E_x = 50MeV$. Spectra have been normalized to the maximum to compare the shapes.

Table 2. Multiplicity of light particles emitted in the decay ^{138}Te at $E_x = 50MeV$ adopting different prescriptions of the isospin effects.

	standard	Z-Zo	$E_{sym}(T)+a(T)$	both effects
M_n	6.03	4.87	4.43	3.42
M_p	0.0001	0.0004	0.0050	0.0057
M_α	0.005	0.005	0.071	0.035

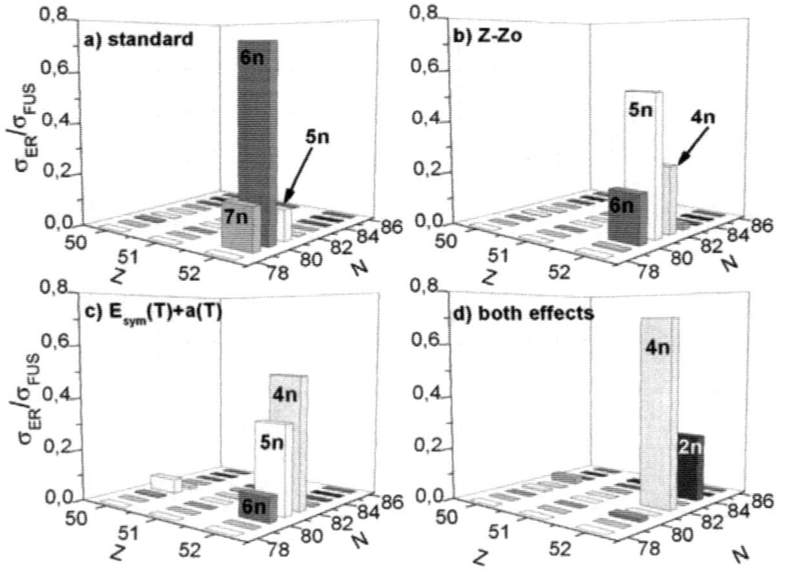

Figure 3. (Color online) Relative evaporation residue cross sections for the reaction ^{134}Sn on ^{4}He, at $E_x = 50MeV$. The calculations have been carried out adopting the prescriptions: a) standard, b) Z-Zo, c) $E_{sym}(T) + a(T)$ and d) inclusion of both isospin effects.

In the four panels of Fig.3 we show the relative cross sections of the evaporation residues obtained using the code Lilita_N97 for the reaction $^{134}Sn + ^{4}He$ at $E_x = 50$ MeV, adopting the prescriptions: a) standard, b) Z-Zo, c) $E_{sym}(T) + a(T)$ and d) both isospin effects.

The standard yield is characterized by having the 6n channel as the most intense one, while Z-Zo produces an enhancement of 5n and 4n channels with a reduction of 6n. Coming to the $E_{sym}(T) + a(T)$ prescription a similar behaviour is observed, but with the 4n channel more intense than the 5n. Combining the two effects d) there is a strong enhancement of the 4n channel and the opening of the 2n channel, all the other channels being depressed. From these results we can conclude that the isospin effects provide significant differences with respect to the standard prescription, and moreover, that evaporation residue yields and particle multiplicities are suitable to disentangle the different effects.

4. Near future experiment with a stable beam

We have explored the possibility to observe clues of isospin effects on fusion-evaporation reactions induced by a stable beam. For this purpose, we have carried out calculations for $^{7}Li+^{58,64}Ni$ at the beam energies 36.6 MeV and 35.6 MeV, respectively to obtain the same excitation energy E_x = 50 MeV and the same maximum angular momentum $J_{MAX} = 19\hbar$. In the Fig.4, the proton center of mass energy spectra for the two systems are superimposed.

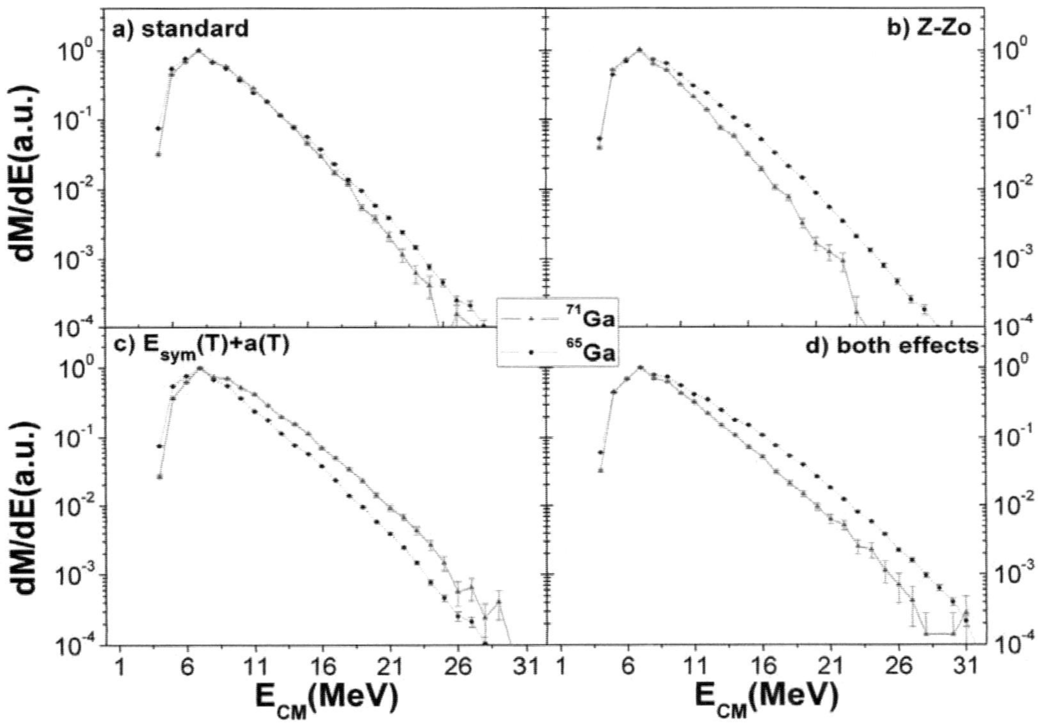

Figure 4. (Color online) Proton energy spectra emitted in the decay of $^{65,71}Ga$ at E_x = 50 MeV. To evidentiate the differences in the shapes, the spectra have been normalized to the maximum. The calculations have been carried out adopting the prescriptions: a) standard, b) Z-Zo, c) $E_{sym}(T)$ + a(T) and d) combining the two previous isospin effects.

As can be seen in Fig.4a, the standard calculations predict essentially no difference, while including both isospin effects a sizeable difference (Fig.4d) is obtained, which can be measured. The effect produced taking into account both isospin dependences comes out to be larger than that due to different values of a parameter (a = A/8 as well as a = A/10, remembering that the standard prescription of Tab.1 assumes $a \sim$ A/9.11). It is interesting to show the predictions for each single effect in order to determine the main responsible of possible experimental differences. Assuming $E_{sym}(T)$ + a(T) we observe an opposite behaviour with respect to the case where both effects are present, as shown in Fig.4c. The high energy tail of the proton spectrum is higher in the decay of ^{71}Ga nuclei than in the ^{65}Ga nuclei. This increase is due to the higher value of the difference N-Z (9 for ^{71}Ga and 3 for ^{65}Ga), that is amplified according to the expression 1.

Looking at Z-Zo effect (Fig.4b), we find a similar behaviour as that with both isospin effects. Due to the higher value of Z-Zo, 1.8 for ^{65}Ga with respect to the value of 0.4 for ^{71}Ga, the proton spectrum from the decay of ^{65}Ga nuclei shows an high energy tail. Therefore, this is an indication of the strong contribution supplied by Z-Zo dependence. In particular, such kind of measure could represent an experimental confirmation for this prescription and, in any case, provides an useful information to improve the adopted parameters. Concerning the spectra the same behaviours have been predicted for alpha particles.

A further observable useful to enlighten the presence of isospin effects is represented by the multiplicity ratios of light charged particles of the two reactions. In Tab.3, it can be observed that the proton multiplicity ratios go from 9 to 26, while the alpha particle ones do not change significantly when the isospin effects are included.

Table 3. Multiplicity ratios of light charged particles emitted in the decay ^{65}Ga and ^{71}Ga at E_x = 50MeV adopting different prescriptions of the isospin effects.

	standard	Z-Zo	E_{sym}(T)+a(T)	both effects
M_p(^{65}Ga)/Mp(^{71}Ga)	9	20	12	26
M_α^{65}Ga)/Ma(^{71}Ga)	2.91	3.26	3.52	3.44

5. Conclusions

Statistical model calculations show that fusion-evaporation reactions are a good probe for studying the level density of neutron-rich nuclei, as those that will be produced by SPES and SPIRAL2 facilities. In particular, light particle energy spectra and multiplicities as well as evaporation residue yields are expected to be good observables in order to enlighten the isospin effects on the level density parameter a and on the symmetry energy allowing to discriminate between the two contributions.

References

[1] Bethe H A 1936 *Phys. Rev.* **50** 332
[2] Al-Quraishi S I et al. 2001 *Phys. Rev.* **C63** 065803
[3] Al-Quraishi S I, Grimes S M, Massey T N and Resler D A 2003 *Phys. Rev.* **C67** 015803
[4] Grimes S M et al. 2008 *AIP Conf. Proc.* **1005** 57
[5] Brondi A et al. 2010 *EPJ Web of Conferences* **2** 04002
[6] See for example Nebbia G et al. 1986 *Phys. Lett.* **B176** 20
Feshbach H 1992 *Theoretical Nuclear Physics-Nuclear Reactions* edited by Wiley J
[7] Bhat M R 1992 in Evaluated Nuclear Structure Data File (ENSDF), *Nuclear Data for Science and Technology*, edited by Qaim S M Springer-Verlag Berlin p.817
[8] Lilita_N97 is an extensively modified version of the original Lilita made by Gomez del Campo J and Stockstad R G, Oak Ridge National Laboratory 1981 *Rep. No TM729 unpublished*
[9] Donati P, Pizzocchero P M, Bortignon P F and Broglia R A 1994 *Phys. Rev. Lett.* **72** 2835
[10] Ormand W E et al. 1989 *Phys. Rev.* **C40** 1510
[11] Dean D J et al. 2002 *Phys. Rev.* **C66** 045802

E0 decay of the 0^+_2 levels in ^{156}Dy and ^{160}Er

2011 J. Phys.: Conf. Ser. 267 012054

(http://iopscience.iop.org/1742-6596/267/1/012054)

E0 decay of the 0_2^+ levels in ^{156}Dy and ^{160}Er

G Lo Bianco[1], D L Balabanski[2], S Nardelli[1], S Das Gupta[1], N Blasi[3], K Gladnishki[4], A Saltarelli[1] and L Fortunato[5]

[1] School of Science and Technology, University of Camerino and INFN - Sez. Perugia, Italy
[2] INRNE, Bulgarian Academy of Sciences, Sofia, Bulgaria
[3] INFN - Sez. Milano, Italy
[4] Faculty of Physics, St. Kliment Ohridski University of Sofia, Bulgaria
[5] ECT*, Trento, Italy

E-mail: balabanski@inrne.bas.bg

Abstract.
The branching between the $E0\ 0_2^+ \rightarrow 0_1^+$ and the $E2\ 0_2^+ \rightarrow 2_1^+$ transitions in ^{156}Dy and ^{160}Er were measured following the ε decay of ^{156}Ho and ^{160}Tm. A potential of the 4th order of the deformation parameter β, the "Lo Bianco potential", was chosen to describe the $U(5) - SU(3)$ first order shape phase transition, covering the whole transitional path. The comparison of the excitation spectra and the measured ratios of reduced transition probabilities, $X(E0/E2)$, with the calculations, indicates that ^{156}Dy is in the spherical region, while ^{160}Er is located in the deformed region, but quite close to the critical point.

1. Introduction

In the geometrical description of collective nuclear motion there are three limiting cases, corresponding to the harmonic vibrator, the symmetrically deformed rotor and the triaxial rotor. Each of them is associated with a particular nuclear shape, spherical, axial-ellipsoidal, and triaxial. The transition from a spherical harmonic vibrator to an axially deformed rotor was described analytically by Iachello, introducing a dynamic symmetry, denoted $X(5)$ [1]. It is interesting in this context to investigate the $E0$ transition strength, in the decay of the first excited 0^+ state, because this quantity is known to have particularly large values in the transitional region between spherical and deformed nuclei [2, 3]. This fact can be explained by strong mixing of states with different deformation [4], or within the framework of the Interacting Boson Model (IBM) [5], by mixing of states with different number of bosons [3].

Electric monopole, $E0$, transitions are forbidden by the γ-decay selection rules and can occur only via the emission of atomic electrons. Electron emission is the only process by which a 0^+ state can decay to another 0^+ state. If the nuclear spin is non-zero electric quadrupole $E2$, and $E0$ transitions compete with each other. The $E0$ transition probability is factorized into electron and nuclear terms [6],

$$W = \Omega \rho^2(E0), \tag{1}$$

where Ω represents all the "non-nuclear" contributions and can be calculated from several models [7, 8, 9]. The nuclear structure information is contained in $\rho^2(E0)$ and can be directly related to different models.

Considering a simple collective geometrical model, Rasmussen estimated the $E0$ transition probability and the dimensionless ratio of the reduced $E0$ and $E2$ transition probabilities [10]

$$X(E0/E2; 0_2^+ \to 0_1^+) = \frac{\rho^2(E0; 0_2^+ \to 0_1^+) \cdot e^2 R_0^4}{B(E2; 0_2^+ \to 2_1^+)} = 4\beta^2, \tag{2}$$

and demonstrated that it is proportional to the deformation, β. This ratio can be compared with the experiment as [11]:

$$X(E0/E2; 0_2^+ \to 0_1^+) = 2.54 \cdot 10^9 A^{4/3} E_\gamma^5 q^2 \alpha_K(E2)/\Omega, \tag{3}$$

where E_γ is the energy of the $E2$ transition in MeV and $\alpha_K(E2)$ is the K-conversion coefficient for $E2$ transition. The quantity q^2 is measured in the experiment as

$$q^2 = \frac{A_{e,K}(E0)}{\epsilon_{e,K}(E0)} \cdot \frac{\epsilon_{e,K}(E2)}{A_{e,K}(E2)}, \tag{4}$$

where $A_{e,K}(E\lambda)$ is the intensity of the K line for the corresponding transition and $\epsilon_{e,K}(E\lambda)$ is the efficiency of the β spectrometer.

We report a study of the $E0$ strength in two nuclei, ^{156}Dy and ^{160}Er. The nucleus ^{156}Dy is considered as a good candidate for $X(5)$ symmetry [12], while ^{160}Er, which has a structural parameter $R_{4/2} = E(4_1^+)/E(2_1^+) = 3.1$ lies, on the transition path between the critical point of the vibrator-rotor shape phase transition and the rigid rotor limit.

2. Experiment

The experiments were performed at the INFN Laboratori Nazionali del Sud (LNS) in Catania. Partial level schemes of ^{156}Dy and ^{160}Er are presented in Fig. 1 and Fig. 2, respectively.

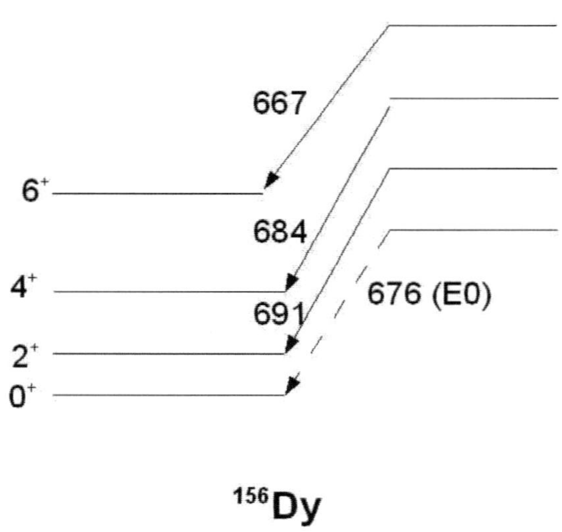

Figure 1. Partial level scheme of ^{156}Dy.

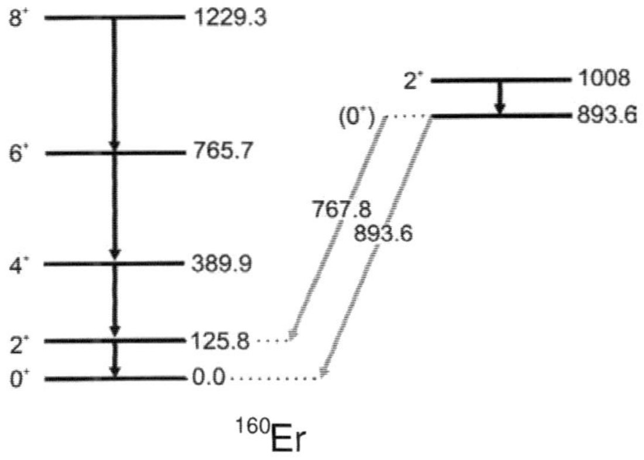

Figure 2. Partial level scheme of ^{160}Er.

Levels in ^{156}Dy were populated by the ^{156}Er $\to$ ^{156}Ho $\to$ ^{156}Dy ϵ decay chain. ^{156}Er was produced by the ^{148}Sm(^{12}C,4n) reaction at 73 MeV. The ^{12}C beam was provided by the LNS tandem accelerator and the target was a self-supporting isotope-enriched foil having a thickness of $\sim$ 0.8 mg/cm^2. Since the ^{156}Er $\to$ ^{156}Ho and the ^{156}Ho $\to$ ^{156}Dy decays have half lives of 19.5 min and 56 min, respectively, the experiment was performed by repeating cycles in which the ^{148}Sm target was irradiated for one hour and, after a 5 min delay, the decay of ^{156}Ho was measured for one hour by detecting off-beam γ rays and conversion electrons. In-beam spectra, produced by the prompt de-excitation of ^{156,157}Er levels, were also collected during the irradiation periods for calibration purposes.

The 0_2^+ state in ^{160}Er was populated through the ^{160}Yb $\rightarrow$ ^{160}Tm $\rightarrow$ ^{160}Er ϵ decay chain. ^{160}Yb was produced by the ^{150}Sm(^{14}N,4n) reaction at 72 MeV. The target was a self-supporting isotope-enriched foil with a thickness of 0.612 mg/cm^2. The half lives of ^{160}Yb and ^{160}Tm are 4.8 min and 9.4 min, respectively. Therefore, a ten-minute on-off beam cycle was used. The data was collected in the beam-off intervals.

This technique has the advantage that the decay chain starts with an even-even nucleus and the β decay passes through low spin states, thus avoiding possible branches through higher spin states in the parent doubly-odd nucleus. As a result, the population of non-yrast states in the daughter nucleus is enhanced and relatively clean spectra are obtained. Another advantage of the technique is that the measurement is done off-beam, which reduces the background of the electron spectra. The γ rays were measured with a coaxial HPGe detector which was positioned at 90° degrees with respect to the beam. Conversion electrons were measured with a mini-orange spectrometer, consisting of a magnetic filter made by permanent Sm-Co magnets, shaped like orange slices, and a 3 mm-thick Si(Li) detector cooled at liquid-nitrogen temperature. The spectrometer at Catania is a replica of the Groningen device [13]. Such spectrometers have a magnetic field which is transverse to the flat trajectories of the electrons. The mini-orange spectrometer was positioned at 135° with respect to the beam, in backward direction. Its efficiency is strongly energy dependent. The transmission window can be varied by changing the number and type of the permanent magnets and by adjusting the distances between the magnets and the target and between the target and the Si(Li) detector.

The efficiency calibration of the electron spectrometer is crucial for the correct treatment of the data. For example, in the ^{160}Er experiment, the ^{124}Sn(^{12}C,4n)^{132}Ba reaction at 60 MeV was used for simultaneous calibration of the HPGe detector and the mini-orange spectrometer, because in this case stretched $E2$ transitions are populated in the desired energy region of $650 - 850$ keV [14]. In the case of ^{156}Dy the calibration was done in two different ways: (i) the lines of the yrast band in ^{156}Er (4n reaction channel) were used, and (ii) the relative efficiency of the mini-orange spectrometer for the 667- and 691-keV lines was derived from the measured electron lines. The efficiency for the $E0$ and the 684-keV transitions was obtained assuming a linear dependence (see Fig. 3 lower panel). The efficiency of the mini-orange spectrometer, ϵ_e, as a function of the electron energy is obtained from the measured areas of the corresponding peaks in the electron (A_e) and the γ-ray spectra (I_γ), *e.g.* in the case of two transitions with energies E_1 and E_2

$$\frac{\epsilon_e(E_1 - B)}{\epsilon_e(E_2 - B)} = \frac{A_e(E_1 - B)}{A_e(E_2 - B)} \times \frac{\alpha(E_2)I_\gamma(E_2)}{\alpha(E_1)I_\gamma(E_1)}, \tag{5}$$

where B is the binding energy of the electron. The theoretical conversion coefficients for the stretched $E2$ are used [15].

The measured γ-ray and conversion electron spectra, revealing the decay of the 0_2^+ states in ^{156}Dy and ^{160}Er are presented in Fig. 3 and Fig. 4, respectively. Transitions belonging to ^{156}Dy and ^{160}Er are indicated with their energies in keV. K- and L-electron lines are indicated by K or L in the conversion electron spectra. Contaminant lines are indicated by $\star$. The arrows in the lower pannels indicate the peak corresponding to the $E0$ transition. In both cases, no peaks at these positions were observed in the γ spectra (indicated by arrows in the upper panels), which sets limits on the degree of contamination.

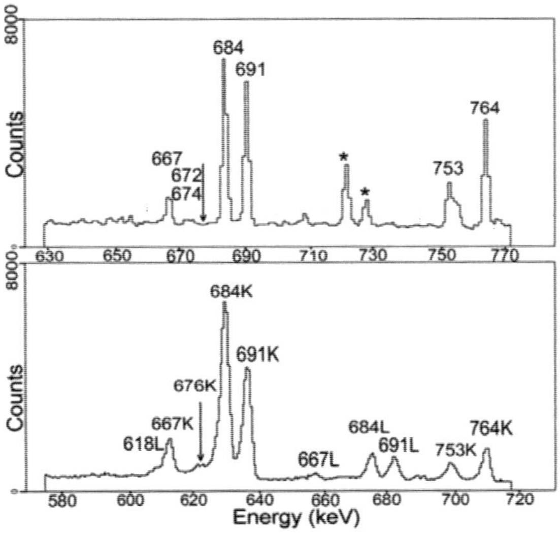

Figure 3. Partial off-beam γ-ray specrum (upper pannel) and conversion electron spectrum (lower pannel), obtained in the ϵ decay of ^{156}Ho.

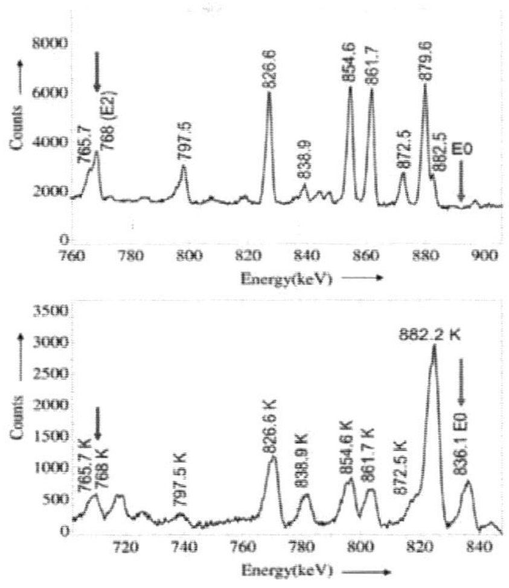

Figure 4. Partial off-beam γ-ray specrum (upper panel) and conversion electron spectrum (lower panel), obtained in the ϵ decay of ^{160}Tm.

3. Results and discussion

The 676-keV $E0$ transition in ^{156}Dy is rather weak (see Fig. 3 lower panel). Therefore, possible sources of contamination were analysed carefully to avoid systematic errors. Four peaks that correspond to the K-electron lines of 667-, 676-, 684- and 691-keV were used in the analysis of the conversion electron spectra. Using the γ-ray intensities from the experiment and the conversion coefficients of Ref. [15], $q^2 = 1.97(70)$ was obtained. Estimating the electronic factor with the method of Kantele [9] as $\Omega_K = 4.05 \cdot 10^{10}$ s^{-1}, the ratio of the reduced transition probabilities is $X(E0/E2; 0_2^+ \to 0_1^+) = 0.045(17)$. The γ-ray intensities, which were measured by Caprio *et al.* [12], can also be used in the analysis, since excited states in ^{156}Dy were populated in exactly the same way in both experiments. They yield $q^2 = 2.17(74)$, in perfect agreement with the value above.

These values can be compared with results from a recent compilation of the $E0$ strength [11], where the authors re-evaluated the existing data. For ^{156}Dy the results are: $q^2 = 3(2)$ and $X(E0/E2; 0_2^+ \to 0_1^+) = 0.08(5)$. The results are in agreement with each other, but with the present experiment, the uncertainty was reduced considerably.

In the case of ^{160}Er more intense $0_2^+ \to 0_1^+$ transition was observed (see Fig. 4b). With the mini-orange efficiency calibration and the γ-ray intensities of the experiment and the conversion coefficients of Ref. [15], $q^2 = 4.3(7)$ was obtained This number should be considered as a preliminary result, since the analysis continues. Estimating the electronic factor with the method of Kantele [9] as $\Omega_K = 7.04 \cdot 10^{10}$ s^{-1}, the ratio of the reduced transition probabilities is $X(E0/E2; 0_2^+ \to 0_1^+) = 0.18(4)$.

The $X(5)$ critical point symmetry [1] is a solution of the Bohr Hamiltonian with a special choice of the potential: $v(\beta, \gamma) = u(\beta) + v(\gamma)$. This potential allows an approximate separation of variables and then the potential in β is chosen as an infinite square well. This choice comes from the fact that, using the coherent state formalisms in the IBM, one can obtain the potential that (only) at the critical point goes as $\sim \beta^4$ and therefore it can be approximated with an infinite square well [16]. In a similar way, for the $U(5) - SU(3)$ first order shape phase transition, a more general potential in β can be introduced, the "Lo Bianco potential", that is parametrized as follows:

$$u(\beta) = V_0(\zeta\beta^4 - 2\zeta\beta_0\beta^3 + (1-\zeta)\beta_0^2\beta^2), \tag{6}$$

with $0 \le \zeta \le 1$. When $\zeta = 0$ there is a spherical minimum, at the critical point with $\zeta = 1/2$ there are two coexisting minima (with a very small bump in between), one in $\beta = 0$ and the other at $\beta = \beta_0$, and at $\zeta = 1$ there is only a unique deformed minimum in $3/2\beta_0$. This potential has the virtue of covering the whole transitional path of the shape phase transition at the price of having three parameters instead of the parameter-free predictions of the $X(5)$ model. The parameters, especially V_0, can be adjusted according to phenomenology to reproduce a subset of low-lying energy levels. An optimization procedure, based on a random walk in the parameter space (similar to Ref. [17]), to find optimal values for the parameters in order to fit the experimental energy levels has been set up. Calculations for several Dy and Er nuclei have been performed. Results for the excitation spectra of ^{160}Er are displayed in Fig. 5.

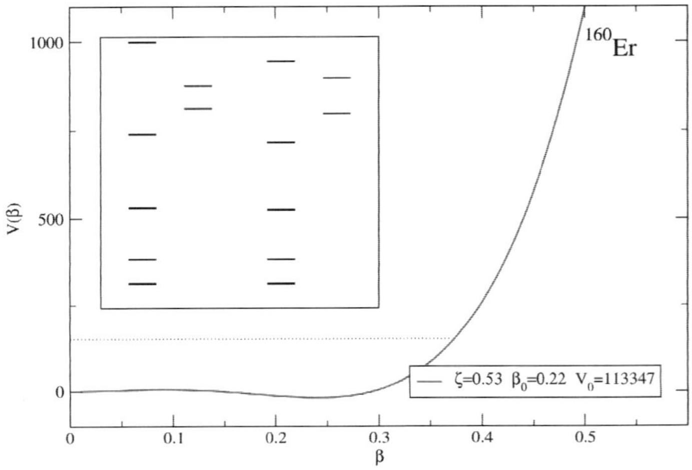

Figure 5. Potential energy (large panel) calculated with the optimization procedure to fit with experimental levels for ^{160}Er. The ground state is indicated by a dotted line. The inner panel contains the experimental (left) and calculated (right) lowest energy levels, 0^+, 2^+, 4^+, 6^+ and 8^+ of the *gs* band, and 0^+ and 2^+ of the β band in reduced energy units.

For the Dy isotopes, in the case of ^{154}Dy a value of $\zeta \approx 0.4$ is found, together with a value of $\beta_0 \approx 0.3$; the interplay of these two parameters makes up a potential with a flat bottom part, but still on the spherical side of the phase transition. From the fits one can see that ^{156}Dy is in the spherical region ($\zeta \approx 0.1$) and ^{158}Dy is just after the critical point ($\zeta = 0.54$), but the deformed minimum is already winning. In fact, although ζ is just after $1/2$, the combination of all three parameters gives a deformed potential and the wave functions are more pronounced at the deformations around the minimum. In the Erbium series of isotopes, ^{158}Er is close to the spherical limit ($\zeta = 0.06$), ^{162}Er is clearly deformed ($\zeta = 0.79$) and ^{160}Er has a value of $\zeta = 0.53$ that locates it quite close to the critical point.

In the framework of the Rasmunssen model [10], for ^{156}Dy the experimental ratio of the reduced transition probabilities, $X(E0/E2)$, was found to correspond to a modest equilibrium deformation, $\beta = 0.11(6)$, while for ^{160}Er a value of $\beta = 0.21(10)$ is obtained.

The matrix elements of the $E0$ operator, $\langle f|T(E0)|i\rangle$, between 0^+ states in ^{156}Dy and ^{160}Er, which are calculated with the present model, are listed in table 1.

Table 1. Matrix elements of the $T(E0)$ operator between 0^+ states with the "Lo Bianco potential".

^{A}Z	$\left\langle 0_1^+\|T(E0)\|0_2^+\right\rangle$	$\left\langle 0_1^+\|T(E0)\|0_3^+\right\rangle$	$\left\langle 0_3^+\|T(E0)\|0_2^+\right\rangle$
^{156}Dy	0.04279	-0.00213	-0.06419
^{160}Er	0.03311	-0.00467	-0.04464

The matrix elements allow a straightforward calculation of $\rho^2(E0)$:

$$\rho^2(E0) = \left(\frac{3}{4\pi}\right)^2 Z^2 |\langle f|T(E0)|i\rangle|^2, \tag{7}$$

which can be compared with the experiment. It turns out that, in the cases where $\rho^2(E0)$ is measured [11], the comparison is not very good. The value calculated in the collective model is about four times bigger than the measured quantity and this is true not only with our potential, but also for the confined beta-soft model [18]. One could wonder whether this is a shortcoming of the collective model, rather than attributing it to details of the potential.

In conclusion, we have measured the ratios of the reduced transition probabilities $X(E0/E2)$ in ^{156}Dy and ^{160}Er and compared them with calculations using a potential, which is a combination of quadratic, cubic and quartic powers of β. The results indicate that ^{156}Dy is in the spherical region, while ^{160}Er is located in the deformed region, but quite close to the critical point for the $U(5) - SU(3)$ first order shape phase transition.

Acknowledgments

We like to give a last farewell to Nanni Lo Bianco, who left us unexpectedly in September 2009. He has been an excellent teacher and collaborator for all of us. This work has been supported by the Italian INFN GAMMA collaboration and by the Bulgarian National Science Fund, contract DID-02/16.

References

[1] Iachello F 2001 *Phys. Rev. Lett.* **87** 052502
[2] Heyde K and Meyer R A 1988 *Phys. Rev.* **C 37** 2170
[3] von Brentano P *et al.* 2004 *Phys. Rev. Lett.* **93** 152502
[4] Wood J L *et al.* 1999 *Nucl. Phys.* **A651** 323
[5] Iachello F and Arima A 1987 *The Interacting Boson Model* (Cambridge: Cambridge University Press)
[6] Church E L and Weneser J 1956 *Phys. Rev.* **103** 1035
[7] Hager R S and Seltzer E C 1968 *Nucl. Data Tabl.* **A4** 1
[8] Bell D A *et al.* 1970 *Can. J. Phys.* **48** 2542
[9] Kantele J 1988 *Nucl. Instr. Meth.* **A271** 625
[10] Rasmussen J O 1960 *Nucl. Phys.* **19** 85
[11] Kibedi T and Spear R H 2005 *At. Data Nucl. Data Tables* **89** 77
[12] Caprio M A *et al.* 2002 *Phys. Rev.* **C 66** 054310
[13] van Klinken J and Wisshak K 1972 *Nucl. Instr. Meth.* **98** 1
[14] Harissopulos S *et al.* 1995 *Phys. Rev.* **C 52** 1796
[15] Kibedi T *et al.* 2008 *Nucl. Instr. Meth.* **A589** 202
[16] Fortunato L 2005 *Eur. Phys. J.* **A 26, s01** 1
[17] Fortunato L 2006 *Phys.Rev.* **C 74** 014310
[18] Bonnet J *et al.* 2009 *Phys.Rev.* **C 79** 034307

Description of giant resonances with Skyrme forces

Description of giant resonances with Skyrme forces

J Kvasil[1], V O Nesterenko[2], P Vesely[1,3], W Kleinig[2,4], and P-G Reinhard[5]

[1]Institute of Particle and Nuclear Physics, Charles University, CZ-18000 Praha 8, Czech Republic

E-mail: kvasil@ipnp.troja.mff.cuni.cz,vesely@ipnp.troja.mff.cuni.cz
[2]Bogoliubov Laboratory of Theoretical Physics, Joint Institute for Nuclear Research, Dubna, Moscow region, 141980, Russia

E-mail: nester@theor.jinr.ru,kleinig@theor.jinr.ru
[3]Department of Physics, Univ. of Jyvaskyla, P.O. Box 35 (YFL), 40014, Jyvaskyla, Finland
[4]Technische Universität Dresden, Institut für Analysis, D-01062, Dresden, Germany
[5]Institut für Theoretische Physik, Universität Erlangen, D-91058, Erlangen, Germany

Abstract. Some representative cases of the description and analysis of giant resonances within the self-consistent separable random-phase-approximation model (SRPA) based on the Skyrme functional are discussed. It is shown that SRPA with SLy6 force well describes the dipole giant resonance in rare-earth and actinide regions. The sensitivity of the E1 strength near the particle thresholds to nuclear deformation is scrutinized. Finally, the open problems of description of spin-flip M1 resonance are discussed.

1. Introduction

Nowadays, the Skyrme Hartree-Fock (SHF) mean-field models represent one of the most widely used theoretical tools for investigation of nuclear structure and dynamics, see e.g. reviews [1, 2]. These models provide high performance in description of ground state properties but description of dynamics is still a developing field, in particular concerning the exploration of giant resonances (GR). In the latter case, the SHF implementations with high accuracy and yet modest computational expense are desirable, especially for investigation of nuclear excitations in heavy spherical and deformed nuclei.

In this connection, the separable random-phase-approximation (SRPA) model with Skyrme forces was recently developed [3, 4, 5] and widely applied to description of electric [3, 4, 7, 8, 9, 10, 11] and magnetic [5, 12, 13] GR. SRPA belongs to self-consistent theories of excitations where both the static mean field and the residual interaction are derived from the same Skyrme energy-density functional [14, 15, 16]. The residual interaction includes all contributions from the functional (including those from the time-odd current and spin densities) as well as the Coulomb (direct and exchange) and pairing (at BCS level) terms. SRPA employs a self-consistent factorization of the residual interaction which considerably reduces the computational expense while maintaining high accuracy. The model is particularly suitable for systematic studies and transparent for the analysis.

In this paper, we present some of our recent results for isovector E1(T=1) and spin-flip M1(T=1) GR [5, 8, 9, 12]. First of all, we demonstrate the high accuracy of SRPA for E1(T=1)

GR in the rare-earth and actinide regions [8]. As a second step, we analyze the impact of nuclear deformation on E1 strength near the neutron and proton thresholds [9]. Being rather weak, this strength is nonetheless important for understanding astrophysical processes [17]. Its shown that the impact of deformation may be negligible or be of both sign, depending on the energy interval between the thresholds and E1(T=1) GR. Finally we inspect the problem of simultaneous description of gross-structure of spin-flip M1(T=1) GR in spherical and deformed nuclei [5, 12, 13]. We show that this problem is still a challenge for Skyrme forces and discuss possible solutions including the implementation of tensor forces.

2. Skyrme-RPA model

Starting point is the Skyrme functional [1, 2]

$$E = \int d\mathbf{r}(\mathcal{H}_{\rm kin} + \mathcal{H}_{\rm Sk}(\rho_q, \tau_q, \mathbf{s}_q, \mathbf{j}_q, \mathbf{J}_q, \mathbf{T}_q) + \mathcal{H}_{\rm pair}(\chi_q) + \mathcal{H}_{\rm C}(\rho_p)) \tag{1}$$

which consists of kinetic, Skyrme, pairing, and Coulomb terms, respectively. It involves time-even (nucleon ρ_q, kinetic-energy τ_q, spin-orbit $\mathbf{J}_q$) and time-odd (current $\mathbf{j}_q$, spin $\mathbf{s}_q$, vector kinetic-energy $\mathbf{T}_q$) densities, as well as the pairing density χ_q. The label q denotes protons and neutrons. The time-odd densities do not contribute to the ground states of even-even nuclei but may be essential for the ground states of odd nuclei [18, 19] and nuclear dynamics including rotation [20] and giant resonances [3, 4, 7, 8, 10, 11]. These densities restore the Galilean invariance of the functional [7, 16, 20] and are important for the spin-flip M1(T=1) GR [3, 4, 5]. The functional (1) includes the tensor contributions $\mathbf{J}^2 = (\mathbf{J}_n + \mathbf{J}_p)^2$ and $\sum_q \mathbf{J}_q^2$, expressed through the squared spin-orbital densities [21], which can be also important for the spin-flip M1(T=1) GR [5, 12, 22].

The SRPA residual interaction is fully determined by the given Skyrme functional and does not add any additional parameters. It is expressed through the second functional derivatives

$$\frac{\delta^2 E}{\delta\rho_{q'}\delta\rho_q}, \frac{\delta^2 E}{\delta\tau_{q'}\delta\rho_q}, \frac{\delta^2 E}{\delta\mathbf{J}_{q'}\delta\rho_q}, \frac{\delta^2 E}{\delta\mathbf{j}_{q'}\delta\mathbf{j}_q}, \tag{2}$$

for electric modes and

$$\frac{\delta^2 E}{\delta\mathbf{j}_{q'}\delta\mathbf{s}_q}, \frac{\delta^2 E}{\delta\mathbf{s}_{q'}\delta\mathbf{s}_q}, \frac{\delta^2 E}{\delta\mathbf{J}_{q'}\delta\mathbf{J}_q}, \frac{\delta^2 E}{\delta\mathbf{T}_{q'}\delta\mathbf{s}_q} \tag{3}$$

for magnetic modes. The isovector E1 and M1 GR are analyzed in terms of strength functions

$$S(E1(M1); E) = \sum_{\nu\neq 0} E_\nu^L |\langle\Psi_\nu|\hat{M}|\Psi_0\rangle|^2 \zeta(E - E_\nu) \tag{4}$$

where Ψ_0 is the ground state, ν runs over the RPA states ($K^\pi = 0^-, 1^-$ for E1 and $K^\pi = 1^+$ for M1) with energies ω_ν and wave functions Ψ_ν. Furthermore, $\zeta(E-E_\nu) = \Delta/[2\pi[(E-E_\nu)^2+\Delta^2/4]]$ is a Lorentz weight with the averaging parameter Δ intended to simulate broadening effects beyond SRPA (escape widths, coupling with complex configurations). The energy weight is used for E1 (L=1) but not for M1 (L=0). The strength function (4) is computed directly, i.e. without calculation of RPA states ν, which additionally reduces the computation expense.

The E1(T=1) GR is computed with the proton and neutron effective charges $e_p^{eff} = N/A$ and $e_n^{eff} = -Z/A$. The spin-flip M1 transition operator is calculated with the proton and neutron spin g-factors $g_s^p = 5.58\varsigma_p$ and $g_s^n = -3.82\varsigma_n$ quenched by ς_p=0.68 and ς_n=0.64. The corresponding isoscalar (T=0) and isovector (T=1) g-factors are $g_s^0 = g_s^p + g_s^n = 1.35$ and $g_s^1 = g_s^p - g_s^n = 6.24$. Note that $g_s^1 \gg g_s^0$, i.e. spin-flip M1 GR is mainly isovector. In the present study the orbital part in M1 operator is omitted.

3. Results and discussion

3.1. E1(T=1) giant resonance

Results of SRPA calculations [8] with the Skyrme force SLy6 [23] for E1(T=1) GR in axially-deformed nuclei of rare-earth and actinide regions are given in Figs. 1 and 2. The Lorentz averaging Δ=2 MeV is used. The figures demonstrate the deformation splitting between the $\mu = 0$ and $\mu = 1$ resonance branches. An excellent agreement with the experimental data is seen. This proves high SRPA accuracy and encourages application of the model to other kinds of nuclear dynamics discussed in the next subsections.

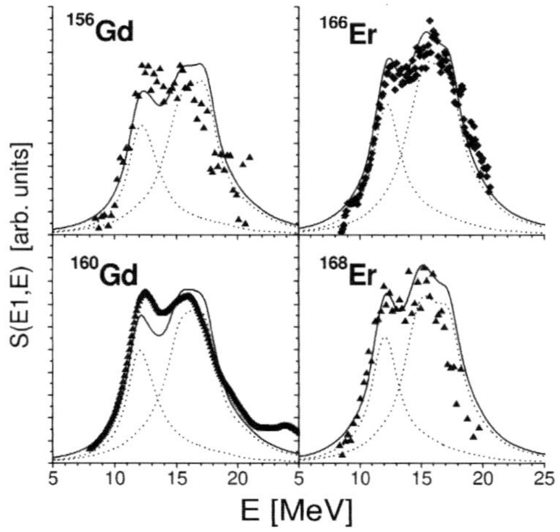

Figure 1. E1(T=1) GR in ^{156,160}Gd and ^{166,168}Er [8]. The calculated strength (solid curve) is compared to the experimental data [24] for total photo-absorption (triangles) and neutron product (rhombus). The GR branches with $\mu = 0$ (left) and $\mu = 1$ (right) are depicted by dotted curves.

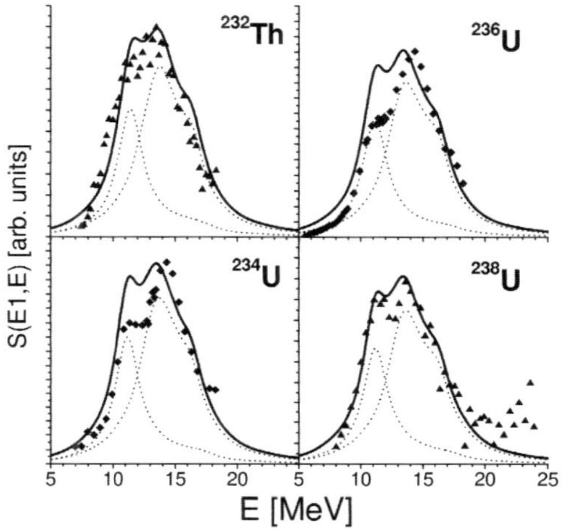

Figure 2. The same as in Fig. 1 for ^{232}Th and ^{234,236,238}U [8].

3.2. Threshold E1 strength

The E1 strength near neutron and proton emission thresholds is a subject of intense research in connection to astrophysical needs, see e.g. [17]. The strength was explored within inconsistent models with Woods-Saxon potential [25, 26] and fully self-consistent approaches, relativistic covariant RPA [27] and Skyrme SRPA [9]. The phenomenological studies claim that nuclear deformation always increases the threshold E1 strength [25, 26]. On the other hand, SRPA calculations with the Skyrme force SLy6 for Mo isotopes [9] show that this is not the case and the deformation impact can vanish (due to the compensation of $\mu = 0$ and $\mu = 1$ resonance contributions) or be of both sign, depending on the proximity of the thresholds to E1(T=1) GR. The similar effect, weakening the pygmy resonance strength due to the deformation, was found in Sn isotopes near the drip line [27]. In the previous SRPA study [9], likely triaxial Mo isotopes were treated as axial. Here we consider a more refine case of the axial prolate nucleus ^{152}Nd.

In Fig. 3 the fraction $\Sigma(E) = \int_{E_{min}}^{E} dES(E1;E)$ of the Thomas-Reiche-Kuhn (TRK) energy-weighted sum rule $\Sigma_{TRK} = \int dES(E1;E) = 60NZ/A$ mb MeV is depicted for the energy interval $[E_{min}, \mathrm{E}]$ with E_{min}=4 MeV. The cases with and without the equilibrium deformation β_2=0.347

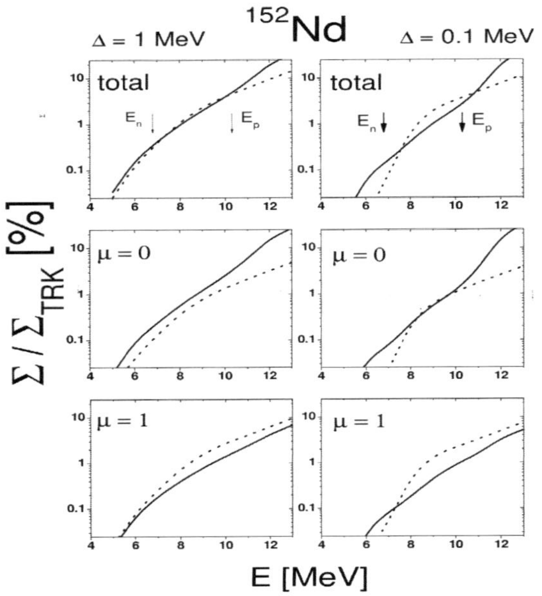

Figure 3. The cumulative E1 strength $\Sigma(E)$ in ^{152}Nd for the large Δ=1 MeV (left) and small Δ=0.1 MeV (right) Lorentz smoothing. The panels give the total E1 strength (upper) and separate contributions from $\mu = 0$ (middle) and $\mu = 1$ (bottom) branches. In every panel the strengths at zero (dash line) and equilibrium β_2=0.347 (solid line) deformations are compared. The neutron E_n=6.8 MeV and proton E_p=10.3 MeV thresholds are approximated by the Fermi level energies in SLy6 mean fields. The logarithmic scale is used to visualize the effects.

are compared. A large and a small Lorentz smoothing are used to embrace the E1(T=1) GR tail (Δ=1 MeV) or local dipole excitations alone (Δ=0.1 MeV).

Fig. 3 shows that near thresholds the influence of deformation on the total E1 strength is negligible for Δ=1 MeV or may be both positive and negative for Δ=0.1 MeV. The effect of deformation is always positive only at $E > 11$ MeV i.e. in close vicinity to the lower resonance branch ($\mu = 0$) at $E_{\mu=0}$=12 MeV. The middle and bottom panels of the figure show that deformation effects for $\mu = 0$ and $\mu = 1$ are of the opposite sign and, to a large extent, compensate each other in the total E1 strength. The splitting due to the prolate deformation down-shifts the $\mu = 0$ and up-shifts the $\mu = 1$ GR branches. As a result, the $\mu = 0$ branch adds strength to the lower threshold region while the $\mu = 1$ one reduces it. Being of the opposite sign, the effects have a similar magnitude since the weaker $\mu = 0$ branch is placed closer to the thresholds while the stronger $\mu = 1$ branch is more remote. Near the resonance the $\mu = 0$ contribution strictly dominates and we always have the increase of the strength as was predicted in [25, 26]. However, the thresholds often lie much below the resonance and then the compensation effect becomes decisive. In this case, the impact of the deformation can be negligible or, depending on the compensation, slightly positive or negative. This result is important for a realistic estimate of the threshold E1 strength in deformed nuclei and subsequent astrophysical conclusions.

3.3. Spin-flip M1 GR

The spin-flip M1 GR is known to be a major source of our knowledge on spin correlations and robust test for the spin-orbit interaction [28]. During the last decades the resonance was widely investigated within various (mainly empirical) models and its basic properties were properly clarified. The availability of self-consistent SHF calculations motivated the revision of the M1 GR studies on new theoretical grounds, see e.g. [29, 30]. The first *systematic* SHF study of this GR was performed within the SRPA [5, 12]. It was shown that none of numerous Skyrme parameterizations is able to describe simultaneously the one-bump gross structure of M1 strength in doubly magic nuclei and two-bump structure in heavy deformed nuclei. That was an alarming message for both spin-flip M1 GR and its Gamow-Teller counterpart.

The problem is illustrated in Fig. 4 where the resonance strength (4) for the forces SkO [31],

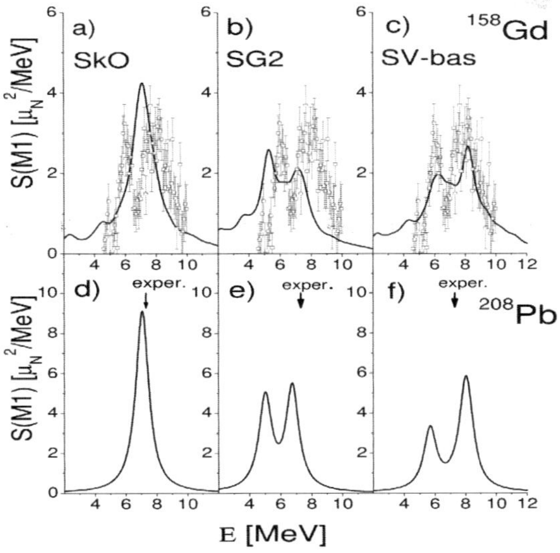

Figure 4. The spin-flip M1 strength in ^{158}Gd and ^{208}Pb for the forces SkO, SG2, and SV-bas. The experimental data are given by boxes with bars for ^{158}Gd [34] and vertical arrows for ^{208}Pb [35]. The Lorentz width Δ=1 MeV is used.

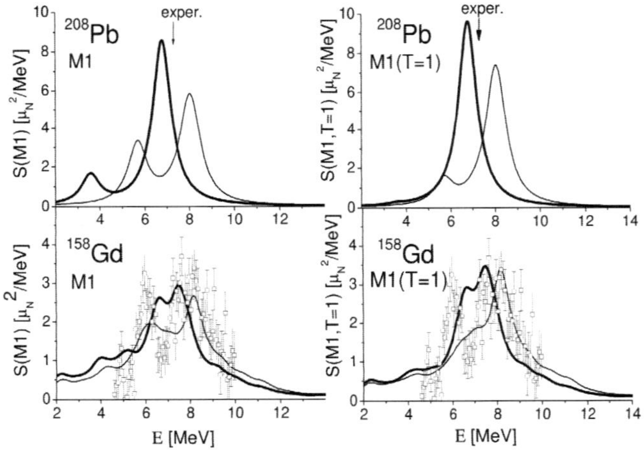

Figure 5. The spin-flip M1 (left) and M1(T=1) (right) strengths in ^{208}Pb and ^{158}Gd for the force SV-bas with (bold solid curve) and without (solid curve) tensor contribution. The experimental data are like in Fig. 4. The Lorentz width is Δ=1 MeV.

SG2 [32], and SV-bas [33] is compared to the experimental data in spherical ^{208}Pb and deformed ^{158}Gd. It is seen that SkO gives one bump in both nuclei, which is good for ^{208}Pb but not for ^{158}Gd. Instead, SG2 and SV-bas always give two-bumps, which is correct for ^{158}Gd but not for ^{208}Pb. The gross structure of spin-flip M1 GR is known to be determined by a fragile balance between the spin correlations (= residual interaction) and spin-orbital splitting (forming the unperturbed proton and neutron peaks) [5, 12, 28]. If the correlations are strong enough then the neutron and proton peaks are mixed and we obtain the one-bump structure. Otherwise, the two-bump proton-neutron structure is preserved. Fig. 4 shows that the given Skyrme parameterizations do not keep the proper balance but distort it in favor of the correlations (SkO) or spin-orbital splitting (SG2 and SV-bas).

As seen from Fig. 5, the description can be improved by replacing the M1 response by the M1(T=1) one. This leads to redistribution of the strength and formation of mainly one-bump structure in ^{208}Pb even for forces like SV-bas [5]. An additional significant improvement can be obtained by including the tensor contribution, see also [22]. Both these two factors have to be in focus of further upgrade of SHF description of spin-flip M1 GR.

4. Conclusions

It was demonstrated that the Skyrme SRPA model performs very well in the description and analysis of various kinds of nuclear giant resonance (GR) excitations. For the threshold E1 strength, a simple compensation scheme was proposed to explain a small deformation effect. The open problems in description of spin-flip M1 GR were discussed. These examples, as well as other SRPA studies of multipole giant resonances [3, 4, 7, 8, 9, 10, 11], show that the model is indeed very effective, accurate, economical and universal theoretical tool.

Acknowledgments

The work was supported by the grants 202/06/0363 and MSM 0021620859 (Ministry of education of the Czech Republic), Heisenberg-Landau (Germany - BLTP JINR) and Votruba - Blokhintsev (Czech Republic - BLTP JINR). P.V. thanks the FIDIPRO support. W.K. and P.-G.R. are grateful for the BMBF support (contracts 06 DD 9052D and 06 ER 9063).

[1] Bender M, Heenen P-H and Reinhard P-G 2003 *Rev. Mod. Phys.* **75** 121
[2] Stone J R and Reinhard P-G 2007 *Prog. Part. Nucl. Phys.* **58** 587
[3] Nesterenko V O, Kvasil J and Reinhard P-G 2002 *Phys. Rev.* C **66** 044307
[4] Nesterenko V O, Kleinig W, Kvasil J, Vesely P, Reinhard P-G and Dolci D S 2006 *Phys. Rev.* C **74** 064306
[5] Vesely P, Kvasil J, Nesterenko V O, Kleinig W, Reinhard P-G and Ponomarev V Yu 2009 *Phys. Rev.* C **80** 031302(R)
[6] Nesterenko V O, Kleinig W, Kvasil J, Vesely P and Reinhard P-G, 2007 *Int. J. Mod. Phys.* E **16**, 624
[7] Nesterenko V O, Kleinig W, Kvasil J, Vesely P and Reinhard P-G, 2008 *Int. J. Mod. Phys.* E **17**, 89
[8] Kleinig W, Nesterenko V O, Kvasil J, Reinhard P-G and Vesely P 2008 *Phys. Rev.* C **78** 044313
[9] Kvasil J, Vesely P, Nesterenko V O, Kleinig W, Reinhard P-G and Frauendorf S 2009 *Int. J. Mod. Phys.* E **18** 975
[10] Nesterenko V O, Kleinig W, Kvasil J, Vesely P, and Reinhard P-G 2007 *Proc. 26th Int. Workshop on Nuclear Theory (Rila, Bulgaria, 2006)* ed. S. Dimitrova (Sofia: INRNE BAS) p.273
[11] Nesterenko V O, Kleinig W, Kvasil J, Vesely P, Reinhard P-G and Novikov A N 2009 *Proc. of 2nd Int. Conf. on Current Problems in Nuclear Physics and Atomic Energy (Kyiv, Ukraine, 2008)*(Kyiv:) p. 108
[12] Nesterenko V O, Kvasil J, Vesely P, Kleinig W, Reinhard P-G and Pomomarev V Yu 2010 *J. Phys. G: Nucl. Part. Phys.* **37** 064034
[13] Nesterenko V O, Kvasil J, Vesely P, Kleinig W and Reinhard P-G 2010 *Int. J. Mod. Phys.* E **19** 558
[14] Skyrme T H R 1956 *Phil. Mag.* **1** 1043
[15] Vautherin D and Brink D M 1972 *Phys. Rev.* C **5** 626
[16] Engel Y M, Brink D M, Goeke K, Krieger S J and Vauterin D 1975 *Nucl. Phys.* A **249** 215
[17] M. Arnold and S. Goriely, *Phys. Rep.* **384**, 1 (2003).
[18] Schunck N, Dobaczewski J, McDonnell J, Moré J, Nazarewicz W, Sarich J and Stoitsov M V 2010 *Phys. Rev.* C **81** 024316
[19] Pototzky K, Erler J, Reinhard P-G and Nesterenko V O 2010 arXiv:1008.0535[nucl-th]; submitted to *Eur. Phys. J. A*
[20] Dobaczewski J and Dudek J 1995 *Phys. Rev.* C **52** 1827
[21] Zow W, Colo G, Ma Zh, Sagawa H and Bortignon P F 2008 *Phys. Rev.* C **77**, 014314
[22] Cao Li-Gang, Colo G, Sagawa H, Bortignon P F and Sciacchitano L 2009 *Phys. Rev.* C **80**, 064304
[23] Chabanat E, Bonche P, Haensel P, Meyer J and Schaeffer R 1997 *Nucl. Phys.* A **627** 710
[24] Varlamov A V, Varlamov V V, Rudenko D S and Stepanov M E 1999 *Atlas of Giant Dipole Resonances*, INDC(NDS)-394, 1999; JANIS database.
[25] Dönau F, Rusev G, Schwengner R, Junghans A R, Schilling K D and Wagner A 2007 *Phys. Rev.* C **76** 014317
[26] Rusev G et al 2009 *Phys. Rev.* C **79** 061303(R)
[27] Peña Artega D, Khan E and Ring P 2009 *Phys. Rev.* C **79** 034311
[28] Harakeh M N and van der Woude A 2001 *Giant Resonances* (Oxford: Clarendon Press)
[29] Sarriguren P, Moya de Guerra E and Nojarov R 1996 *Phys. Rev.* C **54** 690
[30] Hilton R R, Höhenberger W and Ring P 1998 *Eur. Phys. J.* A **1** 257
[31] Reinhard P-G et al 1999 *Phys. Rev.* C **60** 014316
[32] Van Giai N and Sagawa H 1981 *Phys. Lett.* B **106** 379
[33] Klüpfel P, Reinhard P-G, Bürvenich T J and Maruhn J A 2009 *Phys. Rev.* C **79** 034310
[34] Frekers D et al 1990 *Phys. Lett.* B **244** 178.
[35] Laszewski R M, Alarcon R, Dale D S and Hoblit S D 1988 *Phys. Rev. Lett.* **61** 1710

Time-modulation of entangled two-body weak decays with massive neutrinos

Time-modulation of entangled two-body weak decays with massive neutrinos

P. Kienle

Excellence Cluster "Universe" Technische Universität München, Germany

E-mail: paul.kienle@ph.tum.de

Abstract. In recent experiments at the GSI, Darmstadt, time-modulated orbital Electron Capture (EC) decays of H-like $^{140}Pr^{58+}$, $^{142}Pm^{60+}$, and $^{122}I^{52+}$ ions with one electron in the K-shell, coasting in the ESR storage ring with velocity $\beta = 0.71$, were observed. The EC-branches observed using time-resolved Schottky Mass Spectroscopy show exponential decay curves time-modulated with periods $T = 7.06(8)$ s, $T = 7.10(22)$ s, and $T = 6.16(3)$ s and amplitudes $a = 0.18(3)$, $a = 0.23(4)$, and $a = 0.16(2)$ for ^{140}Pr, ^{142}Pm, and ^{122}I decays, respectively. The β^+ branch of ^{142}Pm shows no modulation with $a < 0.03$. As origin of the modulation, we propose quantum beats produced by massive neutrino mass-eigen states emitted in the two body weak decay. From the modulation frequency a value for the difference of the quadratic mass values $m_2^2 - m_1^2 = 2.22(3) \times 10^{-4}$ eV^2 is deduced, which is 2.9 times larger than the value derived by the KamLAND antineutrino oscillation experiment. The origin of the small modulation amplitudes is discussed as the result of partial restoration of the interference terms which are expected to cancel for the usual assumed unitarity of the neutrino flavour mixing matrix.

1. Introduction

The heavy ion accelerator and storage ring facility of the GSI, Darmstadt, offers unique opportunities for weak decay studies of highly ionized atoms [1]. In this report we focus on the first results of studying the time evolution of quasi free two body weak decays. In this way we investigate the properties of massive neutrinos by observing the appearance time of the daughter nucleus, detected with 100% efficiency, and entangled with the massive neutrinos. We focus here on the orbital electron capture of hydrogen like ions, (H-like), with one electron in the K-shell, and observe such decays in a storage ring quasi-free without the complication of such decays in a solid. Our aim was to search for quantum beat like phenomena possibly introduced by mixed massive neutrinos. Its properties can then be studied without observing the neutrino directly due to the quantum-entanglement with the daughter nucleus thus avoiding the small detection efficiency of neutrinos.

For our first experiments [2] we selected allowed Gamow-Teller transitions, GT ($1^+ \rightarrow 0^+$), with strong branches to 0^+ ground-states of stable even-even nuclei, with EC/β^+ branches close to unity, and half-lives in the range of minutes for the observation of beats with modulation times of several seconds and a time resolution in the 100 ms range. Experiments with H-like ^{140}Pr and ^{142}Pm ions fulfill the entire requirements. In addition they have quite different decay energies (Q_{EC}-values) and life-times and thus allow a detailed comparison of the time evolution of the decays with different neutrino energies in different time ranges. Both Q_{EC}-values of several MeV are large enough to be well resolved by measuring the time correlated change of the revolution frequency when the mother-ion decays into the daughter-ion without change of their charge. In a recent experiment we investigated also the ($1^+ \rightarrow 0^+$) GT transition of H-like ^{122}I in order to learn about the A-dependence of the time evolution of such two body decays [3, 4].

2. Experimental results

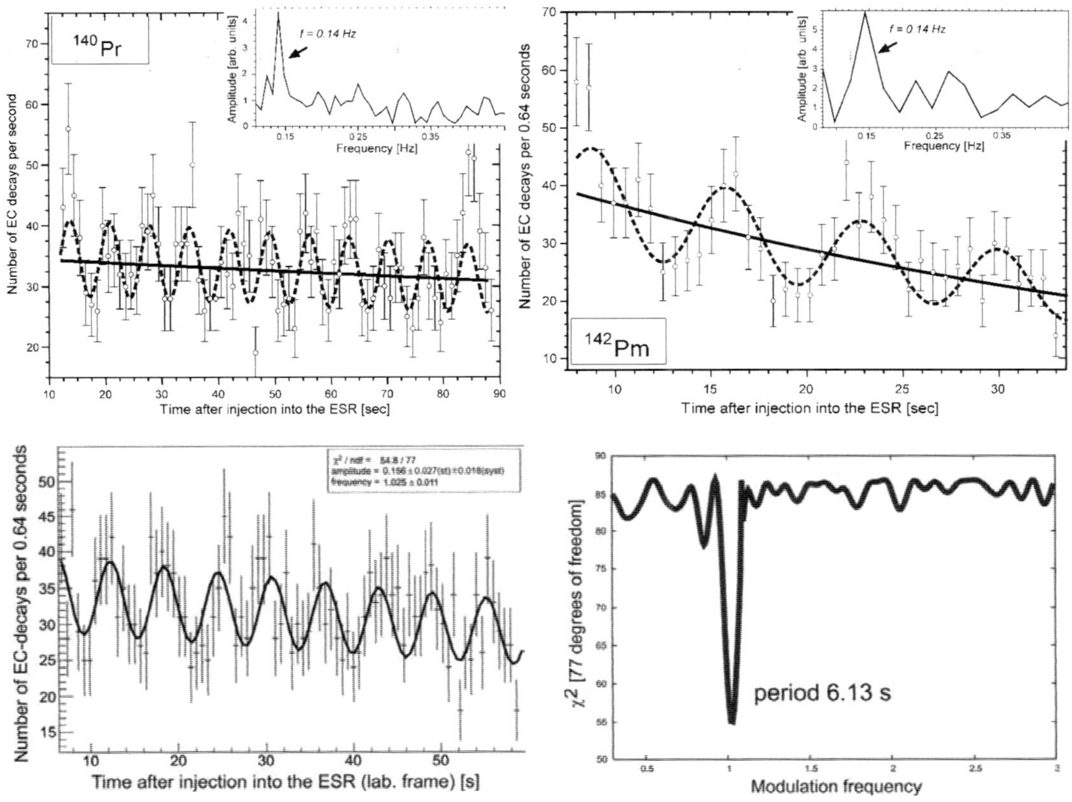

Figure 1. Time dependence of the EC decay rate of ^{140}Pr and ^{142}Pm showing periodic modulations of about 7 s (dashed curve) of the exponential decays (solid curve). The insets show Fourier frequency spectra with maxima at frequencies of $f = 0.14$ s^{-1} [2] (upper two panels). Time dependence of the EC decay rate of ^{122}I with a periodic modulation of about 6 s (lower panel left). $\chi^2(\omega)$ for the fit of the ^{122}I data with a modulated EC decay rate as function of the modulation frequency ω with all other parameters free (lower panel right) [3, 4].

Fig. 1 shows the measured decay spectra for H-like ^{140}Pr and ^{142}Pm ions (upper panels) [2]. A fit with a pure exponential decay function,

$$\frac{dN_{EC}(t)}{dt} = N(0) \cdot \lambda_{EC} \cdot e^{-\lambda t} \quad (1)$$

as indicated with the solid curve, failed to reproduce the data. The expected exponential decay shows a superimposed periodic time modulation. To account for this modulation the data was fitted with the function:

$$\frac{dN_{EC}(t)}{dt} = N(0) \cdot \lambda_{EC}(t) \cdot e^{-\lambda t} \quad (2)$$

with a time dependent decay probability

$$\lambda_{EC}(t) = \lambda_{EC}\left[1 + a\cos(\omega t + \phi)\right] \tag{3}$$

representing a periodic modulation of the partial EC-decay rate $\lambda_{EC}(t)$ with an amplitude a, an angular-frequency ω, and a phase ϕ. For details of the measuring procedure of the decays in the storage ring using time-dependant, single ion Schottky Mass Spectroscopy (SMS) and the results of all fit parameters of both data sets see reference [2] Table 2. Here we quote briefly the main results.

The total decay constants λ obtained from both fitting functions, Eqs. (1) and (2) agree within their error margins for the ^{140}Pr data and for the ^{142}Pm data. All thus deduced decay constants λ agree within two standard errors with the corresponding literature values for neutral atoms corrected for the missing electrons and taking into account the observed "enhancement" of the EC decay of H-like ions [5, 6]. From the angular frequencies 0.890(11) s^{-1} and 0.885(31) s^{-1} we extract the periods of modulation as $T = 7.06(8)$ s and $T = 7.10(22)$ s in the laboratory frame for the decays of ^{140}Pr and ^{142}Pm, respectively. Note that the modulation periods are within their errors the same for both decays and thus independent of the energies of the emitted neutrinos; in contrast they indicate scaling with the mass of the ions. Also the amplitudes $a = 0.18(3)$, ^{140}Pr, and $a = 0.23(4)$, ^{142}Pm, agree within their error margins; the average of both systems is $\langle a \rangle = 0.20(2)$. The results for the phases of the modulation reflect the uncertainties introduced by the difference and variation of the cooling times of the daughter ions as outlined in [2].

In the following we show a still preliminary result gained with the goal to complement the observations so far reported in [2]. In order to test our measuring method and determine whether the observed modulation are characteristic for the two body EC decay, the simultaneously observed decay spectra of the β^+ decay of ^{142}Pm, which is about three time stronger than the EC branch, was analysed [3, 4]. It shows a pure exponential decay, whereas the decay spectrum of the EC branch gained with the same data set showed the expected modulation with angular frequency of 0.90(2) s^{-1} and an amplitude of $a = 0.18(5)$. At a modulation frequency of $\omega = 0.90$ s^{-1} only a vanishing modulation amplitude of $a = 0.03(3)$ is indicated for the β^+ branch [3, 4], as expected theoretically for a continuous neutrino spectrum [7].

In August 2008 we performed a further experiment [3, 4] to study the decay spectrum of the EC decay branch of the $(1^+ - 0^+)$ GT-transition of H-like ^{122}I^{52+} to the ground state of stable ^{122}Te^{52+}. The main purpose of this experiment was studying the A-dependence of the observed modulation frequency. Fig. 1 (Lower panel left) shows an EC decay spectrum of H-like ^{122}I^{52+} of a very recent analysis [3], and a fit of a time modulated exponential decay to the data. It shows a modulation with an angular frequency $\omega = 1.038(6)$ s^{-1}, an amplitude $a = 0.16(2)$. In the lower right panel of Fig. 1 we show the mean square deviation, $\chi^2(\omega)$, of the fit of the data to a modulated decay spectrum eq. (2) as function of the angular frequency ω with all other parameters free. A clear minimum of $\chi^2(\omega)$ is seen at $\omega = 1.038(6)$ s^{-1} with a statistical relevance of about 6σ. So the reduced value of the observed modulation period of $T = 6.13(6)$ s for ^{122}I compared with 7.06(8) s for ^{140}Pr indicates that T scales with the mass number A of the decaying system.

3. Discussion of the results

The periodically modulated exponential EC decay has now been observed for three H-like systems ^{140}Pr, ^{142}Pr, and ^{122}I. The amplitude of the modulation is within the error margin the same in all three systems. Its periods scale with the mass number A of the decaying systems and are independent of their Q_{EC}-values and their half-lives. The three-body β^+ decay of ^{142}Pm is according to preliminary data not modulated with a limit of its amplitude $a = 0.03(3)$ [4]. This result, though preliminary, is most important because it indicates that the modulation is directly connected to the weak decay properties, as it is present for the two-body decay branch only, and absent for the three body decays [7]. It excludes experimental effects and also nuclear properties of the initial state [8, 9]; it thus points to an effect caused by the neutrinos of the EC branch. A further important result is that the time averaged decay probability of the modulated decays is fully consistent with the decay constant of neutral atoms after proper correction [2, 5, 6]. This excludes the influence of possible oscillatory transitions with a period of 7 s between the $F = 1/2$ hyperfine ground state of the H-like system and its $F = 3/2$ excited state which is inert against weak decays. The measured EC/β^+ ratios for H- and He-like decays of ^{140}Pr agrees within 3% with theoretical predictions assuming mass-less neutrinos [5, 6]. This excludes neutrino flavour oscillations as reason for the time modulation of the EC decays because their time averaged decay constants would be reduced in the case of oscillations relative to that of the amplitudes of EC decays of H-like heavy ions.

So what is the origin of these so called "GSI Oscillations ?" Strongly controversial arguments have been put forward, one which connects the observations with quantum beats of flavour mixed massive neutrinos [10, 11, 12] and the others claim that the effect has nothing to do with neutrinos [8, 9, 13, 14, 15]. The latter is based essentially on two arguments that interference terms in the decay probability cannot appear in the present GSI experiment:

i) The final mass-eigenstates of the neutrinos are orthogonal and thus asymptotically being identified by different energies and momenta with the consequence that the interference terms are not observable. We have shown [16] that this assertion does not respect the time differential observation of the development of the decay in our experiments which induces energy and momentum uncertainties leading to an overlap of indistinguishable decay paths and thus to interference in time which may be classified as "neutrino quantum beats" [17].
ii) There is a serious objection against our approach [16], based on the argument that the "neutrinos are not observed" in our experiments. Various authors put forward as a "theorem", that only the squares of the flavour transition amplitudes should be incoherently added in case that the "neutrinos are not observed", and concluded that the observed time modulation of the EC decay has nothing to do with neutrinos. As we have argued experimentally that the observed modulation must originate from the two-body decay involving the emission of massive neutrinos we have proposed possible ways to get around this "non-neutrino theorem" which is based on unitarity of the flavour mixing matrix [18, 19].

In our theory approach, published recently [16] the amplitude $A(m \to d + \nu_e)(t)$ of the EC decay of a H-like mother ion m into a daughter ion d and an electron neutrino ν_e is defined as the sum of the amplitudes of the emission of massive neutrino mass eigenstates $m \to d + \nu_j$ as follows:

$$A(m \to d + \nu_e)(t) = \sum_j U_{ej} A(m \to d + \nu_j)(t) \tag{4}$$

where the coefficients U_{ej} take into account that *only* electron neutrinos contribute to the transition. As a result of time dependent perturbation treatment using wave packets for the daughter ion wave functions for description of their time dependent observation a time modulation of the EC decay probability is predicted in the form

$$\lambda_{EC}(t) = \lambda_{EC}\left[1 + a\cos(\omega_{21}t)\right] \tag{5}$$

with the modulation amplitude a determined by the neutrino flavour mixing angle θ_{12}

$$a = \sin(2\theta_{12}) \tag{6}$$

and the modulation frequency ω_{21} determined by energy and momentum conservation of the decay

$$\omega_{21} = (m_2^2 - m_1^2)/2M_m \tag{7}$$

Note, ω_{21} is equal to the recoil energy differences of the neutrinos with mass m_2 and m_1 emitted by the mother nucleus with mass M_m in the cm system. Note also that, as observed, the modulation frequency scales as the inverse of the mother masses M_m and a numerical value

$$m_2^2 - m_1^2 = 2.12 \times 10^{-4}\ \mathrm{eV}^2 \tag{8}$$

is derived as averaged value from the three decays observed. It is 2.9 larger than the value reported by the KamLAND collaboration for oscillations of antineutrinos from fission products. A possible solution of this difference in terms of neutrino and antineutrino mass corrections of opposite signs induced by virtual lepton-W boson loops in the Coulomb field of heavy nuclei is proposed in [20].

Our approach [16] was criticised recently by Gal [18] and Yazaki [19] who showed that by using the total set of orthogonal neutrino flavour wave function in calculation of the transition probability, and requiring unitarity of the flavour mixing matrix, the interference terms cancel. There are in addition two experimental reasons to use the complete set of neutrino flavour wave functions with definite lepton charges: i) The indication of a finite phase ϕ_{EC} in the measured modulation pattern which is expected to be $\phi_{EC} = 0$ in our approach [16]. ii) The small value of the modulation amplitude which is given in our approach as $a = \sin(2\theta_{12})$. With $\theta_{12} = 34°$ from sun neutrino observations [21] one expects a modulation amplitude $a = 0.93$ which is not observed in our experiments.

Thus we have proposed recently [17] an approach to define the amplitude of EC decays as coherent superposition of the amplitude $A(m \to d\nu_\alpha)(t)$ given by:

$$A(m \to d\nu)(t) = \sum_\alpha e^{-i\phi_\alpha} A(m \to d\nu_\alpha)(t) \tag{9}$$

with the flavour charge index α running over $\alpha = e, \mu$ and τ. In addition we have introduced arbitrary phases ϕ_α to take care of the possible different phase shifts of the lepton flavour components. For the amplitude we are using the following general expression:

$$A(m \to d\nu)(t) = \sum_j \sum_\alpha U_{\alpha j} U_{ej} e^{-i\phi_\alpha} A(m \to d\nu_j)(t) \tag{10}$$

with

$$A(m \to d\nu_j)(t) = -\delta_{M_{F,-\frac{1}{2}}} \sqrt{6M_m} \mathcal{M}_{\mathrm{GT}} \langle \psi_{1s}^{(Z)} \rangle \sqrt{2E_d(\vec{k}_j)E_j(\vec{k}_j)} \frac{e^{i(\Delta E_j - i\varepsilon)t}}{\Delta E_j - i\varepsilon} \Phi_d(\vec{k}_j + \vec{q}_j) \tag{11}$$

with notations as defined in [17].

Using this formalism we get the following interference term in the transition rate as function of the time t:

$$\sum_{l>j} 2\mathrm{Re}\left[\sum_{\beta\alpha} U_{\beta l}^* U_{el}^* e^{i(\phi_\beta - \phi_\alpha)} U_{\alpha l} U_{ej} e^{i\omega_{lj}t}\right] = \frac{1}{\sqrt{2}} \sin 2\theta_{12} (\sin\phi_{\mu e} - \sin\phi_{\tau e}) \sin(\omega_{21}t) \tag{12}$$

which is modulated as $\sin(\omega_{21}t)$ with a reduced amplitude

$$a_{EC} = \frac{1}{\sqrt{2}} \sin 2\theta_{12} (\sin\phi_{\mu e} - \sin\phi_{\tau e}) \tag{13}$$

determined by the mixing angle θ_{12} (For simplicity it was assumed that $\theta_{13} = 0$ and $\theta_{23} = \pi/4$), and the difference of $\sin \phi_{\mu e}$ and $\sin \phi_{\tau e}$. Note that the inference term vanishes for $\phi_{\mu e} = \phi_{\tau e}$ as well as for the special case that both phase differences are zero. The origin of these CP violating phase shifts are so far unknown. They may contain also charged lepton-nucleus Coulomb phase shifts.

Since the GSI experiment did observe modulations with an amplitude $a = 0.20(2)$, Kleinert and Kienle [22] concluded recently that for the partial restoration of the interference term the unitarity of the flavor mixing matrix must be violated. From the observed modulation amplitude a violation of about 10% is derived [22]. A possible origin for such a unitarity violation could be that there are more than three families of leptons in nature and that universality of weak interaction is not valid for all of them, such as for the proposed sterile neutrinos.

More data will be needed to find out which mechanism causes the observed modulation, such as direct comparison of the modulation of bound β decay and EC decay of ^{108}Ag as proposed by us in this context [23]. In such a case the modulation produced by antineutrinos and neutrinos can be compared directly, especially the effect of the Coulomb field which has opposite sign for neutrinos and antineutrinos could be studied.

Acknowledgments

I acknowledge very helpful clarifying discussions with many colleagues, especially with F. Bosch, A. Gal, A. Ivanov, H. Kleinert, Y. Litvinov, and K. Yazaki. This research was partly supported by the DFG cluster of excellence "Origin and Structure of the Universe" of the Technische Universität München.

References

[1] Kienle P and Gross K D 1991 *Naturwissenschaften* **78** 385; Kienle P 2001 *Naturwissenschaften* **88** 313
[2] Litvinov Y N *et al.* 2008 *Phys. Lett.* B **664** 162
[3] Winckler N *et al.* 2008, 2009 *GSI Annual Report*
[4] Kienle P 2009 *Nucl. Phys.* A **827** 510c, and *J. Phys.: Conf. Ser.* **171** 012065
[5] Litvinov Y A *et al.* 2007 *Phys. Rev. Lett.* **99** 261501
[6] Ivanov A N,Faber M,Reda R, and Kienle P 2008 *Phys. Rev.* C **78** 025503
[7] Ivanov A N,Kryshen E L,Pitschmann M, and Kienle P 2008 *Phys. Rev. Lett.* **101** 182501
[8] Giunti C 2008 *Phys. Lett.* B **665** 92
[9] Kienert H *et al. J. Phys: Conf. Ser.* **136** 022049 (*Preprint* hep-ph/0808.2389)
[10] Lipkin H 2009 (*Preprint* hep-ph/0805.0435) and (*Preprint* nucl-th/0908.2039)
[11] Ivanov A N,Reda R, and Kienle P 2008 (*Preprint* nucl-th/0801.2121v5)
[12] Kleinert H and Kienle P 2009 (*Preprint* nucl-th/0803.2938)
[13] Cohen A G,Glashow S L, and Ligeti Z 2009 *Phys. Lett. B* **678** 191 (*Preprint* hep-ph/0810.4602)
[14] Peshkin M 2008 (*Preprint* hep-ph/0804.4891 and hep-ph/0803.0935)
[15] Burkhardt H *et al.* 2008 (*Preprint* hep-ph/0804.1099)
[16] Ivanov A N and Kienle P 2009 *Phys. Rev. Lett.* **103** 062502
[17] Ivanov A N and Kienle P 2009 (*Preprint* nucl-th/0909.1287)
[18] Gal A 2010 (*Preprint* nucl-th/0809.1213)
[19] Yazaki K 2009 private communication
[20] Ivanov A N,Kryshen E L,Pitschmann M, and Kienle P 2009 (*Preprint* nucl-th/0804.1311)
[21] Abe S *et al.* 2008 *Phys. Rev. Lett.* **100** 221803
[22] Kleinert H and Kienle P 2009 (*Preprint* nucl-th/0803.2938v4)
[23] Faber M *et al.* to be published

Doorway states, the super-radiant mechanism and nuclear reactions

2011 J. Phys.: Conf. Ser. 267 012057

(http://iopscience.iop.org/1742-6596/267/1/012057)

Doorway states, the super-radiant mechanism and nuclear reactions

N. Auerbach

School of Physics and Astronomy,

Tel Aviv University, Tel Aviv 69978, Israel

E-mail: auerbach@post.tau.ac.il

Abstract. In the 1950s the possibility of forming a ''super-radiant'' (SR) state in a gas of atoms confined to a volume of a size smaller than the wave length of radiation was suggested by Dicke. The atoms, with two levels, are coupled through their common radiation field. This indirect coupling leads to a redistribution of lifetimes among unstable intrinsic states. A strongly decaying SR state is created at the expense of the rest of the states of the system. Recently the connection of this mechanism to the notion of doorway states in low-energy nuclear reactions, was pointed out. The conditions for appearance of such doorways in nuclear physics processes are discussed.

1. Introduction

In the early 1950s the possibility of forming a ''super-radiant'' (SR) state in a gas of atoms confined to a volume of a size smaller than the wave length of radiation was suggested by Dicke [1]. In the absence of a direct interaction, the atoms are coupled through their common radiation field. This indirect interaction through the continuum leads to a redistribution of lifetimes among unstable intrinsic states. A rapidly decaying SR state is created at the expense of the rest of the states of the system that are ''robbed'' of their decay probability and become narrow. This mechanism has a general origin and analogous phenomena should appear in many quantum systems when quasi-bound states are strongly coupled through common decay channels. The SR approach has been used in many different fields. For example, it was applied in chemistry [2], atomic physics [3], condensed matter physics [4, 5], intermediate energy nuclear physics [6, 7], in particle physics [8], in the theory of nuclear reactions [5, 9, 10]. In the next section we will develop the SR formalism for some simple situations in order to arrive in a heuristic fashion to the idea of a super-radiant state and its connection to the concept of a doorway.

2. The effective complex Hamiltonian.

Following the projection formalism [6,11], let us divide the Hilbert space of nuclear states into two parts, the $\{Q\}$-subspace involving complicated many-body states $|q\rangle$, and the subspace $\{P\}$ of open channels $|c\rangle$. We use the notations Q and P for the corresponding projection operators onto the above subspaces. The wave function of the total system,

$$|\Psi\rangle = Q|\Psi\rangle + P|\Psi\rangle \tag{1}$$

satisfies the Schrödinger equation

$$H|\Psi\rangle = E|\Psi\rangle \tag{2}$$

that can be written as a set of coupled equations,

$$(E - H_{QQ})Q|\Psi\rangle = H_{QP}\, P|\Psi\rangle \tag{3}$$

and

$$(E - H_{PP})P|\Psi\rangle = H_{PQ}Q|\Psi\rangle \tag{4}$$

where we use the notations $H_{AB} = AHB$. Eliminating the part $P|\Psi\rangle$, we obtain an equation in the Q-space,

$$(E - \tilde{H}_{QQ})Q|\Psi\rangle = 0 \tag{5}$$

with the effective Hamiltonian

$$\tilde{H}_{QQ} = H_{QQ} + H_{QP}\frac{1}{E^{(+)} - H_{PP}}H_{PQ} \tag{6}$$

Here $E^{(+)} \equiv E + i0$ contains the infinitesimal imaginary term $+i0$ ensuring correct asymptotic conditions for the continuum wave functions. The second term of the effective Hamiltonian (6) contains a real and imaginary part of the propagator

$$G^{(+)}(E) = \frac{1}{E^{(+)} - H_{PP}} \tag{7}$$

resulting from the principal value and the delta-function $\delta(E - H_{PP})$ (on-shell contributions from channels c open at energy E), respectively. The imaginary part of the effective Hamiltonian is $-(i/2)W$, with

$$W = 2\pi\sum_c H_{QP}|c\rangle\langle c|H_{PQ} \tag{8}$$

Thus, the effective Hamiltonian (7) in Q-space is anti-Hermitian,

$$\tilde{H}_{QQ} = \overline{H}_{QQ} - \frac{i}{2}W \tag{9}$$

where $\overline{H}_{QQ}$ is a symmetric and real matrix that includes, apart from the original Hamiltonian in the Q-space- H_{QQ}, the principal value contribution of the PQ-coupling. The second part is anti-Hermitian. The eigenvalues of $\tilde{H}$, $\tilde{E} = E - (i/2)\Gamma$ are complex poles of the scattering matrix corresponding to the resonances in the cross sections.

To demonstrate in a simple way the role of the anti-Hermitian term we assume that only one channel is open. Then the matrix W, Eq. (8), has a completely separable form,

$$\langle q|W|q'\rangle = 2\pi A_q^c A_{q'}^{c*} \tag{10}$$

where:

$$A_q^c = \langle q|H_{QP}|c\rangle \tag{11}$$

The rank of the matrix W is 1, so that all the eigenvalues of this matrix are zero except one that has the value equal to the trace of the matrix:

$$\Gamma_s = \sum_q \langle q|W|q\rangle = 2\pi\sum_q |A_q^c|^2 \equiv \sum_q \Gamma_q^{\uparrow} \tag{12}$$

with $\Gamma_q^{\uparrow}$ denoting the escape width of the individual levels before the W-matrix is diagonalized. The special unstable state with width Γ_s is often referred to as the *super-*

radiant (SR), in analogy to the Dicke coherent state [1] of a set of two-level atoms coupled through the common radiation field. Here the coherence is generated by the *common decay channel.* The stable states are trapped and decoupled from the continuum.

In the more general case of N intrinsic states and N_c open channels with $N_c << N$ the super-radiant mechanism survives if the mean level spacing D of internal states and their decay widths $\Gamma_q^{\uparrow}$ obey:

$$\kappa^c = \frac{\Gamma_q^{\uparrow}}{D} > 1 \tag{13}$$

In this case we have N_c broad states, while the rest of states $N - N_c$ become very narrow.

3. Doorways

It is often the case that only a subset of intrinsic states $\{Q\}$ connects directly to the $\{P\}\}$ space of channels. The rest of states in $\{Q\}$ will connect to $\{P\}$ only when they obtain admixtures of these selected states of the first type. The special states directly coupled to continuum are the *doorways*, $|d\rangle$. They form the doorway subspace $\{D\}$ within $\{Q\}$, and the corresponding projection operator will be denoted here as *D*. The remaining states in {*Q*} will be denoted as $|\tilde{q}\rangle$ and the subspace as $\{\tilde{Q}\}$.

The full Hamiltonian can be decomposed the following way:

$$\begin{aligned} H = & \left(H_{\tilde{Q}\tilde{Q}} + H_{DD} + H_{\tilde{Q}D} + H_{D\tilde{Q}}\right) \\ & + \left(H_{PP} + H_{DP} + H_{PD}\right) \end{aligned} \tag{14}$$

Note that the terms $H_{P\tilde{Q}}$ and $H_{\tilde{Q}P}$ are missing because in accordance with the doorway hypothesis they are very small. Also note that diagonalizing the operator in the upper line of (14) would give back the states $|q\rangle$ with the components of $|d\rangle$ mixed with $|\tilde{q}\rangle$ states. The two last items in the above equation couple the doorway states, and therefore all the $|q\rangle$ states to the open channels.

3.1 *The case of a single doorway*

We start our discussion with the case when there is only one important doorway state $|d\rangle$. The matrix elements of the effective operator W in the intrinsic space are now given by:

$$\langle q|W|q'\rangle = 2\pi \sum_{c=1}^{N_c} \langle q|H_{DP}|c\rangle\langle c|H_{PD}|q'\rangle \tag{15}$$

with the doorway assumption:

$$\langle q|H_{DP}|c\rangle = \langle q|d\rangle\langle d|H_{DP}|c\rangle \tag{16}$$

where $\langle q|d\rangle$ is the amplitude of the admixture of the doorway into the $|q\rangle$ state. Eq. (16) becomes:

$$\langle q|W|q'\rangle = 2\pi \sum_{c=1}^{N_c} \langle q|H_{DP}|c\rangle\langle c|H_{PD}|q'\rangle \tag{17}$$

with the doorway assumption:

$$\langle q|H_{DP}|c\rangle = \langle q|d\rangle\langle d|H_{DP}|c\rangle \tag{18}$$

where $\langle q|d\rangle$ is the amplitude of the admixture of the doorway into the $|q\rangle$ state. Eq. (16) becomes:

$$\langle q|W|q'\rangle = 2\pi\langle q|d\rangle\langle d|q'\rangle\sum\left|\langle d|H_{DP}|c\rangle\right|^2 \tag{19}$$

again we have separable matrix elements of the matrix W, this time irrespective of the number of open channels $-N_c$.

As discussed above, the criterion of validity of the SR mechanism is that the average spacing between the levels in $\{Q\}$ is smaller than the decay width of such a state "before" the SR mechanism takes effect. This can be expressed, in the case of doorways, the following way. Consider the *spreading width*, $\Gamma_d^{\downarrow}$, of the doorway state representing the fragmentation of $|d\rangle$ into compound states $|\tilde{q}\rangle$. If N_q is the number of compound states in the interval covered by the spreading width, their average energy spacing is:

$$\bar{D} \approx \frac{\Gamma_d^{\downarrow}}{N_q} \tag{20}$$

Before the SR mechanism is turned on, the average decay width of a typical $|q\rangle$ state is:

$$\Gamma_q^{\uparrow} = 2\pi\sum_c\left|\langle q|H_{QP}|c\rangle\right|^2 \tag{21}$$

that can be estimated as:

$$\Gamma_q^{\uparrow} = \frac{\Gamma_s}{N_q} \tag{22}$$

Thus:

$$\frac{\Gamma_q^{\uparrow}}{D_q} \approx \frac{\Gamma_s}{\Gamma_d^{\downarrow}} \approx \frac{\Gamma_d^{\uparrow}}{\Gamma_d^{\downarrow}} \tag{23}$$

We conclude that the requirement for the SR doorway mechanism to be valid can be formulated as:

$$\frac{\Gamma_d^{\uparrow}}{\Gamma_d^{\uparrow}} > 1 \tag{24}$$

4. Examples

4.1 Isobaric Analog States

The isobaric analog state (IAS), $|A\rangle$, is obtained from the parent state with certain isospin T [12] by acting with the isospin lowering o T_- that changes a neutron into a proton:

$$|A\rangle = const\cdot T_-|\pi\rangle \tag{25}$$

In a compound nucleus, the IAS is surrounded by many compound states $|q\rangle$ of lower isospin $T_< = T-1$. The Coulomb interaction violates the isospin symmetry fragmenting the strength of the IAS over many states $|q\rangle$ giving rise to the spreading width $\Gamma_A^{\downarrow}$ of the IAS. If located above thresholds, the IAS can also decay into several continuum channels that gives rise to

the decay width $\Gamma_A^{\uparrow}$. In medium and heavy mass nuclei the condition the in Eq. (24) is satisfied. For example in ^{208}Pb the spreading width is about 80 keV while the escape width is about 160 keV, [12], thus $\Gamma_A^{\uparrow} / \Gamma_A^{\downarrow} \approx 2$.
The SR mechanism is therefore relevant to this case providing a straightforward explanation why the IAS appears as a single resonance with the decay width given by that of $|A\rangle$:

$$\Gamma_A^{\uparrow} = 2\pi \left| \langle A | H_{QP} | P \rangle \right|^2 \tag{26}$$

4.2 Giant resonances

Giant resonances in nuclei (or atomic clusters) present another example of similar physics [13, 14]. Usually, the giant resonances are discussed in terms of p–h configurations with identical spin-parity quantum numbers. The coherent residual interactions form a correlated state that carries much of the transition strength of a corresponding multipole operator. The giant resonances are mostly located in the particle continuum decaying via particle emission to the ground and excited states in the daughter nucleus.
Since the p–h giant resonance $|G\rangle$ is surrounded usually by a dense spectrum of 2p–2h and more complex configurations, the residual strong interaction will mix $|G\rangle$ with this background. Each of the resulting states denoted as $|b\rangle$ will contain the admixture $\langle G|b\rangle$ of the giant resonance. The mixed states $|b\rangle$ couple to the continuum. If we assume that the dominant coupling is that through the admixture of the giant state, then $|G\rangle$ serves as a doorway, and the matrix W is separable, given by:

$$\langle b|W|b'\rangle = \langle b|G\rangle\langle G|b'\rangle \sum_{c=1}^{N_c} \langle G|V|c\rangle\langle c|V|G\rangle \tag{27}$$

The matrix W is again of rank one and the non-zero eigenvalue is:

$$\Gamma_G^{\uparrow} = 2\pi \sum_c \left| \langle G|V|c\rangle \right|^2 \tag{28}$$

This holds under the assumption that the energy spread of the background states $|b\rangle$ is small compared to their decay width:

$$\Gamma_b^{\uparrow} = 2\pi \sum_c \left| \langle b|V|c\rangle \right|^2 \tag{29}$$

As before, this condition can be expressed in this case as:

$$\Gamma_G^{\uparrow} / \Gamma_G^{\downarrow} > 1 \tag{30}$$

the spreading width of the giant resonance is smaller than its total decay width. When this condition is not satisfied, so that the energy intervals between the 2p–2h states are larger than their decay widths $\Gamma_b^{\uparrow}$, the situation of a single decay peak for the giant resonance might not hold. Still one could expect some bunching of background states into groups and have spacing within the group smaller than their decay widths $\Gamma_b^{\uparrow}$. Each such group then can be treated separately using the SR mechanism and appear as a single peak in the decay (or excitation) curve. Then one will observe intermediate structure resonances in the GR energy domain.
I some continuum RPA calculations it was found that high lying giant resonances have substantial decay widths [14] larger than the spreading widths. For example the isovector monopole resonances calculated in heavy nuclei had decay widths of the order of 8-10 MeV, which is larger than the spreading width which is of the order of several MeV. In these cases the condition Eq. [29] for super-radiance is satisfied.

5. **Conclusions**

We have considered few examples of well known nuclear physics phenomena and described the formation of doorways in the framework of the SR mechanism in these phenomena. Because of the generic nature of this mechanism it would be interesting to extend this approach to other mesoscopic systems. In the physics of cold atoms [15] in a trap, the coupling of the internal states to the continuum via the Feshbach resonance influences profoundly the system affecting not only the imaginary part of the effective Hamiltonian but also the real part. This enables one to manipulate the properties of the system by changing the sign and/or the strength of the two-body interaction. The Feshbach resonance in this case serves as a doorway. The traditional theory of intermediate structure [16, 17] considers usually statistical sequences in time of multi-step particle interactions with possible pre-compound (or pre-equilibrium) decay to the continuum. The present formulation emphasizes complementary aspects of underlying physics, namely those that follow from the strict quantum-mechanical description of complicated many-body dynamics generated by the mean field and residual interactions. The approach based on the effective non-Hermitian Hamiltonian predicts the existence of different scales in the energy dependence of observed cross sections. These scales are defined by the intrinsic dynamics that combines collective (coherent) and chaotic (incoherent) features. The new collectivity responsible for the emergence of the doorways, under certain physical conditions, is due to the factorized nature of corresponding terms in the effective Hamiltonian. This separability is related to the properties of unitarity of the scattering matrix. In this way we obtain a supplementary description of many-body dynamics.

We should emphasize that the SR approach with respect to doorways is not in any way contradicting the more conventional approach to the notion of doorways [11, 12]. It presents a different view of the same physical phenomenon, stressing the collective nature of the decay width. The SR formalism in the present application is analogous to the calculation of the collective states in nuclear structure. However in the SR mechanism one deals with the imaginary part of the effective Hamiltonian and not with the real part. This analogy can bridge nuclear structure and nuclear reactions, especially when dealing with weakly bound exotic nuclei. In certain situations the SR mechanism may provide an understanding of the appearance of unexpectedly narrow resonances. As a consequence of the creation of a SR resonance the widths of the other resonances coupled to the same channel, become narrow.

This work was done in collaboration with V. Zelevinsky.

References

[1] Dicke R H 1954 *Phys. Rev.* **93** 99.
[2] Pavlov-Verevkin V B 1988 *Phys. Lett.* A **129** 168.
[3] Flambaum V V, Gribakina A A, Gribakin G F 1996 *Phys. Rev.* A **54** 2066.
[4] Pichugin K, Schanz H, Šeba P, 2001 *Phys. Rev.* E **64** 056227.
[5] Volya A, Zelevinsky V, in: Zelevinsky V (Ed.), *Nuclei and Mesoscopic Physics,* in: *AIP Conference Proceedings*, 2005 vol. 777, p. 229.
[6] Auerbach N, 1994 *Phys. Rev. C* **50** 1606.
[7] Auerbach N, Zelevinsky V, 2002 *Phys. Rev.* C **65** 034601.
[8] Auerbach N, Zelevinsky V, Volya A, 2004 *Phys. Lett.* B **590** 45.
[9]. Sokolov V V, Zelevinsky V, 1989 *Nucl. Phys.* A **504** 562.
[10] Rotter I, 1991 *Rep. Prog. Phys.* **54** 635.
[11] Feshbach H, 1958 *Ann. Phys.* **5** 357; Feshbach H, 1962 *Ann. Phys.* **19** 287.
[12] Auerbach N, Hüfner J, Kerman A K, Shakin C M, 1972 *Rev. Mod. Phys.* **44** 48.
[13] Bortignon P F, Bracco A, Broglia R A, 1998 *Giant Resonances: Nuclear Structure at Finite Temperature,* Harwood Academic, Amsterdam.
[14] Auerbach N and Klein A , 1983 *Nucl. Phys.* A **375** 77.
[15] Smith H, Pethick C J, 2001 *Bose–Einstein Condensation in Dilute Gases,* Cambridge Univ. Press, Cambridge.
[16] Wood R M, Borchers R R, Barschall H H, 1965 *Nucl. Phys.* **71** 529.
[17] Feshbach H, Kerman A K, Koonin S, 1980 *Ann. Phys. (N.Y.)* **125** 429

Novel nuclear structure aspects of the O-decay

2011 J. Phys.: Conf. Ser. 267 012058

(http://iopscience.iop.org/1742-6596/267/1/012058)

Novel nuclear structure aspects of the $0\nu\beta\beta$-decay

J. Menéndez[*,a,b], **A. Poves**[*], **E. Caurier**[†] **and F. Nowacki**[†]

[*]Departamento de Física Teórica, and IFT, UAM-CSIC, Universidad Autónoma de Madrid, 28049-Madrid, Spain
[a]Institut für Kernphysik, Technische Universität Darmstadt, 64289 Darmstadt, Germany
[b]ExtreMe Matter Institute EMMI, GSI Helmholtzzentrum für Schwerionenforschung GmbH, 64291 Darmstadt, Germany
[†]IPHC, IN2P3-CNRS/Université Louis Pasteur, 67037-Strasbourg France.

E-mail: `alfredo.poves@uam.es`

Abstract. We explore the influence of the deformation on the nuclear matrix elements of the neutrinoless double beta decay (NME), concluding that the difference in deformation -or more generally in the amount of quadrupole correlations- between parent and grand daughter nuclei quenches strongly the decay. We correlate these differences with the seniority structure of the nuclear wave functions. In this context, we examine the present discrepancies between the NME's obtained in the framework of the Interacting Shell Model and the Quasiparticle RPA. In our view, part of the discrepancy can be due to the limitations of the spherical QRPA in treating nuclei which have strong quadrupole correlations. We surmise that the NME's in a basis of generalized seniority are approximately model independent, *i. e.* they are "universal".

1. Introduction

The double beta decay is a rare weak process which takes place between two even-even isobars when the single beta decay is energetically forbidden or hindered by large spin difference. The two neutrino beta decay is a second order weak process —the reason of its low rate—, and has been measured in a few nuclei. The $0\nu\beta\beta$ decay is analog but requires neutrinos to be Majorana fermions. With the exception of one unconfirmed claim [1], it has never been observed, and currently there is a number of experiments either taking place or expected for the near future —see e.g. ref. [2]— devoted to detect this process and to set up firmly the nature of neutrinos. Furthermore, the $0\nu\beta\beta$ decay is also sensitive to the absolute scale of the neutrino mass, and hence to the mass hierarchy. Since the half-life of the decay is determined, together with the masses, by the nuclear matrix element for the process, its knowledge is essential to predict the most favorable decays and, once detection is achieved, to settle the neutrino mass scale and hierarchy.

Two different methods were traditionally used to calculate the NME's for $0\nu\beta\beta$ decays, the quasiparticle random-phase approximation and the shell model in large valence spaces (ISM). The QRPA has produced results for most of the possible emitters since long [3, 4, 5]. The ISM, that was limited to a few cases till recently [6], can nowadays describe (or will do it shortly) all the experimentally relevant decays but one, the decay of ^{150}Nd. Other approaches, that share a common prescription for the transition operator (including higher order corrections), and for the treatment of the short range correlations (SRC) and the finite size effects, are the Interacting Boson Model [7], and the Projected Hartree Fock Bogolyuvov method [8].

The expression for the half-life of the $0\nu\beta\beta$ decay can be written as [9]:

$$\left(T_{1/2}^{0\nu\beta\beta}\left(0^{+}\rightarrow 0^{+}\right)\right)^{-1}=G_{01}\left|M^{0\nu\beta\beta}\right|^{2}\left(\frac{\langle m_{\beta\beta}\rangle}{m_e}\right)^{2}, \quad (1)$$

where $\langle m_{\beta\beta}\rangle = \left|\sum_k U_{ek}^2 m_k\right|$ is the effective neutrino mass, a combination of the neutrino mass eigenvalues m_k. U is the neutrino mixing matrix and G_{01} is a kinematic factor dependent on the charge, mass and available energy of the process. $M^{0\nu\beta\beta}$ is the nuclear matrix element of the neutrinoless double beta decay operator, which has Fermi, Gamow-Teller and Tensor components. The kinematic factor G_{01} depends on the value of the coupling constant g_A. In addition, some calculations use different values of r_0 in the formula $R=r_0\ A^{1/3}$. It is therefore convenient to define:

$$M'^{\,0\nu\beta\beta}=\left(\frac{g_A}{1.25}\right)^{2}\left(\frac{1.2}{r_0}\right)M^{0\nu\beta\beta} \quad (2)$$

In this way the theoretical $M'^{\,0\nu\beta\beta}$'s are directly comparable among them irrespective of the values of g_A and r_0 employed in their calculation, since they share a common G_{01} factor —the one computed with $g_A = 1.25$ and r_0=1.2 fm. Thus, the translation of the $M'^{\,0\nu\beta\beta}$'s into half-lives is transparent.

2. Pairing and Quadrupole; The Influence of Deformation

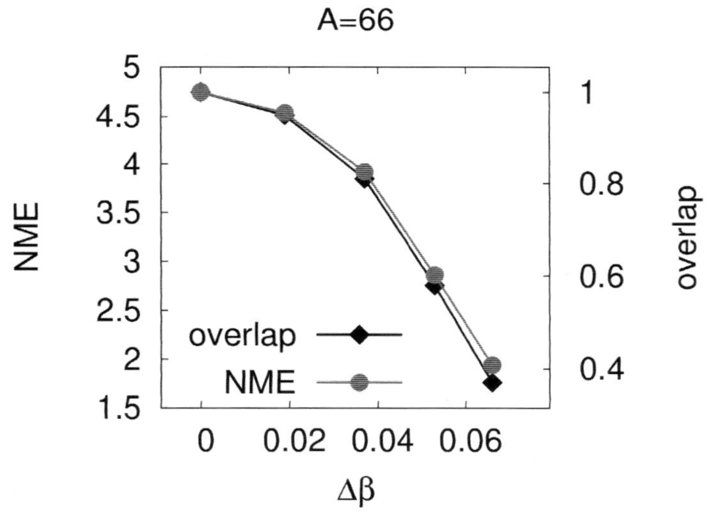

Figure 1. ^{66}Ge → ^{66}Se NME, $M'^{0\nu}$, as a function of the difference in deformation induced by the extra quadrupole interaction added to ^{66}Se.

An important issue regarding the $0\nu\beta\beta$ decay is the role of the correlations; pairing that drives the nucleus toward a superfluid state and quadrupole that favors deformed intrinsic shapes. It has been show recently that the $2\nu\beta\beta$ is hindered by the difference in deformation between the initial and final nuclei [10, 11]. For the neutrinoless mode, the calculations [6] indicate that the pairing interaction favors the decay and that, consequently, the truncations in seniority, which quench the pair breaking action of the quadrupole correlations, produce an overestimation of the values of the NME's. On the other hand, the NME's are also reduced when the parent and grand-daughter nuclei have different deformations [12, 13].

We have chosen to study the (unphysical) transition between the mirror nuclei ^{66}Ge and ^{66}Se in order to have a clearer view of the effect of the deformation in the NME's. This transition has the peculiarity that the wave functions of the initial and final nuclei are identical (provided Coulomb effects are neglected) and consequently it is easier to disentangle the contributions of the $0\nu\beta\beta$ operator and the nuclear wave functions to the NME. The calculations are carried out in the valence space r3g with the effective interaction gcn28:50. The SRC are modeled by a Jastrow factor with the Spencer and Miller parametrization [14], although it has been shown recently that, once the finite size of the nucleon has been taken into account by a dipole form factor, softer options are more realistic [15, 16]

To increase the deformation of a given nucleus we add to the effective interaction a term $\lambda\ Q \cdot Q$. Fig. 1 shows the results when the final nucleus has been artificially deformed by adding an extra quadrupole-quadrupole term. Notice in the first place that for λ=0 both nuclei are deformed with $\beta \sim 0.2$. In spite of that, the NME is a factor of two larger than the values obtained for the A=76 and A=82 decays in the same valence space and with the same interaction. Hence, even if the two A=66 partners are deformed, the fact that their wave functions are identical enhances the decay. Nevertheless, the NME is still far from its expected value in the superfluid limit (NME~8). The figure shows that the reduction of the NME as the difference in deformation increases is very pronounced. For the values of λ between 0.0 and 0.2, the difference in deformation parameter between parent and grand daughter grows from zero to about 0.1. In addition, the NME follows closely the overlap between the wave function of one nucleus obtained with λ=0 and the wave function of the same nucleus obtained with $\lambda \neq 0$. This means that, if we write the final wave function as: $|\Psi\rangle = a|\Psi_0\rangle + b|\Psi_{qq}\rangle$, the $0\nu\beta\beta$ operator does not connect Ψ_0 and Ψ_{qq}. This behavior of the NME's with respect to the difference of deformation between parent and grand daughter is common to all the transitions between mirror nuclei that we have studied (A=50, A=110) and to more realistic cases like the A=82 decay that we have examined in detail in [17]. Therefore we can submit that this is a robust result.

3. The NME's and the seniority structure of the nuclear wave functions

We can also analyze the results of the preceding section in terms of the seniority structure of the wave functions of parent and grand daughter nuclei. Indeed when $\Delta\beta$=0 both ^{66}Ge and ^{66}Se have identical wave functions. The probabilities of the components of different seniority are given in table 1. It is seen that changing β from 0.22 (mildly deformed) to 0.30 (strongly deformed) increases drastically the amount of high seniority components in the wave function, provoking a seniority mismatch between the decaying and the final nuclei. This leads to very large cancelations of the nuclear matrix elements of the decay, as shown also in table 1.

Table 1. The seniority structure of the wave functions in the A=66 mirror decay

	$s = 0$	$s = 4$	$s = 6$	$s = 8$	$s = 10$
$\Delta\beta$=0	39	43	7	10	1
$\Delta\beta$=0.08	6	32	21	31	10
	$M^{0\nu}_F$	$M^{0\nu}_{GT}$	$M^{0\nu}_T$	$M'^{0\nu}$	
$\Delta\beta$=0	-2.02	3.95	0.08	5.16	
$\Delta\beta$=0.08	-0.76	1.65	0.02	2.12	

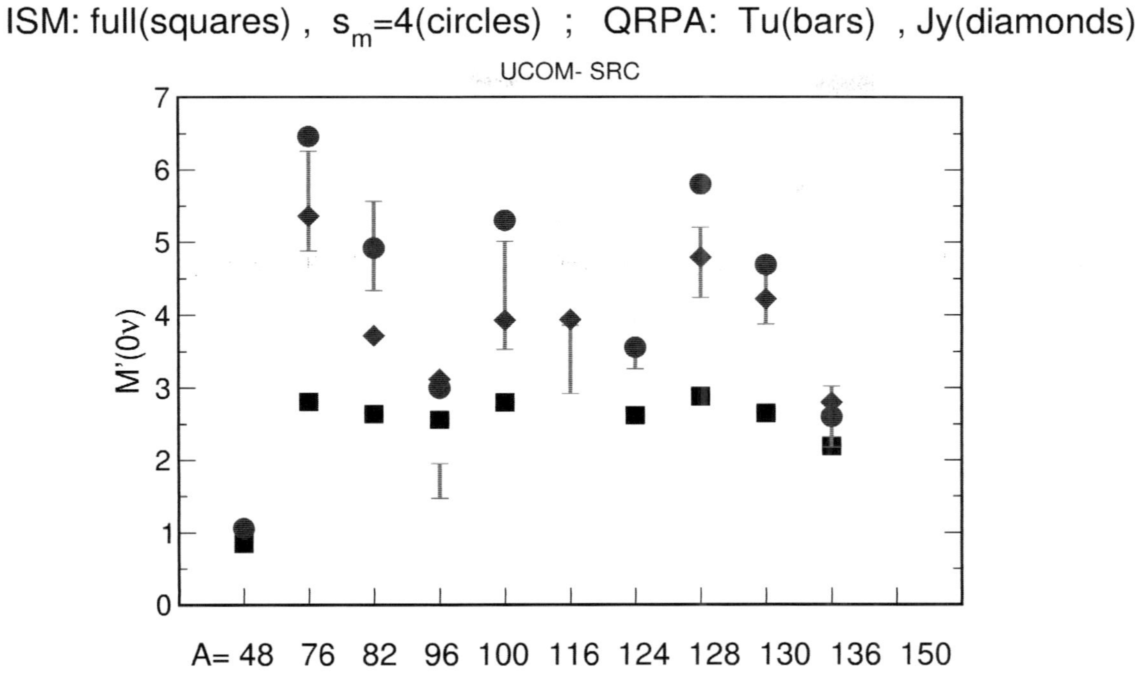

Figure 2. The neutrinoless double beta decay nuclear matrix elements $M'^{0\nu\beta\beta}$ for ISM and QRPA calculations treating the SRC with the UCOM approach. Tu, QRPA results from ref. [18] and Jy, QRPA results from refs. [3, 4]. The ISM results for A=96 and A=100 are preliminary

Coming back to the physically relevant decays, we compare in figure 2 the ISM and QRPA NME's. In both approaches, the SRC are taken into account in the UCOM framework [19] and g_A=1.25 is adopted. We have discussed elsewhere that the discrepancies between both approaches show the following trends: when the nuclei that participate in the decay have a low level of quadrupole correlations, as in the decays of ^{96}Zr, ^{124}Sn and ^{136}Xe, the calculations tend to agree. On the contrary, when the correlations are large, the QRPA in a spherical basis seems not to be able to capture them fully. As the effect of the correlations is to reduce the NME's, the QRPA produces NME's that are too large in ^{76}Ge, ^{82}Se, ^{100}Mo, ^{128}Te, and ^{130}Te. Indeed, when the ISM calculations are truncated to maximum seniority s_m=4, which is the leading order of the ground state correlations in the QRPA (corresponding to the two quasi-particle contribution), they follow closely the QRPA results, as can be seen also in figure 2. Notice that only when the ISM calculations are converged at this level of truncation the two approaches do produce similar NME's.

We compare in table 2 the seniority structure of the wave functions of the ISM and QRPA, in some of the cases for which the latter are available [20]. It is seen that the differences are important and share a common trend: in the QRPA, the seniority structure of parents and grand daughters is much more similar than in the ISM. According to what we have seen in the A=66 case, this is bound to produce larger NME's in the QRPA than in the ISM, as it is actually the case. To make this statement quantitative, we have developed the ISM matrix elements in a basis of generalized seniority

Table 2. The seniority structure of the wave functions in the ISM and QRPA

	$s=0$	$s=4$	$s=6$	$s=8$	$s=10$	$s=12$	$s=14$	$s=16$
					ISM			
^{48}Ca	97	3	-	-	-	-	-	-
^{48}Ti	59	36	4	1	-	-	-	-
^{76}Ge	43	41	7	8	1	-	-	-
^{76}Se	26	41	11	16	4	1	-	-
^{82}Se	50	39	10	1	-	-	-	-
^{82}Kr	44	41	6	8	1	-	-	-
^{128}Te	70	26	3	1	-	-	-	-
^{128}Xe	37	41	9	10	2	-	-	-
					QRPA			
^{76}Ge	55	33	-	10	-	2	-	-
^{76}Se	59	31	-	8	-	2	-	-
^{82}Se	56	32	-	9	-	2	-	-
^{82}Kr	54	34	-	11	-	2	-	-
^{128}Te	52	34	-	11	-	3	-	-
^{128}Xe	40	37	-	17	-	5	-	1

$$M_{F,GT,T} = \sum_{\alpha,\beta} A_{\nu_i(\alpha)} B_{\nu_f(\beta)} \langle \nu_f(\beta) | O_{F,GT,T} | \nu_i(\alpha) \rangle$$

where the A's and B's are the amplitudes of the different seniority components of the wave functions of the initial and final nuclei. Obviously, when we plug the ISM amplitudes in this formula, we recover the ISM NME's. But, what shall we obtain if we put the QRPA amplitudes instead? Indeed, we get approximately the QRPA NME's! (5.73 for A=76 and 4.15 for A=82). Therefore as we had anticipated, the seniority mismatch of the initial and final wave functions, which is severely underestimated in the QRPA calculations, explains most of the discrepancy between the two descriptions. In addition, this result strongly suggests that there is some kind of universal behavior in the NME's of the neutrinoless double beta decay when they are computed in a basis of generalized seniority. If this is so, the only relevant difference between the different theoretical approaches would reside in the seniority structure of the wave functions that they produce.

Table 3. The GT NME's of the A=48 decay in the generalized seniority basis

^{48}Ti	$s=0$	$s=4$	$s=6$	$s=8$
^{48}Ca $s=0$	3.95	-3.68	-	-
^{48}Ca $s=4$	0.00	-0.26	0.08	-0.02

A very spectacular example of the cancellation of the NME by the seniority mismatch is provided by the ^{48}Ca decay. In Table 2 we have included also the seniority structures of the two nuclei, and we see that they are very different. If we now examine the values of the matrix elements $\langle \nu_f(\beta) | O_{GT} | \nu_i(\alpha) \rangle$ we find the values listed in Table 3. There are two large matrix elements one diagonal and another off-diagonal of the same size and opposite sign. If the two nuclei were dominated by the seniority zero components one should obtain $M_{GT}\sim 4$. If ^{48}Ti were a bit more deformed, M_{GT} will be essentially zero. The value produced by the KB3 interaction is

0.75 that is more than a factor five reduction with respect to the seniority zero limit. Earlier work on double beta decays in a basis of generalized seniority (limited to s=0 and s=4 components) showing also this kind of cancellations can be found in ref. [21]

Acknowledgments

This work is partly supported by the Spanish Ministry of Ciencia e Innovación under grant FPA2009-13377, by the Comunidad de Madrid (Spain) project HEPHACOS S2009/ESP-1473, by the IN2P3(France)-CICyT(Spain) collaboration agreements, by the DFG through grant SFB 634 [J.M.], and by the Helmholtz Association through the Helmholtz Alliance Program, contract HA216/EMMI "Extremes of Density and Temperature: Cosmic Matter in the Laboratory" [J.M.].

References

[1] Klapdor-Kleingrothaus H V 2004 *Phys. Lett. B* **586** 198–212 (*Preprint* `hep-ph/0404088`)

[2] AvignoneIII F T, Elliott S R and Engel J 2008 *Rev. Mod. Phys.* **80** 481 arXiv:0708.1033 [nucl-ex]

[3] Kortelainen M and Suhonen J 2007 *Phys. Rev. C* **75** 51303 (*Preprint* `arXiv:0705.0469 [nucl-th]`)

[4] Kortelainen M and Suhonen J 2007 *Phys. Rev. C* **76** 24315 (*Preprint* `arXiv:0708.0115 [nucl-th]`)

[5] Rodin V A, Faessler A, Šimkovic F and Vogel P 2007 *Nucl. Phys. A* **793** 213 (*Preprint* `arXiv:0706.4304 [nucl-th]`)

[6] Caurier E, Menéndez J, Nowacki F and Poves A 2008 *Phys. Rev. Lett.* **100** 52503 (*Preprint* `0709.2137`)

[7] Barea J and Iachello F 2009 *Phys. Rev. C* **79** 044301

[8] Chandra R, Chaturvedi K, Rath P K, Raina P K and Hirsch J G 2009 *Eur. Phys. Lett. A* **86** 32001

[9] Doi M, Kotani T and Takasugi E 1985 *Prog. Theor. Phys. Suppl.* **83** 1

[10] Simkovic F, Pacearescu L and Faessler A 2004 *Nucl. Phys. A* **733** 321

[11] Álvarez Rodríguez R, Sarriguren P, Moya de Guerra E, Pacearescu L, Faessler A and Šimkovic F 2004 *Phys. Rev.* **C70** 64309 (*Preprint* `nucl-th/0411039`)

[12] Caurier E, Nowacki F and Poves A 2008 *Eur. Phys. J. A.* **36** 195–200 (*Preprint* `arXiv:0709.0277 [nucl-th]`)

[13] Simkovic F 2007 *Proceedings of the Workshop on the Calculation of Double Beta Decay Matrix Elements, MEDEX07* vol 942 ed Civitarese O, Stekl I and Suhonen J (AIP Conf. Proc.)

[14] Miller G A and Spencer J E 1976 *Ann. Phys.* **100** 562

[15] Engel J and Hagen G 2009 *Phys. Rev. C* **79** 064317 (*Preprint* `0904.1709`)

[16] Simkovic F, Faessler A, Muther H, Rodin V and Stauf M 2009 *Phys. Rev. C* **79** 055501 (*Preprint* `0902.0331`)

[17] Menéndez J, Poves A, Caurier E and Nowacki F 2009 *Proceedings of the International School of Physics Enrico Fermi, Measurements of Neutrino Mass* ed Ferroni F Vissani F and Brofferio C (IOS Press) arXiv:0809.2183

[18] Šimkovic F, Faessler A, Rodin V A, Vogel P and Engel J 2008 *Phys. Rev. C* **77** 45503 (*Preprint* `0710.2055`)

[19] Feldmeier H, Neff T, Roth R and Schnack J 1998 *Nucl. Phys. A* **632** 61–95 (*Preprint* `nucl-th/9709038`)

[20] Escuderos A, Faessler A, Rodin V and Simkovic F 2010 (*Preprint* `arXiv:1001.3519v2 [nucl-th]`)

[21] Engel J, Vogel P, Ji X and Pittel S 1989 *Phys. Lett. B* **225** 5

Nuclear structure, double beta decay and the neutrino mass

2011 J. Phys.: Conf. Ser. 267 012059

(http://iopscience.iop.org/1742-6596/267/1/012059)

Nuclear structure, double beta decay and the neutrino mass

Amand Faessler

Institute für Theoretische Physik der Universität Tübingen, D-72076 Tübingen, Germany

E-mail: amand.faessler@uni-tuebingen.de

Abstract. The Quasi-particle Random Phase Approximation (QRPA), the Shell Model (SM), the angular momentum projected Hartee-Fock-Bogoliubov approach (HFB) and the Interacting Boson Model (IBM)are compared with each other for the determination the neurinoless double beta decay. The different approaches are compared specifically concerning the the angular momenta and parities of the neutron pairs, which are changed into two protons by the $0\nu\beta\beta$-decay.

1. The different Many Body Approaches for the double beta decay

The main purpose of the neutrinoless double beta decay [1, 2, 3, 4] is to discriminate between Majorana and Dirac neutrinos and to obtain a value for the neutrino mass. To give an absolute scale of the mass the transition matrix element is as important as the measured half life.

First the different many body approaches: QRPA [5, 6, 7], SM [8, 10], HFB [19, 20, 14, 15] and the IBM [17] are characterized, which are used to calculate the neutrinoless double beta decay.

In the QRPA one starts from the transformation to Bogoliubov quasiparticles :

$$a_i^\dagger = u_i c_i^\dagger - v_i c_{\bar{i}}. \tag{1}$$

The creation $c_i^\dagger$ and annihilation operators of time reversed single nucleon states $c_{\bar{i}}$ are usually defined with respect to the oscillator wave functions [5]. The single particle energies are calculated with a Woods Saxon potential [5]. The single particle basis can include up to 23 nucleon levels (all single particle states without core up to the $i_{13/2}$ level) for the protons and also for the neutrons.

The excited states $|m\rangle$ with angular momentum J in the intermediate odd-odd mass nucleus are created from the correlated initial and final 0^+ ground states by a proton-neutron phonon creation operator:

$$|m\rangle = Q_m^\dagger|0^+\rangle; \quad \hat{H}Q_m^\dagger|0^+\rangle = E_m Q_m^\dagger|0^+\rangle. \tag{2}$$

$$Q_m^\dagger = \sum_\alpha [X_\alpha^m A_\alpha^\dagger - Y_\alpha^m A_\alpha], \tag{3}$$

which is defined as a linear superposition of creation operators of proton-neutron quasiparticle pairs:

$$A_\alpha^\dagger = [a_i^\dagger a_k^\dagger]_{JM}, \tag{4}$$

$$[A_\alpha, A_\beta^\dagger] = \delta_{\alpha,\beta} + \hat{X}, \tag{5}$$

The "scattering" terms $\hat{X}$ in (5) are put to zero in the QRPA and are included approximately in Renormalized QRPA (RQRPA), which for the first time was employed to neutrinoless double beta decay in the PhD thesis of Schwieger in Tübingen [21]. The RQRPA includes the Pauli principle for the Fermion pairs and reduces the ground state correlations. The many body Hamilton equation (2) yields the algebraic QRPA equation for the coefficients X_α^m and Y_α^m, and the excitation energies E_m.

The inverse $0\nu\beta\beta$ lifetime for the light Majorana neutrino exchange mechanism is given as the product of three factors,

$$\left(T_{1/2}^{0\nu}\right)^{-1} = G^{0\nu} \left|M^{0\nu}\right|^2 m_{\beta\beta}^2 \,, \tag{6}$$

where $G^{0\nu}$ is a calculable phase space factor, $M^{0\nu}$ is the $0\nu\beta\beta$ nuclear matrix element, and $m_{\beta\beta}$ is the (nucleus-independent) "effective Majorana neutrino mass" which, in standard notation [22], reads

$$m_{\beta\beta} = \left|\sum_{l=1}^{3} m_l e^{i\alpha_l} |U_{el}|^2\right| \,, \tag{7}$$

m_l, α_l and U_{el} being the neutrino masses, the two independent ($\alpha_2 - \alpha_1$ and $\alpha_3 - \alpha_1$) Majorana phases and the ν_e mixing matrix elements, respectively.

The expressions for the matrix elements $M^{0\nu}$ and the corresponding $0\nu\beta\beta$ transition operators are given, e.g., in Ref. [5]:

$$M^{(0\nu)} = M_{GT}^{0\nu} - (\frac{g_V}{g_A})^2 M_F^{0\nu} - M_T^{0\nu} \tag{8}$$

The indices indicate GT (Gamow Teller), F (Fermi) and T (Tensor)contributions. The nuclear radius is $R = r_0 A^{1/3}$, and the parameter r_0 is chosen to be $r_0 = 1.2$ fm. Thus, the matrix elements $M^{0\nu}$ are increased by about 10 % as compared with our earlier calculations, where we chose $r_0 = 1.1$ fm.

The SM approach has been applied [10] to neutrinoless double beta decay [8] using the closure relation with an averaged energy denominator. In this way one does not needs to calculate the intermediate states in the odd-odd intermediate nuclei. The quality of the results depends then on the description of the 0^+ ground states in the initial and final nuclei of the double beta decay system, e.g. $^{76}Ge \rightarrow ^{76}Se$. The $0\nu\beta\beta$ transition matrix element (8) simplifies as shown in Ref. [11] equations (5) to (11). Since the number of many body configurations is increasing drastically with the single nucleon basis, one is forced to restrict for mass numbers $A = 76$ and $A = 82$ in the SM to the single particle basis $1p_{3/2}, 0f_{5/2}, 1p_{1/2}$ and $0g_{9/2}$and for the mass region around $A = 130$ to $0g_{7/2}, 1d_{3/2}, 1d_{5/2}, 2s_{1/2}$ and $0h_{11/2}$. This is a very small particle basis compared to the maximal 23 single particle levels included in the QRPA. But the real problem is, that the spin-orbit partners $0f_{7/2}$ and $0g_{7/2}$ are missing [11]. The SM results then automatically violate the Ikeda sum rule [18] while the QRPA satisfies it exactly. The Ikeda sum rule is

$$S_- - S_+ = 3(N - Z), \tag{9}$$

$$S_- = \sum_\mu \langle 0_i^+ | [\sum_k^A (-)^\mu \sigma_{-\mu}(k) t_+(k)] [\sum_l^A \sigma_\mu(l) t_-(l)] | 0_i^+ \rangle, \tag{10}$$

and for S_+ the subscripts at the isospin rising and lowering operators are exchanged.

The ground state correlation in the QRPA are limited to the ring diagrams (see Fig. 1), and thus to seniority 0, 4, 8, 12, ... configurations (or in deformed nuclei to configurations with 0,

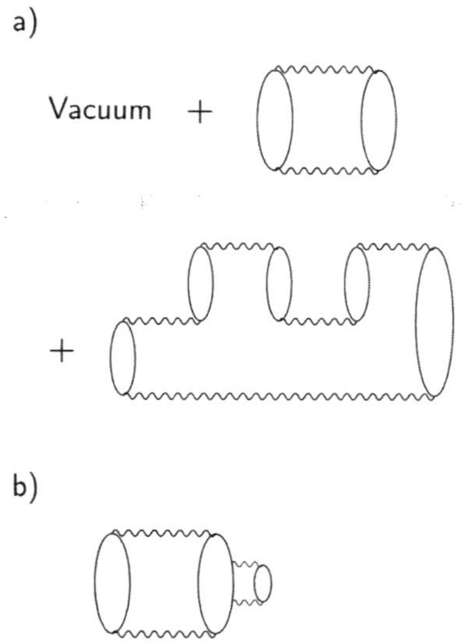

Figure 1. The ring diagrams of type a) are with the "vacuum" included in the ground state correlations in the QRPA. This means that the seniority or in deformed nuclei quasi-particle zero, four, eight, twelve, ... excitations are included in both the QRPA and the SM. But, e.g., diagram b) is not included in the QRPA but is considered in the SM, where seniority 4, 6, 8, 10, 12, ... are included in the ground state correlations [8, 10, 11]. But contributions of seniority six and ten admixtures to the neutrinoless double beta decay probability are suppressed according to the philosophy of the Random Phase Approach, that the ring diagrams are the leading ones (see figure 3).

4, 8, 12, ... quasi-particle excitations), while the shell model can include seniority 0, 4, 6, 8, 10, 12, ... states. This has been discussed in publications of the Strasbourg-Madrid group [8, 11]. But the results [8, 11] show also, that the philosophy of the RPA is approximately correct, that the ring diagrams give the important ground state correlations. The contributions of seniority 6 and 10 are suppressed.

Figure 2 shows the QRPA contributions of different angular momenta of the neutron pairs, which are changed in proton pairs with the same angular momenta. In figure 2 left bar is for ^{82}Se the result for the single nucleon basis $1p_{3/2}, 0f_{5/2}, 1p_{1/2}$ and $0g_{9/2}$ used in the SM (Shell Model). The bar in the middle includes also the two spin-orbit partners $0f_{7/2}$ and $0g_{9/2}$. For the right bar the basis is increased to 9 single nucleon levels for neutrons and also for protons $(0f_{7/2}, 1p_{3/2}, 0f_{5/2}, 1p_{1/2}, 0g_{9/2}, 1d_{5/2}, 2s_{1/2}, 1d_{3/2}, 0g_{7/2})$.

For all results in figure 2 the $2\nu\beta\beta$ is adjusted by the choice of the "renormalisation" factor g_{pp} of the particle-particle strength of the Bonn CD nucleon-nucleon interaction. The radius parameter is chosen to be 1.2 fm.

The QRPA is restricted to the ring diagrams for the ground state correlations, while the Shell Model (SM) takes into account all possible many body states in a given single nucleon basis. The drawback of the SM is the limited basis size. For the light double beta decay systems like ^{76}Ge and ^{82}Se the single particle basis is restricted to the four levels [8, 9, 10, 11] $1p_{3/2}, 0f_{5/2}, 1p_{1/2}, 0g_{9/2}$. Since the spin-orbit partners $0f_{7/2}, 0g_{7/2}$ are are missing the Ikeda Sum Rule (ISR) is violated and one obtains only about 50% (or 60% in ^{128}Te of the the ISR (see

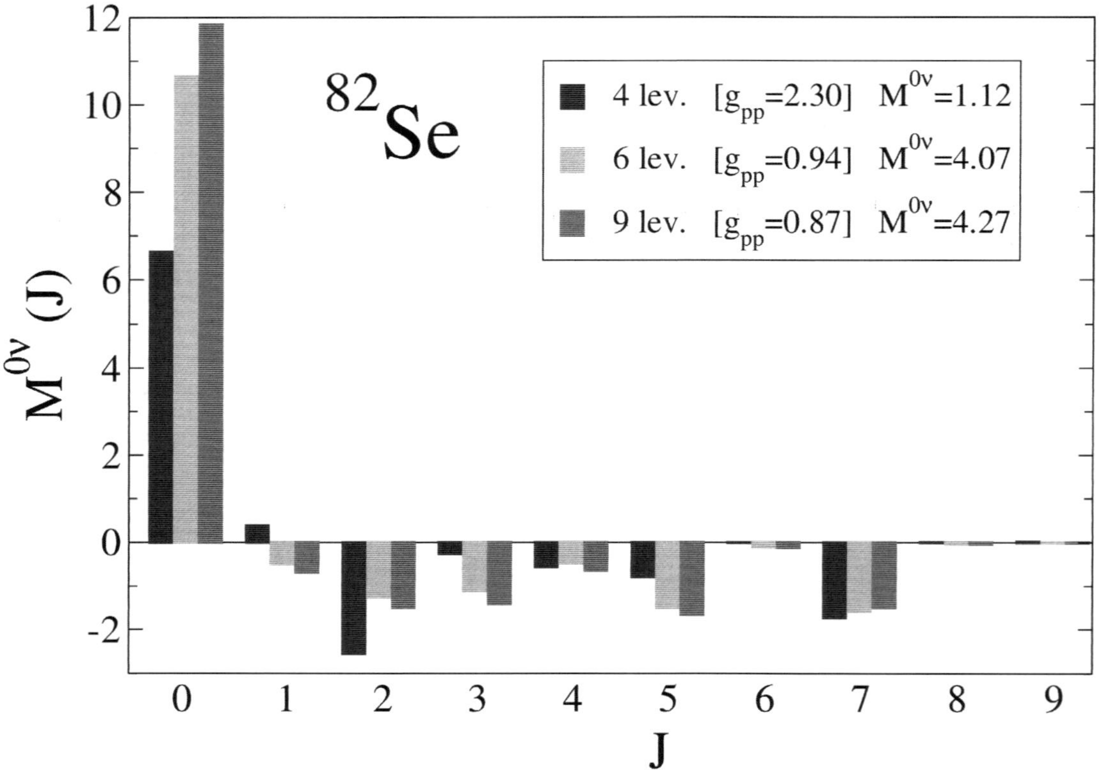

Figure 2. (Color online) Contributions of the transforming neutron pair with different angular momenta J^π to the total $M^{0\nu}$ calculated within the QRPA and different basis sizes for the $0\nu\beta\beta$ decay ^{82}Se$\rightarrow$^{82}Kr. The left bar is calculated with the same basis of four levels used in shell model calculations $1p_{3/2}, 0f_{5/2}, 1p_{1/2}$ and $0g_{9/2}$. The middle bar includes in addition the missing spin-orbit partners $0f_{7/2}$ and $0g_{7/2}$ amounting in total to 6 single-particle levels. This leads to the increase in the neutrinoless matrix element from 1.12 to 4.07. The right bar represents the QRPA result with 9 single particle levels ($1f_{7/2}, 2p_{3/2}, 1f_{5/2}, 2p_{1/2}, 1g_{9/2}, 2d_{5/2}, 3s_{1/2}, 2d_{3/2}, 1g_{7/2}$.). The matrix element gets only slightly increased to 4.27. The spin-orbit partners are essential to fulfill the Ikeda sum rule. In all three QRPA calculations the $2\nu\beta\beta$ is adjusted by the choice of the "renormalisation" factor g_{pp} of the particle-particle strength of the Bonn CD nucleon-nucleon interaction.

table 1). Including the spin-orbit partners into the basis $0f_{7/2}, 1p_{3/2}, 0f_{5/2}, 1p_{1/2}, 0g_{9/2}, 0g_{7/2}$ the ISR is fulfilled.

Table 1 shows also the contributions for seniority zero, for seniority different from zero and for the total matrix element for different sizes of the single nucleon basis for QRPA for $^{82}Se, ^{128}Te, ^{130}Te$ and for the large basis in ^{76}Ge and for the small basis with 4 or 5 levels for the Shell Model [8, 9, 10, 11] in the two nuclei $^{82}Se, ^{128}Te$. In all QRPA calculations the "renormalisation factor" g_{pp} of the particle-particle (pp) force multiplying the pp-nucleon-nucleon matrix elements of the G matrix of the CD Bonn interaction is adjusted to the $2\nu\beta\beta$ decay. In publication [12] the factor g_{pp} has been adjusted together with the axial charge g_A simultaneously to the $2\nu\beta\beta$ decay, to the single β^- decay in the second leg and to electron capture in the first leg for three double beta decay systems . The results for $0\nu\beta\beta$ change with this overconstraint fit less than the general theoretical uncertainties (see figure 3). Only the

1^+ Gamow-Teller state depends in QRPA but not in RQRPA sensitively on g_{pp}, since the 1^+ states in the intermediate nucleus are close to the "phase transition" to "static Gamow-Teller deformations". As a function of the other particle-particle multipole parts of the force the results change only very weakly. Since the neutrinoless double beta decay depends practically only on the natural parity multipoles $0^+, 1^-, 2^+, 3^-, 4^+, \ldots$ and not on 1^+, possible uncertainties of g_{pp}, which affect the 1^+ multipoles, do not affect appreciably the $0\nu\beta\beta$ results.

Table 1. Contribution to the transition matrix elements of the neutrinoless Double Beta Decay for different nuclei for different basis sizes, for QRPA and SM (Shell Model), for seniority zero, for seniority different from zero and for the total value. The last column gives the exhaustion of the Ikeda Sum Rule (ISR) in percent (see eqs. (9), (10)). The basis sets used are: 4 levels: $1p_{3/2}, 0f_{5/2}, 1p_{1/2}, 0g_{9/2}$; 5 levels: $0g_{7/2}, 1d_{5/2}, 2s_{1/2}, 1d_{3/2}, 0h_{11/2}$; 6 levels: $0f_{7/2}, 1p_{3/2}, 0f_{5/2}, 1p_{1/2}, 0g_{9/2}, 0g_{7/2}$; 7 levels: $0g_{7/2}, 1d_{5/2}, 2s_{1/2}, 1d_{3/2}, 0h_{11/2}, 0g_{9/2}, 0h_{9/2}$; 9 levels: $0f_{7/2}, 1p_{3/2}, 0f_{5/2}, 1p_{1/2}, 0g_{9/2}, 1d_{5/2}, 2s_{1/2}, 1d_{3/2}, 0g_{7/2}$;13 levels: All levels of $N = 3$ (p, f) and the $N = 4$ (s, d, g) shells and $0h_{11/2}, 0h_{9/2}, 1f_{7/2}$ and $1f_{5/2}$. Four levels (for: $^{76}Ge,^{82}Se$) and five levels (for: $^{100}Mo,^{128}Te,^{130}Te$) correspond to the Shell Model (SM) basis sets [8, 9, 10, 11]. Since the level four and five basis sets do not contain all spin orbit partners the Ikeda Sum Rule (ISR) is violated and only exhausted by 50 and 60 percents for these sets, respectively.

Nucleus	Basis Size	Method	seniority: $s = 0$	seniority: $s \neq 0$	total	Ikeda SR percents
^{76}Ge	9	QRPA	12.5	-7.5	5.0	100
^{82}Se	4	QRPA	6.6	-5.5	1.1	50
^{82}Se	4	SM	7.8	-5.8	2.0	
^{82}Se	6	QRPA	10.7	-6.6	4.1	100
^{82}Se	9	QRPA	11.8	-7.5	4.3	100
^{100}Mo	13	QRPA	16.3	-12.6	3.7	100
^{128}Te	5	QRPA	9.7	-8.3	1.4	60
^{128}Te	5	SM	10.6	-8.4	2.2	
^{128}Te	7	QRPA	13.7	-10.3	3.4	100
^{128}Te	13	QRPA	16.8	-13.0	3.8	100
^{130}Te	5	QRPA	8.8	-7.5	1.3	60
^{130}Te	7	QRPA	12.2	-9.0	3.2	100
^{130}Te	13	QRPA	14.9	-11.1	3.8	100

In the last ten years P. K. Rath and coworkers [14, 15, 16] have published a whole series of papers (see references in Ref. [15]) on the $2\nu\beta\beta$ decay and since 2008 also on the $0\nu\beta\beta$ decay, in which they used a simple pairing plus quadrupole many body Hamiltonian of the Kumar and Baranger type [23] to calculate the neutrinoless double beta decay transition matrix elements with angular momentum projection from a Hartree-Fock-Bogoliubov (HFB) wave function after variation. Schmid [20] did show, that with the assumption of a real Bogoliubov transformation (real coefficient A and B)

$$a_\alpha^\dagger = \sum_{i=1}^{M} A_{i\alpha} c_i^\dagger + B_{i\alpha} c_i \tag{11}$$

and no parity mixing only $0^+, 2^+, 4^+, \ldots.$ nucleon pairs and excited states are allowed. In this approach one can describe rotational bands and can change neutron pairs with angular momentum $0^+, 2^+, 4^+, \ldots$ into a corresponding proton pair with the same angular momentum and parity in neutrinoless double beta decay for ground to ground transition in even mass nuclei.

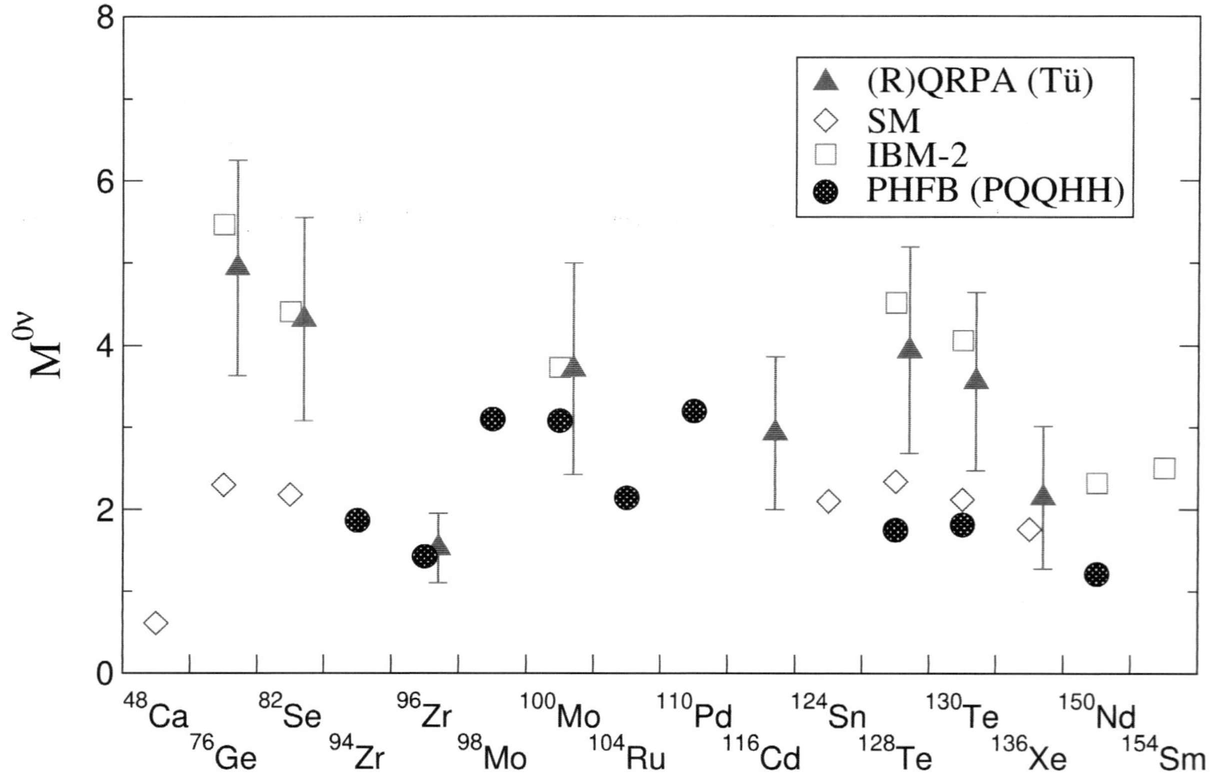

Figure 3. (Color online) Neutrinoless double beta decay transition matrix elements for the different approaches: QRPA [5, 6], the SM [8, 10, 11], the projected HFB method [16] and the IBM [17]. The error bars for the QRPA are calculated as the highest and the lowest values for three different single nucleon basis sets, two different axial charges $g_A = 1.25$ and the quenched value $g_A = 1.00$ and two different treatments of short range correlations (Jastrow-like [27] and the Unitary Correlator Operator Method (UCOM) [28]). The radius parameter is as in this whole work $r_0 = 1.2$ fm.

The projected HFB approach is restricted to contributions of neutron pairs with angular moments $0^+, 2^+, 4^+, \ldots$. The contributions of higher angular momentum neutron to proton pairs $2^+, 4^+, \ldots$ are reduced almost to zero compared to the QRPA and the SM. The reason for this is obvious: in a spherical nucleus the HFB solution contains only seniority zero and no stronger higher angular momentum pairs. The double beta decay system considered has only a small deformation and thus a projected HFB state is not able to describe an appreciable admixture of higher angular momentum pairs for $0^+ \to 0^+$ transitions as can be seen from the BCS wave function after the first Bloch-Messiah transformation [24]. The higher angular momentum contributions increase drastically with increasing intrinsic quadrupole and hexadecapole deformations of the HFB solution.

The IBM (Interacting Boson Model)calculations [17]use the Surface Delta Interaction (SDI) and can only change 0^+ (S) and 2^+ (D) fermionic pairs from two neutrons into two protons. In the bosonization to higher orders this leads to the creation and annihilation of up to three "s" and "d" boson annihilation and creation operators in Ref. [17]. But all these terms of equation (18) of reference [17] originate from the annihilation of a 0^+ (S) or a 2^+ (D) neutron pair into a corresponding proton pair with the same angular momentum. The higher boson terms try only to fulfill the Fermi commutation relations of the original nucleon pairs up to third order. The IBM can therefore change only a 0^+ or a 2^+ neutron pair into a corresponding proton pair.

2. Conclusions

The Shell Model (SM)[8, 10, 11] due to the restricted single particle basis has a severe handicap. The matrix elements in the ^{76}Ge region are by a factor 2 smaller than the Quasi-particle Random Phase (QRPA)[5, 6, 7] results. With the same restricted basis the QRPA obtains roughly the same results as the SM (see figure 4), but the Ikeda sum rule [18] gets strongly violated due to the spin-orbit partners missing in the SM single particle basis.

The SM can admix seniority $0, 4, 6, 8, 10, \ldots$ configurations to the ground state, while the QRPA is restricted in the ground state to the seniority or in deformed nuclei to quasi-particle $0, 4, 8, 12, \ldots$ configurations (see figures 1). But the SM results show also that the seniority 6 and 10 configurations, which are not contained in the QRPA ground state, contribute only a little to the total neutrinoless double beta decay matrix element.

The angular momentum projected Hartee-Fock-Bogoliubov (HFB) method [14] is restricted in its scope. With a real Bogoliubov transformation without parity mixing (11) one can only describe neutron pairs with angular momenta and parity $0^+, 2^+, 4^+, 6^+, \ldots$ changing into two protons for ground state-to-ground state transitions. The restriction for the Interacting Boson Model (IBM) [17] is even more severe: one is restricted to 0^+ and 2^+ neutron pairs changing into two protons.

A comparison of the $0\nu\beta\beta$ transition matrix elements for the different many body methods: QRPA, SM, HFB and IBM are shown in Fig. 3.

If one assumes, that the contested measurement of Klapdor et. al. [29] is correct, one obtains with our matrix elements a double beta decay neutrino mass of $< m_{\beta\beta} >= 0.24 \pm 0.03$ $[eV]$.

Acknowledgments

This work was done in collaboration with Alberto Escuderos, Vadim Rodin and Fedor Simkovic. The author acknowledge support of the Deutsche Forschungsgemeinschaft within both the SFB TR27 "Neutrinos and Beyond" and Graduiertenkolleg GRK683.

References

[1] Doi M, Kotani T and Tagasuki E 1985 *Prog. Theor. Phys.* **83** 1
[2] Tomoda T 1991 *Rep. Prog. Phys.* **54** 53
[3] Faessler A and Šimkovic F 1998 *J. Phys.* G **24** 2139
[4] Avignone III F T, Elliott S R, and Engel J 2008 *Rev. Mod. Phys.* **80** 481
[5] Rodin V A, Faessler A, Šimkovic F, and Vogel P 2006 *Nucl. Phys.* A **766** 107; *Erratum-ibid.* 2007 **793** 213; Šimkovic F, Faessler A, Müther H, Rodin V A and Stauf M 2009 *Phys. Rev.* C **79** 055501
[6] Šimkovic F, Faessler A, Rodin V, Vogel P and Engel J 2008 *Phys. Rev.* C **77** 045503
[7] Kortelainen M and Suhonen J, 2007 *Phys. Rev.* C **75** 051303(R); *ibidem* 2007 **76** 024315
[8] Caurier E, Nowacki F, Menendez J and Poves A 2008 *Phys. Rev. Lett.* **100** 052503
[9] Caurier E, Nowacki F and Poves A 2008 *Eur. Phys.* **A36** 195
[10] Caurier E, Martínez-Pinedo G, Nowacki F, Poves A and Zuker A P 2005 *Rev. Mod. Phys.* **77** 427
[11] Menendez J, Poves A, Caurier E and Nowacki F 2009 *Nucl. Phys.* **A818** 139
[12] Faessler A, Fogli G L, Lisi E, Rodin V, Rotunno A M and Šimkovic F 2008 *J. Phys.* **G35** 075104
[13] Tomoda T, Faessler A, Schmid K W and Gruemmer F 1986 *Nucl. Phys.* **A452** 591
[14] Chaturvedi K, Chandra R, Rath P K, Raina P K and Hirsch J G 2008 *Phys. Rev.* C **78** 054302
[15] Rath P K, Chandra R, Chaturvedi K, Raina P K and Hirsch J G 2009 *Phys. Rev.* C **80** 044303
[16] Chandra R, Chaturvedi K, Rath P K, Raina P K and Hirsch J G 2009 *Europhys. Lett.* **86** 32001
[17] Barea J and Iachello F 2009 *Phys. Rev.* C **79** 044301
[18] Ikeda K 1964 *Prog. Theor. Phys.* **31** 434

[19] Schmid K W, Gruemmer F and Faessler A 1984 *Phys. Rev.* C **29** 291; 1984 *Phys. Rev.* C **29** 308; Schmid K W and Gruemmer F 1987 *Rep. Prog. Phys.* **50** 731
[20] Schmid K W 2004 *Prog. Part. Nucl. Phys.* **52** 565
[21] Schwieger J and Faessler A 1998 *Nucl. Phys. Proc. Suppl.* **66** 191; Schwieger J, Šimkovic F, Faessler A and Kaminski W A 1998 *Phys. Rev.* C **57** 1738
[22] Review of Particle Properties, C. Amsler *et al.* [Particle Data Group] 2008 *Phys. Lett.* B **667** 1
[23] Baranger M and Kumar K 1968 *Nucl. Phys.* **A110** 490
[24] Bloch C and Messiah A 1962 *Nucl. Phys.* **39** 95
[25] Schmid K W 2010 to be published
[26] Berger J F, Girod M and Gogny D 1984 *Nucl. Phys.* **A428** 23c
[27] Miller G A and Spencer J E 1976 *Ann. Phys. (NY)* **100** 562
[28] Feldmeier H, Neff T, Roth T and Schnack J 1998 *Nucl. Phys.* **A632** 61
[29] Klapdor-Kleingrothaus H V, Krivosheina I V 2006 *Mod. Phys. Lett.* **A21** 1547

Lattice animals, fractality and criticality in hadronic and partonic systems

2011 J. Phys.: Conf. Ser. 267 012060

(http://iopscience.iop.org/1742-6596/267/1/012060)

Lattice animals, fractality and criticality in hadronic and partonic systems

L.G. Moretto, J.B. Elliot, P.T. Lake and L. Phair
Nuclear Science Division, Lawrence Berkeley National Laboratory
Berkeley, CA 94720 USA

E-mail: lgmoretto@lbl.gov

Abstract. The cluster description of near coexistence phases (e.g. Fisher theory) requires an evaluation of cluster surface entropy. This surface degeneracy can be estimated with lattice models where clusters appear. The maximum probability lies near the maximum cluster surface. At low temperatures, clusters are forced to be nearly spherical by the surface energy and the associated Boltzmann factor. At higher temperatures and near criticality, the fractal dimension of clusters changes so that clusters become fractal. In the MIT bag model, where there is no surface energy, bags are always fractal.

1. Introduction

Phase transitions have a way to catch the attention of our collective scientific unconscious. The dramatic changes in structure, density and other parameters at times overwhelm our best judgment. We know but forget that, at least for first order transitions, these transitions can be described, and predicted, on the basis of the thermodynamics characterization of each individual infinite phase with the requirement of equal chemical potential. It is a case of myth hiding triviality.

The advent of mesoscopic physics made the waters murkier by introducing the additional complication of finiteness. Now finite size effects transformed, in the eyes of some, the general thermodynamics of infinite systems into an infinity of thermodynamics, one for each finite system.

In nuclear physics we should have known better. On one hand, we are used to finite nuclei, on the other we understand the danger of long range forces, such as the Coulomb force. Eventually, it was learned how to move elegantly from finite to infinite system trough the liquid drop model $A^{-1/3}$ expansion.

In this way, to do justice to our forefathers, some properties of $T = 0$ symmetric nuclear matter were quickly determined from the volume term of the liquid drop expansion. The guess of the existence of a dilute gas-like phase at $T > 0$ should have been obvious as neutrons started "evaporating" under the hands of experimentalists [1, 2].

2. Alleged signals of phase transitions

This should have been enough to predict and, to a large extent, characterize the liquid vapor phase transition and the attendant phase diagram. Unfortunately, the advent of heavy ion reactions and their complexity spurred experimentalists to look, in the foggy environment of the theory and data, for "signals" of phase transitions.

Plateaus in caloric curves were first held to be such signals [3, 4], forgetting that these plateaus are just due to keeping the pressure constant, which was not obvious here.

Similarly, spurred by numerical calculations in finite systems, others saw in negative heat capacities the desired signals [5, 6, 7]. Unfortunately "negative" heat capacities can be produced at command in infinite systems and even in a single phase [8].

Bimodality, another finite size effect, was also considered as a possible signal of a phase transition [9].

It may be a fair assessment to say that all these experimental signals did not lead to any hint of a phase diagram, nor not to a characterization of the phases involved in the alleged phase transitions.

3. Thermodynamic frugality

Yet much more progress could have been achieved much earlier and with very little work. The Clapeyron equation shows the way, giving in differential form the p, T coexistence line:

$$\left.\frac{dp}{dT}\right|_{coex} = \frac{\Delta H_m}{T\,\Delta V_m}. \tag{1}$$

Assuming $\Delta V_m = V_m^{vap}$ and ideal behavior of the gas we obtain

$$\frac{d\ln p}{dT} = \frac{\Delta H_m}{T^2}, \tag{2}$$

and taking for $\Delta H_m = -a_v + T$,

$$p = f(T)\exp\left(-\frac{a_v}{T}\right). \tag{3}$$

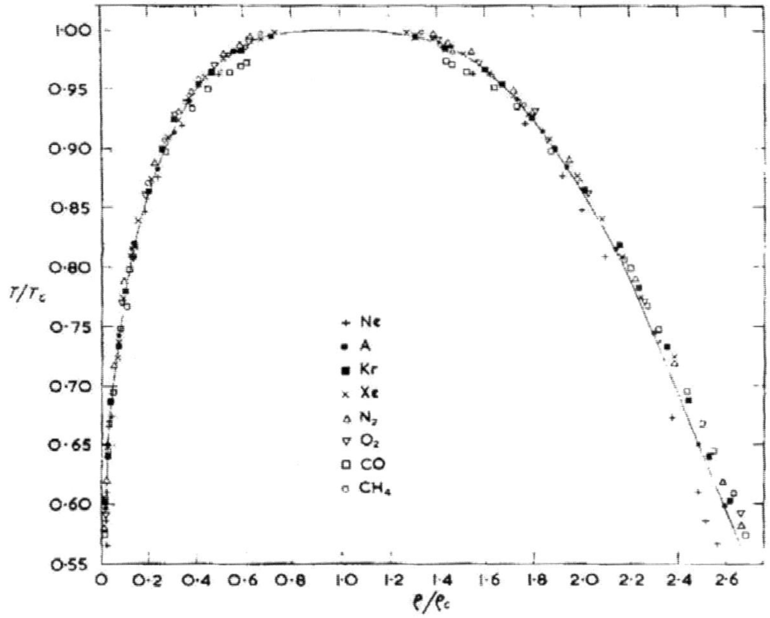

Figure 1. Reduced temperature-density phase diagram for several fluids.

Thus much was implicitly known and could have been made explicit before 1940.

Compare now the above equation with the expression used by Guggenheim [10] to fit the data in figure 1 for a variety of Van der Waals systems:

$$p = p_c \exp\left[-\frac{\Delta H_0}{T}\left(1 - \frac{T}{T_c}\right)\right], \tag{4}$$

where p_c and T_c are the critical pressure and temperature respectively. The two equations are obviously equivalent. So, in order to obtain the full phase diagram of nuclear matter we must obtain the critical parameters from experiments.

4. Critical parameters

In order to obtain such information we must rely on the properties of the gas phase (vapor) in equilibrium with nuclear matter at $T > 0$. We do not have access to such a system. However we have the possibility of studying the particles and clusters "evaporated" from hot finite nuclei. Three problems are to be overcome:

1) No vapor is ever in equilibrium with a nucleus, but emission (evaporation) occurs in vacuum [11].
2) Coulomb distorts the relative abundance of fragments through the Coulomb barrier [12].
3) The finiteness of nuclei affects the rate of emission of fragments [13].

We have shown that:

a) We can easily correct the distortion of rates and abundances due to Coulomb effects [12].
b) We can correct for finite size effects with the complement approach [11].
c) The corrected rates can be transformed in the concentrations of the clusters in equilibrium [13].

Having reached this point, we attempt the characterization of the phase coexistence line through the Fisher cluster model, its formula having been modified to account for finite size effect [13]:

$$\begin{aligned} n(A,T) &= q_0 A^{-\tau} \exp\left(\frac{\epsilon c_0 A^{\sigma}}{T}\right) \text{ (infinite system)}, \\ n(A,A_0,T) &= q_0 \left(\frac{A(A_0 - A)}{A_0}\right)^{-\tau} \exp\left(\frac{\epsilon c_0 [A^{\sigma} + (A_0 - A)^{\sigma} - A_0^{\sigma}]}{T}\right) \text{ (finite system)}, \end{aligned} \tag{5}$$

where $n(A,T)$ and $n(A,A_0,T)$ are the clusters abundances, A_0 is the emitter size, τ and σ are critical exponents, and $\epsilon = \frac{T_c - T}{T_c}$, where T_c is the critical temperature.

This approach can be tested on a finite size Ising Model as well as on a finite size Lennard Jones system and it allows one to recover accurately the critical parameters [13]. In figure 2 we show the Fisher plot, scaled fragment abundancies versus the surface energy divided by the temperature, for a variety of energies, nuclear systems and fragments [14]. The excellent scaling observed is equivalent to a coexistence line and allows us to obtain the critical parameters for infinite nuclear matter.

5. Partonic transitions and the bag model

The encouraging results described above lead us to consider another phase transition that has been studied and looked for in the past decades: the hadronic to partonic phase transition. We concentrate here on the simplest model, the bag model [15], which describes the transition in terms of two vacua: the hadronic vacuum in which the hadrons live, and the partonic vacuum

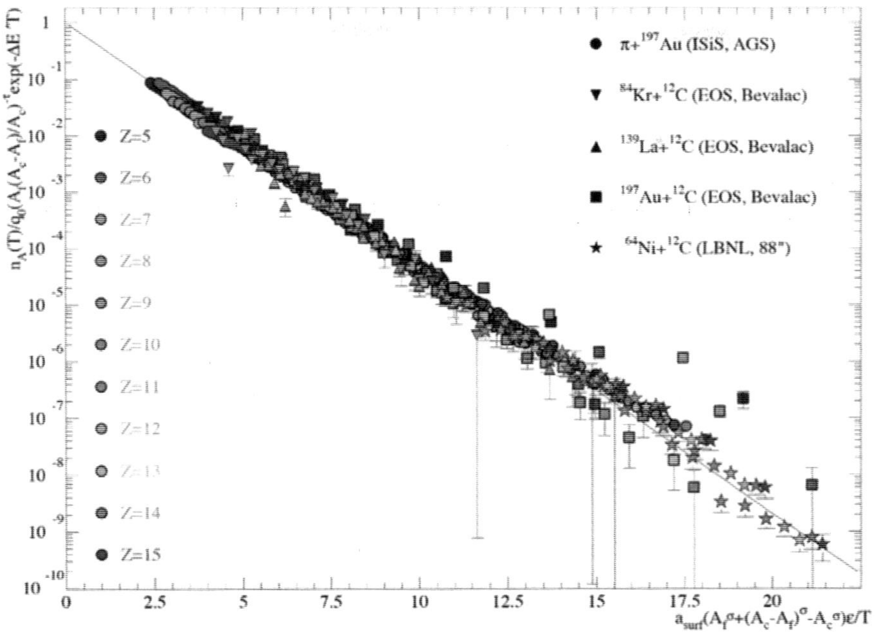

Figure 2. Generalized Fisher scaling for several nuclear systems.

in which quarks and gluons move unconfined. A finite volume of the partonic vacuum, a bag, is confined by a bag wall which exerts a constant pressure, B, on it and which must be balanced by the partonic bag content. In the case of no conserved charges, as in a thermal gas of the various quarks and gluons [15],

$$p = g\frac{\pi^2}{90}T_H^4 = B, \tag{6}$$

where g is the number of partonic degrees of freedom and B is the bag pressure. This leads to a universal, unique bag temperature:

$$T_H = \left(B\frac{90}{g\pi^2}\right)^{1/4}. \tag{7}$$

We have shown that the bag at equilibrium can neither be cooled nor heated [16, 17]. Although its total energy can be varied by increasing or decreasing the size of the bag, the energy density remains constant:

$$\epsilon = g\frac{\pi^2}{30}T_H^4 + B. \tag{8}$$

It follows that the entropy of the bag is extensive and proportional to its mass, m:

$$S = \frac{m}{T_H}, \tag{9}$$

which leads to a level density or bag degeneracy

$$\rho = e^{m/T_H}. \tag{10}$$

This extraordinary result implies that the bag is a perfect thermostat, which will impart its unique temperature T_H to any system in equilibrium with it [16, 17]. Let us consider now an amount of energy, E, to be shared between a bag and an ideal gas in which the mass of the particles of degeneracy $g(m)$ is derived also from the total energy. The total level density, $P(E, \epsilon)$, is the product of the gas and bag level densities,

$$P(E,\epsilon) = \rho_H(E-\epsilon)\rho_{iv}(\epsilon) = g(m)\frac{V^N}{N!\Gamma\left(\frac{3}{2}N\right)}\left(\frac{2\pi m}{h}\right)^{\frac{3}{2}N}\epsilon^{\left(\frac{3}{2}N-1\right)}\exp\left(\frac{E-mN-\epsilon}{T_H}\right). \quad (11)$$

Finding the average energy per particle from the above probability distribution, we obtain:

$$\frac{\epsilon}{N} = \frac{3}{2}T_H. \quad (12)$$

In other words, the gas is isothermal with the bag. Furthermore, the gas density is:

$$\frac{N}{V} = g(m)\left(\frac{2\pi m T_H}{h}\right)^{3/2} e^{-m/T_H}. \quad (13)$$

This means that the concentration of the gas is volume independent. It is a saturated vapor in equilibrium with a bag, like a saturated vapor in equilibrium with its liquid. This is the phase coexistence we were looking for. At temperatures lower than T_H we have a gas of those hadronic particles whose degeneracy is $g < e^{m/T_H}$.

At $T = T_H$ there is the appearance of bags with $g = e^{m/T_H}$ in equilibrium with the gas. As we attempt to increase the temperature by feeding energy to the system, the temperature stays fixed at T_H and the bag(s) grows in size until all the space is occupied by the bag. No higher temperature can be achieved without imposing an external pressure, which would force the disappearance of the vapor. Thus coexistence is signaled by fixed composition of the vapor hadronic constituents at fixed temperature T_H [16, 17].

6. A gas of bags

One may wonder whether a gas of bags (resonance gas) can exist. If it does, what are its properties? First of all, it is clear that a bag is intrinsically indifferent to fragmentation or coalescence:

$$\exp\left(\frac{m}{T_H}\right) = \exp\left(\sum\frac{m_i}{T_H}\right), \text{ with } M = \sum m_i. \quad (14)$$

What about the translational degrees of freedom? The answer is given by the equation 13, with $g(m) = e^{m/T_H}$:

$$\frac{N}{V} = \left(\frac{2\pi m T_H}{h}\right)^{3/2}. \quad (15)$$

The gas of bags has the above mass distribution, which has a maximum at $m = \infty$. The bags in the gas tend to coalesce into the largest possible bag. Is there criticality in the system described so far, and if so what are the critical parameters?

7. Criticality, surface entropy and fractal dimension

We preface the answer to this question with an apparently trivial question: What is the most probable shape of a bag? Should the bag have surface energy, then the bag would tend to be compact and near spherical. But in the bag model considered here, all terms are extensive, meaning proportional to the bag volume. There is no $V^{2/3}$ term that would signify the presence of a surface energy. So what is the shape of the bag? We can give a precise answer in a lattice model. Occupied sites can cluster into objects, called animals, of different sizes. Let us consider a given animal of size A. In how many shapes can it exist? The logarithm of this number is the shape entropy. In three dimensions, the surface entropy has a maximum very near the maximum possible surface. This corresponds to a highly dendritic, fractal shape and not to a sphere-like shape. We would conclude that the bag's most probable shape is fractal. What does this have to do with criticality? In Fisher's model, criticality is wholly dependent on surface entropy. We can rewrite Fisher's formula, as follows:

$$n(A) = g(A) \exp\left(-\frac{c_0 A^\sigma}{T_c}\right), \tag{16}$$

where $\sigma \approx 2/3$ and $\ln[g(A)]$ is the surface entropy. Fisher guesses

$$g(A) = A^{-\tau} \exp(k\, A^\sigma) = A^{-\tau} \exp\left(\frac{c_0\, A^\sigma}{T_c}\right). \tag{17}$$

This guess is apparently strange because for lattice animals $\ln[g(A)] \approx kA$. In order to understand this, let us use equation 16 and the actual $g(A)$ to calculate an effective σ at various temperatures. Figure 3 and 4 show that at low temperatures the Boltzmann factor kills the highly fractal shapes, retaining only the compact ones. Accordingly, $\sigma \approx 2/3$. As we approach the critical temperature, there is a rapid transition from to $\sigma \approx 2/3$ to $\sigma \approx 1$. The clusters go fractal. This would suggest that the bag(s), which are inherently fractal due to the lack of surface energy, portray critical features at their unique temperature T_H. A rather embarrassing mix of 1st order and criticality. One may wonder whether one can cure this problem by introducing an *ad hoc* surface term to the bag model. This is equivalent to introducing an additional volume dependent pressure to the bag constant,

$$p = \frac{1}{3} f(T) - B + \frac{2}{3} a_s V^{-1/3} = 0. \tag{18}$$

This gives a T as an implicit function of the bag volume, V. More specifically, the smaller the bag, the larger its temperature. This is a rather disastrous result because a gas of bags of various sizes cannot be isothermal, and the only way out is to coalesce the bags into a single one.

8. Conclusion

There is a liquid-gas phase transition for symmetric, uncharged infinite nuclear matter, and the relevant coexistence curve has been extracted from the analysis of experimental cluster distributions. The bag model supports a 1st order phase transition between non-Hagedorn hadrons and a bag at a constant, unique temperature T_H. A gas of bags is entropically unstable towards coalescence. The lack of surface energy entropically drives the bag towards fractal shapes, thus simulating criticality. An *ad hoc* introduction of surface energy makes the bags non-isothermal.

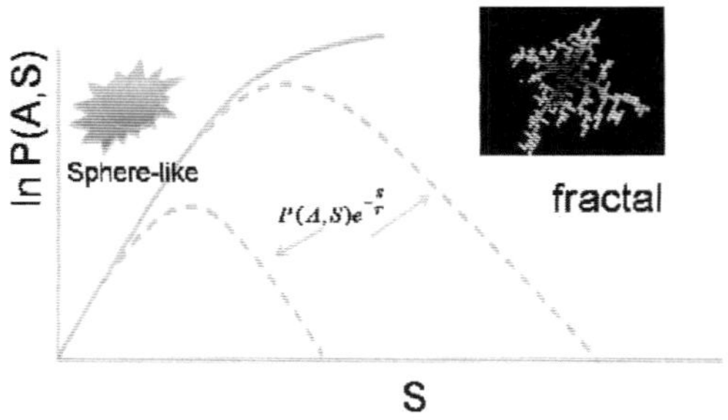

Figure 3. Probability of a lattice animal to have a given surface, S, multiplied by Boltzmann factors at different temperatures.

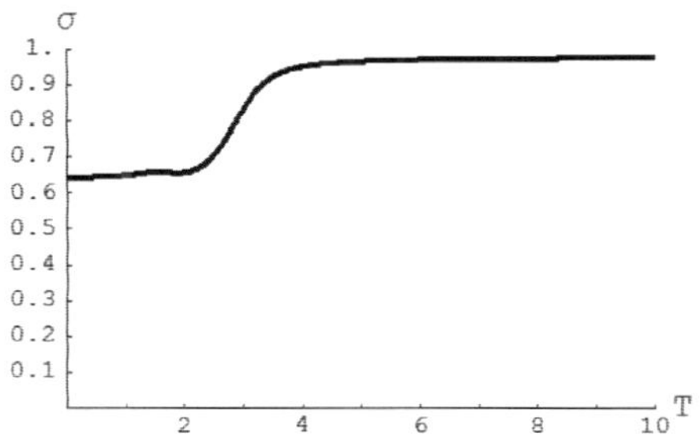

Figure 4. Effective surface dimensionality, σ, for lattice animals on a cubic grid as a function of temperature.

References

[1] Weisskopf V 1937 *Phys. Rev.* **52** 295

[2] Le Couteur K J 1950 *Proc. Phys. Soc.* A **63** 259

[3] Pochodzalla J *et al* 1995 *Phys. Rev. Lett.* **75** 1040

[4] Natowitz J B, Wada R, Hagel K, Keutgen T, Murray M, Makeev A, Qin L, Smith P and Hamilton C 2002 *Phys. Rev.* C **65** 034618

[5] D'Agostino M *et al* 2000 *Phys. Lett.* B **473** 219

[6] Gross D H E 1990 *Rep. Prog. Phys.* **53** 605

[7] Gross D H E 1997 *Phys. Rep.* **279** 119

[8] Moretto L G, Elliott J B, Phair L and Wozniak G J 2002 *Phys. Rev.* C **66** 041601

[9] Chomaz P, Gulminelli F and Duflot V 2001 *Phys. Rev.* E **64** 046114

[10] Guggenheim E A 1945 *J. Chem. Phys.* **13** 253

[11] Moretto L G, Elliott J B and Phair L 2005 *Phys. Rev.* C **72** 064605

[12] Moretto L G, Elliott J B and Phair L 2003 *Phys. Rev.* C **68** 061602

[13] Moretto L G, Bugaev K A, Elliott J B, Ghetti R, Helgesson J and Phair L 2005 *Phys. Rev. Lett.* **94** 202701

[14] Moretto L G, Elliott J B and Phair L 2004 *Prog. Part. Nucl. Phys.* **53** 101 and references therein

[15] Chodos A, Jaffe R L, Johnson K, Thorn C B and Weisskopf V F 1974 *Phys. Rev.* D **9** 3471

[16] Moretto L G, Bugaev K A, Elliott J B and Phair L 2006 *Europhys. Lett.* **76** 402

[17] Moretto L G, Phair L, Bugaev K A and Elliott J B 2006 *PoS* **CPOD2006** 037

Recent results in quantum chaos and its applications to atomic nuclei

2011 J. Phys.: Conf. Ser. 267 012061

(http://iopscience.iop.org/1742-6596/267/1/012061)

Recent results in quantum chaos and its applications to atomic nuclei

J. M. G. Gómez[1], E. Faleiro[2], L. Muñoz[2], R. A. Molina[3] A. Relaño[1], J. Retamosa[1]

[1] Departamento de Física Atómica, Molecular y Nuclear, Universidad Complutense de Madrid, E-28040 Madrid, Spain

[2] Departamento de Física Aplicada, E. U. I. T. Industrial, Universidad Politécnica de Madrid, E-28012 Madrid, Spain

[3] Instituto de Estructura de la Materia, CSIC, E-28006 Madrid, Spain

E-mail: gomezk@nucl.fis.ucm.es

Abstract. A survey of chaotic dynamics in atomic nuclei is presented, using on the one hand standard statistics of quantum chaos studies, and on the other a new approach based on time series analysis methods. The study of shell-model spectra in the pf shell shows that nuclear chaos is strongly isopin dependent and increases with excitation energy. On the other hand, it is found that chaotic quantum systems exhibit $1/f$ noise and regular systems exhibit $1/f^2$ behaviour. It is shown that the time series approach can be used to calculate quite accurately the fraction of missing levels and the existence of mixed symmetries in experimental level spectra.

1. Introduction

The understanding of quantum chaos has greatly advanced during the last two decades. It is well known that there is a clear relationship between the energy level fluctuation properties of a quantum system and the large time scale behaviour of its classical analogue. The spectral fluctuations of a quantum system whose classical analogue is fully integrable are well described by Poisson statistics, i. e. the successive energy levels are not correlated [1]. On the contrary the fluctuation properties of generic quantum systems, which are fully chaotic, coincide with those of random matrix theory (RMT) [2]. Comprehensive reviews can be found in [3, 4], and more recent developments are reviewed in [5].

The information on regular and chaotic nuclear motion available from the analysis of experimental data is rather limited, because it requires the knowledge of sufficiently large pure sequences, i.e. consecutive level samples all with the same quantum numbers (J, π, T) in a given nucleus. The situation is clear above the one-nucleon emission threshold, where a large number of neutron and proton $J^{\pi} = 1/2^{+}$ resonances are identified. The agreement between this Nuclear Data Ensemble (NDE) [6] and the GOE predictions is excellent. In the low energy domain, however, it is rather difficult (if not impossible) to get large enough pure sequences, and for this reason the conclusions are less clear [7, 8]. In order to get deeper understanding of what happens in the low energy region one can use the shell model with configuration mixing. Most shell-model studies on chaotic motion have been done in *sd*-shell nuclei. We present here calculations in pf-shell nuclei, where the configuration space is much larger, and therefore we can study the dependence of chaos on excitation energy and isospin. We present as well a different approach to quantum chaos based on traditional methods of time series analysis. The energy

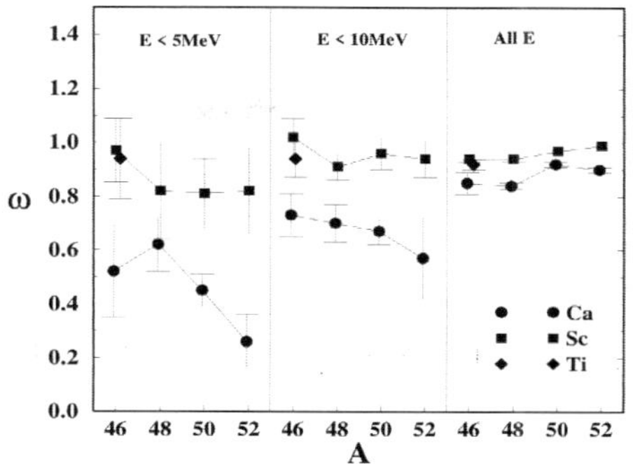

Figure 1. Brody parameter ω for the $A = 46$, 48, 50 and 52 Ca and Sc isotopes and ^{46}Ti, using all the energy levels up to 5 MeV and 10 MeV over the yrast line, and the full spectrum.

spectrum can be considered as a discrete signal, and the sequence of energy levels as a time series. As we shall see, examination of the power spectrum of energy level fluctuations reveals very accurate power laws for completely regular or completely chaotic Hamiltonian quantum systems. It turns out that chaotic systems have $1/f$ noise, in contrast to the brown noise of regular systems [9, 10, 11]. We also present applications of spectral fluctuations to the analysis of imperfect experimental spectra in nuclei [12].

2. Quantum chaos in pf−shell nuclei

To study the isospin dependence of nuclear chaos, we have performed a detailed comparative analysis of spectral properties in the $A = 46$, 48, 50 and 52 Ca and Sc isotopes and in ^{46}Ti [13]. The $T = T_z$ states are considered in all cases. The spectra are obtained by shell-model calculations in the pf shell using the KB3 interaction [14].

The nearest neighbour spacing distribution $P(s)$ has been studied including all the levels up to 5 MeV and 10 MeV above the yrast line, and without any cutoff. The unfolding is performed for each $J^\pi, T = T_z$ pure sequence separately and then the unfolded spacings are gathered into a single set for each nucleus to get better statistics. The values of the Brody parameter ω are displayed in Fig. 1 separated in three sub-panels according to the energy cutoff. Up to 5 MeV above the yrast line, Ca isotopes show spectral fluctuations intermediate between those of regular and chaotic systems, except ^{52}Ca which essentially is a regular system. On the contrary, all the Sc isotopes and ^{46}Ti are very close to GOE fluctuations. For a given A, the big differences between Sc and Ca isotopes must be due to the residual two-body interaction, because the single particle energies are the same in both cases. We argue that the neutron-neutron interaction is much weaker than the neutron-proton interaction and thus the central field motion is less affected in Ca isotopes, where all the valence particles are neutrons. Another interesting feature observed in Fig. 1 is that in Ca isotopes ω shows a strong fall from $A = 48$ to $A = 52$, where $\omega = 0.25$. This astonishing result means that the two-body interaction is almost unable to perturb the single-particle motion in the low energy levels of ^{52}Ca. The ω value for ^{46}Ti has been included in Fig. 1 to show that, at all energies, ω reaches its maximum in ^{46}Sc. Therefore, replacing a single neutron by a proton in Ca isotopes causes a transition from a quasi-regular to a chaotic regime. It is remarkable that this transition takes place abruptly at all excitation energies in all the isotopes. A second replacement of a neutron by a proton does not seem to produce appreciable effects on the $P(s)$ statistic.

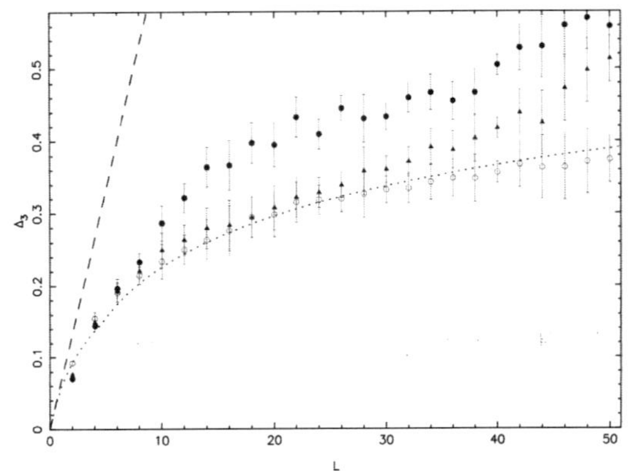

Figure 2. Average Δ_3 for all the $J^\pi = 0^+, T = T_z$ levels of ^{46}Ca (dots), ^{46}Sc (squares) and ^{46}Ti (diamonds). The dotted and dashed curves represent the GOE and Poisson Δ_3 values.

We have studied the behaviour of the $\Delta_3(L)$ statistic as well. The results given below concern the full spectrum and not only the low energy region. Fig. 2 shows $\Delta_3(L)$ values for $L \leq 50$, using the $J^\pi = 0^+$, $T = T_z$ levels of ^{46}Ca, ^{46}Sc and ^{46}Ti. Of the three nuclei, only ^{46}Ti follows the GOE line, at least until $L = 50$. For ^{46}Sc the Δ_3 is close to GOE predictions up to a certain separation value, $L_{sep} \simeq 30$ where it upbends from the GOE curve. In ^{46}Ca the upbending starts at a smaller value $L_{sep} \simeq 10$. The upbending from the GOE curve and a linear growth of this statistic reveals a departure from the chaotic regime. The Δ_3 behaviour clearly shows a strong isospin dependence in the $A = 46$ nuclei, with chaoticity increasing as T decreases. This happens not only from Ca to Sc, but also from Sc to Ti. The same phenomenon is observed for other J values.

Summarising, there exists a clear excitation energy and isospin dependence in the chaoticity degree of nuclear motion. It is observed not only in the ground state region, but along the whole spectrum. When the full spectrum is taken into account, the $P(s)$ distribution is not very sensitive to the isospin dependence, but the effect is clearly seen in the Δ_3 statistic.

3. Time series approach to quantum chaos

A new approach to study spectral fluctuations in quantum systems has been recently presented. [9] The basic idea is to consider the energy spectrum as a time series where energy plays the role of time. Using this formal analogy spectral fluctuations can be studied by means of techniques borrowed from time series analysis. We can characterise the spectral fluctuations by the statistic δ_n defined by $\delta_n = \sum_{i=1}^{n}(s_i - \langle s \rangle)$, $n = 1, 2, \ldots N-1$, where $s_i = \epsilon_{i+1} - \epsilon_i$ is the spacing between two consecutive unfolded energy levels, and N is the total number of levels. Since $< s >= 1$, the δ_n function represents the deviation of the unfolded excitation energy from its mean value n.

One of the simplest methods to analyse correlations in a time series is the study of its power spectrum, which provides information on the correlations at all time scales. The power spectrum $S(k)$ of a discrete and finite series δ_n is given by $S(k) = \left|\hat{\delta}_k\right|^2$, where $\hat{\delta}_k$ is the Fourier transform of δ_n,

$$\hat{\delta}_k = \frac{1}{N}\sum_n \delta_n \exp\left(\frac{2\pi i k n}{N}\right). \quad (1)$$

The first systems studied by this method were the spectra of atomic nuclei calculated using

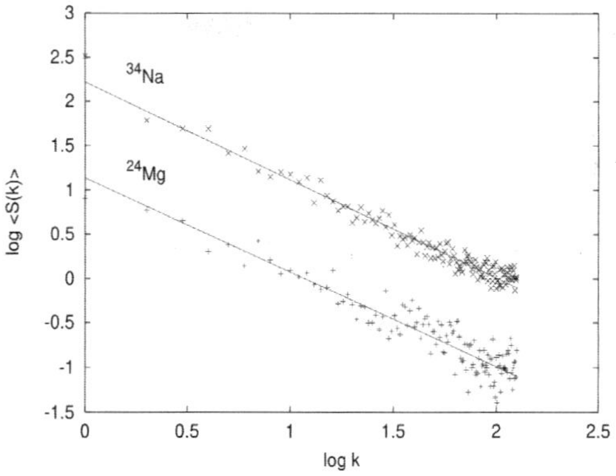

Figure 3. Average power spectrum of the δ_n function for ^{24}Mg and ^{34}Na, using 25 sets of 256 levels from the high level density region. The plots are displaced to avoid overlapping.

the shell model with configuration mixing [9]. Fig. 3 shows the results for a typical stable *sd*-shell nucleus, ^{24}Mg, and for a very exotic nucleus, ^{34}Na, in the *sd* proton and *pf* neutron shells. Clearly, the power spectrum of δ_n follows closely a power law. We may assume the simple functional form $\langle S(k)\rangle \sim 1/k^{\alpha}$. A least-squares fit to the data of Fig. 3 gives $\alpha = 1.11 \pm 0.03$ for ^{34}Na, and $\alpha = 1.06 \pm 0.05$ for ^{24}Mg. These results raised the question of whether there is a general relationship between quantum chaos and the power spectrum of the δ_n fluctuations of the system. A similar numerical calculation performed for the three classical random matrix ensembles GOE, GUE and GSE showed that the power spectrum $\langle S(k)\rangle$ follows quite accurately a power law of type $1/k$ for all of them. Therefore, the spectral fluctuations of chaotic quantum systems described by the δ_n function exhibit $1/f$ noise [9], independently of the symmetries of the system. As is well known, the existence of $1/f$ noise is a remarkable and very ubiquitous property of many complex systems in nature and in social sciences. The origin of the $1/f$ noise in the time fluctuations of these systems is an important open problem. In the case of quantum systems, an exact and complete proof of the $1/f$ noise behaviour seems to be extremely difficult. However, it can be theoretically studied in semiclassical systems or random matrix ensembles, where the mathematical tractability of these systems may help to understand the origin of the $1/f$ noise in chaotic quantum systems.

We have recently shown that the power spectrum of δ_n for fully chaotic or integrable systems can be written in terms of the ensemble form factor $K(\tau)$ as follows [10],

$$\langle S(k)\rangle_\beta = \frac{N^2}{4\pi^2}\left\{\frac{K\left(\frac{k}{N}\right)-1}{k^2}+\frac{K\left(1-\frac{k}{N}\right)-1}{(N-k)^2}\right\}+\frac{1}{4\sin^2\left(\frac{\pi k}{N}\right)}+\Delta,\quad k=1,2,\cdots,N-1,\quad N\gg 1, \tag{2}$$

where β is the repulsion parameter of RMT ensembles and takes the values $\beta = 1$ for GOE, $\beta = 2$ for GUE, and $\beta = 4$ for GSE. Here $\Delta = 0$ for integrable systems and $\Delta = -1/12$ for chaotic systems. This equation, together with the appropriate values of $K(\tau)$, gives explicit expressions of $\langle S(k)\rangle$ for specific ensembles or systems.

When $k \ll N$ the first term of Eq. (2) becomes dominant and we can write $\langle S(k)\rangle_\beta = \dfrac{N}{2\beta\pi^2 k}$

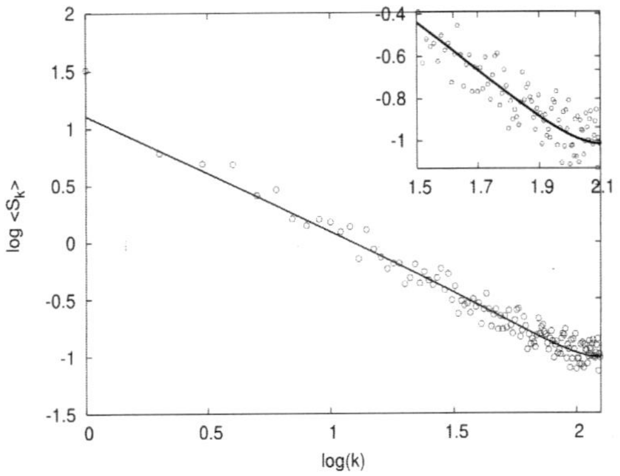

Figure 4. Numerical average power spectrum of the δ_q function for ^{34}Na, calculated using 25 sets of 256 consecutive levels from the high level density region, compared to the parameter free theoretical values (solid line) for GOE.

for chaotic systems and $\langle S(k)\rangle_\beta = \dfrac{N^2}{4\pi^2 k^2}$ for integrable ones. These expressions show that, for small frequencies, the excitation energy fluctuations exhibit $1/f$ noise in chaotic systems and $1/f^2$ noise in integrable systems. As we shall see below, these power laws are also approximately valid through almost the whole frequency domain, due to partial cancellation of higher order terms. Only near $k = N/2$ the effect of these terms becomes appreciable.

We have compared the analytical expression (2) to numerical results obtained for different RMT ensembles, shell model spectra of atomic nuclei and chaotic and regular quantum billiards. In all cases the agreement is excellent. As an example Fig. 4 displays the theoretical values of $\langle S(k)\rangle$ compared with the shell-model results for ^{34}Na. It can be seen that Eq. (2) gives $1/k$ behaviour up to quite high values of the frequency counter k, and the agreement with the numerical power spectrum of ^{34}Na is excellent.

4. Applications

The time series approach to quantum chaos turns out to be quite fruitful to analyse imperfect spectra, i.e., quantum spectra with missing levels or mixed symmetries, which would induce misleading results about the chaotic or regular features of the spectra. It it is possible to generalise Eq. (2) to a more complex situation in which every observed level sequence contains l pure sequences, each with a partial density η_i (the probability that a given level belong to the i-th sequence) and a fraction of observed levels φ_i (levels pertaining to the i-th sequence are observed with probability φ_i). Then the power spectrum of δ_n becomes [12]

$$\langle S(k)\rangle = \frac{N^2}{4\pi^2}\sum_{i=1}^{l}\eta_i\varphi_i\left[\frac{K_i\left(\frac{\varphi_i k}{N\eta_i}\right)-1}{k^2}+\frac{K_i\left(\frac{\varphi_i (N-k)}{N\eta_i}\right)-1}{(N-k)^2}\right]+\frac{1}{4\sin^2(\pi k/N)}+\langle\varphi\rangle^2\Delta, \quad (3)$$

where $K_i(\tau)$ is the original spectral form factor of the i-th pure level sequence and $\langle\varphi^2\rangle = \sum_{i=1}^{l}\eta_i\varphi_i^2$. Clearly in the limit of a very large number of mixed symmetries or a very small fraction of observed levels one obtains again Poisson statistics. To apply (3) in practical cases one needs to assume a form of the spectral form factor $K_i(\tau)$. To see how well the method works in realistic situations [12], let us consider the shell-model spectrum of ^{24}Mg, which exhibits $1/f$ noise (see Fig. 4).

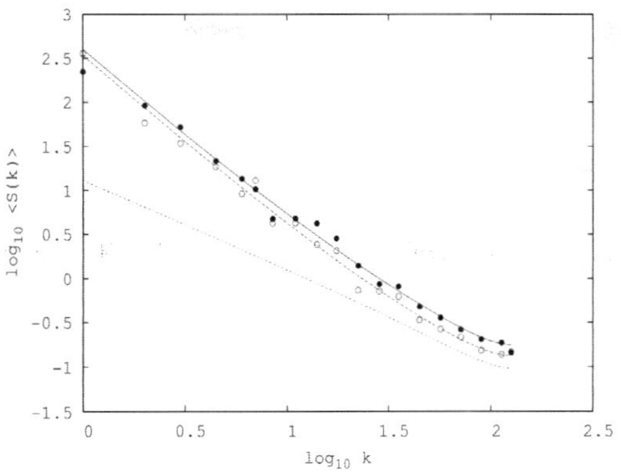

Figure 5. Numerical power spectrum of δ_n with 20% missing levels ($\varphi = 0.8$) using $J = 3$ and $J = 4$ pure sequences of length $N = 256$ (open circles) and incomplete mixed sequences of the same length (filled circles). The solid and dashed lines correspond to Eq (3) with the best fit l and φ values. The theoretical curve for GOE (dotted line) is also shown.

Now Fig 5 shows the power spectrum of δ_n when 20% of the levels are missing ($\varphi = 0.8$), for the case of pure $J^\pi = 3^+$ and $J^\pi = 4^+$ sequences, and for the case of a mixed sequence of $J^\pi = 3^+$ and $J^\pi = 4^+$ states. Using (3), the parameters l and φ are calculated by means of a least-squares fit to the numerical power spectrum. In the case of pure J sequences, we obtain $\varphi = 0.80 \pm 0.03, l = 1.1 \pm 0.3$. Thus we see that the fit detects that the sequences belong to a single J value, and the fraction of observed levels is accurately reproduced. In the case of mixed $J^\pi = 3^+$ and $J^\pi = 4^+$ states, the fit gives $\varphi = 0.77 \pm 0.03, l = 2.1 \pm 0.4$. Therefore it detects hat there are two mixed sequences ($l = 2$) and the fraction of observed levels is also very close to the real 80% fraction. In general, the agreement between the results of the fit and the actual parameters used to generate the level sequences ranges from good to remarkable. Thus we conclude that this method can be very useful to extract relevant statistical information from experimental spectra.

References

[1] Berry M V and Tabor M 1977 *Proc. R. Soc. London* A **356** 375
[2] Bohigas O, Giannoni M J and Schmit C 1984 *Phys. Rev. Lett.* **52** 1
[3] Guhr T, Müller-Groeling A and Weidenmüller H A 1998 *Phys. Rep.* **299** 189
[4] H. J. Stöckmann, *Quantum Chaos*, (Cambridge University Press, Cambridge, 1999)
[5] Gómez J M G, Kar K, Kota V K B, Molina R A, Relaño A and Retamosa J, *Phys. Rep.* to be published
[6] Haq R U, Pandey A and Bohigas O 1982 *Phys. Rev. Lett.* **48** 1086
[7] Shriner jr J F, Mitchell G E and von Egidy T 1991 *Z. Phys. A* **338**, 309
[8] Garret J D, German J R, Courtney L and Espino J M, *in Future Directions in Nuclear Physics, eds. J. Dudek and B. Haas (A. I. P., N.Y., 1992)* p. 345
[9] Relaño A, Gómez J M G, Molina R A, Retamosa J and Faleiro E 2002 *Phys. Rev. Lett.* **89** 244102
[10] Faleiro E, Gómez J M G, Molina R A, Muñoz L, Relaño A and Retamosa J 2004 *Phys. Rev. Lett.* **93** 244101
[11] Gómez J M G, Relaño A, Retamosa J, Faleiro E, Salasnich L, Vranicar M and Robnik M 2005 *Phys. Rev. Lett.* **94** 084101
[12] Molina R A, Retamosa J, Muñoz L, Relaño A and Faleiro E 2007 *Phys. Lett.* B **644** 25
[13] Molina R A, Gómez J M G and Retamosa J 2000 *Phys. Rev.* C **63** 014311
[14] Poves A and Zuker A 1981 *Phys. Rep.* **70** 4

Statistical significance of theoretical predictions: A new dimension in nuclear structure theories (I)

2011 J. Phys.: Conf. Ser. 267 012062

(http://iopscience.iop.org/1742-6596/267/1/012062)

Statistical significance of theoretical predictions: A new dimension in nuclear structure theories (I)

J DUDEK[1], **B SZPAK**[2], **M-G PORQUET**[3] **and B FORNAL**[2]

[1]Institut Pluridisciplinaire Hubert Curien Département de Recherches Subatomiques UMR7178, IN2P3-CNRS and Université de Strasbourg, 23 rue du Loess B.P. 28, F-67037 Strasbourg Cedex 2, France

[2]The Niewodniczański Institute of Nuclear Physics, Polish Academy of Sciences, ul. Radzikowskiego 152, Pl-31-342 Kraków, Poland

[3]Centre de Spectrométrie Nucléaire et de Spectrométrie de Masse, IN2P3-CNRS and Université Paris-Sud, Bâtiments 104 et 108, F-91405 Orsay Campus, France

E-mail: `Jerzy.Dudek@IReS.in2p3.fr`, `Bartlomiej.Szpak@ifj.edu.pl`, `Marie-Genevieve.Porquet@csnsm.in2p3.fr`, `Bogdan.Fornal@ifj.edu.pl`

Abstract.

In this and the follow-up article we briefly discuss what we believe represents one of the most serious problems in contemporary nuclear structure: the question of statistical significance of parametrizations of nuclear microscopic Hamiltonians and the implied predictive power of the underlying theories. In the present *Part I*, we introduce the main lines of reasoning of the so-called *Inverse Problem Theory*, an important sub-field in the contemporary Applied Mathematics, here illustrated on the example of the Nuclear Mean-Field Approach.

1. Introduction

In societies with long history, within their intellectual or political development there arrive issues that become a common secret: *subjects whose discussing is rather avoided* despite the fact that the problems behind are both important and disturbing and everybody knows about their existence. Similarly, in scientific communities certain themes become kind-of-embarrassing and we may even develop a tendency of systematically avoiding the issue. The nuclear physics community with over 100 years of existence has produced already some such delicate subjects. One of them is the actual subject of this article: it has to do with (possibly un-) stable parametrizations of the nuclear structure Hamiltonians, the avoided issue being parametrization's statistical significance and the Hamiltonian's (likely missing) predictive power.

The first signs of theory's missing predictive power are: a) Quickly growing in the literature a number of alternative (thus generally incompatible) parametrizations of the same Hamiltonian and b) Diverging predictions from various parametrizations when extrapolating to nuclear areas that are experimentally unknown at a given time. To illustrate the item a) above: Our community has produced to date over 130 various parametrizations of the effective Hamiltonian of Skyrme Hartree-Fock mean field theory. Moreover, in other microscopic theories parametrizations are sometimes accepted that vary from one narrow nuclear range to another or even from one nucleus to the next. Manifestations of the same problem can be found in yet another important microscopic method: the nuclear shell-model. To illustrate the origin of the problem b) above, let us quote the authors [1] who remark with strong criticism:

> *"Unfortunately, many practitioners of parameter estimation never proceed beyond determining the numerical values of the parameter fit. They deem a fit acceptable if a graph of data and model 'looks good'. This approach is known as chi-by-the-eye. Luckily, its practitioners get what they deserve"*

[i.e. - 'they' get a 'statistical nonsense', see the above reference for explanatory comments].

The question of predictive power of physical theories and statistical significance of theoretical predictions is a vast and quickly growing field of mathematics, so far applied mainly in domains of science other than nuclear physics. These applications range from geology and geophysics to social and political sciences, image modelling, medicine etc., etc. Within the volume of this presentation we will be able to touch upon a small number of selected issues and illustrations belonging to the underlying sub-field of Applied Mathematics: the *Inverse Problem Theory*[1] one of the most important technical tools of the general *Inference Theory*. The former deals with the question of a mathematically correct extracting the mathematical model's adjustable parameters and determining their statistical significance. As for the latter, it has to do with the mathematical treatment necessary to obtain the statistically significant information about *all the data* through theoretical modelling - while having only *very limited sample of experimental data at our disposal* - as an actual information-input.

2. Statistical significance of a Hamiltonian and predictive power of the theory

Examining a physical system we usually wish to obtain 'the full truth', the latter mathematically represented by the solutions of the eigen-value problem $\hat{H}\,\psi_n = e_n\,\psi_n$ in which $\hat{H} = \hat{H}^{true}$ is supposed to contain *the complete* information. Since we believe in the objective existence of studied physical objects and in an ever-lasting improvement of our understanding, the adjective 'complete' signifies here both: already known and yet undiscovered. It then follows that the true solutions $\{e_n, \psi_n\}$ remain asymptotic in time devoted to the research, rather fictitious, ideal and to an extent unknown objects and that, to the same extent, the desired *full truth* remains unknown to us. To say: *unknown* - is an expression of the lack of knowledge synonymous with ignorance, here imposed by nature rather than human laziness. Humans do not know any efficient cure against deliberate laziness, but they did invent a cure against forced ignorance. The solution: *Parametrize our ignorance*; more precisely: estimate by all available means which answer is more-, and which less-likely the right solution. In other words we will need to find relative probability of what we think the right answer is! In Applied Mathematics, the ignorance is usually represented by a random variable X, quantified by its probability distribution $P_X = P_X(x)$, x called 'realisation' of the random variable variable X. In short: The way to progress will be to learn diminishing the effects of ignorance through parametrizing it in terms of probabilities.

To be able to illustrate the formalism with the realistic situations, we will focus on the simple case of nuclear phenomenological mean-field theory, otherwise our considerations are general.

[1] The term 'Inverse Problem Theory' of Applied Mathematics has not much in common with its traditional meaning in quantum mechanics, where for a long time it referred to finding the form of the potential in the Schrödinger equation out of bound states and scattering cross-sections. When we took as a somewhat arbitrary measure of intellectual activity the number of hits in *Google search* to compare two fields, the *nuclear physics* and the *inverse problem*, on Sunday, May 16, 2010, at 10:30 AM, we found

"Nuclear Physics" → 3 200 000 results (about 100 years of active history)
"Inverse Problem" → 510 000 results (about 20 years of active history)

2.1. Theoretical and experimental errors as sources of uncertainties: Probability of the truth
At the origin of our ignorance lie *errors* and here we distinguish between their two sources: a) Often discussed experimental errors, and, much more important and almost never examined: b) The theory errors originating from the fact that *our many-body theories are incomplete*:

$$\text{Theory} \rightarrow \boxed{e_n = e_n^{true} + \delta e_n^{error}(p)} \quad \& \quad \boxed{\varepsilon_n = \varepsilon_n^{true} + \delta\varepsilon_n^{err}} \leftarrow \text{Experiment} \tag{1}$$

Above, both the theoretical and experimental energies, e_n and ε_n, respectively, are random variables characterised by probability distributions $P_n^{th.}(e_n)$ and $P_n^{exp.}(\varepsilon_n)$. Errors propagate to the theory predictions through parameter fits

$$\chi^2(p) \sim \sum w_n \big[\underbrace{(\varepsilon_n^{true} + \delta\varepsilon_n^{err})}_{\textbf{Experiment}} - \underbrace{(e_n^{true} + \delta e_n^{err})}_{\textbf{Theory}} \big]^2 \rightarrow \frac{\partial \chi^2}{\partial p} = 0. \tag{2}$$

The final theory predictions will depend on an appropriate convolution between the error probabilities $P^{th.}(e)$ and $P^{exp.}(\varepsilon)$ leading to the Hamiltonian-parameter probability distributions $P(p)$ and, through the latter, to the error probability distribution of the final predictions for observables. In other words, we suggest an extended strategy in constructing theoretical models:

> *Physical theories should provide not only the numerical predictions but also probabilities with which they are believed to give right answers when compared to experimental data. Studying these probabilities opens a new dimension*[2] *in modelling nuclear phenomena.*

A new formulation (or re-formulation) of theories will require a thorough investigation of the exact forms of the theory's error probability distributions $P^{th.}(e)$. We cannot offer such a full solution at this time, the whole field of research being at its initial stage. Instead, we will focus in this text on the underlying mathematical aspects and in the following one [3] on the illustrations with the help of the mean-field single-particle spectra for ^{208}Pb and ^{132}Sn nuclei and a phenomenological Woods-Saxon toy-model Hamiltonian.

Consider an example of a spherical mean-field Hamiltonian:

$$H_{mf} = H_{mf}(\hat{r}, \hat{p}, \hat{s}; \{p\}); \quad \{p\} \leftrightarrow \text{parameters}, \tag{3}$$

where $\hat{r}$, $\hat{p}$ and $\hat{s}$ denote the operators of position, linear momentum and spin, respectively. In looking for an as *adequate approach as possible* in searching for Hamiltonian parameters we need some guidelines. Our choice: The mean-field Hamiltonian should first of all describe optimally the mean field degrees of freedom[3], i.e. the single-particle wave functions and energies $\{\varphi_\nu, e_\nu\}$.

It would have been totally incorrect to think that the so-called experimental single-particle energies are some elementary, pure-mean-field *observables* ready to compare with the mean-field calculations: they are in fact complex model-dependent objects whose extraction involves the spectroscopic factors, the knowledge of the one-particle transfer cross-sections modelled using non-trivial theoretical input, cf. [2]. We believe that the theoretical analysis leading to the Hamiltonian parameters through the 'experimental single-particle levels' is nevertheless simpler as compared to what is needed to extract the same information from nuclear masses.

[2] Obviously, the word *dimension* should not be taken literally but rather as a nickname for a new direction of research combining the application of advanced methods of probability theories with a new type of physics tasks: modelling the impact of errors caused by missing elements of theories, for instance missing terms in the underlying Hamiltonians - as well as constructing the underlying probability distributions ('theory of theoretical errors').

[3] It is well known that some authors include the nuclear masses to fit the HF Hamiltonian parameters. Such a choice has necessarily a negative impact on the predictive power of the model. This is obvious since the mean-field Hamiltonian represents only a part of physical reality whereas the true masses are contributed by the complete, true Hamiltonian. Consequently, in the spirit of the 'chi-by-the-eye method', the Hamiltonian's parameters are forced to compensate for the absence of certain unknown fragments of this true Hamiltonian - the task that is impossible to keep mathematically under control since the true nuclear Hamiltonian is, as usual, not known.

2.2. Parameters of Hamiltonians and ill-posedeness of the χ^2-minimisation problem
The Hamiltonian's parameter adjustment usually corresponds to a χ^2-minimisation

$$\min_{\{p_k\}} \left\{ \chi^2(\{p\}) = \sum_{j=1}^{n_d} W_j \left[e_j^{exp} - e_j^{th}(\{p\}) \right]^2 \right\} \rightarrow \frac{\partial \chi^2}{\partial p_k} = 0, \text{ for } k = 1 \ldots n_p, \tag{4}$$

where n_d denotes the number of data points, n_p - number of model parameters and $\{W_i\}$ - the physicist-defined set of weight factors. In practical realisations of the inverse problem theory one usually introduces a linearization procedure after simplifying the notation $e_j^{th}(\{p\}) \rightarrow f_j(p)$,

$$f_j(p) \approx f_j(p_0) + \sum_{k=1}^{n_p} \left(\frac{\partial f_j}{\partial p_k} \right) \Big|_{p=p_0} (p_k - p_{0,k}). \tag{5}$$

Introducing further abbreviations

$$J_{jk} \equiv \sqrt{W_j} \left(\frac{\partial f_j}{\partial p_k} \right) \Big|_{p=p_0} \quad \text{and} \quad b_j = \sqrt{W_j} \left[e_j^{exp} - f_j(p_0) \right] \tag{6}$$

one may show that the minimised function takes the form

$$\chi^2(p) = \sum_{j=1}^{n_d} \left[\sum_{k=1}^{n_p} J_{jk} \cdot (p_k - p_{0,k}) - b_j \right]^2. \tag{7}$$

It then follows that the relation within the sought parameters p and the input data b is linear

$$\frac{\partial \chi^2}{\partial p_k} = 0 \quad \leftrightarrow \quad (J^T J) \cdot (p - p_0) = J^T b \quad \rightarrow \quad (p - p_0) = (J^T J)^{-1} J^T b. \tag{8}$$

In Applied Mathematics one usually changes the wording and notation: the set of parameters $\{p\}$ being called 'causes' will be denoted x and the resulting quantities, in our case a set of energies $\{e\}$, called 'effects', is equivalently hidden in b. Observe that generally J is an $n_d \times n_p$ dimensional, i.e., a non-trivially rectangular rather than square matrix since we tend to make the numbers of parameters as small as possible and the number of data points as large as possible and thus $x \in \mathbb{R}^{n_d}$, and $b \in \mathbb{R}^{n_p}$. This is also why it is convenient to use a non-typical notation in (8) that allows to involve explicitly the square matrix $(J^T J)$.

Thanks to the linearized representation in (8) the problem of instabilities of the solutions to the χ^2-minimisation, also called *ill-conditioning*, can be very conveniently posed using the so-called Singular-Value Decomposition theorem applied to J. The latter can always be decomposed as:

$$J = U \cdot D \cdot V^T \text{ with } U \in \mathbb{R}^{n_d \times n_d}, \quad V \in \mathbb{R}^{n_p \times n_p}, \quad D \in \mathbb{R}^{n_d \times n_p}, \tag{9}$$

where diagonal matrix has a form $D = \text{diag}\{d_1, d_2, \ldots d_{min(n_d,n_p)}\}$. Formally the solution x is

$$x = (J^T J)^{-1} J^T b \text{ with } (J^T J)^{-1} J^T = V \cdot D^{-1} \cdot U^T \tag{10}$$

where

$$D^{-1} = \text{diag}\{\tfrac{1}{d_1}, \tfrac{1}{d_2}, \ldots \tfrac{1}{d_p}; 0, 0, \ldots 0\}. \tag{11}$$

Using the Singular-Value Decomposition one shows, within schematic but an obvious notation, that $(J^T J)^{-1} = V \frac{1}{d^2} V^T$. Independently one can derive the expression for the correlation matrix

$$\langle (p_i - \langle p_i \rangle) \cdot (p_j - \langle p_j \rangle) \rangle = \chi^2(p)\, t^2_{\alpha/2, n_d - n_p} \, (J^T J)^{-1}_{ij} \tag{12}$$

where $t_{\alpha/2,n_d-n_p}$ is the usual Student t-probability distribution function. After having defined next the two standard quantities, viz. the so-called unbiased estimator Δ_{rms} for the root-mean-square deviation and the variance-covariance matrix M, respectively, by

$$\Delta^2_{rms} \equiv \frac{\chi^2}{n_d - n_p} \quad \text{and} \quad M \equiv \Delta^2_{rms}(J^T J)^{-1} \tag{13}$$

we obtain the estimates for the 100%-α confidence interval for the least-square estimator $\hat{p}_i$ of the parameter p_i

$$p_i \in [\hat{p}_i - t_{\alpha/2,\nu}\sqrt{M_{ii}}, \hat{p}_i + t_{\alpha/2,\nu}\sqrt{M_{ii}}] \quad \text{with} \quad \nu \equiv n_d - n_p \tag{14}$$

As it is well known, some diagonal elements of the D-matrix can become close to zero or actually vanish - and this property depends generally both on the properties of the actual Hamiltonian as well as on the selection of the data points (in other words: on the experimentalist). We thus arrive at the following important conclusion:

> *If one or more diagonal values $d_k \to 0$, it follows that $(J^T J)^{-1}$ tends to infinity and consequently the correlation matrix and the confidence intervals of all parameters diverge*[4]. *The theory predictions loose stability - thus the theory in question looses its all predictive power - and yet* 'the graph of the solution will usually still look good'- *except that the latter property is now dependent on the experimentalist who measured the data or on the theorist who selected some data points for his/her tests.*

In short: full statistical in-significance referred to as statistical nonsense [and yet - cf. footnote[4]]. [The reader is here referred to the recent publications [4, 5] addressing very similar problems.]

2.3. Elementary tests of stability: Single-point vs. Monte-Carlo approaches, Overfitting

The calculation schematised in (12)-(14) can be performed using just one single minimisation run - and information of this type is referred to as 'single point' or local. Its advantage is relative simplicity but it has also a disadvantage of not representing any global features of the problem as for instance the sensitivity of various observables (such as e.g. single-particle energies) with respect to specific uncertainties in function of the Hamiltonian parameters or, what is related, the fact that some calculated levels may be much more sensitive to uncertainties in inverse-problem input data points as compared to some other levels. Information of this type is illustrated in figure 1 showing the positions of various levels varying in function of the fit-input energy of a given level: here $e_{2g_{9/2}} \in [-2, +2]$ MeV. Let us emphasise that the information of this type may be very precious if we wish to model with precision the positions of certain levels: illustrations like the one in figure 1 show the 'strength of the coupling' thus indicating which levels will need to be measured with precision if we wish to model a particular class of other levels with precision.

Another way of testing the global features of the model *with respect to the data sample at hand* involves the Monte-Carlo simulations. Within the latter approach we assume a certain probability distribution representing an uncertainty of the input data, for instance we suppose that each input level has e.g., Gaussian energy-position probability $P_i(e) \sim \exp[-(e-e_i)^2)/2\sigma_i^2]$ (or impose, deduce, derive ... an alternative probability distribution). One of the first possible applications is to test for an *over-fitting* - a mechanism influencing the model's predictive power.

[4] This does not imply that the theory is in any fundamental trouble since according to the existing methods the negative effects of small or vanishing diagonal (singular) values can be kept easily under control. The problem is: *To bother doing the necessary in place of the chi-by-the-eye approach.*

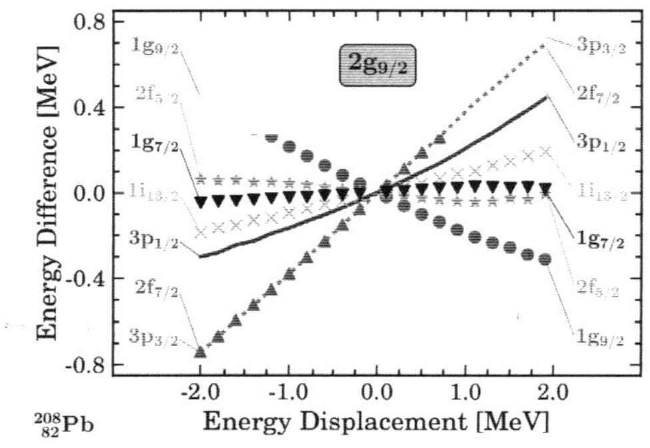

Figure 1. Global correlations in response of the model to various minimisation-input data, are represented by the variation of several (in principle all) single particle levels of the system as function of a single input point, in this example - the experimental $e_{2g_{9/2}}$-level energy. At each of the latter energies the fit of the Hamiltonian parameters is repeated and new parameter set and thus new energies are obtained. Observe a presence of levels nearly insensitive to the variation in input chosen (here: $e_{2g_{9/2}}$) - but there is a richness of possibilities varying from level to level.

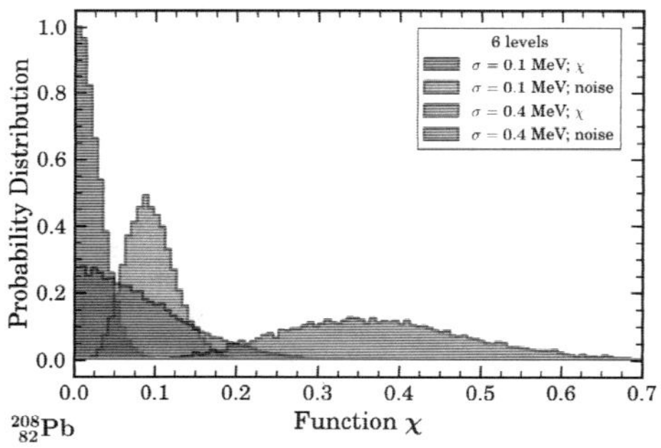

Figure 2. An illustration of the over-fitting of a Hamiltonian studied using Monte-Carlo approach. With the fit-input spectra sampled according to the Gaussian distributions (referred to as 'noise') discussed in the text, here: $\sigma_i = \sigma$, the fitting is repeated 10 000 times, the histograms giving occurrences of any given χ are produced representing related probabilities. The over-fitting takes place when the data points are not sufficiently constraining and it is possible to find as most probable the $\chi^2 = 0$ solution starting from *any input.*

3. Summary and Conclusions

We have discussed statistical significance of theory predictions and of the parametrizations of the underlying Hamiltonians. We introduced selected mathematical elements of the *Inverse Problem Theory* adapted to the application to nuclear Hamiltonians. We formulated the problem of instability of the Hamiltonian's parameter adjustment related to the loss of model's predictive power [for illustrations cf. follow-up in [3]]. As an important element of the future evolution in nuclear theories we suggest a reformulation of their strategic constructions such that the theory *will provide also the probability of the validity of its actual prediction.*

Acknowledgements

This work is supported by the Polish Ministry of Science and Higher Education, under contract N-N202-263238, by LEA COPIGAL and by COPIN-IN2P3 collaboration agreement No.06-126.

References

[1] Press W H, Teukolsky S A, Vetterling W T and Flannery B P 1992 *Numerical Recipes in FORTRAN 77: The Art of Scientific Computing* (Cambridge: Cambridge University Press)

[2] Dudek J et al., 2010 *J. Phys. G: Nucl. Part. Phys.* **37** 064031

[3] Szpak B, Dudek J, Porquet M-G and Fornal B, 2010 *J. Phys.: Conference Series* **This Volume**

[4] Toivanen J et al., 2008 *Phys. Rev.* **C 78**, *034306*

[5] Reinhard P-G and Nazarewicz W, 2010 *Phys. Rev.* **C 81**, *051303(R)*; Kortelainen M et al., 2010 *Phys. Rev.* **C 82**, *024313*

Statistical significance of theoretical predictions: A new dimension in nuclear structure theories (II)

2011 J. Phys.: Conf. Ser. 267 012063

(http://iopscience.iop.org/1742-6596/267/1/012063)

Statistical significance of theoretical predictions: A new dimension in nuclear structure theories (II)

B SZPAK[1], J DUDEK[2], M-G PORQUET[3] and B FORNAL[1]

[1]The Niewodniczański Institute of Nuclear Physics, Polish Academy of Sciences, ul. Radzikowskiego 152, Pl-31-342 Kraków, Poland

[2]Institut Pluridisciplinaire Hubert Curien, Département de Recherches Subatomiques UMR7178, IN2P3-CNRS and Université de Strasbourg, 23 rue du Loess B.P. 28, F-67037 Strasbourg Cedex 2, France

[3]Centre de Spectrométrie Nucléaire et de Spectrométrie de Masse, IN2P3-CNRS and Université Paris-Sud, Bâtiments 104 et 108, F-91405 Orsay Campus, France

E-mail: `Bartlomiej.Szpak@ifj.edu.pl`, `Jerzy.Dudek@IReS.in2p3.fr`, `Marie-Genevieve.Porquet@csnsm.in2p3.fr`, `Bogdan.Fornal@ifj.edu.pl`

Abstract.
We present examples and illustrate associated tests of statistical significance of the parameter fitting procedures in the nuclear mean-field context using a phenomenological toy-model based on the spherical Woods-Saxon Hamiltonian. We calculate the variance-covariance matrix and compare the standard estimates of the confidence intervals (and more generally - the uncertainty distributions) based on the standard χ^2-fitting as opposed to a more general Monte-Carlo simulations. We give arguments for the superiority of the latter approach.

1. Introduction

In [1] further on referred to as (I) we have introduced the problem of uncertainties of theoretical predictions as well as those of the parametrizations of the theory Hamiltonians. Such uncertainties originate not only from the experimental errors but first of all from the inexact character of the theoretical models. These considerations brought us to the notion of the *probability distributions* accompanying the usual numerical theory predictions. In other words: Within the new formulation, the theories are expected to provide not only the pure numerical prediction (as it was the case so far) but also the probability with which we might expect this prediction to represent the real physical system under considerations. At present we are not able to provide the full solution of this type of the problem (the whole approach is under construction); instead we will focus on illustrating the functioning of the most elementary mathematical tools that will be used in the future to fully solve the problem.

2. Elements of the formalism - a short summary

Suppose we have found the minimum of the $\chi^2(p)$ as in (4) or (7) in (I). We may introduce the so-called unbiased estimator for the root-mean-square deviation

$$\Delta^2_{rms} \equiv \frac{\chi^2}{n_d - n_p} \tag{1}$$

where as before n_d is the number of data points and n_p the number of fitted parameters. In probability theory one shows that Δ^2_{rms} has the properties of the so-called χ^2 probability-distribution with $\nu \equiv n_d - n_p$ degrees of freedom. With the help of the Jacobian matrix introduced in (6) of (I), a number of statistical properties of a system can be obtained using the variance-covariance matrix M and the correlation matrix ρ defined by

$$M = \Delta^2_{rms}(J^T J)^{-1} \quad \text{and} \quad \rho_{ij} = M_{ij}/\sqrt{M_{ii}M_{jj}}\,. \tag{2}$$

Consider $(M_{ii})^{1/2}$, the standard error for the ith parameter. The $100\% - \alpha$ confidence interval for the least-square estimator $\hat{p}_i$ of the parameter p_i is given as usual by:

$$p_i \in (\hat{p}_i - t_{\alpha/2,\nu}\sqrt{M_{ii}}, \hat{p}_i + t_{\alpha/2,\nu}\sqrt{M_{ii}})\,, \tag{3}$$

where $t_{\alpha/2,\nu}$ denotes Student $t-$distribution. In the following examples we are interested in 90% confidence intervals thus setting $t_{\alpha/2,\nu} \to t_{0.05,\nu}$ in (3).

Alternatively, one can obtain the variance-covariance matrix and the confidence intervals for the model parameters using Monte-Carlo techniques. Assuming that the value of σ for each experimental data point is given by the Δ_{rms}(obtained from a fit to a set of experimental data) we add the gaussian random noise with variance equal to Δ_{rms} to each experimental point and repeat the minimization thousands of times. The Monte-Carlo method has a number of advantages over the method based on single least-square fit, cf. [2].

The knowledge of full variance-covariance matrix is essential for obtaining information about the first elements of the predictive power of a model: the confidence intervals for parameters and the detection of the possible parameter correlations [for the importance of the latter cf. (I)]. Under certain simplifying assumptions that are not specified here the Probability Distribution Function of the model parameters can expressed by a multivariate Gaussian distribution:

$$P(p) \propto \exp\big[-\tfrac{1}{2}(p-\hat{p})M^{-1}(p-\hat{p})\big]\,, \tag{4}$$

where M^{-1} is the inverse of the variance-covariance matrix and $\hat{p}$ is the estimation of p obtained in a least-square fit. This distribution can be sampled using a Monte-Carlo method to obtain the mean and confidence intervals for parameters of the Hamiltonian and/or physical observables.

3. Illustrations: Single-particle levels within a toy-model

To illustrate our considerations we will use a phenomenological Woods-Saxon Hamiltonian whose parameters are fitted to a set of neutron single particle levels in ^{208}Pb. In this schematic presentation we assume equal weights for all experimental levels. The toy-model Hamiltonian is $H = T + V_c + V_{so}$ where the central potential V_c and the spin-orbit one, V_{so}, are defined by

$$V_c \equiv \frac{V_0}{1+\exp[(r-R_0)/a_0]} \quad \text{and} \quad V_{so} \equiv \frac{1}{r}\frac{d}{dr}\left\{\frac{V_0^{so}}{1+\exp[(r-R_0^{so})/a_0^{so}]}\right\}. \tag{5}$$

It turns out that the solutions to the Schrödinger equation depend very regularly on diffuseness of the spin-orbit close to the χ^2-minimum and to simplify the otherwise schematic illustrations we set $a_0^{so} = 0.6$ fm. Accordingly, the 5-dimensional parameter vector takes the form $\{p\} \equiv \{V_0, r_0, a_0, V_0^{so}, r_0^{so}\}$. The experimental single-particle states are taken from [2]; they are $\{1h_{9/2}, 2f_{7/2}, 1i_{13/2}, 3p_{3/2}, 2f_{5/2}, 3p_{1/2}, 2g_{9/2}, 1i_{11/2}, 1j_{15/2}, 3d_{5/2}, 4s_{1/2}, 2g_{7/2}, 3d_{3/2}\}$ with the corresponding experimental energies $\{\varepsilon\}$/MeV given by

$$\{-10.784, -10.3, -9.8, -8.27, -7.94, -7.37, -3.94, -3.160, -2.517, -2.37, -1.9, -1.44, -1.4\}.$$

An essential element of the analysis consists in evaluating the stability of the obtained solution. For instance when increasing the number of experimental data points we may be able to find out whether the parameters-solutions of the χ^2-minimisation change a lot (or just to the contrary whether they are already stabilised) and/or if the theoretical energies lie systematically close to the corresponding experimental values. However, such an option is seldom at our disposal. Obtaining each new experimental point usually requires a non-trivial instrumental effort so that the experimental points are never in excess. Therefore the only easy option consists in *decreasing* the number of experimental data points taken to the fit and verifying the stability properties in this way.

Formally, by decreasing the number of experimental data points we approach the situation of having too few such points to constrain the number of parameters that is fixed a priori - in this way we approach the limit of model's *ill-posedeness*. For the test purposes we can make ourselves independent of the limitation imposed by the actual experimental situation: we may generate the theoretical spectrum using a reasonable (i.e. close to the empirical values) set of parameters and then try to re-obtain the Hamiltonian parameters using an arbitrarily selected number of those theoretical levels treated as experimental. We will use 7 data points ($1i_{13/2}, 3p_{3/2}, 2f_{5/2}, 3p_{1/2}, 2g_{9/2}, 1i_{11/2}, 1j_{15/2}$) to approach an ill-posed problem. The set of 13 levels represents the realistic scenario (this is the actually known number of experimental neutron levels in ^{208}Pb). By using 21 levels we will be able to estimate how the knowledge of extra levels can constrain the confidence intervals for parameters and observables. Figure 1 illustrates the

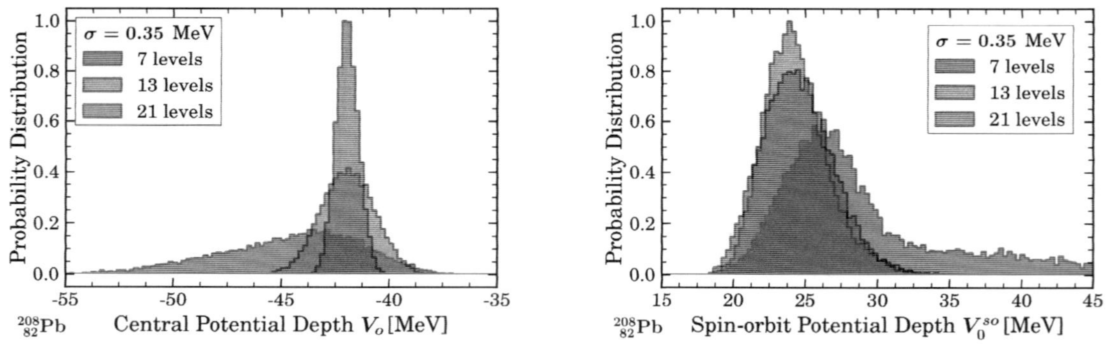

Figure 1. The Probability Distribution Function for the V_0 (left) and V_0^{so} (right) parameters as obtained from the Monte-Carlo simulation assuming common Gaussian random noise with $\sigma = 0.35$ MeV.

'speed' with which an increase in the number of data points constrains more and more the parameters - here the central potential depth and the spin-orbit strength parameters. Observe a characteristic difference: the fitting of spin-orbit strength parameter is influenced by the double-valued structure in the χ^2-minima - see below and also in (I) - the distributions for 13 and 21 levels remain similar in that case.

Indeed, fitting to 13 levels results in two local minima which differ mainly in terms of r_0^{so} parameter, see Table 1; the $r_0^{so} = 1.231$ fm solution is referred to as 'non-compact' and the one with $r_0^{so} = 0.924$ fm as 'compact'. Although the 'non-compact' solution has a smaller value of Δ_{rms} one should remember that the value obtained from a single fit to a certain set of experimental data represents only one single point on the x-axis of the χ^2-distribution[1] which describes Δ_{rms} estimator. Therefore the difference of 0.076 keV between the two results may be a single-point effect which has no statistical significance. This hypothesis is confirmed by a Monte-Carlo simulation tests performed independently which indicates that both solutions are almost equally probable. In the case of fit to 7 levels there also exist two local minima, the 'compact' solution has a smaller value of Δ_{rms} estimator in this case.

Table 1. The values of the Woods-Saxon Hamiltonian parameters as obtained in the fit to 7 or 13 neutron single particle levels in ^{208}Pb. The confidence intervals are defined as $2t_{0.05,\nu}\sqrt{M_{ii}}$.

Quantity	V_0	r_0	a_0	V_0^{so}	r_0^{so}
Non-compact solution $\Delta_{rms} = 0.349$ *MeV (13 levels)*					
Mean value	−42.025	1.320	0.694	24.781	1.231
Standard error	1.217	0.023	0.031	2.639	0.050
Confidence interval	2.264	0.043	0.058	4.907	0.093
Compact solution $\Delta_{rms} = 0.425$ *MeV (13 levels)*					
Mean value	−41.653	1.328	0.670	24.007	0.924
Standard error	1.494	0.029	0.036	2.075	0.035
Confidence interval	2.777	0.053	0.066	3.859	0.066
Compact solution $\Delta_{rms} = 0.151$ *MeV (7 levels)*					
Mean value	−42.923	1.294	0.616	26.692	0.890
Standard error	0.912	0.018	0.027	1.030	0.021
Confidence interval	2.664	0.052	0.079	3.007	0.061

The calculated Jacobian matrix allows to estimate the confidence intervals for the model parameters. In the case of the fit to 13 levels the confidence intervals calculated with the relations (2) and (3) agree with the results of Monte-Carlo calculations - compare the widths of Probability Density Functions presented in Figure 1 with the confidence intervals widths from Table 1. In the case of ill-posed problems the variance-covariance matrix can be unreliable and so will be the confidence intervals. The fit performed to 7 levels results in $\Delta_{rms} = 0.151$ MeV which is more then a factor of 2 smaller then in the fit employing 13 levels. This suggest an effect called *over-fitting* of parameters: a small change in one of experimental data may strongly influence the resulting values of optimal parameters. Observe that the confidence intervals obtained in the fit to 7 levels are of the same size as those obtained in the fit to 13 levels, an abnormal situation for the case of the stable fit. Indeed, the Monte-Carlo simulations performed with $\Delta_{rms} = 0.35$ MeV give confidence intervals which are an order of magnitude bigger in the case of 7 levels used in the fit. This observation supports the thesis about the superiority of the Monte-Carlo tests that contain a global-type information as opposed to a single-point test based on (2).

The results obtained with (2) for fits with 7 and 13 levels are given in the lower left and upper right 'triangles' of Table 2, respectively. One can read from the table a very strong

[1] We wish to make the reader attentive to the existence of two meanings of the term 'chi-square': a. The χ^2-probability distribution studied in the domain of probabilities (as needed in this paragraph) and b. The χ^2 as the minimised function.

Table 2. The correlation matrix for the Woods-Saxon Hamiltonian parameters. The upper right 'triangle' presents the correlation coefficients for 'compact' solution for 13 levels whereas the lower left 'triangle' represents the same information for the 'compact' solution with 7 levels.

	V_0	r_0	a_0	V_0^{so}	r_0^{so}
V_0		0.990	−0.028	0.044	0.341
r_0	0.991		0.056	0.068	0.323
a_0	0.609	0.689		0.218	−0.123
V_0^{so}	−0.161	−0.146	0.110		−0.588
r_0^{so}	0.784	0.787	0.526	−0.547	

correlation between the depth and radius parameters for both central and spin-orbit potentials; these are roughly independent of the number of data points used in the fit. On the contrary, the correlation coefficient between the r_0 and a_0 parameters decreases from 0.743 to 0.092 when the number of experimental levels changes from 7 to 13. These results are confirmed by Monte-Carlo calculations. Figure 2 presents indeed strong correlations between V_O-r_0 and V_0^{so}-r_0^{so} pairs of parameters, here for the 13 levels in the fit.

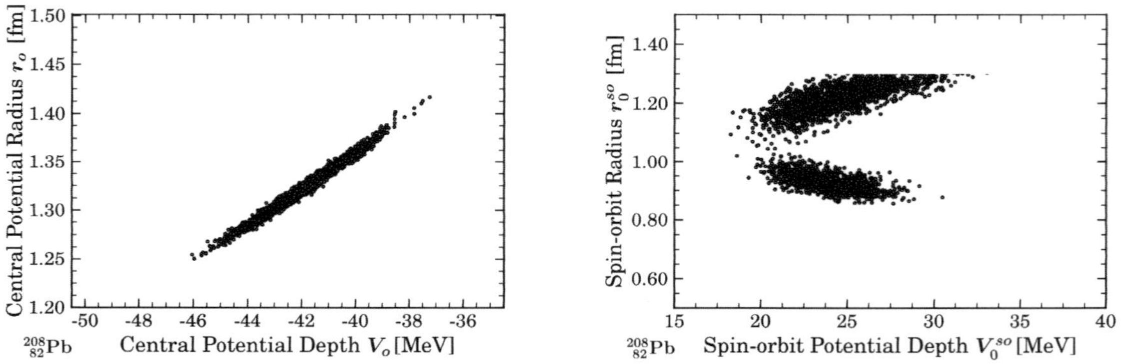

Figure 2. Left: The correlation between the V_0 and r_0 parameters as obtained from the Monte-Carlo simulation assuming $\sigma = 0.35$ MeV in the fit to 13 levels. Right: Same as on the left but for the V_0^{so} and r_0^{so} parameters. Observe the 'compact' and 'non-compact' correlations coexisting independently in the second case. The diagrams contain 10 000 points each.

The proper estimation of a variance-covariance matrix is essential for the reliable predictions of observables and analysis of a predictive power of a model. The importance of correlations between the parameters can be illustrated by comparing the predictions which utilise full variance-covariance matrix with the results of calculations when only diagonal elements (standard errors of parameters) are used. The results of these two kinds of calculations are presented in Figure 3. In the presented calculations the variance-covariance matrix obtained from (2) is sampled with Monte-Carlo method. One clearly sees that neglecting the correlation can significantly increase the variance of predictions.

In the case of a well-posed problem the predictions obtained from Monte-Carlo sampling of variance-covariance matrix and Monte-Carlo simulation should be similar. Indeed, comparison of Figures 3 and 4 shows that this is the case in the analysed example. Complications arise

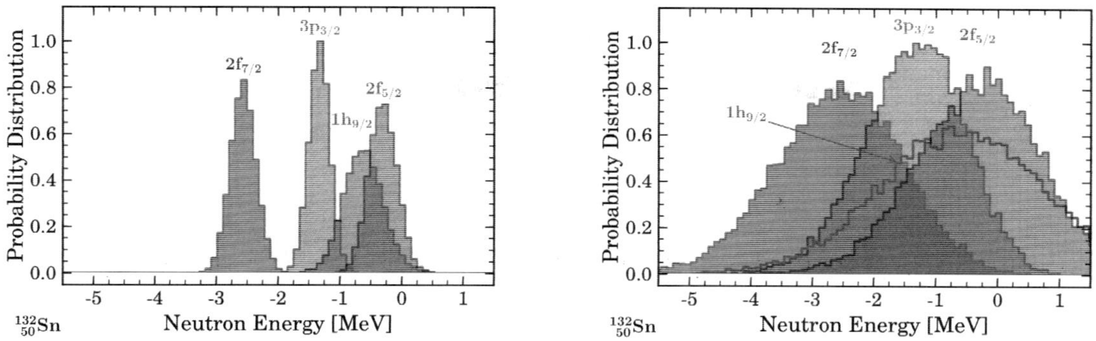

Figure 3. Left: The Probability Density Function for the neutron levels in ^{132}Sn obtained from the extrapolation of parameters given in the first part of Table 1. Right: same as on the left assuming no correlation between the Hamiltonian parameters

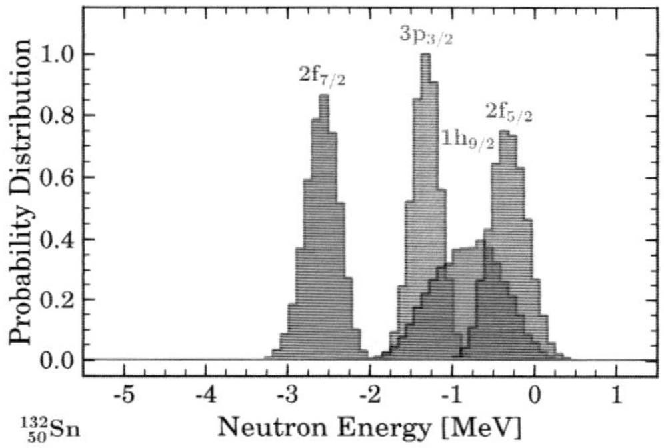

Figure 4. The Probability Density Function for the neutron levels in ^{132}Sn obtained from the extrapolation of parameters obtained in the Monte-Carlo simulation assuming common Gaussian noise with $\sigma = 0.35$ MeV and the fit to 13 levels.

when multiple local χ^2-minima exist. The variance-covariance matrix is calculated only in one minimum while in Monte-Carlo analysis all local minima are taken into account. This explains the different widths of $1h_{9/2}$ levels whose position strongly depends on the spin-orbit parameters. The reader may wish to compare these with more simulation results in [3] and [4].

4. Summary

In this article we have presented and discussed a few elements of the analysis of the statistical significance in determination of the parameters of theory Hamiltonians using a toy-model in the form of a phenomenological (Woods-Saxon) nuclear mean field. Analyses using variance-covariance matrix and the Monte-Carlo simulations were compared pointing out to the superiority of the latter.

Acknowledgements

This work is supported by the Polish Ministry of Science and Higher Education, under contract N-N202-263238, by the LEA COPIGAL and by the COPIN-IN2P3 agreement No.06-126.

References

[1] Dudek J, Szpak B, Porquet M-G and Fornal B 2010 *J. Phys.: Conference Series* **This Issue**

[2] Dudek J, Szpak B, Porquet M-G, Molique H, Rybak K and Fornal B 2010 *J. Phys. G: Nucl. Part. Phys.* **37** 064031

[3] Dudek J, Rybak K, Szpak B, Porquet M-G, Molique H, and Fornal B 2010 *Int. J. Mod. Phys. E* **19** 652

[4] Szpak B, Dudek J, Porquet M-G, Rybak K, Molique H, and Fornal B 2010 *Int. J. Mod. Phys. E* **19** 665

AUTHOR INDEX

AUTHOR INDEX

AUTHOR INDEX